Graduate Texts in Mathematics   176

Graduate Texts in Mathematics

Graduate Texts in Mathematics bridge the gap between passive study and creative understanding, offering graduate-level introductions to advanced topics in mathematics. The volumes are carefully written as teaching aids and highlight characteristic features of the theory. Although these books are frequently used as textbooks in graduate courses, they are also suitable for individual study.

More information about this series at http://www.springer.com/series/136

John M. Lee

Introduction to Riemannian Manifolds

Second Edition

 Springer

John M. Lee
Department of Mathematics
University of Washington
Seattle, WA, USA

ISSN 0072-5285   ISSN 2197-5612   (electronic)
Graduate Texts in Mathematics
ISBN 978-3-030-80106-9   ISBN 978-3-319-91755-9   (eBook)
https://doi.org/10.1007/978-3-319-91755-9

Library of Congress Control Number: 2018943719

Mathematics Subject Classification (2010): 53-01, 53C20, 53B20

Originally published with title "Riemannian Manifolds: An Introduction to Curvature"
1st edition: © Springer-Verlag New York, Inc. 1997
2nd edition: © Springer International Publishing AG 2018, First softcover printing 2021

This Springer imprint is published by the registered company Springer Nature Switzerland AG
The registered company address is: Gewerbestrasse 11, 6330 Cham, Switzerland

Preface

Riemannian geometry is the study of manifolds endowed with *Riemannian metrics*, which are, roughly speaking, rules for measuring lengths of tangent vectors and angles between them. It is the most "geometric" branch of differential geometry. Riemannian metrics are named for the great German mathematician Bernhard Riemann (1826–1866).

This book is designed as a textbook for a graduate course on Riemannian geometry for students who are familiar with the basic theory of smooth manifolds. It focuses on developing an intimate acquaintance with the geometric meaning of curvature, and in particular introducing many of the fundamental results that relate the local geometry of a Riemannian manifold to its global topology (the kind of results I like to call "local-to-global theorems," as explained in Chapter 1). In so doing, it introduces and demonstrates the uses of most of the main technical tools needed for a careful study of Riemannian manifolds.

The book is meant to be introductory, not encyclopedic. Its coverage is reasonably broad, but not exhaustive. It begins with a careful treatment of the machinery of metrics, connections, and geodesics, which are the indispensable tools in the subject. Next comes a discussion of Riemannian manifolds as metric spaces, and the interactions between geodesics and metric properties such as completeness. It then introduces the Riemann curvature tensor, and quickly moves on to submanifold theory in order to give the curvature tensor a concrete quantitative interpretation.

The first local-to-global theorem I discuss is the Gauss–Bonnet theorem for compact surfaces. Many students will have seen a treatment of this in undergraduate courses on curves and surfaces, but because I do not want to assume such a course as a prerequisite, I include a complete proof.

From then on, all efforts are bent toward proving a number of fundamental local-to-global theorems for higher-dimensional manifolds, most notably the Killing–Hopf theorem about constant-curvature manifolds, the Cartan–Hadamard theorem about nonpositively curved manifolds, and Myers's theorem about positively curved ones. The last chapter also contains a selection of other important local-to-global theorems.

Many other results and techniques might reasonably claim a place in an introductory Riemannian geometry book, but they would not fit in this book without drastically broadening its scope. In particular, I do not treat the Morse index theorem, Toponogov's theorem, or their important applications such as the sphere theorem; Hodge theory, gauge theory, minimal surface theory, or other applications of elliptic partial differential equations to Riemannian geometry; or evolution equations such as the Ricci flow or the mean curvature flow. These important topics are for other, more advanced, books.

When I wrote the first edition of this book twenty years ago, a number of superb reference books on Riemannian geometry were already available; in the intervening years, many more have appeared. I invite the interested reader, after reading this book, to consult some of those for a deeper treatment of some of the topics introduced here, or to explore the more esoteric aspects of the subject. Some of my favorites are Peter Petersen's admirably comprehensive introductory text [Pet16]; the elegant introduction to comparison theory by Jeff Cheeger and David Ebin [CE08] (which was out of print for a number of years, but happily has been reprinted by the American Mathematical Society); Manfredo do Carmo's much more leisurely treatment of the same material and more [dC92]; Barrett O'Neill's beautifully integrated introduction to pseudo-Riemannian and Riemannian geometry [O'N83]; Michael Spivak's classic multivolume tome [Spi79], which can be used as a textbook if plenty of time is available, or can provide enjoyable bedtime reading; the breathtaking survey by Marcel Berger [Ber03], which richly earns the word "panoramic" in its title; and the "Encyclopaedia Britannica" of differential geometry books, *Foundations of Differential Geometry* by Shoshichi Kobayashi and Katsumi Nomizu [KN96]. At the other end of the spectrum, Frank Morgan's delightful little book [Mor98] touches on most of the important ideas in an intuitive and informal way with lots of pictures—I enthusiastically recommend it as a prelude to this book. And there are many more to recommend: for example, the books by Chavel [Cha06], Gallot/Hulin/Lafontaine [GHL04], Jost [Jos17], Klingenberg [Kli95], and Jeffrey Lee [LeeJeff09] are all excellent in different ways.

It is not my purpose to replace any of these. Instead, I hope this book fills a niche in the literature by presenting a selective introduction to the main ideas of the subject in an easily accessible way. The selection is small enough to fit (with some judicious cutting) into a single quarter or semester course, but broad enough, I hope, to provide any novice with a firm foundation from which to pursue research or develop applications in Riemannian geometry and other fields that use its tools.

This book is written under the assumption that the student already knows the fundamentals of the theory of topological and smooth manifolds, as treated, for example, in my two other graduate texts [LeeTM, LeeSM]. In particular, the student should be conversant with general topology, the fundamental group, covering spaces, the classification of compact surfaces, topological and smooth manifolds, immersions and submersions, submanifolds, vector fields and flows, Lie brackets and Lie derivatives, tensors, differential forms, Stokes's theorem, and the basic theory of Lie groups. On the other hand, I do not assume any previous acquaintance with Riemannian metrics, or even with the classical theory of curves and surfaces in $\mathbb{R}^3$. (In this subject, anything proved before 1950 can be considered "classical"!)

Although at one time it might have been reasonable to expect most mathematics students to have studied surface theory as undergraduates, many current North American undergraduate math majors never see any differential geometry. Thus the fundamentals of the geometry of surfaces, including a proof of the Gauss–Bonnet theorem, are worked out from scratch here.

The book begins with a nonrigorous overview of the subject in Chapter 1, designed to introduce some of the intuitions underlying the notion of curvature and to link them with elementary geometric ideas the student has seen before. Chapter 2 begins the course proper, with definitions of Riemannian metrics and some of their attendant flora and fauna. Here I also introduce pseudo-Riemannian metrics, which play a central role in Einstein's general theory of relativity. Although I do not attempt to provide a comprehensive introduction to pseudo-Riemannian geometry, throughout the book I do point out which of the constructions and theorems of Riemannian geometry carry over easily to the pseudo-Riemannian case and which do not.

Chapter 3 describes some of the most important "model spaces" of Riemannian and pseudo-Riemannian geometry—those with lots of symmetry—with a great deal of detailed computation. These models form a sort of leitmotif throughout the text, serving as illustrations and testbeds for the abstract theory as it is developed.

Chapter 4 introduces connections, together with some fundamental constructions associated with them such as geodesics and parallel transport. In order to isolate the important properties of connections that are independent of the metric, as well as to lay the groundwork for their further study in arenas that are beyond the scope of this book, such as the Chern–Weil theory of characteristic classes and the Donaldson and Seiberg–Witten theories of gauge fields, connections are defined first on arbitrary vector bundles. This has the further advantage of making it easy to define the induced connections on tensor bundles. Chapter 5 investigates connections in the context of Riemannian (and pseudo-Riemannian) manifolds, developing the Levi-Civita connection, its geodesics, the exponential map, and normal coordinates. Chapter 6 continues the study of geodesics, focusing on their distance-minimizing properties. First, some elementary ideas from the calculus of variations are introduced to prove that every distance-minimizing curve is a geodesic. Then the Gauss lemma is used to prove the (partial) converse—that every geodesic is locally minimizing.

Chapter 7 unveils the first fully general definition of curvature. The curvature tensor is motivated initially by the question whether all Riemannian metrics are "flat" (that is, locally isometric to the Euclidean metric). It turns out that the failure of parallel transport to be path-independent is the primary obstruction to the existence of a local isometry. This leads naturally to a qualitative interpretation of curvature as the obstruction to flatness. Chapter 8 is an investigation of submanifold theory, leading to the definition of sectional curvatures, which give curvature a more quantitative geometric interpretation.

The last four chapters are devoted to the development of some of the most important global theorems relating geometry to topology. Chapter 9 gives a simple moving-frames proof of the Gauss–Bonnet theorem, based on a careful treatment of Hopf's rotation index theorem (often known by its German name, the *Umlaufsatz*). Chapter 10 has a largely technical nature, covering Jacobi fields, conjugate points, the second variation formula, and the index form for later use in comparison theorems. Chapter 11 introduces comparison theory, using a simple comparison theorem for matrix Riccati equations to prove the fundamental fact that bounds on curvature lead to bounds (in the opposite direction) on the size of Jacobi fields, which in turn lead to bounds on many fundamental geometric quantities, such as distances, diameters, and volumes. Finally, in Chapter 12 comes the denouement: proofs of some of the most important local-to-global theorems illustrating the ways in which curvature and topology affect each other.

Exercises and Problems

This book contains many questions for the reader that deserve special mention. They fall into two categories: "exercises," which are integrated into the text, and "problems," grouped at the end of each chapter. Both are essential to a full understanding of the material, but they are of somewhat different characters and serve different purposes.

The exercises include some background material that the student should have seen already in an earlier course, some proofs that fill in the gaps from the text, some simple but illuminating examples, and some intermediate results that are used in the text or the problems. They are, in general, elementary, but they are *not optional*—indeed, they are integral to the continuity of the text. They are chosen and timed so as to give the reader opportunities to pause and think over the material that has just been introduced, to practice working with the definitions, and to develop skills that are used later in the book. I recommend that students stop and do each exercise as it occurs in the text, or at least convince themselves that they know what is involved in the solution of each one, before going any further.

The problems that conclude the chapters are generally more difficult than the exercises, some of them considerably so, and should be considered a central part of the book by any student who is serious about learning the subject. They not only introduce new material not covered in the body of the text, but they also provide the student with indispensable practice in using the techniques explained in the text, both for doing computations and for proving theorems. If the result of a problem is used in an essential way in the text, or in a later problem, the page where it is used is noted at the end of the problem statement. Instructors might want to present some of these problems in class if more than a semester is available.

At the end of the book there are three appendices that contain brief reviews of background material on smooth manifolds, tensors, and Lie groups. I have omitted most of the proofs, but included references to other books where they may be found. The results are collected here in order to clarify what results from topology and smooth manifold theory this book will draw on, and also to establish definitions and conventions that are used throughout the book. I recommend that most readers at least glance through the appendices *before* reading the rest of the book, and consider consulting the indicated references for any topics that are unfamiliar.

About the Second Edition

This second edition, titled *Introduction to Riemannian Manifolds*, has been adapted from my earlier book *Riemannian Manifolds: An Introduction to Curvature*, Graduate Texts in Mathematics 176, Springer 1997.

For those familiar with the first edition, the first difference you will notice about this edition is that it is considerably longer than the first. To some extent, this is due to the addition of more thorough explanations of some of the concepts. But a much more significant reason for the increased length is the addition of many topics that were not covered in the first edition. Here are some of the most important ones: a somewhat expanded treatment of pseudo-Riemannian metrics, together with more consistent explanations of which parts of the theory apply to them; a more detailed treatment of which homogeneous spaces admit invariant metrics; a new treatment of general distance functions and semigeodesic coordinates; introduction of the Weyl tensor and the transformation laws for various curvatures under conformal changes of metric; derivation of the variational equations for hypersurfaces that minimize area with fixed boundary or fixed enclosed volume; an introduction to symmetric spaces; and a treatment of the basic properties of the cut locus. Most importantly, the entire treatment of comparison theory has been revamped and expanded based on Riccati equations, and a handful of local-to-global theorems have been added that were not present in the first edition: Cartan's torsion theorem, Preissman's theorem, Cheng's maximal diameter theorem, Milnor's theorem on polynomial growth of the fundamental group, and Synge's theorem. I hope these will make the book much more useful.

I am aware, though, that one of the attractions of the first edition for some readers was its brevity. For those who would prefer a more streamlined path toward the main local-to-global theorems in Chapter 12, here are topics that can be omitted on a first pass through the book without essential loss of continuity.

- *Chapter 2*: Other generalizations of Riemannian metrics
- *Chapter 3*: Other homogeneous Riemannian manifolds and model pseudo-Riemannian manifolds
- *Chapter 5*: Tubular neighborhoods, Fermi coordinates, and Euclidean and non-Euclidean geometries
- *Chapter 6*: Distance functions and semigeodesic coordinates
- *Chapter 7*: The Weyl tensor and curvatures of conformally related metrics
- *Chapter 8*: Computations in semigeodesic coordinates, minimal hypersurfaces, and constant-mean-curvature hypersurfaces
- *Chapter 9*: The entire chapter
- *Chapter 10*: Locally symmetric spaces and cut points
- *Chapter 11*: Günther's volume comparison theorem and the Bishop–Gromov volume comparison theorem
- *Chapter 12*: All but the theorems of Killing–Hopf, Cartan–Hadamard, and Myers

In addition to the major changes listed above, there are thousands of minor ones throughout the book. Of course, I have attempted to correct all of the mistakes that I became aware of in the first edition. Unfortunately, I surely have not been able to avoid introducing new ones, so if you find anything that seems amiss, please let me know by contacting me through the website listed below. I will keep an updated list of corrections on that website.

I have also adjusted my notation and terminology to be consistent with my two other graduate texts [LeeSM, LeeTM] and hopefully to be more consistent with commonly accepted usage. Like those books, this one now has a notation index just before the subject index, and it uses the same typographical conventions: mathematical terms are typeset in ***bold italics*** when they are officially defined; exercises in the text are indented, numbered consecutively with the theorems, and marked with the special symbol ▶ to make them easier to find; the ends of numbered examples are marked with the symbol //; and the entire book is now set in Times Roman, supplemented by the MathTime Professional II mathematics fonts created by Personal TEX, Inc.

Acknowledgements

I owe an unpayable debt to the authors of the many Riemannian geometry books I have used and cherished over the years, especially the ones mentioned above—I have done little more than rearrange their ideas into a form that seems handy for teaching. Beyond that, I would like to thank my Ph.D. advisor, Richard Melrose, who many years ago introduced me to differential geometry in his eccentric but thoroughly enlightening way; my colleagues Judith Arms, Yu Yuan, and Jim Isenberg, who have provided great help in sorting out what topics should be included; and all of the graduate students at the University of Washington who have suffered with amazing grace through the many flawed drafts of both editions of this book and have provided invaluable feedback, especially Jed Mihalisin, David Sprehn, Collin Litterell, and Maddie Burkhart. And my deepest gratitude goes to Ina Mette of Springer-Verlag (now at the AMS), who first convinced me to turn my lecture notes into a book; without her encouragement, I would never have become a textbook author.

Finally, I would like to dedicate this book to the memory of my late colleague Steve Mitchell, who by his sparkling and joyful example taught me more about teaching and writing than anyone.

Seattle, Washington, USA   John M. Lee
June 2018   www.math.washington.edu/∼lee/

Contents

Chapter 1
What Is Curvature?

If you have spent some time studying modern differential geometry, with its intricate web of manifolds, submanifolds, vector fields, Lie derivatives, tensor fields, differential forms, orientations, and foliations, you might be forgiven for wondering what it all has to do with geometry. In most people's experience, geometry is concerned with properties such as distances, lengths, angles, areas, volumes, and curvature. These concepts, however, are often barely mentioned in typical beginning graduate courses in smooth manifold theory.

The purpose of this book is to introduce the theory of *Riemannian manifolds:* these are smooth manifolds equipped with Riemannian metrics (smoothly varying choices of inner products on tangent spaces), which allow one to measure geometric quantities such as distances and angles. This is the branch of differential geometry in which "geometric" ideas, in the familiar sense of the word, come to the fore. It is the direct descendant of Euclid's plane and solid geometry, by way of Gauss's theory of curved surfaces in space, and it is a dynamic subject of contemporary research.

The central unifying theme in current Riemannian geometry research is the notion of curvature and its relation to topology. This book is designed to help you develop both the tools and the intuition you will need for an in-depth exploration of curvature in the Riemannian setting. Unfortunately, as you will soon discover, an adequate development of curvature in an arbitrary number of dimensions requires a great deal of technical machinery, making it easy to lose sight of the underlying geometric content. To put the subject in perspective, therefore, let us begin by asking some very basic questions: What is curvature? What are some important theorems about it? In this chapter, we explore these and related questions in an informal way, without proofs. The "official" treatment of the subject begins in Chapter 2.

The Euclidean Plane

To get a sense of the kinds of questions Riemannian geometers address and where these questions came from, let us look back at the very roots of our subject. The

© Springer International Publishing AG 2018

J. M. Lee, *Introduction to Riemannian Manifolds*, Graduate Texts in Mathematics 176, https://doi.org/10.1007/978-3-319-91755-9_1

treatment of geometry as a mathematical subject began with Euclidean plane geometry, which you probably studied in secondary school. Its elements are points, lines, distances, angles, and areas; and its most fundamental relationship is ***congruence***—two plane figures are congruent if one can be transformed into the other by a ***rigid motion of the plane***, which is a bijective transformation from the plane to itself that preserves distances. Here are a couple of typical theorems.

Theorem 1.1 (Side-Side-Side). *Two Euclidean triangles are congruent if and only if the lengths of their corresponding sides are equal.*

Theorem 1.2 (Angle-Sum Theorem). *The sum of the interior angles of a Euclidean triangle is π.*

As trivial as they may seem, these theorems serve to illustrate two major types of results that permeate the study of geometry; in this book, we call them "classification theorems" and "local-to-global theorems."

The side-side-side (SSS) theorem is a *classification theorem.* Such a theorem tells us how to determine whether two mathematical objects are equivalent (under some appropriate equivalence relation). An ideal classification theorem lists a small number of computable invariants (whatever "small" may mean in a given context), and says that two objects are equivalent if and only if all of these invariants match. In this case the equivalence relation is congruence, and the invariants are the three side lengths.

The angle-sum theorem is of a different sort. It relates a local geometric property (angle measure) to a global property (that of being a three-sided polygon or triangle). Most of the theorems we study in this book are of this type, which, for lack of a better name, we call *local-to-global theorems.*

After proving the basic facts about points and lines and the figures constructed directly from them, one can go on to study other figures derived from the basic elements, such as circles. Two typical results about circles are given below; the first is a classification theorem, while the second is a local-to-global theorem. (It may not be obvious at this point why we consider the second to be a local-to-global theorem, but it will become clearer soon.)

Theorem 1.3 (Circle Classification Theorem). *Two circles in the Euclidean plane are congruent if and only if they have the same radius.*

Theorem 1.4 (Circumference Theorem). *The circumference of a Euclidean circle of radius R is $2\pi R$.*

If we want to continue our study of plane geometry beyond figures constructed from lines and circles, sooner or later we have to come to terms with other curves in the plane. An arbitrary curve cannot be completely described by one or two numbers such as length or radius; instead, the basic invariant is *curvature*, which is defined using calculus and is a function of position on the curve.

Formally, the ***curvature*** of a plane curve γ is defined to be $\kappa(t) = |\gamma''(t)|$, the length of the acceleration vector, when γ is given a unit-speed parametrization.

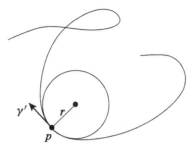

Fig. 1.1: Osculating circle

(Here and throughout this book, the word "curve" refers to a parametrized curve, not a set of points. Typically, a curve will be defined as a smooth function of a real variable t, with a prime representing an ordinary derivative with respect to t.)

Geometrically, the curvature has the following interpretation. Given a point $p = \gamma(t)$, there are many circles tangent to γ at p—namely, those circles whose velocity vector at p is the same as that of γ when both are given unit-speed parametrizations; these are the circles whose centers lie on the line that passes through p and is orthogonal to $\gamma'(p)$. Among these circles, there is exactly one unit-speed parametrized circle whose acceleration vector at p is the same as that of γ; it is called the **osculating circle** (Fig. 1.1). (If the acceleration of γ is zero, replace the osculating circle by a straight line, thought of as a "circle with infinite radius.") The curvature is then $\kappa(t) = 1/R$, where R is the radius of the osculating circle. The larger the curvature, the greater the acceleration and the smaller the osculating circle, and therefore the faster the curve is turning. A circle of radius R has constant curvature $\kappa \equiv 1/R$, while a straight line has curvature zero.

It is often convenient for some purposes to extend the definition of the curvature of a plane curve, allowing it to take on both positive and negative values. This is done by choosing a continuous unit normal vector field N along the curve, and assigning the curvature a positive sign if the curve is turning toward the chosen normal or a negative sign if it is turning away from it. The resulting function κ_N along the curve is then called the **signed curvature**.

Here are two typical theorems about plane curves.

Theorem 1.5 (Plane Curve Classification Theorem). *Suppose γ and $\tilde{\gamma} \colon [a,b] \to \mathbb{R}^2$ are smooth, unit-speed plane curves with unit normal vector fields N and $\tilde{N}$, and $\kappa_N(t)$, $\kappa_{\tilde{N}}(t)$ represent the signed curvatures at $\gamma(t)$ and $\tilde{\gamma}(t)$, respectively. Then γ and $\tilde{\gamma}$ are congruent by a direction-preserving congruence if and only if $\kappa_N(t) = \kappa_{\tilde{N}}(t)$ for all $t \in [a,b]$.*

Theorem 1.6 (Total Curvature Theorem). *If $\gamma \colon [a,b] \to \mathbb{R}^2$ is a unit-speed simple closed curve such that $\gamma'(a) = \gamma'(b)$, and N is the inward-pointing normal, then*

$$\int_a^b \kappa_N(t)\,dt = 2\pi.$$

The first of these is a classification theorem, as its name suggests. The second is a local-to-global theorem, since it relates the local property of curvature to the global (topological) property of being a simple closed curve. We will prove both of these theorems later in the book: the second will be derived as a consequence of a more general result in Chapter 9 (see Corollary 9.6); the proof of the first is left to Problem 9-12.

It is interesting to note that when we specialize to circles, these theorems reduce to the two theorems about circles above: Theorem 1.5 says that two circles are congruent if and only if they have the same curvature, while Theorem 1.6 says that if a circle has curvature κ and circumference C, then $\kappa C = 2\pi$. It is easy to see that these two results are equivalent to Theorems 1.3 and 1.4. This is why it makes sense to regard the circumference theorem as a local-to-global theorem.

Surfaces in Space

The next step in generalizing Euclidean geometry is to start working in three dimensions. After investigating the basic elements of "solid geometry"—points, lines, planes, polyhedra, spheres, distances, angles, surface areas, volumes—one is led to study more general curved surfaces in space (2-dimensional embedded submanifolds of $\mathbb{R}^3$, in the language of differential geometry). The basic invariant in this setting is again curvature, but it is a bit more complicated than for plane curves, because a surface can curve differently in different directions.

The curvature of a surface in space is described by two numbers at each point, called the *principal curvatures*. We will define them formally in Chapter 8, but here is an informal recipe for computing them. Suppose S is a surface in $\mathbb{R}^3$, p is a point in S, and N is a unit normal vector to S at p.

1. Choose a plane Π passing through p and parallel to N. The intersection of Π with a neighborhood of p in S is a plane curve $\gamma \subseteq \Pi$ containing p (Fig. 1.2).
2. Compute the signed curvature κ_N of γ at p with respect to the chosen unit normal N.
3. Repeat this for *all* normal planes Π. The **principal curvatures of S at p**, denoted by κ_1 and κ_2, are the minimum and maximum signed curvatures so obtained.

Although the principal curvatures give us a lot of information about the geometry of S, they do not directly address a question that turns out to be of paramount importance in Riemannian geometry: Which properties of a surface are *intrinsic*? Roughly speaking, intrinsic properties are those that could in principle be measured or computed by a 2-dimensional being living entirely within the surface. More precisely, a property of surfaces in $\mathbb{R}^3$ is called *intrinsic* if it is preserved by *isometries* (maps from one surface to another that preserve lengths of curves).

To see that the principal curvatures are not intrinsic, consider the following two embedded surfaces S_1 and S_2 in $\mathbb{R}^3$ (Figs. 1.3 and 1.4): S_1 is the square in the xy-

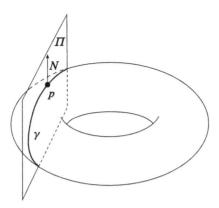

Fig. 1.2: Computing principal curvatures

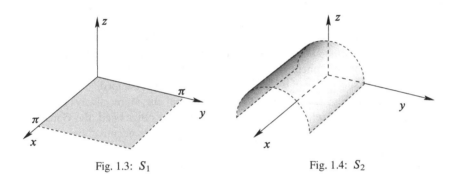

Fig. 1.3: S_1 | Fig. 1.4: S_2

plane where $0 < x < \pi$ and $0 < y < \pi$, and S_2 is the half-cylinder $\{(x, y, z) : z = \sqrt{1 - y^2}, 0 < x < \pi, |y| < 1\}$. If we follow the recipe above for computing principal curvatures (using, say, the downward-pointing unit normal), we find that, since all planes intersect S_1 in straight lines, the principal curvatures of S_1 are $\kappa_1 = \kappa_2 = 0$. On the other hand, it is not hard to see that the principal curvatures of S_2 are $\kappa_1 = 0$ and $\kappa_2 = 1$. However, the map taking $(x, y, 0)$ to $(x, \cos y, \sin y)$ is a diffeomorphism from S_1 to S_2 that preserves lengths of curves, and is thus an isometry.

Even though the principal curvatures are not intrinsic, the great German mathematician Carl Friedrich Gauss made the surprising discovery in 1827 [Gau65] that a particular combination of them is intrinsic. (See also [Spi79, Vol. 2] for an excellent discussion of the details of Gauss's paper.) He found a proof that the product $K = \kappa_1 \kappa_2$, now called the *Gaussian curvature*, is intrinsic. He thought this result was so amazing that he named it *Theorema Egregium*. (This does not mean "totally awful theorem" as its English cognate *egregious* might suggest; a better translation into modern colloquial English might be "totally awesome theorem.")

Fig. 1.5: $K > 0$ Fig. 1.6: $K < 0$

To get a feeling for what Gaussian curvature tells us about surfaces, let us look at a few examples. Simplest of all is any surface that is an open subset of a plane: as we have seen, such a surface has both principal curvatures equal to zero and therefore has constant Gaussian curvature equal to zero. The half-cylinder described above also has $K = \kappa_1 \kappa_2 = 0 \cdot 1 = 0$, as the *Theorema Egregium* tells us it must, being isometric to a square. Another simple example is a sphere of radius R. Every normal plane intersects the sphere in a great circle, which has radius R and therefore curvature $\pm 1/R$ (with the sign depending on whether we choose the outward-pointing or inward-pointing normal). Thus the principal curvatures are both equal to $\pm 1/R$, and the Gaussian curvature is $\kappa_1 \kappa_2 = 1/R^2$. Note that while the signs of the principal curvatures depend on the choice of unit normal, the Gaussian curvature does not: it is always positive on the sphere.

Similarly, any surface that is "bowl-shaped" or "dome-shaped" has positive Gaussian curvature (Fig. 1.5), because the two principal curvatures always have the same sign, regardless of which normal is chosen. On the other hand, the Gaussian curvature of any surface that is "saddle-shaped" (Fig. 1.6) is negative, because the principal curvatures are of opposite signs.

The model spaces of surface theory are the surfaces with constant Gaussian curvature. We have already seen two of them: the Euclidean plane $\mathbb{R}^2$ ($K = 0$), and the sphere of radius R ($K = 1/R^2$). The most important model surface with constant negative Gaussian curvature is called the **hyperbolic plane**, and will be defined in Chapter 3. It is not so easy to visualize because it cannot be realized globally as a smoothly embedded surface in $\mathbb{R}^3$ (see [Spi79, Vol. 3, pp. 373–385] for a proof).

Surface theory is a highly developed branch of geometry. Of all its results, two— a classification theorem and a local-to-global theorem—are generally acknowledged as the most important.

Theorem 1.7 (Uniformization Theorem). *Every connected 2-manifold is diffeomorphic to a quotient of one of the constant-curvature model surfaces described above by a discrete group of isometries without fixed points. Thus every connected 2-manifold has a complete Riemannian metric with constant Gaussian curvature.*

Theorem 1.8 (Gauss–Bonnet Theorem). *Suppose S is a compact Riemannian 2-manifold. Then*

$$\int_S K \, dA = 2\pi \chi(S),$$

where $\chi(S)$ is the Euler characteristic of S.

The uniformization theorem is a classification theorem, because it replaces the problem of classifying surfaces with that of classifying certain discrete groups of isometries of the models. The latter problem is not easy by any means, but it sheds a great deal of new light on the topology of surfaces nonetheless. In Chapter 3, we sketch a proof of the uniformization theorem for the case of compact surfaces.

Although stated here as a geometric-topological result, the uniformization theorem is usually stated somewhat differently and proved using complex analysis. If you are familiar with complex analysis and the complex version of the uniformization theorem, it will be an enlightening exercise after you have finished this book to prove that the complex version of the theorem is equivalent to the one stated here.

The Gauss–Bonnet theorem, on the other hand, is purely a theorem of differential geometry, arguably the most fundamental and important one of all. It relates a local geometric property (the curvature) with a global topological invariant (the Euler characteristic). We give a detailed proof in Chapter 9.

Taken together, these theorems place strong restrictions on the types of metrics that can occur on a given surface. For example, one consequence of the Gauss–Bonnet theorem is that the only compact, connected, orientable surface that admits a metric of strictly positive Gaussian curvature is the sphere. On the other hand, if a compact, connected, orientable surface has nonpositive Gaussian curvature, the Gauss–Bonnet theorem rules out the sphere, and then the uniformization theorem tells us that its universal covering space is topologically equivalent to the plane.

Curvature in Higher Dimensions

We end our survey of the basic ideas of Riemannian geometry by mentioning briefly how curvature appears in higher dimensions. Suppose M is an n-dimensional Riemannian manifold. As with surfaces, the basic geometric invariant is curvature, but curvature becomes a much more complicated quantity in higher dimensions because a manifold may curve in so many different directions.

The first problem we must contend with is that, in general, Riemannian manifolds are not presented to us as embedded submanifolds of Euclidean space. Therefore, we must abandon the idea of cutting out curves by intersecting our manifold with planes, as we did when defining the principal curvatures of a surface in $\mathbb{R}^3$. Instead, we need a more intrinsic way of sweeping out submanifolds. Fortunately, *geodesics*—curves that are the shortest paths between nearby points—are ready-made tools for this and many other purposes in Riemannian geometry. Examples are straight lines in Euclidean space and great circles on a sphere.

The most fundamental fact about geodesics, which we prove in Chapter 4, is that given any point $p \in M$ and any vector v tangent to M at p, there is a unique geodesic starting at p with initial velocity v.

Here is a brief recipe for computing some curvatures at a point $p \in M$.

1. Choose a 2-dimensional subspace Π of the tangent space to M at p.

2. Look at all the geodesics through p whose initial velocities lie in the selected plane Π. It turns out that near p these sweep out a certain 2-dimensional submanifold S_Π of M, which inherits a Riemannian metric from M.

3. Compute the Gaussian curvature of S_Π at p, which the *Theorema Egregium* tells us can be computed from the Riemannian metric that S_Π inherits from M. This gives a number, denoted by $\sec(\Pi)$, called the *sectional curvature* of M at p associated with the plane Π.

Thus the "curvature" of M at p has to be interpreted as a map

$$\sec : \{\text{2-planes in } T_p M\} \to \mathbb{R}.$$

As we will see in Chapter 3, we again have three classes of constant (sectional) curvature model spaces: $\mathbb{R}^n$ with its Euclidean metric (for which $\sec \equiv 0$); the n-sphere of radius R, with the Riemannian metric inherited from $\mathbb{R}^{n+1}$ ($\sec \equiv 1/R^2$); and hyperbolic space of radius R (with $\sec \equiv -1/R^2$). Unfortunately, however, there is as yet no satisfactory uniformization theorem for Riemannian manifolds in higher dimensions. In particular, it is definitely *not* true that every manifold possesses a metric of constant sectional curvature. In fact, the constant-curvature metrics can all be described rather explicitly by the following classification theorem.

Theorem 1.9 (Characterization of Constant-Curvature Metrics). *The complete, connected, n-dimensional Riemannian manifolds of constant sectional curvature are, up to isometry, exactly the Riemannian quotients of the form $\widetilde{M}/\Gamma$, where $\widetilde{M}$ is a Euclidean space, sphere, or hyperbolic space with constant sectional curvature, and Γ is a discrete group of isometries of $\widetilde{M}$ that acts freely on $\widetilde{M}$.*

On the other hand, there are a number of powerful local-to-global theorems, which can be thought of as generalizations of the Gauss–Bonnet theorem in various directions. They are consequences of the fact that positive curvature makes geodesics converge, while negative curvature forces them to spread out. Here (in somewhat simplified form) are two of the most important such theorems.

Theorem 1.10 (Cartan–Hadamard). *Suppose M is a complete, connected Riemannian n-manifold with all sectional curvatures less than or equal to zero. Then the universal covering space of M is diffeomorphic to $\mathbb{R}^n$.*

Theorem 1.11 (Myers). *Suppose M is a complete, connected Riemannian manifold with all sectional curvatures bounded below by a positive constant. Then M is compact and has a finite fundamental group.*

Looking back at the remarks concluding the section on surfaces above, you can see that these last three theorems generalize some of the consequences of the uniformization and Gauss–Bonnet theorems, although not their full strength. It is the primary goal of this book to prove Theorems 1.9, 1.10, and 1.11, among others; it is a primary goal of current research in Riemannian geometry to improve upon them and further generalize the results of surface theory to higher dimensions.

Chapter 2
Riemannian Metrics

In this chapter we officially define Riemannian metrics, and discuss some of the basic computational techniques associated with them. After the definitions, we describe a few standard methods for constructing Riemannian manifolds as submanifolds, products, and quotients of other Riemannian manifolds. Then we introduce some of the elementary geometric constructions provided by Riemannian metrics, the most important of which is the Riemannian distance function, which turns every connected Riemannian manifold into a metric space.

At the end of the chapter, we discuss some important generalizations of Riemannian metrics—most importantly, the pseudo-Riemannian metrics, followed by brief mentions of sub-Riemannian and Finsler metrics.

Before you read this chapter, it would be a good idea to skim through the three appendices after Chapter 12 to get an idea of the prerequisite material that will be assumed throughout this book.

Definitions

Everything we know about the Euclidean geometry of $\mathbb{R}^n$ can be derived from its *dot product*, which is defined for $v = (v^1, \ldots, v^n)$ and $w = (w^1, \ldots, w^n)$ by

$$v \cdot w = \sum_{i=1}^{n} v^i w^i.$$

The dot product has a natural generalization to arbitrary vector spaces. Given a vector space V (which we always assume to be real), an *inner product on V* is a map $V \times V \to \mathbb{R}$, typically written $(v, w) \mapsto \langle v, w \rangle$, that satisfies the following properties for all $v, w, x \in V$ and $a, b \in \mathbb{R}$:

(i) SYMMETRY: $\langle v, w \rangle = \langle w, v \rangle$.
(ii) BILINEARITY: $\langle av + bw, x \rangle = a\langle v, x \rangle + b\langle w, x \rangle = \langle x, av + bw \rangle$.

© Springer International Publishing AG 2018
J. M. Lee, *Introduction to Riemannian Manifolds*, Graduate Texts
in Mathematics 176, https://doi.org/10.1007/978-3-319-91755-9_2

(iii) POSITIVE DEFINITENESS: $\langle v, v \rangle \geq 0$, with equality if and only if $v = 0$.

A vector space endowed with a specific inner product is called an **inner product space**.

An inner product on V allows us to make sense of geometric quantities such as lengths of vectors and angles between vectors. First, we define the **length** or **norm** of a vector $v \in V$ as

$$|v| = \langle v, v \rangle^{1/2}. \tag{2.1}$$

The following identity shows that an inner product is completely determined by knowledge of the lengths of all vectors.

Lemma 2.1 (Polarization Identity). *Suppose $\langle \cdot, \cdot \rangle$ is an inner product on a vector space V. Then for all $v, w \in V$,*

$$\langle v, w \rangle = \tfrac{1}{4} \big(\langle v + w, v + w \rangle - \langle v - w, v - w \rangle \big). \tag{2.2}$$

▶ **Exercise 2.2.** Prove the preceding lemma.

The **angle** between two nonzero vectors $v, w \in V$ is defined as the unique $\theta \in [0, \pi]$ satisfying

$$\cos \theta = \frac{\langle v, w \rangle}{|v| \, |w|}. \tag{2.3}$$

Two vectors $v, w \in V$ are said to be **orthogonal** if $\langle v, w \rangle = 0$, which means that either their angle is $\pi/2$ or one of the vectors is zero. If $S \subseteq V$ is a linear subspace, the set $S^\perp \subseteq V$, consisting of all vectors in V that are orthogonal to every vector in S, is also a linear subspace, called the **orthogonal complement of S**.

Vectors $v_1, \ldots, v_k$ are called **orthonormal** if they are of length 1 and pairwise orthogonal, or equivalently if $\langle v_i, v_j \rangle = \delta_{ij}$ (where δ_{ij} is the *Kronecker delta symbol* defined in Appendix B; see (B.1)). The following well-known proposition shows that every finite-dimensional inner product space has an orthonormal basis.

Proposition 2.3 (Gram–Schmidt Algorithm). *Let V be an n-dimensional inner product space, and suppose $(v_1, \ldots, v_n)$ is any ordered basis for V. Then there is an orthonormal ordered basis $(b_1, \ldots, b_n)$ satisfying the following conditions:*

$$\mathrm{span}(b_1, \ldots, b_k) = \mathrm{span}(v_1, \ldots, v_k) \quad \text{for each } k = 1, \ldots, n. \tag{2.4}$$

Proof. The basis vectors $b_1, \ldots, b_n$ are defined recursively by

$$b_1 = \frac{v_1}{|v_1|}, \tag{2.5}$$

$$b_j = \frac{v_j - \sum_{i=1}^{j-1} \langle v_j, b_i \rangle b_i}{\left| v_j - \sum_{i=1}^{j-1} \langle v_j, b_i \rangle b_i \right|}, \quad 2 \leq j \leq n. \tag{2.6}$$

Because $v_1 \neq 0$ and $v_j \notin \mathrm{span}(b_1, \ldots, b_{j-1})$ for each $j \geq 2$, the denominators are all nonzero. These vectors satisfy (2.4) by construction, and are orthonormal by direct computation. □

If two vector spaces V and W are both equipped with inner products, denoted by $\langle\cdot,\cdot\rangle_V$ and $\langle\cdot,\cdot\rangle_W$, respectively, then a map $F\colon V \to W$ is called a *linear isometry* if it is a vector space isomorphism that preserves inner products: $\langle F(v), F(v')\rangle_W = \langle v, v'\rangle_V$. If V and W are inner product spaces of dimension n, then given any choices of orthonormal bases $(v_1,\dots,v_n)$ for V and $(w_1,\dots,w_n)$ for W, the linear map $F\colon V \to W$ determined by $F(v_i) = w_i$ is easily seen to be a linear isometry. Thus all inner product spaces of the same finite dimension are linearly isometric to each other.

Riemannian Metrics

To extend these geometric ideas to abstract smooth manifolds, we define a structure that amounts to a smoothly varying choice of inner product on each tangent space.

Let M be a smooth manifold. A *Riemannian metric* on M is a smooth covariant 2-tensor field $g \in \mathcal{T}^2(M)$ whose value g_p at each $p \in M$ is an inner product on T_pM; thus g is a symmetric 2-tensor field that is positive definite in the sense that $g_p(v,v) \geq 0$ for each $p \in M$ and each $v \in T_pM$, with equality if and only if $v = 0$. A *Riemannian manifold* is a pair (M,g), where M is a smooth manifold and g is a specific choice of Riemannian metric on M. If M is understood to be endowed with a specific Riemannian metric, we sometimes say "M is a Riemannian manifold."

The next proposition shows that Riemannian metrics exist in great abundance.

Proposition 2.4. *Every smooth manifold admits a Riemannian metric.*

▶ **Exercise 2.5.** Use a partition of unity to prove the preceding proposition.

We will give a number of examples of Riemannian metrics, along with several systematic methods for constructing them, later in this chapter and in the next.

If M is a smooth manifold with boundary, a Riemannian metric on M is defined in exactly the same way: a smooth symmetric 2-tensor field g that is positive definite everywhere. A *Riemannian manifold with boundary* is a pair (M,g), where M is a smooth manifold with boundary and g is a Riemannian metric on M. Many of the results we will discuss in this book work equally well for manifolds with or without boundary, with the same proofs, and in such cases we will state them in that generality. But when the treatment of a boundary would involve additional difficulties, we will generally restrict attention to the case of manifolds without boundary, since that is our primary interest. Many problems involving Riemannian manifolds with boundary can be addressed by embedding into a larger manifold without boundary and extending the Riemannian metric arbitrarily to the larger manifold; see Proposition A.31 in Appendix A.

A Riemannian metric is not the same as a metric in the sense of metric spaces (though, as we will see later in this chapter, the two concepts are related). In this book, when we use the word "metric" without further qualification, it always refers to a Riemannian metric.

Let g be a Riemannian metric on a smooth manifold M with or without boundary. Because g_p is an inner product on $T_p M$ for each $p \in M$, we often use the following angle-bracket notation for $v, w \in T_p M$:

$$\langle v, w \rangle_g = g_p(v, w).$$

Using this inner product, we can define lengths of tangent vectors, angles between nonzero tangent vectors, and orthogonality of tangent vectors as described above. The length of a vector $v \in T_p M$ is denoted by $|v|_g = \langle v, v \rangle_g^{1/2}$. If the metric is understood, we sometimes omit it from the notation, and write $\langle v, w \rangle$ and $|v|$ in place of $\langle v, w \rangle_g$ and $|v|_g$, respectively.

The starting point for Riemannian geometry is the following fundamental example.

Example 2.6 (The Euclidean Metric). The **Euclidean metric** is the Riemannian metric $\bar{g}$ on $\mathbb{R}^n$ whose value at each $x \in \mathbb{R}^n$ is just the usual dot product on $T_x \mathbb{R}^n$ under the natural identification $T_x \mathbb{R}^n \cong \mathbb{R}^n$. This means that for $v, w \in T_x \mathbb{R}^n$ written in standard coordinates $(x^1, \dots, x^n)$ as $v = \sum_i v^i \partial_i|_x$, $w = \sum_j w^j \partial_j|_x$, we have

$$\langle v, w \rangle_{\bar{g}} = \sum_{i=1}^n v^i w^i.$$

When working with $\mathbb{R}^n$ as a Riemannian manifold, we always assume we are using the Euclidean metric unless otherwise specified.   ∥

Isometries

Suppose (M, g) and $(\widetilde{M}, \widetilde{g})$ are Riemannian manifolds with or without boundary. An **isometry from** (M, g) **to** $(\widetilde{M}, \widetilde{g})$ is a diffeomorphism $\varphi \colon M \to \widetilde{M}$ such that $\varphi^* \widetilde{g} = g$. Unwinding the definitions shows that this is equivalent to the requirement that φ be a smooth bijection and each differential $d\varphi_p \colon T_p M \to T_{\varphi(p)} \widetilde{M}$ be a linear isometry. We say (M, g) and $(\widetilde{M}, \widetilde{g})$ are **isometric** if there exists an isometry between them.

A composition of isometries and the inverse of an isometry are again isometries, so being isometric is an equivalence relation on the class of Riemannian manifolds with or without boundary. Our subject, *Riemannian geometry*, is concerned primarily with properties of Riemannian manifolds that are preserved by isometries.

If (M, g) and $(\widetilde{M}, \widetilde{g})$ are Riemannian manifolds, a map $\varphi \colon M \to \widetilde{M}$ is a *local isometry* if each point $p \in M$ has a neighborhood U such that $\varphi|_U$ is an isometry onto an open subset of $\widetilde{M}$.

▶ **Exercise 2.7.** Prove that if (M, g) and $(\widetilde{M}, \widetilde{g})$ are Riemannian manifolds of the same dimension, a smooth map $\varphi \colon M \to \widetilde{M}$ is a local isometry if and only if $\varphi^* \widetilde{g} = g$.

A Riemannian n-manifold is said to be *flat* if it is locally isometric to a Euclidean space, that is, if every point has a neighborhood that is isometric to an open set in

$\mathbb{R}^n$ with its Euclidean metric. Problem 2-1 shows that all Riemannian 1-manifolds are flat; but we will see later that this is far from the case in higher dimensions.

An isometry from (M,g) to itself is called an *isometry of* (M,g). The set of all isometries of (M,g) is a group under composition, called the *isometry group of* (M,g); it is denoted by $\mathrm{Iso}(M,g)$, or sometimes just $\mathrm{Iso}(M)$ if the metric is understood.

A deep theorem of Sumner B. Myers and Norman E. Steenrod [MS39] shows that if M has finitely many components, then $\mathrm{Iso}(M,g)$ has a topology and smooth structure making it into a finite-dimensional Lie group acting smoothly on M. We will neither prove nor use the Myers–Steenrod theorem, but if you are interested, a good source for the proof is [Kob72].

Local Representations for Metrics

Suppose (M,g) is a Riemannian manifold with or without boundary. If $(x^1,\ldots,x^n)$ are any smooth local coordinates on an open subset $U \subseteq M$, then g can be written locally in U as

$$g = g_{ij}\, dx^i \otimes dx^j \tag{2.7}$$

for some collection of n^2 smooth functions g_{ij} for $i, j = 1,\ldots,n$. (Here and throughout the book, we use the Einstein summation convention; see p. 375.) The component functions of this tensor field constitute a matrix-valued function (g_{ij}), characterized by $g_{ij}(p) = \langle \partial_i|_p, \partial_j|_p \rangle$, where $\partial_i = \partial/\partial x^i$ is the ith coordinate vector field; this matrix is symmetric in i and j and depends smoothly on $p \in U$. If $v = v^i \partial_i|_p$ is a vector in T_pM such that $g_{ij}(p)v^j = 0$, it follows that $\langle v,v \rangle = g_{ij}(p)v^i v^j = 0$, which implies $v = 0$; thus the matrix $(g_{ij}(p))$ is always nonsingular. The notation for g can be shortened by expressing it in terms of the symmetric product (see Appendix B): using the symmetry of g_{ij}, we compute

$$
\begin{aligned}
g &= g_{ij}\, dx^i \otimes dx^j \\
&= \tfrac{1}{2}\big(g_{ij}\, dx^i \otimes dx^j + g_{ji}\, dx^i \otimes dx^j\big) \\
&= \tfrac{1}{2}\big(g_{ij}\, dx^i \otimes dx^j + g_{ij}\, dx^j \otimes dx^i\big) \\
&= g_{ij}\, dx^i dx^j.
\end{aligned}
$$

For example, the Euclidean metric on $\mathbb{R}^n$ (Example 2.6) can be expressed in standard coordinates in several ways:

$$\bar{g} = \sum_i dx^i dx^i = \sum_i \big(dx^i\big)^2 = \delta_{ij}\, dx^i dx^j. \tag{2.8}$$

The matrix of $\bar{g}$ in these coordinates is thus $\bar{g}_{ij} = \delta_{ij}$.

More generally, if $(E_1,\ldots,E_n)$ is any smooth local frame for TM on an open subset $U \subseteq M$ and $(\varepsilon^1,\ldots,\varepsilon^n)$ is its dual coframe, we can write g locally in U as

$$g = g_{ij}\, \varepsilon^i \varepsilon^j, \tag{2.9}$$

where $g_{ij}(p) = \langle E_i|_p, E_j|_p \rangle$, and the matrix-valued function (g_{ij}) is symmetric and smooth as before.

A Riemannian metric g acts on smooth vector fields $X, Y \in \mathfrak{X}(M)$ to yield a real-valued function $\langle X, Y \rangle$. In terms of any smooth local frame, this function is expressed locally by $\langle X, Y \rangle = g_{ij} X^i Y^j$ and therefore is smooth. Similarly, we obtain a nonnegative real-valued function $|X| = \langle X, X \rangle^{1/2}$, which is continuous everywhere and smooth on the open subset where $X \neq 0$.

A local frame (E_i) for M on an open set U is said to be an **orthonormal frame** if the vectors $E_1|_p, \ldots, E_n|_p$ are an orthonormal basis for $T_p M$ at each $p \in U$. Equivalently, (E_i) is an orthonormal frame if and only if

$$\langle E_i, E_j \rangle = \delta_{ij},$$

in which case g has the local expression

$$g = (\varepsilon^1)^2 + \cdots + (\varepsilon^n)^2,$$

where $(\varepsilon^i)^2$ denotes the symmetric product $\varepsilon^i \varepsilon^i = \varepsilon^i \otimes \varepsilon^i$.

Proposition 2.8 (Existence of Orthonormal Frames). *Let (M, g) be a Riemannian n-manifold with or without boundary. If (X_j) is any smooth local frame for TM over an open subset $U \subseteq M$, then there is a smooth orthonormal frame (E_j) over U such that* $\operatorname{span}(E_1|_p, \ldots, E_k|_p) = \operatorname{span}(X_1|_p, \ldots, X_k|_p)$ *for each $k = 1, \ldots, n$ and each $p \in U$. In particular, for every $p \in M$, there is a smooth orthonormal frame (E_j) defined on some neighborhood of p.*

Proof. Applying the Gram–Schmidt algorithm to the vectors $(X_1|_p, \ldots, X_n|_p)$ at each $p \in U$, we obtain an ordered n-tuple of rough orthonormal vector fields $(E_1, \ldots, E_n)$ over U satisfying the span conditions. Because the vectors whose norms appear in the denominators of (2.5)–(2.6) are nowhere vanishing, those formulas show that each vector field E_j is smooth. The last statement of the proposition follows by applying this construction to any smooth local frame in a neighborhood of p. $\qquad\square$

Warning: A common mistake made by beginners is to assume that one can find coordinates near p such that the *coordinate frame* (∂_i) is orthonormal. Proposition 2.8 does not show this. In fact, as we will see in Chapter 7, this is possible only when the metric is *flat*, that is, locally isometric to the Euclidean metric.

For a Riemannian manifold (M, g) with or without boundary, we define the **unit tangent bundle** to be the subset $UTM \subseteq TM$ consisting of unit vectors:

$$UTM = \{ (p, v) \in TM : |v|_g = 1 \}. \tag{2.10}$$

Proposition 2.9 (Properties of the Unit Tangent Bundle). *If (M, g) is a Riemannian manifold with or without boundary, its unit tangent bundle UTM is a* *smooth, properly embedded codimension-1 submanifold with boundary in TM, with* $\partial(UTM) = \pi^{-1}(\partial M)$ *(where* $\pi: UTM \to M$ *is the canonical projection). The unit tangent bundle is connected if and only if M is connected, and compact if and only if M is compact.*

▶ **Exercise 2.10.** Use local orthonormal frames to prove the preceding proposition.

Methods for Constructing Riemannian Metrics

Many examples of Riemannian manifolds arise naturally as submanifolds, products, and quotients of other Riemannian manifolds. In this section, we introduce some of the tools for constructing such metrics.

Riemannian Submanifolds

Every submanifold of a Riemannian manifold automatically inherits a Riemannian metric, and many interesting Riemannian metrics are defined in this way. The key fact is the following lemma.

Lemma 2.11. *Suppose $\left(\widetilde{M}, \widetilde{g}\right)$ is a Riemannian manifold with or without boundary, M is a smooth manifold with or without boundary, and $F: M \to \widetilde{M}$ is a smooth map. The smooth 2-tensor field $g = F^*\widetilde{g}$ is a Riemannian metric on M if and only if F is an immersion.*

▶ **Exercise 2.12.** Prove Lemma 2.11.

Suppose $\left(\widetilde{M}, \widetilde{g}\right)$ is a Riemannian manifold with or without boundary. Given a smooth immersion $F: M \to \widetilde{M}$, the metric $g = F^*\widetilde{g}$ is called the ***metric induced by F***. On the other hand, if M is already endowed with a *given* Riemannian metric g, an immersion or embedding $F: M \to \widetilde{M}$ satisfying $F^*\widetilde{g} = g$ is called an ***isometric immersion*** or ***isometric embedding***, respectively. Which terminology is used depends on whether the metric on M is considered to be given independently of the immersion or not.

The most important examples of induced metrics occur on submanifolds. Suppose $M \subseteq \widetilde{M}$ is an (immersed or embedded) submanifold, with or without boundary. The ***induced metric on M*** is the metric $g = \iota^*\widetilde{g}$ induced by the inclusion map $\iota: M \hookrightarrow \widetilde{M}$. With this metric, M is called a ***Riemannian submanifold*** (or ***Riemannian submanifold with boundary***) of $\widetilde{M}$. We always consider submanifolds (with or without boundary) of Riemannian manifolds to be endowed with the induced metrics unless otherwise specified.

If (M, g) is a Riemannian submanifold of $\left(\widetilde{M}, \widetilde{g}\right)$, then for every $p \in M$ and $v, w \in T_p M$, the definition of the induced metric reads

$$g_p(v, w) = \widetilde{g}_p\big(d\iota_p(v), d\iota_p(w)\big).$$

Because we usually identify $T_p M$ with its image in $T_p \widetilde{M}$ under $d\iota_p$, and think of $d\iota_p$ as an inclusion map, what this really amounts to is $g_p(v, w) = \widetilde{g}_p(v, w)$ for $v, w \in T_p M$. In other words, the induced metric g is just the restriction of $\widetilde{g}$ to vectors tangent to M. Many of the examples of Riemannian metrics that we will encounter are obtained in this way, starting with the following.

Example 2.13 (Spheres). For each positive integer n, the unit n-sphere $\mathbb{S}^n \subseteq \mathbb{R}^{n+1}$ is an embedded n-dimensional submanifold. The Riemannian metric induced on $\mathbb{S}^n$ by the Euclidean metric is denoted by $\mathring{g}$ and known as the **round metric** or **standard metric** on $\mathbb{S}^n$.   //

The next lemma describes one of the most important tools for studying Riemannian submanifolds. If $\big(\widetilde{M}, \widetilde{g}\big)$ is an m-dimensional smooth Riemannian manifold and $M \subseteq \widetilde{M}$ is an n-dimensional submanifold (both with or without boundary), a local frame $(E_1, \ldots, E_m)$ for $\widetilde{M}$ on an open subset $\widetilde{U} \subseteq \widetilde{M}$ is said to be **adapted to M** if the first n vector fields $(E_1, \ldots, E_n)$ are tangent to M. In case $\widetilde{M}$ has empty boundary (so that slice coordinates are available), adapted local orthonormal frames are easy to find.

Proposition 2.14 (Existence of Adapted Orthonormal Frames). *Let $\big(\widetilde{M}, \widetilde{g}\big)$ be a Riemannian manifold (without boundary), and let $M \subseteq \widetilde{M}$ be an embedded smooth submanifold with or without boundary. Given $p \in M$, there exist a neighborhood $\widetilde{U}$ of p in $\widetilde{M}$ and a smooth orthonormal frame for $\widetilde{M}$ on $\widetilde{U}$ that is adapted to M.*

▶ **Exercise 2.15.** Prove the preceding proposition. [Hint: Apply the Gram–Schmidt algorithm to a coordinate frame in slice coordinates (see Prop. A.22).]

Suppose $\big(\widetilde{M}, \widetilde{g}\big)$ is a Riemannian manifold and $M \subseteq \widetilde{M}$ is a smooth submanifold with or without boundary in $\widetilde{M}$. Given $p \in M$, a vector $v \in T_p \widetilde{M}$ is said to be **normal to M** if $\langle v, w \rangle = 0$ for every $w \in T_p M$. The space of all vectors normal to M at p is a subspace of $T_p \widetilde{M}$, called the **normal space at p** and denoted by $N_p M = (T_p M)^\perp$. At each $p \in M$, the ambient tangent space $T_p \widetilde{M}$ splits as an orthogonal direct sum $T_p \widetilde{M} = T_p M \oplus N_p M$. A section N of the ambient tangent bundle $T\widetilde{M}|_M$ is called a **normal vector field along M** if $N_p \in N_p M$ for each $p \in M$. The set

$$NM = \coprod_{p \in M} N_p M$$

is called the **normal bundle of M**.

Proposition 2.16 (The Normal Bundle). *If $\widetilde{M}$ is a Riemannian m-manifold and $M \subseteq \widetilde{M}$ is an immersed or embedded n-dimensional submanifold with or without boundary, then NM is a smooth rank-$(m - n)$ vector subbundle of the ambient tangent bundle $T\widetilde{M}|_M$. There are smooth bundle homomorphisms*

$$\pi^\top : T\widetilde{M}|_M \to TM, \qquad \pi^\perp : T\widetilde{M}|_M \to NM,$$

*called the **tangential** and **normal projections**, that for each $p \in M$ restrict to orthogonal projections from $T_p \widetilde{M}$ to $T_p M$ and $N_p M$, respectively.*

Proof. Given any point $p \in M$, Theorem A.16 shows that there is a neighborhood U of p in M that is embedded in $\widetilde{M}$, and then Proposition 2.14 shows that there is a smooth orthonormal frame $(E_1, \ldots, E_m)$ that is adapted to U on some neighborhood $\widetilde{U}$ of p in $\widetilde{M}$. This means that the restrictions of $(E_1, \ldots, E_n)$ to $\widetilde{U} \cap U$ form a local orthonormal frame for M. Given such an adapted frame, the restrictions of the last $m - n$ vector fields $(E_{n+1}, \ldots, E_m)$ to M form a smooth local frame for NM, so it follows from Lemma A.34 that NM is a smooth subbundle.

The bundle homomorphisms $\pi^\top$ and $\pi^\perp$ are defined pointwise as orthogonal projections onto the tangent and normal spaces, respectively, which shows that they are uniquely defined. In terms of an adapted orthonormal frame, they can be written

$$\pi^\top \left(X^1 E_1 + \cdots + X^m E_m \right) = X^1 E_1 + \cdots + X^n E_n,$$
$$\pi^\perp \left(X^1 E_1 + \cdots + X^m E_m \right) = X^{n+1} E_{n+1} + \cdots + X^m E_m,$$

which shows that they are smooth. $\qquad\square$

In case $\widetilde{M}$ is a manifold with boundary, the preceding constructions do not always work, because there is not a fully general construction of slice coordinates in that case. However, there is a satisfactory result in case the submanifold is the boundary itself, using boundary coordinates in place of slice coordinates.

Suppose (M, g) is a Riemannian manifold with boundary. We will always consider ∂M to be a Riemannian submanifold with the induced metric.

Proposition 2.17 (Existence of Outward-Pointing Normal). *If (M, g) is a smooth Riemannian manifold with boundary, the normal bundle to ∂M is a smooth rank-1 vector bundle over ∂M, and there is a unique smooth outward-pointing unit normal vector field along all of ∂M.*

▶ **Exercise 2.18.** Prove this proposition. [Hint: Use the paragraph preceding Prop. B.17 as a starting point.]

Computations on a submanifold $M \subseteq \widetilde{M}$ are usually carried out most conveniently in terms of a ***smooth local parametrization***: this is a smooth map $X \colon U \to \widetilde{M}$, where U is an open subset of $\mathbb{R}^n$ (or $\mathbb{R}^n_+$ in case M has a boundary), such that $X(U)$ is an open subset of M, and such that X, regarded as a map from U into M, is a diffeomorphism onto its image. Note that we can think of X either as a map into M or as a map into $\widetilde{M}$; both maps are typically denoted by the same symbol X. If we put $V = X(U) \subseteq M$ and $\varphi = X^{-1} \colon V \to U$, then (V, φ) is a smooth coordinate chart on M.

Suppose (M, g) is a Riemannian submanifold of $(\widetilde{M}, \widetilde{g})$ and $X \colon U \to \widetilde{M}$ is a smooth local parametrization of M. The coordinate representation of g in these coordinates is given by the following 2-tensor field on U:

$$\left(\varphi^{-1} \right)^* g = X^* g = X^* \iota^* \widetilde{g} = (\iota \circ X)^* \widetilde{g}.$$

Since $\iota \circ X$ is just the map X itself, regarded as a map into $\widetilde{M}$, this is really just $X^*\tilde{g}$. The simplicity of the formula for the pullback of a tensor field makes this expression exceedingly easy to compute, once a coordinate expression for $\tilde{g}$ is known. For example, if M is an immersed n-dimensional Riemannian submanifold of $\mathbb{R}^m$ and $X: U \to \mathbb{R}^m$ is a smooth local parametrization of M, the induced metric on U is just

$$g = X^*\tilde{g} = \sum_{i=1}^{m}(dX^i)^2 = \sum_{i=1}^{m}\left(\sum_{j=1}^{n}\frac{\partial X^i}{\partial u^j}du^j\right)^2 = \sum_{i=1}^{m}\sum_{j,k=1}^{n}\frac{\partial X^i}{\partial u^j}\frac{\partial X^i}{\partial u^k}du^j\,du^k.$$

Example 2.19 (Metrics in Graph Coordinates). If $U \subseteq \mathbb{R}^n$ is an open set and $f: U \to \mathbb{R}$ is a smooth function, then the **graph of f** is the subset $\Gamma(f) = \{(x, f(x)) : x \in U\} \subseteq \mathbb{R}^{n+1}$, which is an embedded submanifold of dimension n. It has a global parametrization $X: U \to \mathbb{R}^{n+1}$ called a **graph parametrization**, given by $X(u) = (u, f(u))$; the corresponding coordinates $(u^1, \ldots, u^n)$ on M are called **graph coordinates**. In graph coordinates, the induced metric of $\Gamma(f)$ is

$$X^*\tilde{g} = X^*\left((dx^1)^2 + \cdots + (dx^{n+1})^2\right) = (du^1)^2 + \cdots + (du^n)^2 + df^2.$$

Applying this to the upper hemisphere of $\mathbb{S}^2$ with the parametrization $X: \mathbb{B}^2 \to \mathbb{R}^3$ given by

$$X(u,v) = \left(u, v, \sqrt{1-u^2-v^2}\right),$$

we see that the round metric on $\mathbb{S}^2$ can be written locally as

$$\begin{aligned}
\mathring{g} = X^*\tilde{g} &= du^2 + dv^2 + \left(\frac{u\,du + v\,dv}{\sqrt{1-u^2-v^2}}\right)^2 \\
&= \frac{(1-v^2)\,du^2 + (1-u^2)\,dv^2 + 2uv\,du\,dv}{1-u^2-v^2}.
\end{aligned}$$ //

Example 2.20 (Surfaces of Revolution). Let H be the half-plane $\{(r,z) : r > 0\}$, and suppose $C \subseteq H$ is an embedded 1-dimensional submanifold. The **surface of revolution** determined by C is the subset $S_C \subseteq \mathbb{R}^3$ given by

$$S_C = \left\{(x,y,z) : \left(\sqrt{x^2+y^2}, z\right) \in C\right\}.$$

The set C is called its **generating curve** (see Fig. 2.1). Every smooth local parametrization $\gamma(t) = (a(t), b(t))$ for C yields a smooth local parametrization for S_C of the form

$$X(t,\theta) = (a(t)\cos\theta, a(t)\sin\theta, b(t)), \tag{2.11}$$

provided that (t,θ) is restricted to a sufficiently small open set in the plane. The t-coordinate curves $t \mapsto X(t,\theta_0)$ are called **meridians**, and the θ-coordinate curves $\theta \mapsto X(t_0,\theta)$ are called **latitude circles**. The induced metric on S_C is

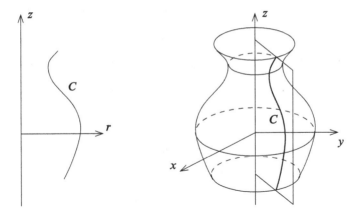

Fig. 2.1: A surface of revolution

$$X^*\overline{g} = d(a(t)\cos\theta)^2 + d(a(t)\sin\theta)^2 + d(b(t))^2$$
$$= (a'(t)\cos\theta\,dt - a(t)\sin\theta\,d\theta)^2$$
$$+ (a'(t)\sin\theta\,dt + a(t)\cos\theta\,d\theta)^2 + (b'(t)dt)^2$$
$$= \big(a'(t)^2 + b'(t)^2\big)dt^2 + a(t)^2 d\theta^2.$$

In particular, if γ is a **unit-speed curve** (meaning that $|\gamma'(t)|^2 = a'(t)^2 + b'(t)^2 \equiv 1$), this reduces to $dt^2 + a(t)^2 d\theta^2$.

Here are some examples of surfaces of revolution and their induced metrics.

- If C is the semicircle $r^2 + z^2 = 1$, parametrized by $\gamma(\varphi) = (\sin\varphi, \cos\varphi)$ for $0 < \varphi < \pi$, then S_C is the unit sphere (minus the north and south poles). The map $X(\varphi, \theta) = (\sin\varphi\cos\theta, \sin\varphi\sin\theta, \cos\varphi)$ constructed above is called the **spherical coordinate parametrization**, and the induced metric is $d\varphi^2 + \sin^2\varphi\,d\theta^2$. (This example is the source of the terminology for meridians and latitude circles.)
- If C is the circle $(r-2)^2 + z^2 = 1$, parametrized by $\gamma(t) = (2 + \cos t, \sin t)$, we obtain a torus of revolution, whose induced metric is $dt^2 + (2 + \cos t)^2 d\theta^2$.
- If C is a vertical line parametrized by $\gamma(t) = (1, t)$, then S_C is the unit cylinder $x^2 + y^2 = 1$, and the induced metric is $dt^2 + d\theta^2$. Note that this means that the parametrization $X: \mathbb{R}^2 \to \mathbb{R}^3$ is an isometric immersion. //

Example 2.21 (The n-Torus as a Riemannian Submanifold). The smooth covering map $X: \mathbb{R}^n \to \mathbb{T}^n$ described in Example A.52 restricts to a smooth local parametrization on any sufficiently small open subset of $\mathbb{R}^n$, and the induced metric is equal to the Euclidean metric in (u^i) coordinates, and therefore the induced metric on $\mathbb{T}^n$ is flat. //

▶ **Exercise 2.22.** Verify the claims in Examples 2.19–2.21.

Riemannian Products

Next we consider products. If (M_1, g_1) and (M_2, g_2) are Riemannian manifolds, the product manifold $M_1 \times M_2$ has a natural Riemannian metric $g = g_1 \oplus g_2$, called the **product metric**, defined by

$$g_{(p_1,p_2)}\big((v_1,v_2),(w_1,w_2)\big) = g_1\big|_{p_1}(v_1,w_1) + g_2\big|_{p_2}(v_2,w_2), \qquad (2.12)$$

where (v_1, v_2) and (w_1, w_2) are elements of $T_{p_1} M_1 \oplus T_{p_2} M_2$, which is naturally identified with $T_{(p_1,p_2)}(M_1 \times M_2)$. Smooth local coordinates $(x^1, \ldots, x^n)$ for M_1 and $(x^{n+1}, \ldots, x^{n+m})$ for M_2 give coordinates $(x^1, \ldots, x^{n+m})$ for $M_1 \times M_2$. In terms of these coordinates, the product metric has the local expression $g = g_{ij} dx^i dx^j$, where (g_{ij}) is the block diagonal matrix

$$(g_{ij}) = \begin{pmatrix} (g_1)_{ab} & 0 \\ 0 & (g_2)_{cd} \end{pmatrix};$$

here the indices a, b run from 1 to n, and c, d run from $n + 1$ to $n + m$. Product metrics on products of three or more Riemannian manifolds are defined similarly.

▶ **Exercise 2.23.** Show that the induced metric on $\mathbb{T}^n$ described in Exercise 2.21 is equal to the product metric obtained from the usual induced metric on $\mathbb{S}^1 \subseteq \mathbb{R}^2$.

Here is an important generalization of product metrics. Suppose (M_1, g_1) and (M_2, g_2) are two Riemannian manifolds, and $f: M_1 \to \mathbb{R}^+$ is a strictly positive smooth function. The **warped product** $M_1 \times_f M_2$ is the product manifold $M_1 \times M_2$ endowed with the Riemannian metric $g = g_1 \oplus f^2 g_2$, defined by

$$g_{(p_1,p_2)}\big((v_1,v_2),(w_1,w_2)\big) = g_1\big|_{p_1}(v_1,w_1) + f(p_1)^2 g_2\big|_{p_2}(v_2,w_2),$$

where $(v_1, v_2), (w_1, w_2) \in T_{p_1} M_1 \oplus T_{p_2} M_2$ as before. (Despite the similarity with the notation for product metrics, $g_1 \oplus f^2 g_2$ is generally not a product metric unless f is constant.) A wide variety of metrics can be constructed in this way; here are just a few examples.

Example 2.24 (Warped Products).

(a) With $f \equiv 1$, the warped product $M_1 \times_f M_2$ is just the space $M_1 \times M_2$ with the product metric.

(b) Every surface of revolution can be expressed as a warped product, as follows. Let H be the half-plane $\{(r, z) : r > 0\}$, let $C \subseteq H$ be an embedded smooth 1-dimensional submanifold, and let $S_C \subseteq \mathbb{R}^3$ denote the corresponding surface of revolution as in Example 2.20. Endow C with the Riemannian metric induced from the Euclidean metric on H, and let $\mathbb{S}^1$ be endowed with its standard metric. Let $f: C \to \mathbb{R}$ be the distance to the z-axis: $f(r, z) = r$. Then Problem 2-3 shows that S_C is isometric to the warped product $C \times_f \mathbb{S}^1$.

(c) If we let ρ denote the standard coordinate function on $\mathbb{R}^+ \subseteq \mathbb{R}$, then the map $\Phi(\rho,\omega) = \rho\omega$ gives an isometry from the warped product $\mathbb{R}^+ \times_\rho \mathbb{S}^{n-1}$ to $\mathbb{R}^n \smallsetminus \{0\}$ with its Euclidean metric (see Problem 2-4).   //

Riemannian Submersions

Unlike submanifolds and products of Riemannian manifolds, which automatically inherit Riemannian metrics of their own, quotients of Riemannian manifolds inherit Riemannian metrics only under very special circumstances. In this section, we see what those circumstances are.

Suppose $\widetilde{M}$ and M are smooth manifolds, $\pi : \widetilde{M} \to M$ is a smooth submersion, and $\widetilde{g}$ is a Riemannian metric on $\widetilde{M}$. By the submersion level set theorem (Corollary A.25), each fiber $\widetilde{M}_y = \pi^{-1}(y)$ is a properly embedded smooth submanifold of $\widetilde{M}$. At each point $x \in \widetilde{M}$, we define two subspaces of the tangent space $T_x\widetilde{M}$ as follows: the **vertical tangent space at x** is

$$V_x = \operatorname{Ker} d\pi_x = T_x\left(\widetilde{M}_{\pi(x)}\right)$$

(that is, the tangent space to the fiber containing x), and the **horizontal tangent space at x** is its orthogonal complement:

$$H_x = (V_x)^\perp.$$

Then the tangent space $T_x\widetilde{M}$ decomposes as an orthogonal direct sum $T_x\widetilde{M} = H_x \oplus V_x$. Note that the vertical space is well defined for every submersion, because it does not refer to the metric; but the horizontal space depends on the metric.

A vector field on $\widetilde{M}$ is said to be a **horizontal vector field** if its value at each point lies in the horizontal space at that point; a **vertical vector field** is defined similarly. Given a vector field X on M, a vector field $\widetilde{X}$ on $\widetilde{M}$ is called a **horizontal lift of** X if $\widetilde{X}$ is horizontal and π-related to X. (The latter property means that $d\pi_x(\widetilde{X}_x) = X_{\pi(x)}$ for each $x \in \widetilde{M}$.)

The next proposition is the principal tool for doing computations on Riemannian submersions.

Proposition 2.25 (Properties of Horizontal Vector Fields). *Let $\widetilde{M}$ and M be smooth manifolds, let $\pi : \widetilde{M} \to M$ be a smooth submersion, and let $\widetilde{g}$ be a Riemannian metric on $\widetilde{M}$.*

(a) *Every smooth vector field W on $\widetilde{M}$ can be expressed uniquely in the form $W = W^H + W^V$, where W^H is horizontal, W^V is vertical, and both W^H and W^V are smooth.*

(b) *Every smooth vector field on M has a unique smooth horizontal lift to $\widetilde{M}$.*

(c) *For every $x \in \widetilde{M}$ and $v \in H_x$, there is a vector field $X \in \mathfrak{X}(M)$ whose horizontal lift $\widetilde{X}$ satisfies $\widetilde{X}_x = v$.*

Proof. Let $p \in \widetilde{M}$ be arbitrary. Because π is a smooth submersion, the rank theorem (Theorem A.15) shows that there exist smooth coordinate charts $(\widetilde{U}, (x^i))$ centered at p and $(U, (u^j))$ centered at $\pi(p)$ in which π has the coordinate representation

$$\pi(x^1, \ldots, x^n, x^{n+1}, \ldots, x^m) = (x^1, \ldots, x^n),$$

where $m = \dim \widetilde{M}$ and $n = \dim M$. It follows that at each point $q \in \widetilde{U}$, the vertical space V_q is spanned by the vectors $\partial_{n+1}|_q, \ldots, \partial_m|_q$. (It probably will not be the case, however, that the horizontal space is spanned by the other n basis vectors.) If we apply the Gram–Schmidt algorithm to the ordered frame $(\partial_{n+1}, \ldots, \partial_m, \partial_1, \ldots, \partial_n)$, we obtain a smooth orthonormal frame $(E_1, \ldots, E_m)$ on $\widetilde{U}$ such that V_q is spanned by $(E_1|_q, \ldots, E_{m-n}|_q)$ at each $q \in \widetilde{U}$. It follows that H_q is spanned by $(E_{m-n+1}|_q, \ldots, E_m|_q)$.

Now let $W \in \mathfrak{X}(\widetilde{M})$ be arbitrary. At each point $q \in \widetilde{M}$, W_q can be written uniquely as a sum of a vertical vector plus a horizontal vector, thus defining a decomposition $W = W^V + W^H$ into rough vertical and horizontal vector fields. To see that they are smooth, just note that in a neighborhood of each point we can express W in terms of a frame $(E_1, \ldots, E_m)$ of the type constructed above as $W = W^1 E_1 + \cdots + W^m E_m$ with smooth coefficients (W^i), and then it follows that $W^V = W^1 E_1 + \cdots + W^{m-n} E_{m-n}$ and $W^H = W^{m-n+1} E_{m-n+1} + \cdots + W^m E_m$, both of which are smooth.

The proofs of (b) and (c) are left to Problem 2-5. □

The fact that every horizontal vector at a point of $\widetilde{M}$ can be extended to a horizontal lift on all of $\widetilde{M}$ (part (c) of the preceding proposition) is highly useful for computations. It is important to be aware, though, that not every horizontal vector field on $\widetilde{M}$ is a horizontal lift, as the next exercise shows.

▶ **Exercise 2.26.** Let $\pi: \mathbb{R}^2 \to \mathbb{R}$ be the projection map $\pi(x, y) = x$, and let W be the smooth vector field $y \partial_x$ on $\mathbb{R}^2$. Show that W is horizontal, but there is no vector field on $\mathbb{R}$ whose horizontal lift is equal to W.

Now we can identify some quotients of Riemannian manifolds that inherit metrics of their own. Let us begin by describing what such a metric should look like.

Suppose $(\widetilde{M}, \widetilde{g})$ and (M, g) are Riemannian manifolds, and $\pi: \widetilde{M} \to M$ is a smooth submersion. Then π is said to be a ***Riemannian submersion*** if for each $x \in \widetilde{M}$, the differential $d\pi_x$ restricts to a linear isometry from H_x onto $T_{\pi(x)}M$. In other words, $\widetilde{g}_x(v, w) = g_{\pi(x)}(d\pi_x(v), d\pi_x(w))$ whenever $v, w \in H_x$.

Example 2.27 (Riemannian Submersions).

(a) The projection $\pi: \mathbb{R}^{n+k} \to \mathbb{R}^n$ onto the first n coordinates is a Riemannian submersion if $\mathbb{R}^{n+k}$ and $\mathbb{R}^n$ are both endowed with their Euclidean metrics.
(b) If M and N are Riemannian manifolds and $M \times N$ is endowed with the product metric, then both projections $\pi_M: M \times N \to M$ and $\pi_N: M \times N \to N$ are Riemannian submersions.

(c) If $M \times_f N$ is a warped product manifold, then the projection $\pi_M : M \times_f N \to M$ is a Riemannian submersion, but π_N typically is not.   //

Given a Riemannian manifold $(\widetilde{M}, \widetilde{g})$ and a surjective submersion $\pi : \widetilde{M} \to M$, it is almost never the case that there is a metric on M that makes π into a Riemannian submersion. It is not hard to see why: for this to be the case, whenever $p_1, p_2 \in \widetilde{M}$ are two points in the same fiber $\pi^{-1}(y)$, the linear maps $(d\pi_{p_i}|_{H_{p_i}})^{-1} : T_y M \to H_{p_i}$ both have to pull $\widetilde{g}$ back to the same inner product on $T_y M$.

There is, however, an important special case in which there is such a metric. Suppose $\pi : \widetilde{M} \to M$ is a smooth surjective submersion, and G is a group acting on $\widetilde{M}$. (See Appendix C for a review of the basic definitions and terminology regarding group actions on manifolds.) We say that the action is **vertical** if every element $\varphi \in G$ takes each fiber to itself, meaning that $\pi(\varphi \cdot p) = \pi(p)$ for all $p \in \widetilde{M}$. The action is **transitive on fibers** if for each $p, q \in \widetilde{M}$ such that $\pi(p) = \pi(q)$, there exists $\varphi \in G$ such that $\varphi \cdot p = q$.

If in addition $\widetilde{M}$ is endowed with a Riemannian metric, the action is said to be an **isometric action** or an **action by isometries**, and the metric is said to be **invariant under G**, if the map $x \mapsto \varphi \cdot x$ is an isometry for each $\varphi \in G$. In that case, provided the action is effective (so that different elements of G define different isometries of $\widetilde{M}$), we can identify G with a subgroup of $\mathrm{Iso}(\widetilde{M}, g)$. Since an isometry is, in particular, a diffeomorphism, every isometric action is an action by diffeomorphisms.

Theorem 2.28. *Let $(\widetilde{M}, \widetilde{g})$ be a Riemannian manifold, let $\pi : \widetilde{M} \to M$ be a surjective smooth submersion, and let G be a group acting on $\widetilde{M}$. If the action is isometric, vertical, and transitive on fibers, then there is a unique Riemannian metric on M such that π is a Riemannian submersion.*

Proof. Problem 2-6.   □

The next corollary describes one important situation to which the preceding theorem applies.

Corollary 2.29. *Suppose $(\widetilde{M}, \widetilde{g})$ is a Riemannian manifold, and G is a Lie group acting smoothly, freely, properly, and isometrically on $\widetilde{M}$. Then the orbit space $M = \widetilde{M}/G$ has a unique smooth manifold structure and Riemannian metric such that π is a Riemannian submersion.*

Proof. Under the given hypotheses, the quotient manifold theorem (Thm. C.17) shows that M has a unique smooth manifold structure such that the quotient map $\pi : \widetilde{M} \to M$ is a smooth submersion. It follows easily from the definitions in that case that the given action of G on $\widetilde{M}$ is vertical and transitive on fibers. Since the action is also isometric, Theorem 2.28 shows that M inherits a unique Riemannian metric making π into a Riemannian submersion.   □

Here is an important example of a Riemannian metric defined in this way. A larger class of such metrics is described in Problem 2-7.

Example 2.30 (The Fubini–Study Metric). Let n be a positive integer, and consider the complex projective space $\mathbb{CP}^n$ defined in Example C.19. That example shows that the map $\pi \colon \mathbb{C}^{n+1} \smallsetminus \{0\} \to \mathbb{CP}^n$ sending each point in $\mathbb{C}^{n+1} \smallsetminus \{0\}$ to its span is a surjective smooth submersion. Identifying $\mathbb{C}^{n+1}$ with $\mathbb{R}^{2n+2}$ endowed with its Euclidean metric, we can view the unit sphere $\mathbb{S}^{2n+1}$ with its round metric $\mathring{g}$ as an embedded Riemannian submanifold of $\mathbb{C}^{n+1} \smallsetminus \{0\}$. Let $p \colon \mathbb{S}^{2n+1} \to \mathbb{CP}^n$ denote the restriction of the map π. Then p is smooth, and it is surjective, because every 1-dimensional complex subspace contains elements of unit norm. We need to show that it is a submersion. Let $z_0 \in \mathbb{S}^{2n+1}$ and set $\zeta_0 = p(z_0) \in \mathbb{CP}^n$. Since π is a smooth submersion, it has a smooth local section $\sigma \colon U \to \mathbb{C}^{n+1}$ defined on a neighborhood U of ζ_0 and satisfying $\sigma(\zeta_0) = z_0$ (Thm. A.17). Let $\nu \colon \mathbb{C}^{n+1} \smallsetminus \{0\} \to \mathbb{S}^{2n+1}$ be the radial projection onto the sphere:

$$\nu(z) = \frac{z}{|z|}.$$

Since dividing an element of $\mathbb{C}^{n+1}$ by a nonzero scalar does not change its span, it follows that $p \circ \nu = \pi$. Therefore, if we set $\tilde{\sigma} = \nu \circ \sigma$, we have $p \circ \tilde{\sigma} = p \circ \nu \circ \sigma = \pi \circ \sigma = \mathrm{Id}_U$, so $\tilde{\sigma}$ is a local section of p. By Theorem A.17, this shows that p is a submersion.

Define an action of $\mathbb{S}^1$ on $\mathbb{S}^{2n+1}$ by complex multiplication:

$$\lambda \cdot \left(z^1, \dots, z^{n+1}\right) = \left(\lambda z^1, \dots, \lambda z^{n+1}\right),$$

for $\lambda \in \mathbb{S}^1$ (viewed as a complex number of norm 1) and $z = \left(z^1, \dots, z^{n+1}\right) \in \mathbb{S}^{2n+1}$. This is easily seen to be isometric, vertical, and transitive on fibers of p. By Theorem 2.28, therefore, there is a unique metric on $\mathbb{CP}^n$ such that the map $p \colon \mathbb{S}^{2n+1} \to \mathbb{CP}^n$ is a Riemannian submersion. This metric is called the ***Fubini–Study metric***; you will have a chance to study its geometric properties in Problems 3-19 and 8-13.   //

Riemannian Coverings

Another important special case of Riemannian submersions occurs in the context of covering maps. Suppose $\left(\widetilde{M}, \tilde{g}\right)$ and (M, g) are Riemannian manifolds. A smooth covering map $\pi \colon \widetilde{M} \to M$ is called a ***Riemannian covering*** if it is a local isometry.

Proposition 2.31. *Suppose $\pi \colon \widetilde{M} \to M$ is a smooth normal covering map, and $\tilde{g}$ is any metric on $\widetilde{M}$ that is invariant under all covering automorphisms. Then there is a unique metric g on M such that π is a Riemannian covering.*

Proof. Proposition A.49 shows that π is a surjective smooth submersion. The automorphism group acts vertically by definition, and Proposition C.21 shows that it acts transitively on fibers when the covering is normal. It then follows from Theorem 2.28 that there is a unique metric g on M such that π is a Riemannian submersion.

Since a Riemannian submersion between manifolds of the same dimension is a local isometry, it follows that π is a Riemannian covering.   □

Proposition 2.32. Suppose $(\widetilde{M}, \widetilde{g})$ is a Riemannian manifold, and Γ is a discrete Lie group acting smoothly, freely, properly, and isometrically on $\widetilde{M}$. Then $\widetilde{M}/\Gamma$ has a unique Riemannian metric such that the quotient map $\pi\colon \widetilde{M} \to \widetilde{M}/\Gamma$ is a normal Riemannian covering.

Proof. Proposition C.23 shows that π is a smooth normal covering map, and Proposition 2.31 shows that $M = \widetilde{M}/\Gamma$ has a unique Riemannian metric such that π is a Riemannian covering.   □

Corollary 2.33. Suppose (M, g) and $(\widetilde{M}, \widetilde{g})$ are connected Riemannian manifolds, $\pi\colon \widetilde{M} \to M$ is a normal Riemannian covering map, and $\Gamma = \operatorname{Aut}_\pi(\widetilde{M})$. Then M is isometric to $\widetilde{M}/\Gamma$.

Proof. Proposition C.20 shows that with the discrete topology, Γ is a discrete Lie group acting smoothly, freely, and properly on $\widetilde{M}$, and then Proposition C.23 shows that $\widetilde{M}/\Gamma$ is a smooth manifold and the quotient map $q\colon \widetilde{M} \to \widetilde{M}/\Gamma$ is a smooth normal covering map. The fact that both π and q are normal coverings implies that Γ acts transitively on the fibers of both maps, so the two maps are constant on each other's fibers. Proposition A.19 then implies that there is a diffeomorphism $F\colon M \to \widetilde{M}/\Gamma$ that satisfies $q \circ F = \pi$. Because both q and π are local isometries, F is too, and because it is bijective it is a global isometry.   □

Example 2.34. The two-element group $\Gamma = \{\pm 1\}$ acts smoothly, freely, properly, and isometrically on $\mathbb{S}^n$ by multiplication. Example C.24 shows that the quotient space is diffeomorphic to the real projective space $\mathbb{RP}^n$ and the quotient map $q\colon \mathbb{S}^n \to \mathbb{RP}^n$ is a smooth normal covering map. Because the action is isometric, Proposition 2.32 shows that there is a unique metric on $\mathbb{RP}^n$ such that q is a Riemannian covering.   //

Example 2.35 (The Open Möbius Band). The *open Möbius band* is the quotient space $M = \mathbb{R}^2/\mathbb{Z}$, where $\mathbb{Z}$ acts on $\mathbb{R}^2$ by $n \cdot (x, y) = (x + n, (-1)^n y)$. This action is smooth, free, proper, and isometric, and therefore M inherits a flat Riemannian metric such that the quotient map is a Riemannian covering. (See Problem 2-8.)   //

▶ **Exercise 2.36.** Let $\mathbb{T}^n \subseteq \mathbb{R}^{2n}$ be the n-torus with its induced metric. Show that the map $X\colon \mathbb{R}^n \to \mathbb{T}^n$ of Example 2.21 is a Riemannian covering.

Basic Constructions on Riemannian Manifolds

Every Riemannian metric yields an abundance of useful constructions on manifolds, besides the obvious ones of lengths of vectors and angles between them. In this section we describe the most basic ones. Throughout this section M is a smooth manifold with or without boundary.

Raising and Lowering Indices

One elementary but important property of Riemannian metrics is that they allow us to convert vectors to covectors and vice versa. Given a Riemannian metric g on M, we define a bundle homomorphism $\hat{g} \colon TM \to T^*M$ by setting

$$\hat{g}(v)(w) = g_p(v, w)$$

for all $p \in M$ and $v, w \in T_pM$. If X and Y are smooth vector fields on M, this yields

$$\hat{g}(X)(Y) = g(X, Y),$$

which implies, first, that $\hat{g}(X)(Y)$ is linear over $C^\infty(M)$ in Y and thus $\hat{g}(X)$ is a smooth covector field by the tensor characterization lemma (Lemma B.6); and second, that the covector field $\hat{g}(X)$ is linear over $C^\infty(M)$ as a function of X, and thus $\hat{g}$ is a smooth bundle homomorphism.

Given a smooth local frame (E_i) and its dual coframe (ε^i), let $g = g_{ij}\,\varepsilon^i\,\varepsilon^j$ be the local expression for g. If $X = X^i E_i$ is a smooth vector field, the covector field $\hat{g}(X)$ has the coordinate expression

$$\hat{g}(X) = \left(g_{ij}X^i\right)\varepsilon^j.$$

Thus the matrix of $\hat{g}$ in any local frame is the same as the matrix of g itself.

Given a vector field X, it is standard practice to denote the components of the covector field $\hat{g}(X)$ by

$$X_j = g_{ij}X^i,$$

so that

$$\hat{g}(X) = X_j\varepsilon^j,$$

and we say that $\hat{g}(X)$ is obtained from X by **lowering an index**. With this in mind, the covector field $\hat{g}(X)$ is denoted by $X^\flat$ and called X **flat**, borrowing from the musical notation for lowering a tone.

Because the matrix (g_{ij}) is nonsingular at each point, the map $\hat{g}$ is invertible, and the matrix of $\hat{g}^{-1}$ is just the inverse matrix of (g_{ij}). We denote this inverse matrix by (g^{ij}), so that $g^{ij}g_{jk} = g_{kj}g^{ji} = \delta^i_k$. The symmetry of g_{ij} easily implies that (g^{ij}) is also symmetric in i and j. In terms of a local frame, the inverse map $\hat{g}^{-1}$ is given by

$$\hat{g}^{-1}(\omega) = \omega^i E_i,$$

where

$$\omega^i = g^{ij}\omega_j. \tag{2.13}$$

If ω is a covector field, the vector field $\hat{g}^{-1}(\omega)$ is called (what else?) ω **sharp** and denoted by $\omega^\sharp$, and we say that it is obtained from ω by **raising an index**. The two inverse isomorphisms $\flat$ and $\sharp$ are known as the **musical isomorphisms**.

Probably the most important application of the sharp operator is to extend the classical gradient operator to Riemannian manifolds. If g is a Riemannian metric on M and $f: M \to \mathbb{R}$ is a smooth function, the **gradient of f** is the vector field $\operatorname{grad} f = (df)^{\sharp}$ obtained from df by raising an index. Unwinding the definitions, we see that $\operatorname{grad} f$ is characterized by the fact that

$$df_p(w) = \langle \operatorname{grad} f|_p, w \rangle \qquad \text{for all } p \in M, w \in T_p M, \tag{2.14}$$

and has the local basis expression

$$\operatorname{grad} f = \left(g^{ij} E_i f \right) E_j.$$

Thus if (E_i) is an orthonormal frame, then $\operatorname{grad} f$ is the vector field whose components are the same as the components of df; but in other frames, this will not be the case.

The next proposition shows that the gradient has the same geometric interpretation on a Riemannian manifold as it does in Euclidean space. If f is a smooth real-valued function on a smooth manifold M, recall that a point $p \in M$ is called a **regular point of f** if $df_p \neq 0$, and a **critical point of f** otherwise; and a level set $f^{-1}(c)$ is called a **regular level set** if every point of $f^{-1}(c)$ is a regular point of f (see Appendix A). Corollary A.26 shows that each regular level set is an embedded smooth hypersurface in M.

Proposition 2.37. *Suppose (M, g) is a Riemannian manifold, $f \in C^{\infty}(M)$, and $\mathcal{R} \subseteq M$ is the set of regular points of f. For each $c \in \mathbb{R}$, the set $M_c = f^{-1}(c) \cap \mathcal{R}$, if nonempty, is an embedded smooth hypersurface in M, and $\operatorname{grad} f$ is everywhere normal to M_c.*

Proof. Problem 2-9. ∎

The flat and sharp operators can be applied to tensors of any rank, in any index position, to convert tensors from covariant to contravariant or vice versa. Formally, this operation is defined as follows: if F is any (k, l)-tensor and $i \in \{1, \dots, k+l\}$ is any covariant index position for F (meaning that the ith argument is a vector, not a covector), we can form a new tensor $F^{\sharp}$ of type $(k+1, l-1)$ by setting

$$F^{\sharp}(\alpha_1, \dots, \alpha_{k+l}) = F(\alpha_1, \dots, \alpha_{i-1}, \alpha_i^{\sharp}, \alpha_{i+1}, \dots, \alpha_{k+l})$$

whenever $\alpha_1, \dots, \alpha_{k+l}$ are vectors or covectors as appropriate. In any local frame, the components of $F^{\sharp}$ are obtained by multiplying the components of F by g^{kl} and contracting one of the indices of g^{kl} with the ith index of F. Similarly, if i is a contravariant index position, we can define a $(k-1, l+1)$-tensor $F^{\flat}$ by

$$F^{\flat}(\alpha_1, \dots, \alpha_{k+l}) = F(\alpha_1, \dots, \alpha_{i-1}, \alpha_i^{\flat}, \alpha_{i+1}, \dots, \alpha_{k+l}).$$

In components, it is computed by multiplying by g_{kl} and contracting.

For example, if A is a mixed 3-tensor given in terms of a local frame by

$$A = A_i{}^j{}_k \, \varepsilon^i \otimes E_j \otimes \varepsilon^k$$

(see (B.6)), we can lower its middle index to obtain a covariant 3-tensor $A^\flat$ with components

$$A_{ijk} = g_{jl} A_i{}^l{}_k.$$

To avoid overly cumbersome notation, we use the symbols $F^\sharp$ and $F^\flat$ without explicitly specifying which index position the sharp or flat operator is to be applied to; when there is more than one choice, we will always stipulate in words what is meant.

Another important application of the flat and sharp operators is to extend the trace operator introduced in Appendix B to covariant tensors. If h is any covariant k-tensor field on a Riemannian manifold with $k \geq 2$, we can raise one of its indices (say the last one for definiteness) and obtain a $(1, k - 1)$-tensor $h^\sharp$. The trace of $h^\sharp$ is thus a well-defined covariant $(k - 2)$-tensor field (see Exercise B.3). We define the *trace of h with respect to g* as

$$\operatorname{tr}_g h = \operatorname{tr}\left(h^\sharp\right).$$

Sometimes we may wish to raise an index other than the last, or to take the trace on a pair of indices other than the last covariant and contravariant ones. In each such case, we will say in words what is meant.

The most important case is that of a covariant 2-tensor field. In this case, $h^\sharp$ is a $(1, 1)$-tensor field, which can equivalently be regarded as an endomorphism field, and $\operatorname{tr}_g h$ is just the ordinary trace of this endomorphism field. In terms of a basis, this is

$$\operatorname{tr}_g h = h_i{}^i = g^{ij} h_{ij}.$$

In particular, in an orthonormal frame this is the ordinary trace of the matrix (h_{ij}) (the sum of its diagonal entries); but if the frame is not orthonormal, then this trace is different from the ordinary trace.

▶ **Exercise 2.38.** If g is a Riemannian metric on M and (E_i) is a local frame on M, there is a potential ambiguity about what the expression (g^{ij}) represents: we have defined it to mean the inverse matrix of (g_{ij}), but one could also interpret it as the components of the contravariant 2-tensor field $g^{\sharp\sharp}$ obtained by raising both of the indices of g. Show that these two interpretations lead to the same result.

Inner Products of Tensors

A Riemannian metric yields, by definition, an inner product on tangent vectors at each point. Because of the musical isomorphisms between vectors and covectors, it is easy to carry the inner product over to covectors as well.

Suppose g is a Riemannian metric on M, and $x \in M$. We can define an inner product on the cotangent space $T_x^* M$ by

$$\langle \omega, \eta \rangle_g = \langle \omega^\sharp, \eta^\sharp \rangle_g.$$

(Just as with inner products of vectors, we might sometimes omit g from the notation when the metric is understood.) To see how to compute this, we just use the basis formula (2.13) for the sharp operator, together with the relation $g_{kl}g^{ki} = g_{lk}g^{ki} = \delta_l^i$, to obtain

$$\langle \omega, \eta \rangle = g_{kl}\left(g^{ki}\omega_i\right)\left(g^{lj}\eta_j\right)$$
$$= \delta_l^i g^{lj}\omega_i \eta_j$$
$$= g^{ij}\omega_i \eta_j.$$

In other words, the inner product on covectors is represented by the inverse matrix $\left(g^{ij}\right)$. Using our conventions for raising and lowering indices, this can also be written

$$\langle \omega, \eta \rangle = \omega_i \eta^i = \omega^j \eta_j.$$

▶ **Exercise 2.39.** Let (M, g) be a Riemannian manifold with or without boundary, let (E_i) be a local frame for M, and let (ε^i) be its dual coframe. Show that the following are equivalent:

(a) (E_i) is orthonormal.
(b) (ε^i) is orthonormal.
(c) $(\varepsilon^i)^\sharp = E_i$ for each i.

This construction can be extended to tensor bundles of any rank, as the following proposition shows. First a bit of terminology: if $E \to M$ is a smooth vector bundle, a **smooth fiber metric on E** is an inner product on each fiber E_p that varies smoothly, in the sense that for any (local) smooth sections σ, τ of E, the inner product $\langle \sigma, \tau \rangle$ is a smooth function.

Proposition 2.40 (Inner Products of Tensors). *Let (M, g) be an n-dimensional Riemannian manifold with or without boundary. There is a unique smooth fiber metric on each tensor bundle $T^{(k,l)}TM$ with the property that if $\alpha_1, \dots, \alpha_{k+l}$, $\beta_1, \dots, \beta_{k+l}$ are vector or covector fields as appropriate, then*

$$\langle \alpha_1 \otimes \cdots \otimes \alpha_{k+l}, \ \beta_1 \otimes \cdots \otimes \beta_{k+l} \rangle = \langle \alpha_1, \beta_1 \rangle \cdot \dots \cdot \langle \alpha_{k+l}, \beta_{k+l} \rangle. \qquad (2.15)$$

With this inner product, if $(E_1, \dots, E_n)$ is a local orthonormal frame for TM and $(\varepsilon^1, \dots, \varepsilon^n)$ is the corresponding dual coframe, then the collection of tensor fields $E_{i_1} \otimes \cdots \otimes E_{i_k} \otimes \varepsilon^{j_1} \otimes \cdots \otimes \varepsilon^{j_l}$ as all the indices range from 1 to n forms a local orthonormal frame for $T^{(k,l)}(T_p M)$. In terms of any (not necessarily orthonormal) frame, this fiber metric satisfies

$$\langle F, G \rangle = g_{i_1 r_1} \cdots g_{i_k r_k} g^{j_1 s_1} \cdots g^{j_l s_l} F^{i_1 \dots i_k}_{j_1 \dots j_l} G^{r_1 \dots r_k}_{s_1 \dots s_l}. \qquad (2.16)$$

If F and G are both covariant, this can be written

$$\langle F, G \rangle = F_{j_1 \dots j_l} G^{j_1 \dots j_l},$$

*where the last factor on the right represents the components of G with all of its
indices raised:*

$$G^{j_1 \cdots j_l} = g^{j_1 s_1} \cdots g^{j_l s_l} G_{s_1 \ldots s_l}.$$

Proof. Problem 2-11. □

The Volume Form and Integration

Another important construction provided by a metric on an oriented manifold is a
canonical volume form.

Proposition 2.41 (The Riemannian Volume Form). *Let (M, g) be an oriented Riemannian n-manifold with or without boundary. There is a unique n-form dV_g on
M, called the **Riemannian volume form**, characterized by any one of the following
three equivalent properties:*

(a) If $(\varepsilon^1, \ldots, \varepsilon^n)$ is any local oriented orthonormal coframe for T^*M, then

$$dV_g = \varepsilon^1 \wedge \cdots \wedge \varepsilon^n.$$

(b) If $(E_1, \ldots, E_n)$ is any local oriented orthonormal frame for TM, then

$$dV_g(E_1, \ldots, E_n) = 1.$$

(c) If $(x^1, \ldots, x^n)$ are any oriented local coordinates, then

$$dV_g = \sqrt{\det(g_{ij})}\, dx^1 \wedge \cdots \wedge dx^n.$$

Proof. Problem 2-12. □

The significance of the Riemannian volume form is that it allows us to integrate
functions on an oriented Riemannian manifold, not just differential forms. If f is
a continuous, compactly supported real-valued function on an oriented Riemannian
n-manifold (M, g) with or without boundary, then $f\, dV_g$ is a compactly supported
n-form. Therefore, the integral $\int_M f\, dV_g$ makes sense, and we define it to be the
integral of f over M. Similarly, if M is compact, the **volume of M** is defined to
be

$$\text{Vol}(M) = \int_M dV_g = \int_M 1\, dV_g.$$

In particular, if $D \subseteq M$ is a **regular domain** (a closed, embedded codimension-0
submanifold with boundary), we can apply these definitions to D with its induced
metric and thereby make sense of the integral of f over D and, in case D is compact, the volume of D.

The notation dV_g is chosen to emphasize the similarity of the integral $\int_M f\, dV_g$
with the standard integral of a function over an open subset of $\mathbb{R}^n$. It is *not* meant
to imply that dV_g is an exact form; in fact, if M is a compact oriented manifold

without boundary, then dV_g is never exact, because its integral over M is positive, and exact forms integrate to zero by Stokes's theorem.

Because there are two conventions in common use for the wedge product (see p. 401), it should be noted that properties (a) and (c) of Proposition 2.41 are the same regardless of which convention is used; but property (b) holds only for the determinant convention that we use. If the Alt convention is used, the number 1 should be replaced by $1/n!$ in that formula.

▶ **Exercise 2.42.** Suppose (M, g) and $(\widetilde{M}, \widetilde{g})$ are oriented Riemannian manifolds, and $\varphi : M \to \widetilde{M}$ is an orientation-preserving isometry. Prove that $\varphi^* dV_{\widetilde{g}} = dV_g$.

For Riemannian hypersurfaces, we have the following important characterization of the volume form on the hypersurface in terms of that of the ambient manifold. If X is a vector field and μ is a differential form, recall that $X \lrcorner \mu$ denotes **interior multiplication** of μ by X (see p. 401).

Proposition 2.43. *Suppose M is a hypersurface in an oriented Riemannian manifold $(\widetilde{M}, \widetilde{g})$ and g is the induced metric on M. Then M is orientable if and only if there exists a global unit normal vector field N for M, and in that case the volume form of (M, g) is given by*

$$dV_g = \left(N \lrcorner dV_{\widetilde{g}} \right)\big|_M. \tag{2.17}$$

Proof. Problem 2-13.   □

When M is not orientable, we can still define integrals of functions, but now we have to use *densities* instead of differential forms (see pp. 405–406).

Proposition 2.44 (The Riemannian Density). *If (M, g) is any Riemannian manifold, then there is a unique smooth positive density μ on M, called the **Riemannian density**, with the property that*

$$\mu(E_1, \ldots, E_n) = 1 \tag{2.18}$$

for every local orthonormal frame (E_i).

▶ **Exercise 2.45.** Prove this proposition by showing that μ can be defined in terms of any local orthonormal frame by
$$\mu = |\varepsilon^1 \wedge \cdots \wedge \varepsilon^n|.$$

Let (M, g) be a Riemannian manifold (with or without boundary). If M is oriented and dV_g is its Riemannian volume form, then its Riemannian density is easily seen to be equal to $|dV_g|$. On the other hand, the Riemannian density is defined whether M is oriented or not. It is customary to denote the Riemannian density by the same notation dV_g that we use for the Riemannian volume form, and to specify when necessary whether the notation refers to a density or a form. In either case, we can define the integral of a compactly supported smooth function $f : M \to \mathbb{R}$ as $\int_M f\, dV_g$. This is to be interpreted as the integral of a density when M is nonorientable; when M is orientable, it can be interpreted either as the integral of a density or as the integral of an n-form (with respect to some choice of orientation), because both give the same result.

The Divergence and the Laplacian

In advanced calculus, you have undoubtedly been introduced to three important differential operators involving vector fields on $\mathbb{R}^3$: the *gradient* (which takes real-valued functions to vector fields), *divergence* (vector fields to functions), and *curl* (vector fields to vector fields). We have already described how the gradient operator can be generalized to Riemannian manifolds (see equation (2.14)); now we can show that the divergence operator also generalizes easily to that setting. Problem 2-27 describes a similar, but more limited, generalization of the curl.

Suppose (M, g) is an oriented Riemannian n-manifold with or without boundary, and dV_g is its volume form. If X is a smooth vector field on M, then $X \lrcorner \, dV_g$ is an $(n-1)$-form. The exterior derivative of this $(n-1)$-form is a smooth n-form, so it can be expressed as a smooth function multiplied by dV_g. That function is called the **divergence of** X, and denoted by div X; thus it is characterized by the following formula:

$$d(X \lrcorner \, dV_g) = (\operatorname{div} X) dV_g. \tag{2.19}$$

Even if M is nonorientable, in a neighborhood of each point we can choose an orientation and define the divergence by (2.19), and then note that reversing the orientation changes the sign of dV_g on both sides of the equation, so div X is well defined, independently of the choice of orientation. In this way, we can define the divergence operator on *any* Riemannian manifold with or without boundary, by requiring that it satisfy (2.19) for any choice of orientation in a neighborhood of each point.

The most important application of the divergence operator is the **divergence theorem**, which you will be asked to prove in Problem 2-22.

Using the divergence operator, we can define another important operator, this one acting on real-valued functions. The **Laplacian** (or **Laplace–Beltrami operator**) is the linear operator $\Delta: C^\infty(M) \to C^\infty(M)$ defined by

$$\Delta u = \operatorname{div}(\operatorname{grad} u). \tag{2.20}$$

(Note that many books, including [LeeSM] and the first edition of this book, define the Laplacian as $-\operatorname{div}(\operatorname{grad} u)$. The main reason for choosing the negative sign is so that the operator will have nonnegative eigenvalues; see Problem 2-24. But the definition we give here is much more common in Riemannian geometry.)

The next proposition gives alternative formulas for these operators.

Proposition 2.46. *Let* (M, g) *be a Riemannian manifold with or without boundary, and let* (x^i) *be any smooth local coordinates on an open set* $U \subseteq M$. *The coordinate representations of the divergence and Laplacian are as follows:*

$$\operatorname{div}\left(X^i\,\frac{\partial}{\partial x^i}\right) = \frac{1}{\sqrt{\det g}}\,\frac{\partial}{\partial x^i}\left(X^i\,\sqrt{\det g}\right),$$

$$\Delta u = \frac{1}{\sqrt{\det g}}\,\frac{\partial}{\partial x^i}\left(g^{ij}\,\sqrt{\det g}\,\frac{\partial u}{\partial x^j}\right),$$

where $\det g = \det(g_{kl})$ *is the determinant of the component matrix of g in these coordinates. On* $\mathbb{R}^n$ *with the Euclidean metric and standard coordinates, these reduce to*

$$\operatorname{div}\left(X^i\,\frac{\partial}{\partial x^i}\right) = \sum_{i=1}^{n} \frac{\partial X^i}{\partial x^i},$$

$$\Delta u = \sum_{i=1}^{n} \frac{\partial^2 u}{(\partial x^i)^2}.$$

Proof. Problem 2-21. $\square$

Lengths and Distances

Perhaps the most important tool that a Riemannian metric gives us is the ability to measure lengths of curves and distances between points. Throughout this section, (M, g) denotes a Riemannian manifold with or without boundary.

Without further qualification, a ***curve*** in M always means a ***parametrized curve***, that is, a continuous map $\gamma \colon I \to M$, where $I \subseteq \mathbb{R}$ is some interval. Unless otherwise specified, we will not worry about whether the interval is bounded or unbounded, or whether it includes endpoints or not. To say that γ is a ***smooth curve*** is to say that it is smooth as a map from the manifold (with boundary) I to M. If I has one or two endpoints and M has empty boundary, then γ is smooth if and only if it extends to a smooth curve defined on some open interval containing I. (If $\partial M \neq \varnothing$, then smoothness of γ has to be interpreted as meaning that each coordinate representation of γ has a smooth extension to an open interval.) A ***curve segment*** is a curve whose domain is a compact interval.

A smooth curve $\gamma \colon I \to M$ has a well-defined velocity $\gamma'(t) \in T_{\gamma(t)}M$ for each $t \in I$. We say that γ is a ***regular curve*** if $\gamma'(t) \neq 0$ for $t \in I$. This implies that γ is an immersion, so its image has no "corners" or "kinks."

We wish to use curve segments as "measuring tapes" to define distances between points in a connected Riemannian manifold. Many aspects of the theory become technically much simpler if we work with a slightly larger class of curve segments instead of just the regular ones. We now describe the appropriate class of curves.

If $[a, b] \subseteq \mathbb{R}$ is a closed bounded interval, a ***partition of*** $[a, b]$ is a finite sequence $(a_0, \ldots, a_k)$ of real numbers such that $a = a_0 < a_1 < \cdots < a_k = b$. Each interval $[a_{i-1}, a_i]$ is called a ***subinterval of the partition***. If M is a smooth manifold with or without boundary, a (continuous) curve segment $\gamma \colon [a, b] \to M$ is said to be *piecewise regular* if there exists a partition $(a_0, \ldots, a_k)$ of $[a, b]$ such that $\gamma|_{[a_{i-1}, a_i]}$ is a regular curve segment (meaning it is smooth with nonvanishing velocity) for $i = 1, \ldots, k$. For brevity, we refer to a piecewise regular curve segment as an *admissible curve*, and any partition $(a_0, \ldots, a_k)$ such that $\gamma|_{[a_{i-1}, a_i]}$ is smooth for each i an *admissible partition for γ*. (There are many admissible partitions for a given admissible curve, because we can always add more points to the partition.) It is also convenient to consider any map $\gamma \colon \{a\} \to M$ whose domain is a single real number to be an admissible curve.

Suppose γ is an admissible curve and $(a_0, \ldots, a_k)$ is an admissible partition for it. At each of the intermediate partition points $a_1, \ldots, a_{k-1}$, there are two one-sided velocity vectors, which we denote by

$$\gamma'(a_i^-) = \lim_{t \nearrow a_i} \gamma'(t),$$
$$\gamma'(a_i^+) = \lim_{t \searrow a_i} \gamma'(t).$$

They are both nonzero, but they need not be equal.

If $\gamma \colon I \to M$ is a smooth curve, we define a *reparametrization of γ* to be a curve of the form $\widetilde{\gamma} = \gamma \circ \varphi \colon I' \to M$, where $I' \subseteq \mathbb{R}$ is another interval and $\varphi \colon I' \to I$ is a diffeomorphism. Because intervals are connected, φ is either strictly increasing or strictly decreasing on I'. We say that $\widetilde{\gamma}$ is a *forward reparametrization* if φ is increasing, and a *backward reparametrization* if it is decreasing.

For an admissible curve $\gamma \colon [a, b] \to M$, we define a *reparametrization of γ* a little more broadly, as a curve of the form $\widetilde{\gamma} = \gamma \circ \varphi$, where $\varphi \colon [c, d] \to [a, b]$ is a homeomorphism for which there is a partition $(c_0, \ldots, c_k)$ of $[c, d]$ such that the restriction of φ to each subinterval $[c_{i-1}, c_i]$ is a diffeomorphism onto its image.

If $\gamma \colon [a, b] \to M$ is an admissible curve, we define the *length of γ* to be

$$L_g(\gamma) = \int_a^b |\gamma'(t)|_g \, dt.$$

The integrand is bounded and continuous everywhere on $[a, b]$ except possibly at the finitely many points where γ is not smooth, so this integral is well defined.

Proposition 2.47 (Properties of Lengths). *Suppose (M, g) is a Riemannian manifold with or without boundary, and $\gamma \colon [a, b] \to M$ is an admissible curve.*

(a) ADDITIVITY OF LENGTH: *If $a < c < b$, then*

$$L_g(\gamma) = L_g\left(\gamma|_{[a,c]}\right) + L_g\left(\gamma|_{[c,b]}\right).$$

(b) PARAMETER INDEPENDENCE: *If $\widetilde{\gamma}$ is a reparametrization of γ, then $L_g(\gamma) = L_g(\widetilde{\gamma})$.*

(c) ISOMETRY INVARIANCE: *If (M, g) and $(\widetilde{M}, \widetilde{g})$ are Riemannian manifolds (with or without boundary) and $\varphi \colon M \to \widetilde{M}$ is a local isometry, then $L_g(\gamma) = L_{\widetilde{g}}(\varphi \circ \gamma)$.*

▶ **Exercise 2.48.** Prove Proposition 2.47.

Suppose $\gamma\colon [a,b] \to M$ is an admissible curve. The **arc-length function of γ** is the function $s\colon [a,b] \to \mathbb{R}$ defined by

$$s(t) = L_g\big(\gamma|_{[a,t]}\big) = \int_a^t |\gamma'(u)|_g \, du.$$

It is continuous everywhere, and it follows from the fundamental theorem of calculus that it is smooth wherever γ is, with derivative $s'(t) = |\gamma'(t)|$.

For this reason, if $\gamma\colon I \to M$ is any smooth curve (not necessarily a curve segment), we define the **speed of γ** at any time $t \in I$ to be the scalar $|\gamma'(t)|$. We say that γ is a **unit-speed curve** if $|\gamma'(t)| = 1$ for all t, and a **constant-speed curve** if $|\gamma'(t)|$ is constant. If γ is a piecewise smooth curve, we say that γ has unit speed if $|\gamma'(t)| = 1$ wherever γ is smooth.

If $\gamma\colon [a,b] \to M$ is a unit-speed admissible curve, then its arc-length function has the simple form $s(t) = t - a$. If, in addition, its parameter interval is of the form $[0,b]$ for some $b > 0$, then the arc-length function is $s(t) = t$. For this reason, a unit-speed admissible curve whose parameter interval is of the form $[0,b]$ is said to be **parametrized by arc length**.

Proposition 2.49. *Suppose (M,g) is a Riemannian manifold with or without boundary.*

(a) *Every regular curve in M has a unit-speed forward reparametrization.*

(b) *Every admissible curve in M has a unique forward reparametrization by arc length.*

Proof. Suppose $\gamma\colon I \to M$ is a regular curve. Choose an arbitrary $t_0 \in I$, and define $s\colon I \to \mathbb{R}$ by

$$s(t) = \int_{t_0}^t |\gamma'(u)|_g \, du.$$

Since $s'(t) = |\gamma'(t)|_g > 0$, it follows that s is a strictly increasing local diffeomorphism, and thus maps I diffeomorphically onto an interval $I' \subseteq \mathbb{R}$. If we let $\varphi = s^{-1}\colon I' \to I$, then $\widetilde{\gamma} = \gamma \circ \varphi$ is a forward reparametrization of γ, and the chain rule gives

$$\big|\widetilde{\gamma}'(t)\big|_g = |\varphi'(s)\gamma'(\varphi(s))|_g = \frac{1}{s'(\varphi(s))}|\gamma'(\varphi(s))|_g = 1.$$

Thus $\widetilde{\gamma}$ is a unit-speed reparametrization of γ.

Now let $\gamma\colon [a,b] \to M$ be an admissible curve. We prove the existence statement in part (b) by induction on the number of smooth segments in an admissible partition. If γ has only one smooth segment, then it is a regular curve segment, and (b) follows by applying (a) in the special case $I = [a,b]$ and choosing $t_0 = a$. Assuming that the result is true for admissible curves with k smooth segments, suppose γ

has an admissible partition $(a_0, \ldots, a_{k+1})$. The inductive hypothesis gives piece-wise regular homeomorphisms $\varphi \colon [0,c] \to [a, a_k]$ and $\psi \colon [0,d] \to [a_k, b]$ such that $\gamma \circ \varphi$ is an arc-length reparametrization of $\gamma|_{[a,a_k]}$ and $\gamma \circ \psi$ is an arc-length reparametrization of $\gamma|_{[a_k,b]}$. If we define $\widetilde{\varphi} \colon [0, c+d] \to [a,b]$ by

$$\widetilde{\varphi}(s) = \begin{cases} \varphi(s), & s \in [0,c], \\ \psi(s-c), & s \in [c, c+d], \end{cases}$$

then $\gamma \circ \widetilde{\varphi}$ is a reparametrization of γ by arc length.

To prove uniqueness, suppose that $\widetilde{\gamma} = \gamma \circ \varphi$ and $\widehat{\gamma} = \gamma \circ \psi$ are both for-ward reparametrizations of γ by arc length. Since both have the same length, it follows that φ and ψ both have the same parameter domain $[0,c]$, and thus both are piecewise regular homeomorphisms from $[0,c]$ to $[a,b]$. If we define $\eta = \varphi^{-1} \circ \psi \colon [0,c] \to [0,c]$, then η is a piecewise regular increasing homeomor-phism satisfying $\widehat{\gamma} = \gamma \circ \varphi \circ \eta = \widetilde{\gamma} \circ \eta$. The fact that both $\widetilde{\gamma}$ and $\widehat{\gamma}$ are of unit speed implies the following equality for all $s \in [0,c]$ except the finitely many values at which $\widetilde{\gamma}$ or η is not smooth:

$$1 = \left|\widehat{\gamma}'(s)\right|_g = \left|\widetilde{\gamma}'(\eta(s))\eta'(s)\right|_g = \left|\widetilde{\gamma}'(\eta(s))\right|_g \eta'(s) = \eta'(s).$$

Since η is continuous and $\eta(0) = 0$, it follows that $\eta(s) = s$ for all $s \in [0,c]$, and thus $\widetilde{\gamma} = \widehat{\gamma}$. $\qquad\square$

The Riemannian Distance Function

We are now in a position to introduce one of the most important concepts from classical geometry into the Riemannian setting: distances between points.

Suppose (M,g) is a connected Riemannian manifold with or without boundary. For each pair of points $p,q \in M$, we define the **Riemannian distance from p to q**, denoted by $d_g(p,q)$, to be the infimum of the lengths of all admissible curves from p to q. The following proposition guarantees that $d_g(p,q)$ is a well-defined nonnegative real number for each $p,q \in M$.

Proposition 2.50. *If M is a connected smooth manifold (with or without boundary), then any two points of M can be joined by an admissible curve.*

Proof. Let $p,q \in M$ be arbitrary. Since a connected manifold is path-connected, p and q can be connected by a continuous path $c \colon [a,b] \to M$. By compactness, there is a partition of $[a,b]$ such that $c([a_{i-1}, a_i])$ is contained in a single smooth coordinate ball (or half-ball in case $\partial M \neq \varnothing$) for each i. Then we may replace each such curve segment by a straight-line path in coordinates, yielding an admissible curve γ between the same points (Fig. 2.2). $\qquad\square$

For convenience, if (M,g) is a disconnected Riemannian manifold, we also let $d_g(p,q)$ denote the Riemannian distance from p to q, provided that p and q lie in the same connected component of M. (See also Problem 2-30.)

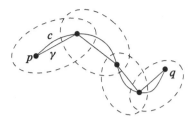

Fig. 2.2: Any two points can be connected by an admissible curve

Proposition 2.51 (Isometry Invariance of the Riemannian Distance Function).
Suppose (M, g) and $\left(\widetilde{M}, \widetilde{g}\right)$ are connected Riemannian manifolds with or without boundary, and $\varphi \colon M \to \widetilde{M}$ is an isometry. Then $d_{\widetilde{g}}(\varphi(x), \varphi(y)) = d_g(x, y)$ for all $x, y \in M$.

▶ **Exercise 2.52.** Prove the preceding proposition.

(Note that unlike lengths of curves, Riemannian distances are not necessarily preserved by *local* isometries; see Problem 2-31.)

We wish to show that the Riemannian distance function turns M into a metric space, whose metric topology is the same as its original manifold topology. To do so, we need the following lemmas.

Lemma 2.53. *Let g be a Riemannian metric on an open subset $W \subseteq \mathbb{R}^n$ or $\mathbb{R}^n_+$, and let $\bar{g}$ denote the Euclidean metric on W. For every compact subset $K \subseteq W$, there are positive constants c, C such that for all $x \in K$ and all $v \in T_x \mathbb{R}^n$,*

$$c|v|_{\bar{g}} \le |v|_g \le C|v|_{\bar{g}}. \tag{2.21}$$

Proof. Define a continuous function $F \colon TW \to \mathbb{R}$ by $F(x, v) = |v|_g = g_x(v, v)^{1/2}$ for $x \in W$ and $v \in T_x \mathbb{R}^n$. Let $K \subseteq W$ be any compact subset, and define $L \subseteq T\mathbb{R}^n$ by

$$L = \left\{ (x, v) \in T\mathbb{R}^n : x \in K, \ |v|_{\bar{g}} = 1 \right\}.$$

Under the canonical identification of $T\mathbb{R}^n$ with $\mathbb{R}^n \times \mathbb{R}^n$, L is equal to the product set $K \times \mathbb{S}^{n-1}$, which is compact. Since F is positive on L, there are positive constants c, C such that

$$c \le |v|_g \le C \quad \text{for all } (x, v) \in L.$$

For each $x \in K$ and $v \in T_x \mathbb{R}^n$ with $v \ne 0$, we can write $v = \lambda \hat{v}$, where $\lambda = |v|_{\bar{g}}$ and $\hat{v} = v/\lambda$. Then $(x, \hat{v}) \in L$, and it follows from the homogeneity of $|\cdot|_g$ that

$$|v|_g = \lambda |\hat{v}|_g \le \lambda C = C|v|_{\bar{g}}.$$

The same inequality holds trivially when $v = 0$. Arguing similarly in the other direction, we conclude that (2.21) holds for all $x \in K$ and $v \in T_x \mathbb{R}^n$. $\qquad \square$

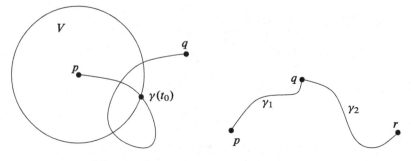

Fig. 2.3: $d_g(p,q) \geq D$ Fig. 2.4: The triangle inequality

The next lemma shows how to transfer this result to Riemannian manifolds.

Lemma 2.54. *Let (M,g) be a Riemannian manifold with or without boundary and let d_g be its Riemannian distance function. Suppose U is an open subset of M and $p \in U$. Then p has a coordinate neighborhood $V \subseteq U$ with the property that there are positive constants C, D satisfying the following inequalities:*

(a) *If $q \in V$, then $d_g(p,q) \leq C d_{\bar{g}}(p,q)$, where $\bar{g}$ is the Euclidean metric in the given coordinates on V.*
(b) *If $q \in M \smallsetminus V$, then $d_g(p,q) \geq D$.*

Proof. Let W be any neighborhood of p contained in U on which there exist smooth coordinates (x^i). Using these coordinates, we can identify W with an open subset of $\mathbb{R}^n$ or $\mathbb{R}^n_+$. Let K be a compact subset of W containing a neighborhood V of p, chosen as follows: If p is an interior point of M, let K be the closed Euclidean ball $\bar{B}_\varepsilon(0)$, with $\varepsilon > 0$ chosen small enough that $K \subseteq W$, and let $V = B_\varepsilon(0)$. If $p \in \partial M$, let K be a closed half-ball $\bar{B}_\varepsilon(0) \cap \mathbb{R}^n_+$ with ε chosen similarly, and let $V = B_\varepsilon(0) \cap \mathbb{R}^n_+$. Let $\bar{g} = \sum_i (dx^i)^2$ denote the Euclidean metric in these coordinates, and let c, C be constants satisfying (2.21). If γ is any admissible curve whose image lies entirely in V, it follows that

$$cL_{\bar{g}}(\gamma) \leq L_g(\gamma) \leq CL_{\bar{g}}(\gamma). \tag{2.22}$$

To prove (a), suppose $q \in V$. Letting $\gamma \colon [a,b] \to V$ be a parametrization of the line segment from p to q, we conclude from (2.22) that

$$d_g(p,q) \leq L_g(\gamma) \leq CL_{\bar{g}}(\gamma) = C d_{\bar{g}}(p,q).$$

To prove (b), suppose $q \in M \smallsetminus V$. If $\gamma \colon [a,b] \to M$ is any admissible curve from p to q, let t_0 denote the infimum of times $t \in [a,b]$ such that $\gamma(t) \notin V$. It follows that $\gamma([a,t_0]) \subseteq K$ and $d_{\bar{g}}(p, \gamma(t_0)) = \varepsilon$ by continuity (Fig. 2.3), so (2.22) implies

$$L_g(\gamma) \geq L_g(\gamma|_{[a,t_0]}) \geq cL_{\bar{g}}(\gamma|_{[a,t_0]}) \geq c\, d_{\bar{g}}(p, \gamma(t_0)) = c\varepsilon.$$

Taking the infimum over all such curves γ, we obtain (b) with $D = c\varepsilon$. $\qquad\square$

Theorem 2.55 (Riemannian Manifolds as Metric Spaces). *Let (M, g) be a connected Riemannian manifold with or without boundary. With the distance function d_g, M is a metric space whose metric topology is the same as the given manifold topology.*

Proof. It is immediate from the definition of d_g that $d_g(p, q) = d_g(q, p) \geq 0$ and $d_g(p, p) = 0$. On the other hand, suppose $p, q \in M$ are distinct. Let $U \subseteq M$ be an open set that contains p but not q, and choose a coordinate neighborhood V of p contained in U and satisfying the conclusion of Lemma 2.54. Then Lemma 2.54(b) shows that $d_g(p, q) \geq D > 0$.

The triangle inequality follows from the fact that an admissible curve from p to q can be combined with one from q to r (possibly changing the starting time of the parametrization of the second) to yield one from p to r whose length is the sum of the lengths of the two given curves (Fig. 2.4). (This is one reason for defining distance using *piecewise regular* curves instead of just regular ones.) This completes the proof that d_g turns M into a metric space.

It remains to show that the metric topology is the same as the manifold topology. Suppose first that $U \subseteq M$ is open in the manifold topology. For each $p \in U$, we can choose a coordinate neighborhood V of p contained in U with positive constants C, D satisfying the conclusions of Lemma 2.54. The contrapositive of part (b) of that lemma says $d_g(p, q) < D \Rightarrow q \in V \subseteq U$, which means that the metric ball of radius D is contained in U. Thus U is open in the metric topology induced by d_g.

On the other hand, suppose U' is open in the metric topology. Given $p \in U'$, choose $\delta > 0$ such that the d_g-metric ball of radius δ around p is contained in U'. Let V be any neighborhood of p that is open in the manifold topology and satisfies the conclusions of Lemma 2.54, with corresponding constants C, D. (We are not claiming that $V \subseteq U'$.) Choose ε small enough that $C\varepsilon < \delta$. Lemma 2.54(a) shows that if q is a point of V such that $d_{\bar{g}}(p, q) < \varepsilon$, then $d_g(p, q) \leq C\varepsilon < \delta$, and thus q lies in the metric ball of radius δ about p, and hence in U'. Since the set $\{q \in V : d_{\bar{g}}(p, q) < \varepsilon\}$ is open in the given manifold topology, this shows that U' is also open in the manifold topology. $\qquad\square$

Thanks to the preceding theorem, it makes sense to apply all the concepts of the theory of metric spaces to a connected Riemannian manifold (M, g). For example, we say that M is (*metrically*) *complete* if every Cauchy sequence in M converges. A subset $A \subseteq M$ is *bounded* if there is a constant C such that $d_g(p, q) \leq C$ for all $p, q \in A$; if this is the case, the *diameter of A* is the smallest such constant:

$$\operatorname{diam}(A) = \sup\{d_g(p, q) : p, q \in A\}.$$

Since every compact metric space is bounded, every compact connected Riemannian manifold with or without boundary has finite diameter. (Note that the unit sphere with the Riemannian distance determined by the round metric has diameter π, not 2, since the Riemannian distance between antipodal points is π. See Problem 6-2.)

Pseudo-Riemannian Metrics

From the point of view of geometry, Riemannian metrics are by far the most important structures that manifolds carry. However, there is a generalization of Riemannian metrics that has become especially important because of its application to physics.

Before defining this generalization, we begin with some linear-algebraic preliminaries. Suppose V is a finite-dimensional vector space, and q is a symmetric covariant 2-tensor on V (also called a *symmetric bilinear form*). Just as for an inner product, there is a linear map $\hat{q} : V \to V^*$ defined by

$$\hat{q}(v)(w) = q(v, w) \text{ for all } v, w \in V.$$

We say that q is *nondegenerate* if $\hat{q}$ is an isomorphism.

Lemma 2.56. *Suppose q is a symmetric covariant 2-tensor on a finite-dimensional vector space V. The following are equivalent:*

(a) *q is nondegenerate.*
(b) *For every nonzero $v \in V$, there is some $w \in V$ such that $q(v, w) \neq 0$.*
(c) *If $q = q_{ij} \varepsilon^i \varepsilon^j$ in terms of some basis (ε^j) for V^*, then the matrix (q_{ij}) is invertible.*

▶ **Exercise 2.57.** Prove the preceding lemma.

Every inner product is a nondegenerate symmetric bilinear form, as is every symmetric bilinear form that is *negative definite* (which is defined by obvious analogy with positive definite). But there are others that are neither positive definite nor negative definite, as we will see below.

We use the term *scalar product* to denote any nondegenerate symmetric bilinear form on a finite-dimensional vector space V, and reserve the term *inner product* for the special case of a positive definite scalar product. A *scalar product space* is a finite-dimensional vector space endowed with a scalar product. When convenient, we will often use a notation like $\langle \cdot, \cdot \rangle$ to denote a scalar product. We say that vectors $v, w \in V$ are *orthogonal* if $\langle v, w \rangle = 0$, just as in the case of an inner product. Given a vector $v \in V$, we define the *norm of v* to be $|v| = |\langle v, v \rangle|^{1/2}$. Note that in the indefinite case, it is possible for a nonzero vector to be orthogonal to itself, and thus to have norm zero. Thus $|v|$ is not technically a norm in the sense defined on page 47 below, but it is customary to call it "the norm of v" anyway.

▶ **Exercise 2.58.** Prove that the polarization identity (2.2) holds for every scalar product.

If $S \subseteq V$ is any linear subspace, the set of vectors in V that are orthogonal to every vector in S is a linear subspace denoted by $S^\perp$.

Lemma 2.59. *Suppose (V, q) is a finite-dimensional scalar product space, and $S \subseteq V$ is a linear subspace.*

(a) $\dim S + \dim S^\perp = \dim V$.
(b) $(S^\perp)^\perp = S$.

Proof. Define a linear map $\Phi \colon V \to S^*$ by $\Phi(v) = \hat{q}(v)|_S$. Note that $v \in \operatorname{Ker}\Phi$ if and only if $q(v,x) = \hat{q}(v)(x) = 0$ for all $x \in S$, so $\operatorname{Ker}\Phi = S^\perp$. If $\varphi \in S^*$ is arbitrary, there is a covector $\tilde{\varphi} \in V^*$ whose restriction to S is equal to φ. (For example, such a covector is easily constructed after choosing a basis for S and extending it to a basis for V.) Since $\hat{q}$ is an isomorphism, there exists $v \in V$ such that $\hat{q}(v) = \tilde{\varphi}$. It follows that $\Phi(v) = \varphi$, and therefore Φ is surjective. By the rank–nullity theorem, the dimension of $S^\perp = \operatorname{Ker}\Phi$ is equal to $\dim V - \dim S^* = \dim V - \dim S$. This proves (a).

To prove (b), note that every $v \in S$ is orthogonal to every element of $S^\perp$ by definition, so $S \subseteq (S^\perp)^\perp$. Because these finite-dimensional vector spaces have the same dimension by part (a), they are equal. $\qquad\square$

An ordered k-tuple $(v_1, \ldots, v_k)$ of elements of V is said to be **orthonormal** if $|v_i| = 1$ for each i and $\langle v_i, v_j \rangle = 0$ for $i \neq j$, or equivalently, if $\langle v_i, v_j \rangle = \pm\delta_{ij}$ for each i and j. We wish to prove that every scalar product space has an orthonormal basis. Note that the usual Gram–Schmidt algorithm does not always work in this situation, because the vectors that appear in the denominators in (2.5)–(2.6) might have vanishing norms. In order to get around this problem, we introduce the following definitions. If (V, q) is a finite-dimensional scalar product space, a subspace $S \subseteq V$ is said to be **nondegenerate** if the restriction of q to $S \times S$ is nondegenerate. An ordered k-tuple of vectors $(v_1, \ldots, v_k)$ in V is said to be **nondegenerate** if for each $j = 1, \ldots, k$, the vectors $(v_1, \ldots, v_j)$ span a nondegenerate j-dimensional subspace of V. For example, every orthonormal basis is nondegenerate.

Lemma 2.60. *Suppose (V, q) is a finite-dimensional scalar product space, and $S \subseteq V$ is a linear subspace. The following are equivalent:*

(a) S is nondegenerate.
(b) $S^\perp$ is nondegenerate.
(c) $S \cap S^\perp = \{0\}$.
(d) $V = S \oplus S^\perp$.

▶ **Exercise 2.61.** Prove the preceding lemma.

Lemma 2.62 (Completion of Nondegenerate Bases). *Suppose (V, q) is an n-dimensional scalar product space, and $(v_1, \ldots, v_k)$ is a nondegenerate k-tuple in V with $0 \leq k < n$. Then there exist vectors $v_{k+1}, \ldots, v_n \in V$ such that $(v_1, \ldots, v_n)$ is a nondegenerate basis for V.*

Proof. Let $S = \operatorname{span}(v_1, \ldots, v_k) \subseteq V$. Because $k < n$, $S^\perp$ is a nontrivial subspace of V, and Lemma 2.60 shows that $S^\perp$ is nondegenerate and $V = S \oplus S^\perp$. By the nondegeneracy of $S^\perp$, there must be a vector in $S^\perp$ with nonzero length, because otherwise the polarization identity would imply that all inner products of pairs of elements of S would be zero. If $v_{k+1} \in S^\perp$ is any vector with nonzero length, then $(v_1, \ldots, v_{k+1})$ is easily seen to be a nondegenerate $(k+1)$-tuple. Repeating this argument for $v_{k+2}, \ldots, v_n$ completes the proof. $\qquad\square$

Proposition 2.63 (Gram–Schmidt Algorithm for Scalar Products). *Suppose*
(V,q) *is an n-dimensional scalar product space. If* (v_i) *is a nondegenerate basis for*
V, *then there is an orthonormal basis* (b_i) *with the property that* $\operatorname{span}(b_1,\dots,b_k) =$
$\operatorname{span}(v_1,\dots,v_k)$ *for each* $k = 1,\dots,n$.

Proof. As in the positive definite case, the basis (b_i) is constructed recursively,
starting with $b_1 = v_1/|v_1|$ and noting that the assumption that v_1 spans a nonde-
generate subspace ensures that $|v_1| \neq 0$. At the inductive step, assuming we have
constructed $(b_1,\dots,b_k)$, we first set

$$z = v_{k+1} - \sum_{i=1}^{k} \frac{\langle v_{k+1}, b_i \rangle}{\langle b_i, b_i \rangle} b_i.$$

Each denominator $\langle b_i, b_i \rangle$ is equal to ± 1, so this defines z as a nonzero element
of V orthogonal to $b_1,\dots,b_k$, and with the property that $\operatorname{span}(b_1,\dots,b_k,z) =$
$\operatorname{span}(v_1,\dots,v_{k+1})$. If $\langle z,z \rangle = 0$, then z is orthogonal to $\operatorname{span}(v_1,\dots,v_{k+1})$, contradicting the nondegeneracy assumption. Thus we can complete the inductive step
by putting $b_{k+1} = z/|z|$.   □

Corollary 2.64. *Suppose* (V,q) *is an n-dimensional scalar product space. There is*
a basis (β^i) *for* V^* *with respect to which q has the expression*

$$q = (\beta^1)^2 + \cdots + (\beta^r)^2 - (\beta^{r+1})^2 - \cdots - (\beta^{r+s})^2, \qquad (2.23)$$

for some nonnegative integers r, s with $r + s = n$.

Proof. Let (b_i) be an orthonormal basis for V, and let (β^i) be the dual basis for V^*.
A computation shows that q has a basis expression of the form (2.23), but perhaps
with the positive and negative terms in a different order. Reordering the basis so that
the positive terms come first, we obtain (2.23).   □

It turns out that the numbers r and s in (2.23) are independent of the choice of
basis. The key to proving this is the following classical result from linear algebra.

Proposition 2.65 (Sylvester's Law of Inertia). *Suppose* (V,q) *is a finite-dimen-*
sional scalar product space. If q has the representation (2.23) *in some basis, then the*
number r is the maximum dimension among all subspaces on which the restriction of
q is positive definite, and thus r and s are independent of the choice of basis.

Proof. Problem 2-33.   □

The integer s in the expression (2.23) (the number of negative terms) is called
the *index of q*, and the ordered pair (r,s) is called the *signature of q*.

Now suppose M is a smooth manifold. A *pseudo-Riemannian metric on M*
(called by some authors a *semi-Riemannian metric*) is a smooth symmetric 2-tensor
field g that is nondegenerate at each point of M, and with the same signature everywhere. Every Riemannian metric is also a pseudo-Riemannian metric.

Proposition 2.66 (Orthonormal Frames for Pseudo-Riemannian Manifolds). *Let* (M, g) *be a pseudo-Riemannian manifold. For each* $p \in M$, *there exists a smooth orthonormal frame on a neighborhood of* p *in* M.

▶ **Exercise 2.67.** Prove the preceding proposition.

▶ **Exercise 2.68.** Suppose (M_1, g_1) and (M_2, g_2) are pseudo-Riemannian manifolds of signatures (r_1, s_1) and (r_2, s_2), respectively. Show that $(M_1 \times M_2, g_1 \oplus g_2)$ is a pseudo-Riemannian manifold of signature $(r_1 + r_2, s_1 + s_2)$.

For nonnegative integers r and s, we define the ***pseudo-Euclidean space of signature*** (r, s), denoted by $\mathbb{R}^{r,s}$, to be the manifold $\mathbb{R}^{r+s}$, with standard coordinates denoted by $\left(\xi^1, \ldots, \xi^r, \tau^1, \ldots, \tau^s\right)$, and with the pseudo-Riemannian metric $\bar{q}^{(r,s)}$ defined by

$$\bar{q}^{(r,s)} = \left(d\xi^1\right)^2 + \cdots + \left(d\xi^r\right)^2 - \left(d\tau^1\right)^2 - \cdots - \left(d\tau^s\right)^2. \qquad (2.24)$$

By far the most important pseudo-Riemannian metrics (other than the Riemannian ones) are the ***Lorentz metrics***, which are pseudo-Riemannian metrics of index 1, and thus signature $(r, 1)$. (Some authors, especially in the physics literature, prefer to use signature $(1, r)$; either one can be converted to the other by multiplying the metric by -1, so there is no significant difference.)

The pseudo-Euclidean metric $\bar{q}^{(r,1)}$ is a Lorentz metric called the ***Minkowski metric***, and the Lorentz manifold $\mathbb{R}^{r,1}$ is called $(\mathbf{r} + \mathbf{1})$-***dimensional Minkowski space***. If we denote standard coordinates on $\mathbb{R}^{r,1}$ by $\left(\xi^1, \ldots, \xi^r, \tau\right)$, then the Minkowski metric is given by

$$\bar{q}^{(r,1)} = \left(d\xi^1\right)^2 + \cdots + \left(d\xi^r\right)^2 - (d\tau)^2. \qquad (2.25)$$

In the special case of $\mathbb{R}^{3,1}$, the Minkowski metric is the fundamental invariant of Albert Einstein's *special theory of relativity*, which can be expressed succinctly by saying that if gravity is ignored, then spacetime is accurately modeled by $(3 + 1)$-dimensional Minkowski space, and the laws of physics have the same form in every coordinate system in which the Minkowski metric has the expression (2.25). The differing physical characteristics of "space" (the ξ^i directions) and "time" (the τ direction) arise from the fact that they are subspaces on which the metric is positive definite and negative definite, respectively. The *general theory of relativity* includes gravitational effects by allowing the Lorentz metric to vary from point to point.

Many, but not all, results from the theory of Riemannian metrics apply equally well to pseudo-Riemannian metrics. Throughout this book, we will attempt to point out which results carry over directly to pseudo-Riemannian metrics, which ones can be adapted with minor modifications, and which ones do not carry over at all. As a rule of thumb, proofs that depend only on the nondegeneracy of the metric tensor, such as properties of the musical isomorphisms and existence and uniqueness of geodesics, work fine in the pseudo-Riemannian setting, while proofs that use positivity in an essential way, such as those involving lengths of curves, do not.

One notable result that does *not* carry over to the pseudo-Riemannian case is Proposition 2.4, about the existence of metrics. For example, the following result characterizes those manifolds that admit Lorentz metrics.

Theorem 2.69. *A smooth manifold M admits a Lorentz metric if and only if it admits a rank-1 tangent distribution (i.e., a rank-1 subbundle of TM).*

Proof. Problem 2-34.   □

With some more sophisticated tools from algebraic topology, it can be shown that every noncompact connected smooth manifold admits a Lorentz metric, and a compact connected smooth manifold admits a Lorentz metric if and only if its Euler characteristic is zero (see [O'N83, p. 149]). It follows that no even-dimensional sphere admits a Lorentz metric, because $\mathbb{S}^{2n}$ has Euler characteristic equal to 2.

For a thorough treatment of pseudo-Riemannian metrics from a mathematical point of view, see the excellent book [O'N83]; a more physical treatment can be found in [HE73].

Pseudo-Riemannian Submanifolds

The theory of submanifolds is only slightly more complicated in the pseudo-Riemannian case. If $\left(\widetilde{M},\widetilde{g}\right)$ is a pseudo-Riemannian manifold, a smooth submanifold $\iota: M \hookrightarrow \widetilde{M}$ is called a ***pseudo-Riemannian submanifold of $\widetilde{M}$*** if $\iota^*\widetilde{g}$ is nondegenerate with constant signature. If this is the case, we always consider M to be endowed with the induced pseudo-Riemannian metric $\iota^*\widetilde{g}$. In the special case in which $\iota^*\widetilde{g}$ is positive definite, M is called a ***Riemannian submanifold***.

The nondegeneracy hypothesis is not automatically satisfied: for example, if $M \subseteq \mathbb{R}^{1,1}$ is the submanifold $\{(\xi, \tau) : \xi = \tau\}$ and $\iota: M \to \mathbb{R}^{1,1}$ is inclusion, then the pullback tensor $\iota^*\overline{q}^{(1,1)}$ is identically zero on M.

For hypersurfaces (submanifolds of codimension 1), the nondegeneracy condition is easy to check. If $M \subseteq \widetilde{M}$ is a smooth submanifold and $p \in M$, then a vector $v \in T_p\widetilde{M}$ is said to be ***normal to M*** if $\langle v, x \rangle = 0$ for all $x \in T_pM$, just as in the Riemannian case. The space of all normal vectors at p is a subspace of $T_p\widetilde{M}$ denoted by N_pM.

Proposition 2.70. *Suppose $\left(\widetilde{M},\widetilde{g}\right)$ is a pseudo-Riemannian manifold of signature (r,s). Let M be a smooth hypersurface in $\widetilde{M}$, and let $\iota: M \hookrightarrow \widetilde{M}$ be the inclusion map. Then the pullback tensor field $\iota^*\widetilde{g}$ is nondegenerate if and only if $\widetilde{g}(v,v) \neq 0$ for every $p \in M$ and every nonzero normal vector $v \in N_pM$. If $\widetilde{g}(v,v) > 0$ for every nonzero normal vector to M, then M is a pseudo-Riemannian submanifold of signature $(r-1,s)$; and if $\widetilde{g}(v,v) < 0$ for every such vector, then M is a pseudo-Riemannian submanifold of signature $(r,s-1)$.*

Proof. Given $p \in M$, Lemma 2.60 shows that T_pM is a nondegenerate subspace of $T_p\widetilde{M}$ if and only if the one-dimensional subspace $(T_pM)^{\perp} = N_pM$ is nondegenerate, which is the case if and only if every nonzero $v \in N_pM$ satisfies $\widetilde{g}(v,v) \neq 0$.

Now suppose $\widetilde{g}(v,v) > 0$ for every nonzero normal vector v. Let $p \in M$ be arbitrary, and let v be a nonzero element of $N_p M$. Writing $n = \dim \widetilde{M}$, we can complete v to a nondegenerate basis $(v, w_2, \ldots, w_n)$ for $T_p \widetilde{M}$ by Lemma 2.62, and then use the Gram–Schmidt algorithm to find an orthonormal basis $(b_1, \ldots, b_n)$ for $T_p \widetilde{M}$ such that $\mathrm{span}(b_1) = N_p M$. It follows that $\mathrm{span}(b_2, \ldots, b_n) = T_p M$. If (β^j) is the dual basis to (b_i), then $\widetilde{g}_p$ has a basis representation of the form $(\beta^1)^2 \pm (\beta^2)^2 \pm \cdots \pm (\beta^n)^2$, with a total of r positive terms and s negative ones, and with a positive sign on the first term $(\beta^1)^2$. Therefore, $\iota^* \widetilde{g}_p = \pm (\beta^2)^2 \pm \cdots \pm (\beta^n)^2$ has signature $(r-1, s)$. The argument for the case $\widetilde{g}(v,v) < 0$ is similar.   □

If $(\widetilde{M}, \widetilde{g})$ is a pseudo-Riemannian manifold and $f \in C^\infty(M)$, then the **gradient of** f is defined as the smooth vector field $\mathrm{grad}\, f = (df)^\sharp$ just as in the Riemannian case.

Corollary 2.71. *Suppose $(\widetilde{M}, \widetilde{g})$ is a pseudo-Riemannian manifold of signature (r,s), $f \in C^\infty(\widetilde{M})$, and $M = f^{-1}(c)$ for some $c \in \mathbb{R}$. If $\widetilde{g}(\mathrm{grad}\, f, \ \mathrm{grad}\, f) > 0$ everywhere on M, then M is an embedded pseudo-Riemannian submanifold of $\widetilde{M}$ of signature $(r-1, s)$; and if $\widetilde{g}(\mathrm{grad}\, f, \ \mathrm{grad}\, f) < 0$ everywhere on M, then M is an embedded pseudo-Riemannian submanifold of $\widetilde{M}$ of signature $(r, s-1)$. In either case, $\mathrm{grad}\, f$ is everywhere normal to M.*

Proof. Problem 2-35.   □

Proposition 2.72 (Pseudo-Riemannian Adapted Orthonormal Frames). *Suppose $(\widetilde{M}, \widetilde{g})$ is a pseudo-Riemannian manifold, and $M \subseteq \widetilde{M}$ is an embedded pseudo-Riemannian or Riemannian submanifold. For each $p \in M$, there exists a smooth orthonormal frame on a neighborhood of p in $\widetilde{M}$ that is adapted to M.*

Proof. Write $m = \dim \widetilde{M}$ and $n = \dim M$, and let $p \in M$ be arbitrary. Proposition 2.66 shows that there is a smooth orthonormal frame $(E_1, \ldots, E_n)$ for M on some neighborhood of p in M. Then by Lemma 2.62, we can find vectors $v_{n+1}, \ldots, v_m \in T_p \widetilde{M}$ such that $(E_1|_p, \ldots, E_n|_p, v_{n+1}, \ldots, v_m)$ is a nondegenerate basis for $T_p \widetilde{M}$. Now extend $v_{n+1}, \ldots, v_m$ arbitrarily to smooth vector fields $V_{n+1}, \ldots, V_m$ on a neighborhood of p in $\widetilde{M}$. By continuity, the ordered m-tuple $(E_1, \ldots, E_n, V_{n+1}, \ldots, V_m)$ will be a nondegenerate frame for $\widetilde{M}$ in some (possibly smaller) neighborhood of p. Applying the Gram–Schmidt algorithm (Prop. 2.63) to this local frame yields a smooth local orthonormal frame $(E_1, \ldots, E_m)$ for $\widetilde{M}$ with the property that $(E_1, \ldots, E_n)$ restricts to a local orthonormal frame for M.   □

The next corollary is proved in the same way as Proposition 2.16.

Corollary 2.73 (Normal Bundle to a Pseudo-Riemannian Submanifold). *Suppose $(\widetilde{M}, \widetilde{g})$ is a pseudo-Riemannian manifold, and $M \subseteq \widetilde{M}$ is an embedded pseudo-Riemannian or Riemannian submanifold. The set of vectors normal to M is a smooth vector subbundle of $T\widetilde{M}|_M$, called the **normal bundle to M**.*   □

Other Generalizations of Riemannian Metrics

Pseudo-Riemannian metrics are obtained by relaxing the positivity requirement for Riemannian metrics. In addition, there are other useful generalizations that result when we relax other requirements. In this section we touch briefly on two of those generalizations. We will not treat these anywhere else in the book, but it is useful to know the definitions.

Sub-Riemannian Metrics

The first generalization arises when we relax the requirement that the metric be defined on the whole tangent space.

A **sub-Riemannian metric** (also sometimes known as a **singular Riemannian metric** or **Carnot–Carathéodory metric**) on a smooth manifold M is a smooth fiber metric on a smooth tangent distribution $S \subseteq TM$ (i.e., a vector subbundle of TM). Since lengths make sense only for vectors in S, the only curves whose lengths can be measured are those whose velocity vectors lie everywhere in S. Therefore, one usually imposes some condition on S that guarantees that any two nearby points can be connected by such a curve. This is, in a sense, the opposite of the Frobenius integrability condition, which would restrict every such curve to lie in a single leaf of a foliation.

Sub-Riemannian metrics arise naturally in the study of the abstract models of real submanifolds of complex space $\mathbb{C}^n$, called *CR manifolds* (short for *Cauchy–Riemann manifolds*). CR manifolds are real manifolds endowed with a tangent distribution $S \subseteq TM$ whose fibers carry the structure of complex vector spaces (with an additional integrability condition that need not concern us here). In the model case of a submanifold $M \subseteq \mathbb{C}^n$, S is the set of vectors tangent to M that remain tangent after multiplication by $i = \sqrt{-1}$ in the ambient complex coordinates. If S is sufficiently far from being integrable, choosing a fiber metric on S results in a sub-Riemannian metric whose geometric properties closely reflect the complex-analytic properties of M as a subset of $\mathbb{C}^n$.

Another motivation for studying sub-Riemannian metrics arises from *control theory*. In this subject, one is given a manifold with a vector field depending on parameters called *controls,* with the goal being to vary the controls so as to obtain a solution curve with desired properties, often one that minimizes some function such as arc length. If the vector field is constrained to be everywhere tangent to a distribution S on the manifold (for example, in the case of a robot arm whose motion is restricted by the orientations of its hinges), then the function can often be modeled as a sub-Riemannian metric and optimal solutions modeled as sub-Riemannian geodesics.

A useful introduction to the geometry of sub-Riemannian metrics is provided in the article [Str86].

Finsler Metrics

Another important generalization arises from relaxing the requirement that norms of vectors be defined in terms of an inner product on each tangent space.

In general, a *norm* on a vector space V is a real-valued function on V, usually written $v \mapsto |v|$, that satisfies

   (i) HOMOGENEITY: $|cv| = |c||v|$ for $v \in V$ and $c \in \mathbb{R}$;
  (ii) POSITIVITY: $|v| \geq 0$ for $v \in V$, with equality if and only if $v = 0$;
 (iii) TRIANGLE INEQUALITY: $|v + w| \leq |v| + |w|$ for $v, w \in V$.

For example, the length function associated with an inner product is a norm.

Now suppose M is a smooth manifold. A *Finsler metric* on M is a continuous function $F : TM \to \mathbb{R}$, smooth on the set of nonzero vectors, whose restriction to each tangent space $T_p M$ is a norm. Again, the norm function associated with a Riemannian metric is a special case.

The inventor of Riemannian geometry himself, Bernhard Riemann, clearly envisaged an important role in n-dimensional geometry for what we now call Finsler metrics; he restricted his investigations to the "Riemannian" case purely for simplicity (see [Spi79, Vol. 2]). However, it was not until the late twentieth century that Finsler metrics began to be studied seriously from a geometric point of view.

The recent upsurge of interest in Finsler metrics has been motivated in part by the fact that two different Finsler metrics appear very naturally in the theory of several complex variables. For certain bounded open sets in $\mathbb{C}^n$ (the ones with smooth, strictly convex boundaries, for example), the *Kobayashi metric* and the *Carathéodory metric* are intrinsically defined, biholomorphically invariant Finsler metrics. Combining differential-geometric and complex-analytic methods has led to striking new insights into both the function theory and the geometry of such domains. We do not treat Finsler metrics further in this book, but you can consult one of the recent books on the subject [AP94, BCS00, JP13].

Problems

2-1. Show that every Riemannian 1-manifold is flat. (*Used on pp. 13, 193.*)

2-2. Suppose V and W are finite-dimensional real inner product spaces of the same dimension, and $F : V \to W$ is any map (not assumed to be linear or even continuous) that preserves the origin and all distances: $F(0) = 0$ and $|F(x) - F(y)| = |x - y|$ for all $x, y \in V$. Prove that F is a linear isometry. [Hint: First show that F preserves inner products, and then show that it is linear.] (*Used on p. 187.*)

2-3. Given a smooth embedded 1-dimensional submanifold $C \subseteq H$ as in Example 2.24(b), show that the surface of revolution $S_C \subseteq \mathbb{R}^3$ with its induced metric is isometric to the warped product $C \times_a \mathbb{S}^1$, where $a : C \to \mathbb{R}$ is the distance to the z-axis.

2-4. Let $\rho\colon \mathbb{R}^+ \to \mathbb{R}$ be the restriction of the standard coordinate function, and let $\mathbb{R}^+ \times_\rho \mathbb{S}^{n-1}$ denote the resulting warped product (see Example 2.24(c)). Define $\Phi\colon \mathbb{R}^+ \times_\rho \mathbb{S}^{n-1} \to \mathbb{R}^n \smallsetminus \{0\}$ by $\Phi(\rho,\omega) = \rho\omega$. Show that Φ is an isometry between the warped product metric and the Euclidean metric on $\mathbb{R}^n \smallsetminus \{0\}$. (*Used on p. 293.*)

2-5. Prove parts (b) and (c) of Proposition 2.25 (properties of horizontal vector fields). (*Used on p. 146.*)

2-6. Prove Theorem 2.28 (if $\pi\colon \widetilde{M} \to M$ is a surjective smooth submersion, and a group acts on $\widetilde{M}$ isometrically, vertically, and transitively on fibers, then M inherits a unique Riemannian metric such that π is a Riemannian submersion).

2-7. For $0 < k < n$, the set $G_k(\mathbb{R}^n)$ of k-dimensional linear subspaces of $\mathbb{R}^n$ is called a **Grassmann manifold** or **Grassmannian**. The group $\mathrm{GL}(n,\mathbb{R})$ acts transitively on $G_k(\mathbb{R}^n)$ in an obvious way, and $G_k(\mathbb{R}^n)$ has a unique smooth manifold structure making this action smooth (see [LeeSM, Example 21.21]).

   (a) Let $V_k(\mathbb{R}^n)$ denote the set of orthonormal ordered k-tuples of vectors in $\mathbb{R}^n$. By arranging the vectors in k columns, we can view $V_k(\mathbb{R}^n)$ as a subset of the vector space $\mathrm{M}(n \times k, \mathbb{R})$ of all $n \times k$ real matrices. Prove that $V_k(\mathbb{R}^n)$ is a smooth submanifold of $\mathrm{M}(n \times k, \mathbb{R})$ of dimension $k(2n - k - 1)/2$, called a **Stiefel manifold**. [Hint: Consider the map $\Phi\colon \mathrm{M}(n \times k, \mathbb{R}) \to \mathrm{M}(k \times k, \mathbb{R})$ given by $\Phi(A) = A^T A$.]

   (b) Show that the map $\pi\colon V_k(\mathbb{R}^n) \to G_k(\mathbb{R}^n)$ that sends a k-tuple to its span is a surjective smooth submersion.

   (c) Give $V_k(\mathbb{R}^n)$ the Riemannian metric induced from the Euclidean metric on $\mathrm{M}(n \times k, \mathbb{R})$. Show that the right action of $O(k)$ on $V_k(\mathbb{R}^n)$ by matrix multiplication on the right is isometric, vertical, and transitive on fibers of π, and thus there is a unique metric on $G_k(\mathbb{R}^n)$ such that π is a Riemannian submersion. [Hint: It might help to note that the Euclidean inner product on $\mathrm{M}(n \times k, \mathbb{R})$ can be written in the form $\langle A, B \rangle = \mathrm{tr}\left(A^T B\right)$.]

   (*Used on p. 82.*)

2-8. Prove that the action of $\mathbb{Z}$ on $\mathbb{R}^2$ defined in Example 2.35 is smooth, free, proper, and isometric, and therefore the open Möbius band inherits a flat Riemannian metric such that the quotient map is a Riemannian covering.

2-9. Prove Proposition 2.37 (the gradient is orthogonal to regular level sets).

2-10. Suppose (M,g) is a Riemannian manifold, $f \in C^\infty(M)$, and $X \in \mathfrak{X}(M)$ is a nowhere-vanishing vector field. Prove that $X = \mathrm{grad}\, f$ if and only if $Xf \equiv |X|_g^2$ and X is orthogonal to the level sets of f at all regular points of f. (*Used on pp. 161, 180.*)

2-11. Prove Proposition 2.40 (inner products on tensor bundles).

2-12. Prove Proposition 2.41 (existence and uniqueness of the Riemannian volume form).

2-13. Prove Proposition 2.43 (characterizing the volume form of a Riemannian hypersurface). [Hint: To prove (2.17), use an adapted orthonormal frame.]

2-14. Suppose $\left(\widetilde{M},\widetilde{g}\right)$ and (M,g) are compact connected Riemannian manifolds, and $\pi:\widetilde{M}\to M$ is a k-sheeted Riemannian covering. Prove that $\mathrm{Vol}\left(\widetilde{M}\right)=k\cdot\mathrm{Vol}(M)$. (*Used on p. 363.*)

2-15. Suppose (M_1,g_1) and (M_2,g_2) are oriented Riemannian manifolds of dimensions k_1 and k_2, respectively. Let $f:M_1\to\mathbb{R}^+$ be a smooth function, and let $g=g_1\oplus f^2g_2$ be the corresponding warped product metric on $M_1\times_f M_2$. Prove that the Riemannian volume form of g is given by

$$dV_g = f^{k_2}dV_{g_1}\wedge dV_{g_2},$$

where f, dV_{g_1}, and dV_{g_2} are understood to be pulled back to $M_1\times M_2$ by the projection maps. (*Used on p. 295.*)

2-16. Let (M,g) be a Riemannian n-manifold. Show that for each $k=1,\dots,n$, there is a unique fiber metric $\langle\cdot,\cdot\rangle_g$ on the bundle $\Lambda^k T^*M$ that satisfies

$$\left\langle\omega^1\wedge\cdots\wedge\omega^k,\eta^1\wedge\cdots\wedge\eta^k\right\rangle_g = \det\left(\left\langle\omega^i,\eta^j\right\rangle_g\right) \tag{2.26}$$

whenever $\omega^1,\dots,\omega^k,\eta^1,\dots,\eta^k$ are covectors at a point $p\in M$. [Hint: Define the inner product locally by declaring the set of k-covectors

$$\left\{\varepsilon^{i_1}\wedge\cdots\wedge\varepsilon^{i_k}\big|_p : i_1<\cdots<i_k\right\}$$

to be an orthonormal basis for $\Lambda^k\left(T_p^*M\right)$ whenever (ε^i) is a local orthonormal coframe for T^*M, and then prove that the resulting inner product satisfies (2.26) and is independent of the choice of frame.]

2-17. Because we regard the bundle $\Lambda^k T^*M$ of k-forms as a subbundle of the bundle $T^k T^*M$ of covariant k-tensors, we have two inner products to choose from on k-forms: the one defined in Problem 2-16, and the restriction of the tensor inner product defined in Proposition 2.40. For this problem, we use the notation $\langle\cdot,\cdot\rangle$ to denote the inner product of Problem 2-16, and $\langle\!\langle\cdot,\cdot\rangle\!\rangle$ to denote the restriction of the tensor inner product.

(a) Using the convention for the wedge product that we use in this book (see p. 400), prove that

$$\langle\cdot,\cdot\rangle_g = \frac{1}{k!}\langle\!\langle\cdot,\cdot\rangle\!\rangle_g.$$

(b) Show that if the Alt convention is used for the wedge product (p. 401), the formula becomes

$$\langle \cdot, \cdot \rangle_g = k! \langle\!\langle \cdot, \cdot \rangle\!\rangle_g.$$

2-18. Let (M, g) be an oriented Riemannian n-manifold.

(a) For each $k = 0, \ldots, n$, show that there is a unique smooth bundle homomorphism $*: \Lambda^k T^*M \to \Lambda^{n-k} T^*M$, called the **Hodge star operator**, satisfying

$$\omega \wedge *\eta = \langle \omega, \eta \rangle_g \, dV_g$$

for all smooth k-forms ω, η, where $\langle \cdot, \cdot \rangle_g$ is the inner product on k-forms defined in Problem 2-16. (For $k = 0$, interpret the inner product as ordinary multiplication.) [Hint: First prove uniqueness, and then define $*$ locally by setting

$$*\left(\varepsilon^{i_1} \wedge \cdots \wedge \varepsilon^{i_k}\right) = \pm \varepsilon^{j_1} \wedge \cdots \wedge \varepsilon^{j_{n-k}}$$

in terms of an orthonormal coframe $\left(\varepsilon^i\right)$, where the indices $j_1, \ldots, j_{n-k}$ are chosen such that $(i_1, \ldots, i_k, j_1, \ldots, j_{n-k})$ is some permutation of $(1, \ldots, n)$.]

(b) Show that $*: \Lambda^0 T^*M \to \Lambda^n T^*M$ is given by $*f = f \, dV_g$.

(c) Show that $**\omega = (-1)^{k(n-k)}\omega$ if ω is a k-form.

2-19. Regard $\mathbb{R}^n$ as a Riemannian manifold with the Euclidean metric and the standard orientation, and let $*$ denote the Hodge star operator defined in Problem 2-18.

(a) Calculate $*dx^i$ for $i = 1, \ldots, n$.

(b) Calculate $*\left(dx^i \wedge dx^j\right)$ in the case $n = 4$.

2-20. Let M be an oriented Riemannian 4-manifold. A 2-form ω on M is said to be **self-dual** if $*\omega = \omega$, and **anti-self-dual** if $*\omega = -\omega$.

(a) Show that every 2-form ω on M can be written uniquely as a sum of a self-dual form and an anti-self-dual form.

(b) On $M = \mathbb{R}^4$ with the Euclidean metric, determine the self-dual and anti-self-dual forms in standard coordinates.

2-21. Prove Proposition 2.46 (the coordinate formulas for the divergence and the Laplacian).

2-22. Suppose (M, g) is a compact Riemannian manifold with boundary.

(a) Prove the following **divergence theorem** for $X \in \mathfrak{X}(M)$:

$$\int_M (\operatorname{div} X) \, dV_g = \int_{\partial M} \langle X, N \rangle_g \, dV_{\hat{g}},$$

where N is the outward unit normal to ∂M and $\hat{g}$ is the induced metric on ∂M. [Hint: Prove it first in the case that M is orientable, and then apply that case to the orientation covering of M (Prop. B.18).]

(b) Show that the divergence operator satisfies the following product rule for $u \in C^\infty(M)$ and $X \in \mathfrak{X}(M)$:

$$\operatorname{div}(uX) = u \operatorname{div} X + \langle \operatorname{grad} u, X \rangle_g,$$

and deduce the following "integration by parts" formula:

$$\int_M \langle \operatorname{grad} u, X \rangle_g \, dV_g = \int_{\partial M} u \langle X, N \rangle_g \, dV_{\hat g} - \int_M u \operatorname{div} X \, dV_g.$$

What does this say when M is a compact interval in $\mathbb{R}$?

(Used on p. 149.)

2-23. Let (M, g) be a compact Riemannian manifold with or without boundary. A function $u \in C^\infty(M)$ is said to be **harmonic** if $\Delta u = 0$, where Δ is the Laplacian defined on page 32.

(a) Prove **Green's identities**:

$$\int_M u \Delta v \, dV_g = \int_{\partial M} u \, Nv \, dV_{\hat g} - \int_M \langle \operatorname{grad} u, \operatorname{grad} v \rangle_g \, dV_g,$$

$$\int_M (u \Delta v - v \Delta u) \, dV_g = \int_{\partial M} (u \, Nv - v \, Nu) \, dV_{\hat g},$$

where N is the outward unit normal vector field on ∂M and $\hat g$ is the induced metric on ∂M.

(b) Show that if M is connected, $\partial M \neq \varnothing$, and u, v are harmonic functions on M whose restrictions to ∂M agree, then $u \equiv v$.

(c) Show that if M is connected and $\partial M = \varnothing$, then the only harmonic functions on M are the constants, and every smooth function u satisfies $\int_M \Delta u \, dV_g = 0$.

(Used on pp. 149, 223.)

2-24. Let (M, g) be a compact Riemannian manifold (without boundary). A real number λ is called an **eigenvalue of M** if there exists a smooth function u on M, not identically zero, such that $-\Delta u = \lambda u$. In this case, u is called an **eigenfunction** corresponding to λ.

(a) Prove that 0 is an eigenvalue of M, and that all other eigenvalues are strictly positive.

(b) Show that if u and v are eigenfunctions corresponding to distinct eigenvalues, then $\int_M uv \, dV_g = 0$.

(Used on p. 149.)

2-25. Let (M, g) be a compact connected Riemannian n-manifold with nonempty boundary. A number $\lambda \in \mathbb{R}$ is called a **Dirichlet eigenvalue for M** if there exists a smooth real-valued function u on M, not identically zero, such that $-\Delta u = \lambda u$ and $u|_{\partial M} = 0$. Similarly, λ is called a **Neumann eigenvalue** if there exists such a u satisfying $-\Delta u = \lambda u$ and $Nu|_{\partial M} = 0$, where N is the outward unit normal.

(a) Show that every Dirichlet eigenvalue is strictly positive.
(b) Show that 0 is a Neumann eigenvalue, and all other Neumann eigenvalues are strictly positive.

2-26. DIRICHLET'S PRINCIPLE: Suppose (M,g) is a compact connected Riemannian n-manifold with nonempty boundary. Prove that a function $u \in C^\infty(M)$ is harmonic if and only if it minimizes $\int_M |\operatorname{grad} u|^2 \, dV_g$ among all smooth functions with the same boundary values. [Hint: For any function $f \in C^\infty(M)$ that vanishes on ∂M, expand $\int_M |\operatorname{grad}(u + \varepsilon f)|^2 \, dV_g$ and use Problem 2-22.]

2-27. Suppose (M,g) is an oriented Riemannian 3-manifold.

(a) Define $\beta: TM \to \Lambda^2 T^*M$ by $\beta(X) = X \lrcorner dV_g$. Show that β is a smooth bundle isomorphism, and thus we can define the *curl* of a vector field $X \in \mathfrak{X}(M)$ by

$$\operatorname{curl} X = \beta^{-1} d(X^\flat).$$

(b) Show that the following diagram commutes:

$$
\begin{array}{ccccccc}
C^\infty(M) & \xrightarrow{\operatorname{grad}} & \mathfrak{X}(M) & \xrightarrow{\operatorname{curl}} & \mathfrak{X}(M) & \xrightarrow{\operatorname{div}} & C^\infty(M) \\
\downarrow{\scriptstyle \operatorname{Id}} & & \downarrow{\scriptstyle \flat} & & \downarrow{\scriptstyle \beta} & & \downarrow{\scriptstyle *} \\
\Omega^0(M) & \xrightarrow{d} & \Omega^1(M) & \xrightarrow{d} & \Omega^2(M) & \xrightarrow{d} & \Omega^3(M),
\end{array}
\qquad (2.27)
$$

where $*(f) = f \, dV_g$, and use this to prove that $\operatorname{curl}(\operatorname{grad} f) = 0$ for every $f \in C^\infty(M)$, and $\operatorname{div}(\operatorname{curl} X) = 0$ for every $X \in \mathfrak{X}(M)$.

(c) Compute the formula for the curl in standard coordinates on $\mathbb{R}^3$ with the Euclidean metric.

2-28. Let (M,g) be an oriented Riemannian manifold and let $*$ denote its Hodge star operator (Problem 2-18). Show that for every $X \in \mathfrak{X}(M)$,

$$X \lrcorner dV_g = *(X^\flat),$$
$$\operatorname{div} X = *d*(X^\flat),$$

and, when $\dim M = 3$,

$$\operatorname{curl} X = (*d(X^\flat))^\sharp,$$

where the curl of a vector field is defined as in Problem 2-27.

2-29. Let (M,g) be a compact oriented Riemannian n-manifold. For $1 \le k \le n$, define a map $d^*: \Omega^k(M) \to \Omega^{k-1}(M)$ by $d^*\omega = (-1)^{n(k+1)+1} *d*\omega$, where $*$ is the Hodge star operator defined in Problem 2-18. Extend this definition to 0-forms by defining $d^*\omega = 0$ for $\omega \in \Omega^0(M)$.

(a) Show that $d^* \circ d^* = 0$.

(b) Show that the formula

$$(\omega, \eta) = \int_M \langle \omega, \eta \rangle_g \, dV_g$$

defines an inner product on $\Omega^k(M)$ for each k, where $\langle \cdot, \cdot \rangle_g$ is the inner product on forms defined in Problem 2-16.

(c) Show that $(d^*\omega, \eta) = (\omega, d\eta)$ for all $\omega \in \Omega^k(M)$ and $\eta \in \Omega^{k-1}(M)$.

2-30. Suppose (M, g) is a (not necessarily connected) Riemannian manifold. Show that there is a distance function d on M that induces the given topology and restricts to the Riemannian distance on each component of M. (*Used on p. 187.*)

2-31. Suppose (M, g) and $(\widetilde{M}, \widetilde{g})$ are connected Riemannian manifolds, and $\varphi \colon M \to \widetilde{M}$ is a local isometry. Show that $d_{\widetilde{g}}(\varphi(x), \varphi(y)) \leq d_g(x, y)$ for all $x, y \in M$. Give an example to show that equality need not hold. (*Used on p. 37.*)

2-32. Let (M, g) be a Riemannian manifold and $\gamma \colon [a, b] \to M$ a smooth curve segment. For each continuous function $f \colon [a, b] \to \mathbb{R}$, we define the **integral of f with respect to arc length**, denoted by $\int_\gamma f \, ds$, by

$$\int_\gamma f \, ds = \int_a^b f(t) |\gamma'(t)|_g \, dt.$$

(a) Show that $\int_\gamma f \, ds$ is independent of parametrization in the following sense: if $\varphi \colon [c, d] \to [a, b]$ is a diffeomorphism, then

$$\int_a^b f(t) |\gamma'(t)|_g \, dt = \int_c^d \widetilde{f}(u) |\widetilde{\gamma}'(u)|_g \, du,$$

where $\widetilde{f} = f \circ \varphi$ and $\widetilde{\gamma} = \gamma \circ \varphi$.

(b) Suppose now that γ is a smooth embedding, so that $C = \gamma([a, b])$ is an embedded submanifold with boundary in M. Show that

$$\int_\gamma f \, ds = \int_C (f \circ \gamma^{-1}) \, d\widetilde{V},$$

where $d\widetilde{V}$ is the Riemannian volume element on C associated with the induced metric and the orientation determined by γ.

(*Used on p. 273.*)

2-33. Prove Proposition 2.65 (Sylvester's law of inertia).

2-34. Prove Theorem 2.69 (existence of Lorentz metrics), as follows.

  (a) For sufficiency, assume that $D \subseteq TM$ is a 1-dimensional distribution, and choose any Riemannian metric g on M. Show that locally it is possible to choose a g-orthonormal frame (E_i) and dual coframe (ε^i) such that E_1 spans D; and then show that the Lorentz metric $-(\varepsilon^1)^2 + (\varepsilon^2)^2 + \cdots + (\varepsilon^n)^2$ is independent of the choice of frame.

  (b) To prove necessity, suppose that g is a Lorentz metric on M, and let g_0 be any Riemannian metric. Show that for each $p \in M$, there are exactly two g_0-unit vectors $v_0, -v_0$ on which the function $v \mapsto g(v, v)$ takes its minimum among all unit vectors in $T_p M$, and use Lagrange multipliers to conclude that there exists a number $\lambda(p) < 0$ such that $g(v_0, w) = \lambda(p)g_0(v_0, w)$ for all $w \in T_p M$. You may use the following standard result from perturbation theory: *if U is an open subset of $\mathbb{R}^n$ and $A \colon U \to \mathrm{GL}(n, \mathbb{R})$ is a smooth matrix-valued function such that $A(x)$ is symmetric and has exactly one negative eigenvalue for each $x \in U$, then there exist smooth functions $\lambda \colon U \to (-\infty, 0)$ and $X \colon U \to \mathbb{R}^n \smallsetminus \{0\}$ such that $A(x)X(x) = \lambda(x)X(x)$ for all $x \in U$.*

2-35. Prove Corollary 2.71 (about level sets in pseudo-Riemannian manifolds). (*Used on p. 63.*)

Chapter 3
Model Riemannian Manifolds

Before we delve into the general theory of Riemannian manifolds, we pause to give it some substance by introducing a variety of "model Riemannian manifolds" that should help to motivate the general theory. These manifolds are distinguished by having a high degree of symmetry.

We begin by describing the most symmetric model spaces of all—Euclidean spaces, spheres, and hyperbolic spaces. We analyze these in detail, and prove that each one has a very large isometry group: not only is there an isometry taking any point to any other point, but in fact one can find an isometry taking any orthonormal basis at one point to any orthonormal basis at any other point. As we will see in Chapter 8, this has strong consequences for the curvatures of these manifolds.

After introducing these very special models, we explore some more general classes of Riemannian manifolds with symmetry—the invariant metrics on Lie groups, homogeneous spaces, and symmetric spaces.

At the end of the chapter, we give a brief introduction to some analogous models in the pseudo-Riemannian case. For the particular case of Lorentz manifolds, these are the Minkowski spaces, de Sitter spaces, and anti-de Sitter spaces, which are important model spaces in general relativity.

Symmetries of Riemannian Manifolds

The main feature of the Riemannian manifolds we are going to introduce in this chapter is that they are all highly symmetric, meaning that they have large groups of isometries.

Let (M, g) be a Riemannian manifold. Recall that $\mathrm{Iso}(M, g)$ denotes the set of all isometries from M to itself, which is a group under composition. We say that (M, g) is a *homogeneous Riemannian manifold* if $\mathrm{Iso}(M, g)$ acts transitively on M, which is to say that for each pair of points $p, q \in M$, there is an isometry $\varphi \colon M \to M$ such that $\varphi(p) = q$.

© Springer International Publishing AG 2018
J. M. Lee, *Introduction to Riemannian Manifolds*, Graduate Texts
in Mathematics 176, https://doi.org/10.1007/978-3-319-91755-9_3

The isometry group does more than just act on M itself. For every $\varphi \in \text{Iso}(M,g)$, the global differential $d\varphi$ maps TM to itself and restricts to a linear isometry $d\varphi_p \colon T_p M \to T_{\varphi(p)} M$ for each $p \in M$.

Given a point $p \in M$, let $\text{Iso}_p(M,g)$ denote the isotropy subgroup at p, that is, the subgroup of $\text{Iso}(M,g)$ consisting of isometries that fix p. For each $\varphi \in \text{Iso}_p(M,g)$, the linear map $d\varphi_p$ takes $T_p M$ to itself, and the map $I_p \colon \text{Iso}_p(M,g) \to \text{GL}(T_p M)$ given by $I_p(\varphi) = d\varphi_p$ is a representation of $\text{Iso}_p(M,g)$, called the **isotropy representation**. We say that M is **isotropic at p** if the isotropy representation of $\text{Iso}_p(M,g)$ acts transitively on the set of unit vectors in $T_p M$. If M is isotropic at every point, we say simply that M is **isotropic**.

There is an even stronger kind of symmetry than isotropy. Let $O(M)$ denote the set of all orthonormal bases for all tangent spaces of M:

$$O(M) = \coprod_{p \in M} \{\text{orthonormal bases for } T_p M\}.$$

There is an induced action of $\text{Iso}(M,g)$ on $O(M)$, defined by using the differential of an isometry φ to push an orthonormal basis at p forward to an orthonormal basis at $\varphi(p)$:

$$\varphi \cdot (b_1,\ldots,b_n) = \big(d\varphi_p(b_1),\ldots,d\varphi_p(b_n)\big). \tag{3.1}$$

We say that (M,g) is **frame-homogeneous** if this induced action is transitive on $O(M)$, or in other words, if for all $p,q \in M$ and choices of orthonormal bases at p and q, there is an isometry taking p to q and the chosen basis at p to the one at q. (Warning: Some authors, such as [Boo86, dC92, Spi79], use the term *isotropic* to refer to the property we have called frame-homogeneous.)

Proposition 3.1. *Let (M,g) be a Riemannian manifold.*

(a) *If M is isotropic at one point and it is homogeneous, then it is isotropic.*
(b) *If M is frame-homogeneous, then it is homogeneous and isotropic.*

Proof. Problem 3-3. $\square$

A homogeneous Riemannian manifold looks geometrically the same at every point, while an isotropic one looks the same in every direction. It turns out that an isotropic Riemannian manifold is automatically homogeneous; however, a Riemannian manifold can be isotropic at one point without being isotropic (for example, the paraboloid $z = x^2 + y^2$ in $\mathbb{R}^3$ with the induced metric); homogeneous without being isotropic anywhere (for example, the Berger metrics on $\mathbb{S}^3$ discussed in Problem 3-10 below); or homogeneous and isotropic without being frame-homogeneous (for example, the Fubini–Study metrics on complex projective spaces discussed in Example 2.30). The proofs of these claims will have to wait until we have developed the theories of geodesics and curvature (see Problems 6-18, 8-5, 8-16, and 8-13).

As mentioned in Chapter 1, the Myers–Steenrod theorem shows that $\text{Iso}(M,g)$ is always a Lie group acting smoothly on M. Although we will not use that result, in many cases we can identify a smooth Lie group action that accounts for at least some of the isometry group, and in certain cases we will be able to prove that it is the entire isometry group.

Euclidean Spaces

The simplest and most important model Riemannian manifold is of course *n-dimensional Euclidean space*, which is just $\mathbb{R}^n$ with the Euclidean metric $\bar{g}$ given by (2.8).

Somewhat more generally, if V is any n-dimensional real vector space endowed with an inner product, we can set $g(v,w) = \langle v,w \rangle$ for any $p \in V$ and any $v,w \in T_pV \cong V$. Choosing an orthonormal basis $(b_1,\ldots,b_n)$ for V defines a basis isomorphism from $\mathbb{R}^n$ to V that sends $(x^1,\ldots,x^n)$ to $x^i b_i$; this is easily seen to be an isometry of (V,g) with $(\mathbb{R}^n,\bar{g})$, so all n-dimensional inner product spaces are isometric to each other as Riemannian manifolds.

It is easy to construct isometries of the Riemannian manifold $(\mathbb{R}^n,\bar{g})$: for example, every orthogonal linear transformation $A: \mathbb{R}^n \to \mathbb{R}^n$ preserves the Euclidean metric, as does every translation $x \mapsto b + x$. It follows that every map of the form $x \mapsto b + Ax$, formed by first applying the orthogonal map A and then translating by b, is an isometry.

It turns out that the set of all such isometries can be realized as a Lie group acting smoothly on $\mathbb{R}^n$. Regard $\mathbb{R}^n$ as a Lie group under addition, and let $\theta: O(n) \times \mathbb{R}^n \to \mathbb{R}^n$ be the natural action of $O(n)$ on $\mathbb{R}^n$. Define the *Euclidean group* $E(n)$ to be the semidirect product $\mathbb{R}^n \rtimes_\theta O(n)$ determined by this action: this is the Lie group whose underlying manifold is the product space $\mathbb{R}^n \times O(n)$, with multiplication given by $(b,A)(b',A') = (b + Ab', AA')$ (see Example C.12). It has a faithful representation given by the map $\rho: E(n) \to GL(n+1,\mathbb{R})$ defined in block form by

$$\rho(b,A) = \begin{pmatrix} A & b \\ 0 & 1 \end{pmatrix},$$

where b is considered an $n \times 1$ column matrix.

The Euclidean group acts on $\mathbb{R}^n$ via

$$(b,A) \cdot x = b + Ax. \tag{3.2}$$

Problem 3-1 shows that when $\mathbb{R}^n$ is endowed with the Euclidean metric, this action is isometric and the induced action on $O(\mathbb{R}^n)$ is transitive. (Later, we will see that this is in fact the full isometry group of $(\mathbb{R}^n,\bar{g})$—see Problem 5-11—but we do not need that fact now.) Thus each Euclidean space is frame-homogeneous.

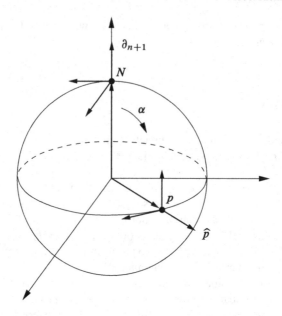

Fig. 3.1: Transitivity of $O(n+1)$ on $O(\mathbb{S}^n(R))$

Spheres

Our second class of model Riemannian manifolds comes in a family, with one for each positive real number. Given $R > 0$, let $\mathbb{S}^n(R)$ denote the sphere of radius R centered at the origin in $\mathbb{R}^{n+1}$, endowed with the metric $\mathring{g}_R$ (called the **round metric of radius R**) induced from the Euclidean metric on $\mathbb{R}^{n+1}$. When $R = 1$, it is the round metric on $\mathbb{S}^n$ introduced in Example 2.13, and we use the notation $\mathring{g} = \mathring{g}_1$.

One of the first things one notices about the spheres is that like Euclidean spaces, they are highly symmetric. We can immediately write down a large group of isometries of $\mathbb{S}^n(R)$ by observing that the linear action of the orthogonal group $O(n+1)$ on $\mathbb{R}^{n+1}$ preserves $\mathbb{S}^n(R)$ and the Euclidean metric, so its restriction to $\mathbb{S}^n(R)$ acts isometrically on the sphere. (Problem 5-11 will show that this is the full isometry group.)

Proposition 3.2. *The group $O(n+1)$ acts transitively on $O(\mathbb{S}^n(R))$, and thus each round sphere is frame-homogeneous.*

Proof. It suffices to show that given any $p \in \mathbb{S}^n(R)$ and any orthonormal basis (b_i) for $T_p\mathbb{S}^n(R)$, there is an orthogonal map that takes the "north pole" $N = (0,\dots,0,R)$ to p and the basis $(\partial_1,\dots,\partial_n)$ for $T_N\mathbb{S}^n(R)$ to (b_i).

To do so, think of p as a vector of length R in $\mathbb{R}^{n+1}$, and let $\hat{p} = p/R$ denote the unit vector in the same direction (Fig. 3.1). Since the basis vectors (b_i) are tangent to the sphere, they are orthogonal to $\hat{p}$, so $(b_1,\dots,b_n,\hat{p})$ is an orthonormal basis for $\mathbb{R}^{n+1}$. Let α be the matrix whose columns are these basis vectors. Then $\alpha \in O(n+1)$, and by elementary linear algebra, α takes the standard basis vectors $(\partial_1, \ldots, \partial_{n+1})$ to $(b_1, \ldots, b_n, \hat{p})$. It follows that $\alpha(N) = p$. Moreover, since α acts linearly on $\mathbb{R}^{n+1}$, its differential $d\alpha_N : T_N\mathbb{R}^{n+1} \to T_p\mathbb{R}^{n+1}$ is represented in standard coordinates by the same matrix as α itself, so $d\alpha_N(\partial_i) = b_i$ for $i = 1, \ldots, n$, and α is the desired orthogonal map. $\square$

Another important feature of the round metrics—one that is much less evident than their symmetry—is that they bear a certain close relationship to the Euclidean metrics, which we now describe. Two metrics g_1 and g_2 on a manifold M are said to be *conformally related* (or *pointwise conformal* or just *conformal*) to each other if there is a positive function $f \in C^\infty(M)$ such that $g_2 = fg_1$. Given two Riemannian manifolds (M, g) and $(\widetilde{M}, \widetilde{g})$, a diffeomorphism $\varphi : M \to \widetilde{M}$ is called a *conformal diffeomorphism* (or a *conformal transformation*) if it pulls $\widetilde{g}$ back to a metric that is conformal to g:

$$\varphi^*\widetilde{g} = fg \text{ for some positive } f \in C^\infty(M).$$

Problem 3-6 shows that conformal diffeomorphisms are the same as angle-preserving diffeomorphisms. Two Riemannian manifolds are said to be *conformally equivalent* if there is a conformal diffeomorphism between them.

A Riemannian manifold (M, g) is said to be *locally conformally flat* if every point of M has a neighborhood that is conformally equivalent to an open set in $(\mathbb{R}^n, \bar{g})$.

▶ **Exercise 3.3.** (a) Show that for every smooth manifold M, conformality is an equivalence relation on the set of all Riemannian metrics on M.
　　(b) Show that conformal equivalence is an equivalence relation on the class of all Riemannian manifolds.

▶ **Exercise 3.4.** Suppose g_1 and $g_2 = fg_1$ are conformally related metrics on an oriented n-manifold. Show that their volume forms are related by $dV_{g_2} = f^{n/2}dV_{g_1}$.

A conformal equivalence between $\mathbb{R}^n$ and $\mathbb{S}^n(R)$ minus a point is provided by *stereographic projection* from the north pole. This is the map $\sigma : \mathbb{S}^n(R) \smallsetminus \{N\} \to \mathbb{R}^n$ that sends a point $P \in \mathbb{S}^n(R) \smallsetminus \{N\}$, written $P = (\xi^1, \ldots, \xi^n, \tau)$, to $u = (u^1, \ldots, u^n) \in \mathbb{R}^n$, where $U = (u^1, \ldots, u^n, 0)$ is the point where the line through N and P intersects the hyperplane $\{(\xi, \tau) : \tau = 0\}$ in $\mathbb{R}^{n+1}$ (Fig. 3.2). Thus U is characterized by the fact that $(U - N) = \lambda(P - N)$ for some nonzero scalar λ. Writing $N = (0, R)$, $U = (u, 0)$, and $P = (\xi, \tau) \in \mathbb{R}^{n+1} = \mathbb{R}^n \times \mathbb{R}$, we obtain the system of equations

$$\begin{aligned} u^i &= \lambda \xi^i, \\ -R &= \lambda(\tau - R). \end{aligned} \tag{3.3}$$

Solving the second equation for λ and plugging it into the first equation, we get the following formula for stereographic projection from the north pole of the sphere of radius R:

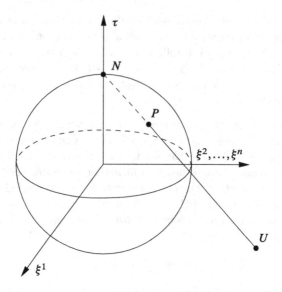

Fig. 3.2: Stereographic projection

$$\sigma(\xi, \tau) = u = \frac{R\xi}{R - \tau}. \tag{3.4}$$

It follows from this formula that σ is defined and smooth on all of $\mathbb{S}^n(R) \smallsetminus \{N\}$. The easiest way to see that it is a diffeomorphism is to compute its inverse. Solving the two equations of (3.3) for τ and ξ^i gives

$$\xi^i = \frac{u^i}{\lambda}, \qquad \tau = R\frac{\lambda - 1}{\lambda}. \tag{3.5}$$

The point $P = \sigma^{-1}(u)$ is characterized by these equations and the fact that P is on the sphere. Thus, substituting (3.5) into $|\xi|^2 + \tau^2 = R^2$ gives

$$\frac{|u|^2}{\lambda^2} + R^2\frac{(\lambda - 1)^2}{\lambda^2} = R^2,$$

from which we conclude

$$\lambda = \frac{|u|^2 + R^2}{2R^2}.$$

Inserting this back into (3.5) gives the formula

$$\sigma^{-1}(u) = (\xi, \tau) = \left(\frac{2R^2 u}{|u|^2 + R^2}, R\frac{|u|^2 - R^2}{|u|^2 + R^2} \right), \tag{3.6}$$

which by construction maps $\mathbb{R}^n$ back to $\mathbb{S}^n(R) \smallsetminus \{N\}$ and shows that σ is a diffeomorphism.

Proposition 3.5. *Stereographic projection is a conformal diffeomorphism between* $\mathbb{S}^n(R) \smallsetminus \{N\}$ *and* $\mathbb{R}^n$.

Proof. The inverse map σ^{-1} is a smooth parametrization of $\mathbb{S}^n(R) \smallsetminus \{N\}$, so we can use it to compute the pullback metric. Using the usual technique of substitution to compute pullbacks, we obtain the following coordinate representation of $\overset{\circ}{g}_R$ in stereographic coordinates:

$$(\sigma^{-1})^* \overset{\circ}{g}_R = (\sigma^{-1})^* \overline{g} = \sum_j \left(d\left(\frac{2R^2 u^j}{|u|^2 + R^2} \right) \right)^2 + \left(d\left(R \frac{|u|^2 - R^2}{|u|^2 + R^2} \right) \right)^2.$$

If we expand each of these terms individually, we get

$$d\left(\frac{2R^2 u^j}{|u|^2 + R^2} \right) = \frac{2R^2 du^j}{|u|^2 + R^2} - \frac{4R^2 u^j \sum_i u^i du^i}{(|u|^2 + R^2)^2};$$

$$d\left(R \frac{|u|^2 - R^2}{|u|^2 + R^2} \right) = \frac{2R \sum_i u^i du^i}{|u|^2 + R^2} - \frac{2R(|u|^2 - R^2) \sum_i u^i du^i}{(|u|^2 + R^2)^2}$$

$$= \frac{4R^3 \sum_i u^i du^i}{(|u|^2 + R^2)^2}.$$

Therefore,

$$(\sigma^{-1})^* \overset{\circ}{g}_R = \frac{4R^4 \sum_j (du^j)^2}{(|u|^2 + R^2)^2} - \frac{16R^4 \left(\sum_i u^i du^i \right)^2}{(|u|^2 + R^2)^3} + \frac{16R^4 |u|^2 \left(\sum_i u^i du^i \right)^2}{(|u|^2 + R^2)^4}$$

$$+ \frac{16R^6 \left(\sum_i u^i du^i \right)^2}{(|u|^2 + R^2)^4}$$

$$= \frac{4R^4 \sum_j (du^j)^2}{(|u|^2 + R^2)^2}.$$

In other words,

$$(\sigma^{-1})^* \overset{\circ}{g}_R = \frac{4R^4}{(|u|^2 + R^2)^2} \overline{g}, \tag{3.7}$$

where $\overline{g}$ now represents the Euclidean metric on $\mathbb{R}^n$, and so σ is a conformal diffeomorphism. $\square$

Corollary 3.6. *Each sphere with a round metric is locally conformally flat.*

Proof. Stereographic projection gives a conformal equivalence between a neighborhood of any point except the north pole and Euclidean space; applying a suitable rotation and then stereographic projection (or stereographic projection from the south pole), we get such an equivalence for a neighborhood of the north pole as well. $\square$

Hyperbolic Spaces

Our third class of model Riemannian manifolds is perhaps less familiar than the other two. For each $n \geq 1$ and each $R > 0$ we will define a frame-homogeneous Riemannian manifold $\mathbb{H}^n(R)$, called **hyperbolic space of radius R**. There are four equivalent models of the hyperbolic spaces, each of which is useful in certain contexts. In the next theorem, we introduce all of them and show that they are isometric.

Theorem 3.7. *Let n be an integer greater than 1. For each fixed $R > 0$, the following Riemannian manifolds are all mutually isometric.*

(a) (HYPERBOLOID MODEL) $\mathbb{H}^n(R)$ *is the submanifold of Minkowski space $\mathbb{R}^{n,1}$ defined in standard coordinates $\left(\xi^1,\ldots,\xi^n,\tau\right)$ as the "upper sheet" $\{\tau > 0\}$ of the two-sheeted hyperboloid $\left(\xi^1\right)^2 + \cdots + \left(\xi^n\right)^2 - \tau^2 = -R^2$, with the induced metric*

$$\breve{g}_R^1 = \iota^* \bar{q},$$

where $\iota\colon \mathbb{H}^n(R) \to \mathbb{R}^{n,1}$ is inclusion, and $\bar{q} = \bar{q}^{(n,1)}$ is the Minkowski metric:

$$\bar{q} = \left(d\xi^1\right)^2 + \cdots + \left(d\xi^n\right)^2 - (d\tau)^2. \tag{3.8}$$

(b) (BELTRAMI–KLEIN MODEL) $\mathbb{K}^n(R)$ *is the ball of radius R centered at the origin in $\mathbb{R}^n$, with the metric given in coordinates $(w^1,\ldots,w^n)$ by*

$$\breve{g}_R^2 = R^2 \frac{(dw^1)^2 + \cdots + (dw^n)^2}{R^2 - |w|^2} + R^2 \frac{\left(w^1\,dw^1 + \cdots + w^n\,dw^n\right)^2}{\left(R^2 - |w|^2\right)^2}. \tag{3.9}$$

(c) (POINCARÉ BALL MODEL) $\mathbb{B}^n(R)$ *is the ball of radius R centered at the origin in $\mathbb{R}^n$, with the metric given in coordinates $(u^1,\ldots,u^n)$ by*

$$\breve{g}_R^3 = 4R^4 \frac{(du^1)^2 + \cdots + (du^n)^2}{(R^2 - |u|^2)^2}.$$

(d) (POINCARÉ HALF-SPACE MODEL) $\mathbb{U}^n(R)$ *is the upper half-space in $\mathbb{R}^n$ defined in coordinates $(x^1,\ldots,x^{n-1},y)$ by $\mathbb{U}^n(R) = \{(x,y) : y > 0\}$, endowed with the metric*

$$\breve{g}_R^4 = R^2 \frac{(dx^1)^2 + \cdots + (dx^{n-1})^2 + dy^2}{y^2}.$$

Proof. Let $R > 0$ be given. We need to verify that $\mathbb{H}^n(R)$ is actually a Riemannian submanifold of $\mathbb{R}^{n,1}$, or in other words that $\breve{g}_R^1$ is positive definite. One way to do this is to show, as we will below, that it is the pullback of $\breve{g}_R^2$ or $\breve{g}_R^3$ (both of which are manifestly positive definite) by a diffeomorphism. Alternatively, here is a direct proof using some of the theory of submanifolds of pseudo-Riemannian manifolds developed in Chapter 1.

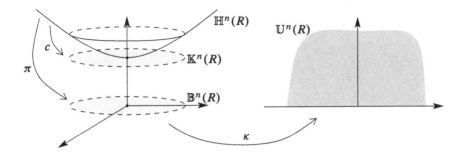

Fig. 3.3: Isometries among the hyperbolic models

Note that $\mathbb{H}^n(R)$ is an open subset of a level set of the smooth function $f : \mathbb{R}^{n,1} \to \mathbb{R}$ given by $f(\xi, \tau) = (\xi^1)^2 + \cdots + (\xi^n)^2 - \tau^2$. We have

$$df = 2\xi^1 d\xi^1 + \cdots + 2\xi^n d\xi^n - 2\tau d\tau,$$

and therefore the gradient of f with respect to $\bar{q}$ is given by

$$\operatorname{grad} f = 2\xi^1 \frac{\partial}{\partial \xi^1} + \cdots + 2\xi^n \frac{\partial}{\partial \xi^n} + 2\tau \frac{\partial}{\partial \tau}. \tag{3.10}$$

Direct computation shows that

$$\bar{q}(\operatorname{grad} f, \operatorname{grad} f) = 4 \left(\sum_i (\xi^i)^2 - \tau^2 \right),$$

which is equal to $-4R^2$ at points of $\mathbb{H}^n(R)$. Thus it follows from Corollary 2.71 that $\mathbb{H}^n(R)$ is a pseudo-Riemannian submanifold of signature $(n, 0)$, which is to say it is Riemannian.

We will show that all four Riemannian manifolds are mutually isometric by defining isometries $c : \mathbb{H}^n(R) \to \mathbb{K}^n(R)$, $\pi : \mathbb{H}^n(R) \to \mathbb{B}^n(R)$, and $\kappa : \mathbb{B}^n(R) \to \mathbb{U}^n(R)$ (shown schematically in Fig. 3.3).

We begin with a geometric construction of a diffeomorphism called **central projection** from the hyperboloid to the ball,

$$c : \mathbb{H}^n(R) \to \mathbb{K}^n(R),$$

which turns out to be an isometry between the two metrics given in (a) and (b). For any $P = (\xi^1, \ldots, \xi^n, \tau) \in \mathbb{H}^n(R) \subseteq \mathbb{R}^{n,1}$, set $c(P) = w \in \mathbb{K}^n(R)$, where $W = (w, R) \in \mathbb{R}^{n,1}$ is the point where the line from the origin to P intersects the hyperplane $\{(\xi, \tau) : \tau = R\}$ (Fig. 3.4). Because W is characterized as the unique scalar multiple of P whose last coordinate is R, we have $W = RP/\tau$, and therefore

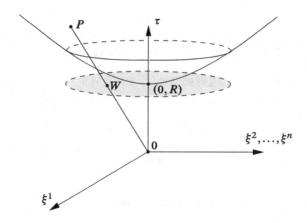

Fig. 3.4: Central projection from the hyperboloid to the Beltrami–Klein model

c is given by the formula

$$c(\xi, \tau) = \frac{R\xi}{\tau}. \tag{3.11}$$

The relation $|\xi|^2 - \tau^2 = -R^2$ guarantees that $|c(\xi, \tau)|^2 = R^2(1 - R^2/\tau^2) < R^2$, so c maps $\mathbb{H}^n(R)$ into $\mathbb{K}^n(R)$. To show that c is a diffeomorphism, we determine its inverse map. Let $w \in \mathbb{K}^n(R)$ be arbitrary. The unique positive scalar λ such that the point $(\xi, \tau) = \lambda(w, R)$ lies on $\mathbb{H}^n(R)$ is characterized by $\lambda^2 |w|^2 - \lambda^2 R^2 = -R^2$, and therefore

$$\lambda = \frac{R}{\sqrt{R^2 - |w|^2}}.$$

It follows that the following smooth map is an inverse for c:

$$c^{-1}(w) = (\xi, \tau) = \left(\frac{Rw}{\sqrt{R^2 - |w|^2}}, \frac{R^2}{\sqrt{R^2 - |w|^2}} \right). \tag{3.12}$$

Thus c is a diffeomorphism. To show that it is an isometry between $\breve{g}_R^1$ and $\breve{g}_R^2$, we use the fact that $\breve{g}_R^1$ is the metric induced from $\bar{q}$, analogously to the computation we did for stereographic projection above. With (ξ, τ) defined by (3.12), we have

$$d\xi^i = \frac{R\,dw^i}{\sqrt{R^2 - |w|^2}} + \frac{Rw^i \sum_j w^j \, dw^j}{\left(R^2 - |w|^2 \right)^{3/2}},$$

$$d\tau = \frac{R^2 \sum_j w^j \, dw^j}{\left(R^2 - |w|^2 \right)^{3/2}}.$$

It is then straightforward to compute that $\left(c^{-1} \right)^* \breve{g}_R^1 = \sum_i (d\xi^i)^2 - (d\tau)^2 = \breve{g}_R^2$.

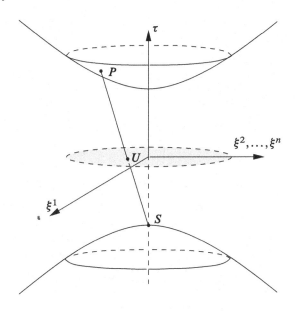

Fig. 3.5: Hyperbolic stereographic projection

Next we describe a diffeomorphism

$$\pi : \mathbb{H}^n(R) \to \mathbb{B}^n(R)$$

from the hyperboloid to the ball, called ***hyperbolic stereographic projection***, which is an isometry between the metrics of (a) and (c). Let $S \in \mathbb{R}^{n,1}$ denote the point $S = (0, \dots, 0, -R)$. For any $P = (\xi^1, \dots, \xi^n, \tau) \in \mathbb{H}^n(R) \subseteq \mathbb{R}^{n,1}$, set $\pi(P) = u \in \mathbb{B}^n(R)$, where $U = (u, 0) \in \mathbb{R}^{n,1}$ is the point where the line through S and P intersects the hyperplane $\{(\xi, \tau) : \tau = 0\}$ (Fig. 3.5). The point U is characterized by $(U - S) = \lambda(P - S)$ for some nonzero scalar λ, or

$$\begin{aligned} u^i &= \lambda \xi^i, \\ R &= \lambda(\tau + R). \end{aligned} \qquad (3.13)$$

These equations can be solved in the same manner as in the spherical case to yield

$$\pi(\xi, \tau) = u = \frac{R\xi}{R + \tau},$$

which takes its values in $\mathbb{B}^n(R)$ because $|\pi(\xi, \tau)|^2 = R^2(\tau^2 - R^2)(\tau^2 + R^2) < R^2$. A computation similar to the ones before shows that the inverse map is

$$\pi^{-1}(u) = (\xi, \tau) = \left(\frac{2R^2 u}{R^2 - |u|^2}, R\frac{R^2 + |u|^2}{R^2 - |u|^2} \right).$$

We will show that $(\pi^{-1})^* \breve{g}_R^1 = \breve{g}_R^3$. The computation proceeds just as in the spherical case, so we skip over most of the details:

$$
(\pi^{-1})^* \breve{g}_R^1 = \sum_j \left(d \left(\frac{2R^2 u^j}{R^2 - |u|^2} \right) \right)^2 - \left(d \left(R \frac{R^2 + |u|^2}{R^2 - |u|^2} \right) \right)^2
$$
$$
= \frac{4R^4 \sum_j (du^j)^2}{(R^2 - |u|^2)^2}
$$
$$
= \breve{g}_R^3.
$$

Next we consider the Poincaré half-space model, by constructing an explicit diffeomorphism

$$
\kappa \colon \mathbb{U}^n(R) \to \mathbb{B}^n(R).
$$

In this case it is more convenient to write the coordinates on the ball as $(u,v) = (u^1, \dots, u^{n-1}, v)$. In the 2-dimensional case, κ is easy to write down in complex notation $w = u + iv$ and $z = x + iy$. It is a variant of the classical **Cayley transform**:

$$
\kappa(z) = w = iR \frac{z - iR}{z + iR}. \tag{3.14}
$$

Elementary complex analysis shows that this is a complex-analytic diffeomorphism taking $\mathbb{U}^2(R)$ onto $\mathbb{B}^2(R)$. Separating z into real and imaginary parts, we can also write this in real terms as

$$
\kappa(x,y) = (u,v) = \left(\frac{2R^2 x}{|x|^2 + (y+R)^2}, R \frac{|x|^2 + |y|^2 - R^2}{|x|^2 + (y+R)^2} \right). \tag{3.15}
$$

This same formula makes sense in any dimension n if we interpret x to mean $(x^1, \dots, x^{n-1})$, and it is easy to check that it maps the upper half-space $\{y > 0\}$ into the ball of radius R. A direct computation shows that its inverse is

$$
\kappa^{-1}(u,v) = (x,y) = \left(\frac{2R^2 u}{|u|^2 + (v-R)^2}, R \frac{R^2 - |u|^2 - v^2}{|u|^2 + (v-R)^2} \right),
$$

so κ is a diffeomorphism, called the **generalized Cayley transform**. The verification that $\kappa^* \breve{g}_R^3 = \breve{g}_R^4$ is basically a long calculation, and is left to Problem 3-4. $\qquad\square$

We often use the generic notation $\mathbb{H}^n(R)$ to refer to any one of the Riemannian manifolds of Theorem 3.7, and $\breve{g}_R$ to refer to the corresponding metric; the special case $R = 1$ is denoted by $(\mathbb{H}^n, \breve{g})$ and is called simply **hyperbolic space**, or in the 2-dimensional case, the **hyperbolic plane**.

Because all of the models for a given value of R are isometric to each other, when analyzing them geometrically we can use whichever model is most convenient for the application we have in mind. The next corollary is an example in which the Poincaré ball and half-space models serve best.

Corollary 3.8. *Each hyperbolic space is locally conformally flat.*

Proof. In either the Poincaré ball model or the half-space model, the identity map gives a global conformal equivalence with an open subset of Euclidean space. □

The examples presented so far might give the impression that most Riemannian manifolds are locally conformally flat. This is far from the truth, but we do not yet have the tools to prove it. See Problem 8-25 for some explicit examples of Riemannian manifolds that are not locally conformally flat.

The symmetries of $\mathbb{H}^n(R)$ are most easily seen in the hyperboloid model. Let $O(n, 1)$ denote the group of linear maps from $\mathbb{R}^{n,1}$ to itself that preserve the Minkowski metric, called the $(n + 1)$-*dimensional Lorentz group*. Note that each element of $O(n, 1)$ preserves the hyperboloid $\{\tau^2 - |\xi|^2 = R^2\}$, which has two components determined by $\tau > 0$ and $\tau < 0$. We let $O^+(n, 1)$ denote the subgroup of $O(n, 1)$ consisting of maps that take the $\tau > 0$ component of the hyperboloid to itself. (This is called the **orthochronous Lorentz group**, because physically it represents coordinate changes that preserve the forward time direction.) Then $O^+(n, 1)$ preserves $\mathbb{H}^n(R)$, and because it preserves $\bar{q}$ it acts isometrically on $\mathbb{H}^n(R)$. (Problem 5-11 will show that this is the full isometry group.) Recall that $O(\mathbb{H}^n(R))$ denotes the set of all orthonormal bases for all tangent spaces of $\mathbb{H}^n(R)$.

Proposition 3.9. *The group* $O^+(n, 1)$ *acts transitively on* $O(\mathbb{H}^n(R))$, *and therefore* $\mathbb{H}^n(R)$ *is frame-homogeneous.*

Proof. The argument is entirely analogous to the proof of Proposition 3.2, so we give only a sketch. Suppose $p \in \mathbb{H}^n(R)$ and (b_i) is an orthonormal basis for $T_p\mathbb{H}^n(R)$. Identifying $p \in \mathbb{R}^{n,1}$ with an element of $T_p\mathbb{R}^{n,1}$ in the usual way, we can regard $\hat{p} = p/R$ as a $\bar{q}$-unit vector in $T_p\mathbb{R}^{n,1}$, and (3.10) shows that it is a scalar multiple of the $\bar{q}$-gradient of the defining function f and thus is orthogonal to $T_p\mathbb{H}^n(R)$ with respect to $\bar{q}$. Thus $(b_1, \ldots, b_n, b_{n+1} = \hat{p})$ is a $\bar{q}$-orthonormal basis for $\mathbb{R}^{n,1}$, and $\bar{q}$ has the following expression in terms of the dual basis (β^j):

$$\bar{q} = (\beta^1)^2 + \cdots + (\beta^n)^2 - (\beta^{n+1})^2.$$

Thus the matrix whose columns are $(b_1, \ldots, b_{n+1})$ is an element of $O^+(n, 1)$ sending $N = (0, \ldots, 0, R)$ to p and ∂_i to b_i (Fig. 3.6). □

Invariant Metrics on Lie Groups

Lie groups provide us with another large class of homogeneous Riemannian manifolds. (See Appendix C for a review of the basic facts about Lie groups that we will use.)

Let G be a Lie group. A Riemannian metric g on G is said to be **left-invariant** if it is invariant under all left translations: $L_\varphi^* g = g$ for all $\varphi \in G$. Similarly, g is **right-invariant** if it is invariant under all right translations, and **bi-invariant** if it is

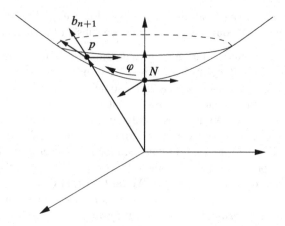

Fig. 3.6: Frame homogeneity of $\mathbb{H}^n(R)$

both left- and right-invariant. The next lemma shows that left-invariant metrics are easy to come by.

Lemma 3.10. *Let G be a Lie group and let $\mathfrak{g}$ be its Lie algebra of left-invariant vector fields.*

(a) *A Riemannian metric g on G is left-invariant if and only if for all $X, Y \in \mathfrak{g}$, the function $g(X, Y)$ is constant on G.*

(b) *The restriction map $g \mapsto g_e \in \Sigma^2 \left(T_e^* G \right)$ together with the natural identification $T_e G \cong \mathfrak{g}$ gives a bijection between left-invariant Riemannian metrics on G and inner products on $\mathfrak{g}$.*

▶ **Exercise 3.11.** Prove the preceding lemma.

Thus all we need to do to construct a left-invariant metric is choose any inner product on $\mathfrak{g}$, and define a metric on G by applying that inner product to left-invariant vector fields. Right-invariant metrics can be constructed in a similar way using right-invariant vector fields. Since a Lie group acts transitively on itself by either left or right translation, every left-invariant or right-invariant metric is homogeneous.

Much more interesting are the bi-invariant metrics, because, as you will be able to prove later (Problems 7-13 and 8-17), their curvatures are intimately related to the structure of the Lie algebra of the group. But bi-invariant metrics are generally much rarer than left-invariant or right-invariant ones; in fact, some Lie groups have no bi-invariant metrics at all (see Problems 3-12 and 3-13). Fortunately, there is a complete answer to the question of which Lie groups admit bi-invariant metrics, which we present in this section.

We begin with a proposition that shows how to determine whether a given left-invariant metric is bi-invariant, based on properties of the *adjoint representation* of the group. Recall that this is the representation $\mathrm{Ad} \colon G \to \mathrm{GL}(\mathfrak{g})$ given by $\mathrm{Ad}(\varphi) = (C_\varphi)_* \colon \mathfrak{g} \to \mathfrak{g}$, where $C_\varphi \colon G \to G$ is the automorphism defined by conjugation: $C_\varphi(\psi) = \varphi \psi \varphi^{-1}$. See Appendix C for more details.

Proposition 3.12. *Let G be a Lie group and $\mathfrak{g}$ its Lie algebra. Suppose g is a left-invariant Riemannian metric on G, and let $\langle \cdot, \cdot \rangle$ denote the corresponding inner product on $\mathfrak{g}$ as in Lemma 3.10. Then g is bi-invariant if and only if $\langle \cdot, \cdot \rangle$ is invariant under the action of $\mathrm{Ad}(G) \subseteq \mathrm{GL}(\mathfrak{g})$, in the sense that $\langle \mathrm{Ad}(\varphi)X, \mathrm{Ad}(\varphi)Y \rangle = \langle X, Y \rangle$ for all $X, Y \in \mathfrak{g}$ and $\varphi \in G$.*

Proof. We begin the proof with some preliminary computations. Suppose g is left-invariant and $\langle \cdot, \cdot \rangle$ is the associated inner product on $\mathfrak{g}$. Let $\varphi \in G$ be arbitrary, and note that C_φ is the composition of left multiplication by φ followed by right multiplication by φ^{-1}. Thus for every $X \in \mathfrak{g}$, left-invariance implies $(R_{\varphi^{-1}})_* X = (R_{\varphi^{-1}})_* (L_\varphi)_* X = (C_\varphi)_* X = \mathrm{Ad}(\varphi)X$. Therefore, for all $\psi \in G$ and $X, Y \in \mathfrak{g}$, we have

$$\left((R_{\varphi^{-1}})^* g\right)_\psi (X_\psi, Y_\psi) = g_{\psi \varphi^{-1}} \left(\left((R_{\varphi^{-1}})_* X\right)_{\psi \varphi^{-1}}, \left((R_{\varphi^{-1}})_* Y\right)_{\psi \varphi^{-1}}\right)$$
$$= g_{\psi \varphi^{-1}} \left((\mathrm{Ad}(\varphi)X)_{\psi \varphi^{-1}}, (\mathrm{Ad}(\varphi)Y)_{\psi \varphi^{-1}}\right)$$
$$= \langle \mathrm{Ad}(\varphi)X, \mathrm{Ad}(\varphi)Y \rangle.$$

Now assume that $\langle \cdot, \cdot \rangle$ is invariant under $\mathrm{Ad}(G)$. Then the expression on the last line above is equal to $\langle X, Y \rangle = g_\psi (X_\psi, Y_\psi)$, which shows that $(R_{\varphi^{-1}})^* g = g$. Since this is true for all $\varphi \in G$, it follows that g is bi-invariant.

Conversely, assuming that g is bi-invariant, we have $(R_{\varphi^{-1}})^* g = g$ for each $\varphi \in G$, so the above computation yields

$$\langle X, Y \rangle = g_\psi (X_\psi, Y_\psi) = \left((R_{\varphi^{-1}})^* g\right)_\psi (X_\psi, Y_\psi) = \langle \mathrm{Ad}(\varphi)X, \mathrm{Ad}(\varphi)Y \rangle,$$

which shows that $\langle \cdot, \cdot \rangle$ is $\mathrm{Ad}(G)$-invariant. $\square$

In order to apply the preceding proposition, we need a lemma about finding invariant inner products on vector spaces. Recall from Appendix C that for every finite-dimensional real vector space V, $\mathrm{GL}(V)$ denotes the Lie group of all invertible linear maps from V to itself. If H is a subgroup of $\mathrm{GL}(V)$, an inner product $\langle \cdot, \cdot \rangle$ on V is said to be **H-invariant** if $\langle hx, hy \rangle = \langle x, y \rangle$ for all $x, y \in V$ and $h \in H$.

Lemma 3.13. *Suppose V is a finite-dimensional real vector space and H is a subgroup of $\mathrm{GL}(V)$. There exists an H-invariant inner product on V if and only if H has compact closure in $\mathrm{GL}(V)$.*

Proof. Assume first that there exists an H-invariant inner product $\langle \cdot, \cdot \rangle$ on V. This implies that H is contained in the subgroup $\mathrm{O}(V) \subseteq \mathrm{GL}(V)$ consisting of linear isomorphisms of V that are orthogonal with respect to this inner product. Choosing an orthonormal basis of V yields a Lie group isomorphism between $\mathrm{O}(V)$ and $\mathrm{O}(n) \subseteq \mathrm{GL}(n, \mathbb{R})$ (where $n = \dim V$), so $\mathrm{O}(V)$ is compact; and the closure of H is a closed subset of this compact group, and thus is itself compact.

Conversely, suppose H has compact closure in $\mathrm{GL}(V)$, and let K denote the closure. A simple limiting argument shows that K is itself a subgroup, and thus it is a Lie group by the closed subgroup theorem (Thm. C.8). Let $\langle \cdot, \cdot \rangle_0$ be an arbitrary inner product on V, and let μ be a right-invariant volume form on K (for example, the volume form of some right-invariant metric on K). For fixed $x, y \in V$, define a smooth function $f_{x,y} \colon K \to \mathbb{R}$ by $f_{x,y}(k) = \langle kx, ky \rangle_0$. Then define a new inner product $\langle \cdot, \cdot \rangle$ on V by

$$\langle x, y \rangle = \int_K f_{x,y}\, \mu.$$

It follows directly from the definition that $\langle \cdot, \cdot \rangle$ is symmetric and bilinear over $\mathbb{R}$. For each nonzero $x \in V$, we have $f_{x,x} > 0$ everywhere on K, so $\langle x, x \rangle > 0$, showing that $\langle \cdot, \cdot \rangle$ is indeed an inner product.

To see that it is invariant under K, let $k_0 \in K$ be arbitrary. Then for all $x, y \in V$ and $k \in K$, we have

$$\begin{aligned}
f_{k_0 x, k_0 y}(k) &= \langle k k_0 x, k k_0 y \rangle_0 \\
&= f_{x,y} \circ R_{k_0}(k),
\end{aligned}$$

where $R_{k_0} \colon K \to K$ is right translation by k_0. Because μ is right-invariant, it follows from diffeomorphism invariance of the integral that

$$\begin{aligned}
\langle k_0 x, k_0 y \rangle &= \int_K f_{k_0 x, k_0 y}\, \mu \\
&= \int_K \left(f_{x,y} \circ R_{k_0} \right) \mu \\
&= \int_K R_{k_0}^* \left(f_{x,y}\, \mu \right) \\
&= \int_K f_{x,y}\, \mu = \langle x, y \rangle.
\end{aligned}$$

Thus $\langle \cdot, \cdot \rangle$ is K-invariant, and it is also H-invariant because $H \subseteq K$. $\square$

Theorem 3.14 (Existence of Bi-invariant Metrics). *Let G be a Lie group and $\mathfrak{g}$ its Lie algebra. Then G admits a bi-invariant metric if and only if $\mathrm{Ad}(G)$ has compact closure in $\mathrm{GL}(\mathfrak{g})$.*

Proof. Proposition 3.12 shows that there is a bi-invariant metric on G if and only if there is an $\mathrm{Ad}(G)$-invariant inner product on $\mathfrak{g}$, and Lemma 3.13 in turn shows that the latter is true if and only if $\mathrm{Ad}(G)$ has compact closure in $\mathrm{GL}(\mathfrak{g})$. $\square$

The most important application of the preceding theorem is to compact groups.

Corollary 3.15. *Every compact Lie group admits a bi-invariant Riemannian metric.*

Proof. If G is compact, then $\mathrm{Ad}(G)$ is a compact subgroup of $\mathrm{GL}(\mathfrak{g})$ because $\mathrm{Ad} \colon G \to \mathrm{GL}(\mathfrak{g})$ is continuous. $\square$

Another important application is to prove that certain Lie groups do not admit bi-invariant metrics. One way to do this is to note that if $\mathrm{Ad}(G)$ has compact closure in $\mathrm{GL}(\mathfrak{g})$, then every orbit of $\mathrm{Ad}(G)$ must be a bounded subset of $\mathfrak{g}$ with respect to any choice of norm, because it is contained in the image of the compact set $\mathrm{Ad}(G)$ under a continuous map of the form $\varphi \mapsto \varphi(X_0)$ from $\mathrm{GL}(\mathfrak{g})$ to $\mathfrak{g}$. Thus if one can find an element $X_0 \in \mathfrak{g}$ and a subset $S \subseteq G$ such that the elements of the form $\mathrm{Ad}(\varphi)X$ are unbounded in $\mathfrak{g}$ for $\varphi \in S$, then there is no bi-invariant metric.

Here are some examples.

Example 3.16 (Invariant Metrics on Lie Groups).

(a) Every left-invariant metric on an abelian Lie group is bi-invariant, because the adjoint representation is trivial. Thus the Euclidean metric on $\mathbb{R}^n$ and the flat metric on $\mathbb{T}^n$ of Example 2.21 are both bi-invariant.

(b) If a metric g on a Lie group G is left-invariant, then the induced metric on every Lie subgroup $H \subseteq G$ is easily seen to be left-invariant. Similarly, if g is bi-invariant, then the induced metric on H is bi-invariant.

(c) The Lie group $\mathrm{SL}(2,\mathbb{R})$ (the group of 2×2 real matrices of determinant 1) admits many left-invariant metrics (as does every positive-dimensional Lie group), but no bi-invariant ones. To see this, recall that the Lie algebra of $\mathrm{SL}(2,\mathbb{R})$ is isomorphic to the algebra $\mathfrak{sl}(2,\mathbb{R})$ of trace-free 2×2 matrices, and the adjoint representation is given by $\mathrm{Ad}(A)X = AXA^{-1}$ (see Example C.10). If we let $X_0 = \left(\begin{smallmatrix} 0 & 1 \\ 0 & 0 \end{smallmatrix}\right) \in \mathfrak{sl}(2,\mathbb{R})$ and $A_c = \left(\begin{smallmatrix} c & 0 \\ 0 & 1/c \end{smallmatrix}\right) \in \mathrm{SL}(2,\mathbb{R})$ for $c > 0$, then $\mathrm{Ad}(A_c)X_0 = \left(\begin{smallmatrix} 0 & c^2 \\ 0 & 0 \end{smallmatrix}\right)$, which is unbounded as $c \to \infty$. Thus the orbit of X_0 is not contained in any compact subset, which implies that there is no bi-invariant metric on $\mathrm{SL}(2,\mathbb{R})$. A similar argument shows that $\mathrm{SL}(n,\mathbb{R})$ admits no bi-invariant metric for any $n \geq 2$. In view of (b) above, this shows also that $\mathrm{GL}(n,\mathbb{R})$ admits no bi-invariant metric for $n \geq 2$. (Of course, $\mathrm{GL}(1,\mathbb{R})$ does admit bi-invariant metrics because it is abelian.)

(d) With $\mathbb{S}^3$ regarded as a submanifold of $\mathbb{C}^2$, the map

$$(w, z) \mapsto \begin{pmatrix} w & z \\ -\bar{z} & \bar{w} \end{pmatrix} \tag{3.16}$$

gives a diffeomorphism from $\mathbb{S}^3$ to $\mathrm{SU}(2)$. Under the inverse of this map, the round metric on $\mathbb{S}^3$ pulls back to a bi-invariant metric on $\mathrm{SU}(2)$, as Problem 3-10 shows.

(e) Let $\mathfrak{o}(n)$ denote the Lie algebra of $\mathrm{O}(n)$, identified with the algebra of skew-symmetric $n \times n$ matrices, and define a bilinear form on $\mathfrak{o}(n)$ by

$$\langle A, B \rangle = \mathrm{tr}(A^T B).$$

This is an Ad-invariant inner product, and thus determines a bi-invariant Riemannian metric on $\mathrm{O}(n)$ (see Problem 3-11).

(f) Let $\mathbb{U}^n$ be the upper half-space as defined in Theorem 3.7. We can regard $\mathbb{U}^n$ as a Lie group by identifying each point $(x, y) = (x^1, \ldots, x^{n-1}, y) \in \mathbb{U}^n$ with an invertible $n \times n$ matrix as follows:

$$(x, y) \quad \longleftrightarrow \quad \begin{pmatrix} I_{n-1} & 0 \\ x^T & y \end{pmatrix},$$

where I_{n-1} is the $(n-1) \times (n-1)$ identity matrix. Then the hyperbolic metric $\breve{g}_R^4$ is left-invariant on $\mathbb{U}^n$ but not right-invariant (see Problem 3-12).

(g) For $n \geq 1$, the $(2n+1)$-dimensional **Heisenberg group** is the Lie subgroup $H_n \subseteq \mathrm{GL}(n+2, \mathbb{R})$ defined by

$$H_n = \left\{ \begin{pmatrix} 1 & x^T & z \\ 0 & 1 & y \\ 0 & 0 & 1 \end{pmatrix} : x, y \in \mathbb{R}^n, \ z \in \mathbb{R} \right\},$$

where x and y are treated as column matrices. These are the simplest examples of **nilpotent Lie groups**, meaning that the series of subgroups $G \supseteq [G, G] \supseteq [G, [G, G]] \supseteq \cdots$ eventually reaches the trivial subgroup (where for any subgroups $G_1, G_2 \subseteq G$, the notation $[G_1, G_2]$ means the subgroup of G generated by all elements of the form $x_1 x_2 x_1^{-1} x_2^{-1}$ for $x_i \in G_i$). There are many left-invariant metrics on H_n, but no bi-invariant ones, as Problem 3-13 shows.

(h) Our last example is a group that plays an important role in the classification of 3-manifolds. Let Sol denote the following 3-dimensional Lie subgroup of $\mathrm{GL}(3, \mathbb{R})$:

$$\mathrm{Sol} = \left\{ \begin{pmatrix} e^z & 0 & x \\ 0 & e^{-z} & y \\ 0 & 0 & 1 \end{pmatrix} : x, y, z \in \mathbb{R} \right\}.$$

This group is the simplest nonnilpotent example of a **solvable Lie group**, meaning that the series of subgroups $G \supseteq [G, G] \supseteq [[G, G], [G, G]] \supseteq \cdots$ eventually reaches the trivial subgroup. Like the Heisenberg groups, Sol admits left-invariant metrics but not bi-invariant ones (Problem 3-14).   //

In fact, John Milnor showed in 1976 [Mil76] that the only Lie groups that admit bi-invariant metrics are those that are isomorphic to direct products of compact groups and abelian groups.

Other Homogeneous Riemannian Manifolds

There are many homogeneous Riemannian manifolds besides the frame-homogeneous ones and the Lie groups with invariant metrics. To identify other examples, it is natural to ask the following question: If M is a smooth manifold endowed with a smooth, transitive action by a Lie group G (called a **homogeneous G-space** or just a **homogeneous space**), is there a Riemannian metric on M that is invariant under the group action?

The next theorem gives a necessary and sufficient condition for existence of an invariant Riemannian metric that is usually easy to check.

Theorem 3.17 (Existence of Invariant Metrics on Homogeneous Spaces). *Suppose G is a Lie group and M is a homogeneous G-space. Let p_0 be a point in M, and let $I_{p_0}: G_{p_0} \to \mathrm{GL}\left(T_{p_0} M\right)$ denote the isotropy representation at p_0. There exists a G-invariant Riemannian metric on M if and only if $I_{p_0}\left(G_{p_0}\right)$ has compact closure in $\mathrm{GL}\left(T_{p_0} M\right)$.*

Proof. Assume first that g is a G-invariant metric on M. Then the inner product g_{p_0} on $T_{p_0} M$ is invariant under the isotropy representation, so it follows from Lemma 3.13 that $I_{p_0}(G_{p_0})$ has compact closure in $\mathrm{GL}\left(T_{p_0} M\right)$.

Conversely, assume that $I_{p_0}\left(G_{p_0}\right)$ has compact closure in $\mathrm{GL}\left(T_{p_0} M\right)$. Lemma 3.13 shows that there is an inner product g_{p_0} on $T_{p_0}(M)$ that is invariant under the isotropy representation. For arbitrary $p \in M$, we define an inner product g_p on $T_p M$ by choosing an element $\varphi \in G$ such that $\varphi(p) = p_0$ and setting

$$g_p = \left(d\varphi_p\right)^* g_{p_0}.$$

If φ_1, φ_2 are any two such elements of G, then $\varphi_1 = h\varphi_2$ with $h = \varphi_1 \varphi_2^{-1} \in G_{p_0}$, so

$$\left(d\varphi_1|_p\right)^* g_{p_0} = \left(d(h\varphi_2)_p\right)^* g_{p_0} = \left(d\varphi_2|_p\right)^* \left(dh_{p_0}\right)^* g_{p_0} = \left(d\varphi_2|_p\right)^* g_{p_0},$$

showing that g is well defined as a rough tensor field on M. An easy computation shows that g is G-invariant, so it remains only to show that it is smooth.

The map $\pi: G \to M$ given by $\pi(\psi) = \psi \cdot p_0$ is a smooth surjection because the action is smooth and transitive. Given $\varphi \in G$, if we let $\theta_\varphi: M \to M$ denote the map $p \mapsto \varphi \cdot p$ and $L_\varphi: G \to G$ the left translation by φ, then the map π satisfies

$$\pi \circ L_\varphi(\psi) = (\varphi\psi) \cdot p_0 = \varphi \cdot (\psi \cdot p_0) = \theta_\varphi \circ \pi(\psi), \tag{3.17}$$

so it is equivariant with respect to these two actions. Thus it is a submersion by the equivariant rank theorem (Thm. C.14).

Define a rough 2-tensor field τ on G by $\tau = \pi^* g$. (It will typically not be positive definite, because $\tau_e(v, w) = 0$ if either v or w is tangent to the isotropy group G_{p_0} and thus in the kernel of $d\pi_e$.) For all $\varphi \in G$, (3.17) implies

$$L_\varphi^* \tau = L_\varphi^* \pi^* g = \left(\pi \circ L_\varphi\right)^* g = \left(\theta_\varphi \circ \pi\right)^* g = \pi^* \theta_\varphi^* g = \pi^* g = \tau,$$

where the next-to-last equality follows from the G-invariance of g. Thus τ is a left-invariant tensor field on G. Every basis $(X_1, \ldots, X_n)$ for the Lie algebra of G forms a smooth global left-invariant frame for G, and with respect to such a frame the components $\tau(X_i, X_j)$ are constant; thus τ is a smooth tensor field on G.

For each $p \in M$, the fact that π is a surjective smooth submersion implies that there exist a neighborhood U of p and a smooth local section $\sigma: U \to G$ (Thm. A.17). Then

$$g|_U = (\pi \circ \sigma)^* g = \sigma^* \pi^* g = \sigma^* \tau,$$

showing that g is smooth on U. Since this holds in a neighborhood of each point, g is smooth. $\qquad\square$

The next corollary, which follows immediately from Theorem 3.17, addresses the most commonly encountered case. (Other necessary and sufficient conditions for the existence of invariant metrics are given in [Poo81, 6.58–6.59].)

Corollary 3.18. *If a Lie group G acts smoothly and transitively on a smooth manifold M with compact isotropy groups, then there exists a G-invariant Riemannian metric on M.* □

▶ **Exercise 3.19.** Suppose G is a Lie group and M is a homogeneous G-space that admits at least one g-invariant metric. Show that for each $p \in M$, the map $g \mapsto g_p$ gives a bijection between G-invariant metrics on M and $I_p(G_p)$-invariant inner products on $T_p M$.

Locally Homogeneous Riemannian Manifolds

A Riemannian manifold (M, g) is said to be **locally homogeneous** if for every pair of points $p, q \in M$ there is a Riemannian isometry from a neighborhood of p to a neighborhood of q that takes p to q. Similarly, we say that (M, g) is **locally frame-homogeneous** if for every $p, q \in M$ and every pair of orthonormal bases (v_i) for $T_p M$ and (w_i) for $T_q M$, there is an isometry from a neighborhood of p to a neighborhood of q that takes p to q, and whose differential takes v_i to w_i for each i.

Every homogeneous Riemannian manifold is locally homogeneous, and every frame-homogeneous one is locally frame-homogeneous. Every proper open subset of a homogeneous or frame-homogeneous Riemannian manifold is locally homogeneous or locally frame-homogeneous, respectively. More interesting examples arise in the following way.

Proposition 3.20. *Suppose $\bigl(\widetilde{M}, \widetilde{g}\bigr)$ is a homogeneous Riemannian manifold, (M, g) is a Riemannian manifold, and $\pi: \widetilde{M} \to M$ is a Riemannian covering. Then (M, g) is locally homogeneous. If $\bigl(\widetilde{M}, \widetilde{g}\bigr)$ is frame-homogeneous, then (M, g) is locally frame-homogeneous.*

▶ **Exercise 3.21.** Prove this proposition.

Locally homogeneous Riemannian metrics play an important role in classification theorems for manifolds, especially in low dimensions. The most fundamental case is that of compact 2-manifolds, for which we have the following important theorem.

Theorem 3.22 (Uniformization of Compact Surfaces). *Every compact, connected, smooth 2-manifold admits a locally frame-homogeneous Riemannian metric, and a Riemannian covering by the Euclidean plane, hyperbolic plane, or round unit sphere.*

Sketch of proof. The proof relies on the topological classification of compact surfaces (see, for example, [LeeTM, Thms. 6.15 and 10.22]), which says that every

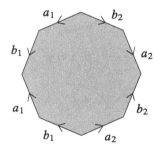

Fig. 3.7: A connected sum of tori

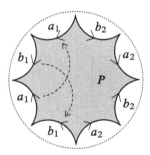

Fig. 3.8: Constructing a Riemannian covering

connected compact surface is homeomorphic to a sphere, a connected sum of one or more tori, or a connected sum of one or more projective planes. The crux of the proof is showing that each of the model surfaces on this list has a metric that admits a Riemannian covering by one of the model frame-homogeneous manifolds, and therefore is locally frame-homogeneous by Proposition 3.20. We consider each model surface in turn.

The 2-sphere: $\mathbb{S}^2$, of course, has its round metric, and the identity map is a Riemannian covering.

The 2-torus: Exercise 2.36 shows that the flat metric on $\mathbb{T}^2$ described in Example 2.21 admits a Riemannian covering by $(\mathbb{R}^2, \bar{g})$.

A connected sum of $n \geq 2$ copies of $\mathbb{T}^2$: It is shown in [LeeTM, Example 6.13] that such a surface is homeomorphic to a quotient of a regular $4n$-sided polygonal region by side identifications indicated schematically by the sequence of labels $a_1 b_1 a_1^{-1} b_1^{-1} \ldots a_n b_n a_n^{-1} b_n^{-1}$ (Fig. 3.7 illustrates the case $n = 2$). Let $G \subseteq GL(2, \mathbb{C})$ be the following subgroup:

$$G = \left\{ \begin{pmatrix} \alpha & \beta \\ \bar{\beta} & \bar{\alpha} \end{pmatrix} : \alpha, \beta \in \mathbb{C}, \ |\alpha|^2 - |\beta|^2 > 0 \right\}. \tag{3.18}$$

Problem 3-8 shows that G acts transitively and isometrically on the Poincaré disk with its hyperbolic metric. It is shown in [LeeTM] that for each $n \geq 2$, there is a discrete subgroup $\Gamma_n \subseteq G$ such that the quotient map $\mathbb{B}^2 \to \mathbb{B}^2/\Gamma_n$ is a covering map, and $\mathbb{B}^2/\Gamma_n$ is homeomorphic to a connected sum of n tori. The group is found by first identifying a compact region $P \subseteq \mathbb{B}^2$ bounded by a "regular geodesic polygon," which is a union of $4n$ congruent circular arcs making interior angles that all measure exactly $\pi/2n$, so that $4n$ of them fit together locally to fill out a neighborhood of a point (see Fig. 3.8). (The name "geodesic polygon" reflects the fact that these circular arcs are segments of geodesics with respect to the hyperbolic metric, as we will see in Chapter 5.) Then Γ_n is the group generated by certain elements of G that take an edge with a label a_n or b_n to the other edge with the same label. Because Γ_n acts isometrically, it follows that such a connected sum admits a locally

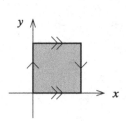

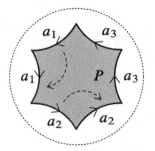

Fig. 3.9: The Klein bottle   Fig. 3.10: Connected sum of three projective planes

frame-homogeneous metric and a Riemannian covering by the Poincaré disk. For details, see the proofs of Theorems 12.29 and 12.30 of [LeeTM].

The projective plane: Example 2.34 shows that $\mathbb{R}\mathbb{P}^2$ has a metric that admits a 2-sheeted Riemannian covering by $(\mathbb{S}^2, \overset{\circ}{g})$.

A connected sum of two copies of $\mathbb{R}\mathbb{P}^2$: It is shown in [LeeTM, Lemma 6.16] that $\mathbb{R}\mathbb{P}^2 \# \mathbb{R}\mathbb{P}^2$ is homeomorphic to the **Klein bottle**, which is the quotient space of the unit square $[0,1] \times [0,1]$ by the equivalence relation generated by the following relations (see Fig. 3.9):

$$(x,0) \sim (x,1) \qquad \text{for } 0 \le x \le 1,$$
$$(0,y) \sim (1,1-y) \quad \text{for } 0 \le y \le 1. \tag{3.19}$$

Let E(2) be the Euclidean group in two dimensions as defined earlier in this chapter, and let $\Gamma \subseteq E(2)$ be the subgroup defined by

$$\Gamma = \left\{ (b,A) \in E(2) : b = (k,l) \text{ with } k,l \in \mathbb{Z}, \text{and } A = \begin{pmatrix} 1 & 0 \\ 0 & (-1)^k \end{pmatrix} \right\}. \tag{3.20}$$

It turns out that Γ acts freely and properly on $\mathbb{R}^2$, and every Γ-orbit has a representative in the unit square such that two points in the square are in the same orbit if and only if they satisfy the equivalence relation generated by the relations in (3.19). Thus the orbit space $\mathbb{R}^2/\Gamma$ is homeomorphic to the Klein bottle, and since the group action is isometric, it follows that the Klein bottle inherits a flat, locally frame-homogeneous metric and the quotient map is a Riemannian covering. Problem 3-18 asks you to work out the details.

A connected sum of $n \ge 3$ *copies of* $\mathbb{R}\mathbb{P}^2$: Such a surface is homeomorphic to a quotient of a regular $2n$-sided polygonal region by side identifications according to $a_1 a_1 a_2 a_2 \ldots a_n a_n$ [LeeTM, Example 6.13]. As in the case of a connected sum of tori, there is a compact region $P \subseteq \mathbb{B}^2$ bounded by a $2n$-sided regular geodesic polygon whose interior angles are all π/n (see Fig. 3.10), and there is a discrete group of isometries that realizes the appropriate side identifications and yields a quotient homeomorphic to the connected sum. The new ingredient here is that because such a connected sum is not orientable, we must work with the full group of isometries of $\mathbb{B}^2$, not just the (orientation-preserving) ones determined by elements of G; but otherwise the argument is essentially the same as the one for connected sums of tori. The details can be found in [Ive92, Section VII.1].

There is one remaining step. The arguments above show that each compact topological 2-manifold possesses a smooth structure and a locally frame-homogeneous Riemannian metric, which admits a Riemannian covering by one of the three frame-homogeneous model spaces. However, we started with a *smooth* compact 2-manifold, and we are looking for a Riemannian metric that is smooth with respect to the given smooth structure. To complete the proof, we appeal to a result by James Munkres [Mun56], which shows that any two smooth structures on a 2-manifold are related by a diffeomorphism; thus after pulling back the metric by this diffeomorphism, we obtain a locally frame-homogeneous metric on M with its originally given smooth structure.   □

Locally homogeneous metrics also play a key role in the classification of compact 3-manifolds. In 1982, William Thurston made a conjecture about the classification of such manifolds, now known as the ***Thurston geometrization conjecture***. The conjecture says that every compact, orientable 3-manifold can be expressed as a connected sum of compact manifolds, each of which either admits a Riemannian covering by a homogeneous Riemannian manifold or can be cut along embedded tori so that each piece admits a finite-volume locally homogeneous Riemannian metric. An important ingredient in the analysis leading up to the conjecture was his classification of all simply connected homogeneous Riemannian 3-manifolds that admit finite-volume Riemannian quotients. Thurston showed that there are exactly eight such manifolds (see [Thu97] or [Sco83] for a proof):

- $\mathbb{R}^3$ with the Euclidean metric
- $\mathbb{S}^3$ with a round metric
- $\mathbb{H}^3$ with a hyperbolic metric
- $\mathbb{S}^2 \times \mathbb{R}$ with a product of a round metric and the Euclidean metric
- $\mathbb{H}^2 \times \mathbb{R}$ with a product of a hyperbolic metric and the Euclidean metric
- The Heisenberg group H_1 of Example 3.16(g) with a left-invariant metric
- The group Sol of Example 3.16(h) with a left-invariant metric
- The universal covering group of $SL(2, \mathbb{R})$ with a left-invariant metric

The Thurston geometrization conjecture was proved in 2003 by Grigori Perelman. The proof is described in several books [BBBMP, KL08, MF10, MT14].

Symmetric Spaces

We end this section with a brief introduction to another class of Riemannian manifolds with abundant symmetry, called *symmetric spaces*. They turn out to be intermediate between frame-homogeneous and homogeneous Riemannian manifolds.

Here is the definition. If (M, g) is a Riemannian manifold and $p \in M$, a ***point reflection at p*** is an isometry $\varphi \colon M \to M$ that fixes p and satisfies $d\varphi_p = -\mathrm{Id} \colon T_p M \to T_p M$. A Riemannian manifold (M, g) is called a (***Riemannian***) ***symmetric space*** if it is connected and for each $p \in M$ there exists a point reflection at

p. (The modifier "Riemannian" is included to distinguish such spaces from other kinds of symmetric spaces that can be defined, such as *pseudo-Riemannian symmetric spaces* and *affine symmetric spaces*; since we will be concerned only with Riemannian symmetric spaces, we will sometimes refer to them simply as "symmetric spaces" for brevity.)

Although we do not yet have the tools to prove it, we will see later that every Riemannian symmetric space is homogeneous (see Problem 6-19). More generally, (M, g) is called a **(Riemannian) locally symmetric space** if each $p \in M$ has a neighborhood U on which there exists an isometry $\varphi \colon U \to U$ that is a point reflection at p. Clearly every Riemannian symmetric space is locally symmetric.

The next lemma can be used to facilitate the verification that a given Riemannian manifold is symmetric.

Lemma 3.23. *If (M, g) is a connected homogeneous Riemannian manifold that possesses a point reflection at one point, then it is symmetric.*

Proof. Suppose (M, g) satisfies the hypothesis, and let $\varphi \colon M \to M$ be a point reflection at $p \in M$. Given any other point $q \in M$, by homogeneity there is an isometry $\psi \colon M \to M$ satisfying $\psi(p) = q$. Then $\widetilde{\varphi} = \psi \circ \varphi \circ \psi^{-1}$ is an isometry that fixes q. Because $d\psi_p$ is linear, it commutes with multiplication by -1, so

$$d\widetilde{\varphi}_q = d\psi_p \circ \left(-\mathrm{Id}_{T_p M}\right) \circ d\left(\psi^{-1}\right)_q = \left(-\mathrm{Id}_{T_q M}\right) \circ d\psi_p \circ d\left(\psi^{-1}\right)_q$$
$$= -\mathrm{Id}_{T_q M}.$$

Thus $\widetilde{\varphi}$ is a point reflection at q. □

Example 3.24 (Riemannian Symmetric Spaces).

(a) Suppose (M, g) is any connected frame-homogeneous Riemannian manifold. Then for each $p \in M$, we can choose an orthonormal basis (b_i) for $T_p M$, and frame homogeneity guarantees that there is an isometry $\varphi \colon M \to M$ that fixes p and sends (b_i) to $(-b_i)$, which implies that $d\varphi_p = -\mathrm{Id}$. Thus every frame-homogeneous Riemannian manifold is a symmetric space. In particular, all Euclidean spaces, spheres, and hyperbolic spaces are symmetric.

(b) Suppose G is a connected Lie group with a bi-invariant Riemannian metric g. If we define $\Phi \colon G \to G$ by $\Phi(x) = x^{-1}$, then it is straightforward to check that $d\Phi_e(v) = -v$ for every $v \in T_e G$, from which it follows that $d\Phi_e^*(g_e) = g_e$. To see that Φ is an isometry, let $p \in G$ be arbitrary. The identity $q^{-1} = (p^{-1}q)^{-1} p^{-1}$ for all $q \in G$ implies that $\Phi = R_{p^{-1}} \circ \Phi \circ L_{p^{-1}}$, and therefore it follows from bi-invariance of g that

$$(\Phi^* g)_p = d\Phi_p^* g_{p^{-1}} = d(L_{p^{-1}})_p^* \circ d\Phi_e^* \circ d(R_{p^{-1}})_e^* g_{p^{-1}} = g_p.$$

Therefore Φ is an isometry of g and hence a point reflection at e. Lemma 3.23 then implies that (G, g) is a symmetric space.

(c) The complex projective spaces introduced in Example 2.30 and the Grassmann manifolds introduced in Problem 2-7 are all Riemannian symmetric spaces (see Problems 3-19 and 3-20).

(d) Every product of Riemannian symmetric spaces is easily seen to be a symmetric space when endowed with the product metric. A symmetric space is said to be **irreducible** if it is not isometric to a product of positive-dimensional symmetric spaces. //

Model Pseudo-Riemannian Manifolds

The definitions of the Euclidean, spherical, and hyperbolic metrics can easily be adapted to give analogous classes of frame-homogeneous pseudo-Riemannian manifolds.

The first example is one we have already seen: the **pseudo-Euclidean space of signature** (r,s) is the pseudo-Riemannian manifold $\left(\mathbb{R}^{r,s}, \overline{q}^{(r,s)}\right)$, where $\overline{q}^{(r,s)}$ is the pseudo-Riemannian metric defined by (2.24).

There are also pseudo-Riemannian analogues of the spherical and hyperbolic metrics. For nonnegative integers r and s and a positive real number R, we define the **pseudosphere** $\left(\mathbb{S}^{r,s}(R), \overset{\circ}{q}_{R}^{(r,s)}\right)$ and the **pseudohyperbolic space** $\left(\mathbb{H}^{r,s}(R), \overset{\vee}{q}_{R}^{(r,s)}\right)$ as follows. As manifolds, $\mathbb{S}^{r,s}(R) \subseteq \mathbb{R}^{r+1,s}$ and $\mathbb{H}^{r,s}(R) \subseteq \mathbb{R}^{r,s+1}$ are defined by

$$\mathbb{S}^{r,s}(R) = \left\{(\xi,\tau) : \left(\xi^{1}\right)^{2} + \cdots + \left(\xi^{r+1}\right)^{2} - \left(\tau^{1}\right)^{2} - \cdots - \left(\tau^{s}\right)^{2} = R^{2}\right\},$$

$$\mathbb{H}^{r,s}(R) = \left\{(\xi,\tau) : \left(\xi^{1}\right)^{2} + \cdots + \left(\xi^{r}\right)^{2} - \left(\tau^{1}\right)^{2} - \cdots - \left(\tau^{s+1}\right)^{2} = -R^{2}\right\}.$$

The metrics are the ones induced from the respective pseudo-Euclidean metrics: $\overset{\circ}{q}_{R}^{(r,s)} = \iota^{*}\overline{q}^{(r+1,s)}$ on $\mathbb{S}^{r,s}(R)$, and $\overset{\vee}{q}_{R}^{(r,s)} = \iota^{*}\overline{q}^{(r,s+1)}$ on $\mathbb{H}^{r,s}(R)$.

Theorem 3.25. *For all r, s, and R as above, $\mathbb{S}^{r,s}(R)$ and $\mathbb{H}^{r,s}(R)$ are pseudo-Riemannian manifolds of signature (r,s).*

Proof. Problem 3-22. □

It turns out that these pseudo-Riemannian manifolds all have the same degree of symmetry as the three classes of model Riemannian manifolds introduced earlier. For pseudo-Riemannian manifolds, though, it is necessary to modify the definition of frame homogeneity slightly. If (M,g) is a pseudo-Riemannian manifold of signature (r,s), let us say that an orthonormal basis for some tangent space $T_{p}M$ is in **standard order** if the expression for g_{p} in terms of the dual basis (ε^{i}) is $(\varepsilon^{1})^{2} + \cdots + (\varepsilon^{r})^{2} - (\varepsilon^{r+1})^{2} - \cdots - (\varepsilon^{r+s})^{2}$, with all positive terms coming before the negative terms. With this understanding, we define $O(M)$ to be the set of all standard-ordered orthonormal bases for all tangent spaces to M, and we say that (M,g) is **frame-homogeneous** if the isometry group acts transitively on $O(M)$.

Theorem 3.26. *All pseudo-Euclidean spaces, pseudospheres, and pseudohyperbolic spaces are frame-homogeneous.*

Proof. Problem 3-23. □

In the particular case of signature $(n, 1)$, the Lorentz manifolds $\left(\mathbb{S}^{n,1}(R), \overset{\circ}{q}_R^{(n,1)}\right)$ and $\left(\mathbb{H}^{n,1}(R), \overset{\smile}{q}_R^{(n,1)}\right)$ are called *de Sitter space of radius R* and *anti-de Sitter space of radius R*, respectively.

Problems

3-1. Show that (3.2) defines a smooth isometric action of E(n) on $(\mathbb{R}^n, \bar{g})$, and the induced action on O($\mathbb{R}^n$) is transitive. (*Used on p. 57.*)

3-2. Prove that the metric on $\mathbb{RP}^n$ described in Example 2.34 is frame-homogeneous. (*Used on p. 145*)

3-3. Prove Proposition 3.1 (about homogeneous and isotropic Riemannian manifolds).

3-4. Complete the proof of Theorem 3.7 by showing that $\kappa^* \overset{\smile}{g}_R^3 = \overset{\smile}{g}_R^4$.

3-5. (a) Prove that $\left(\mathbb{S}^n(R), \overset{\circ}{g}_R\right)$ is isometric to $\left(\mathbb{S}^n, R^2\overset{\circ}{g}\right)$ for each $R > 0$.

(b) Prove that $\left(\mathbb{H}^n(R), \overset{\smile}{g}_R\right)$ is isometric to $\left(\mathbb{H}^n, R^2\overset{\smile}{g}\right)$ for each $R > 0$.

(c) We could also have defined a family of metrics on $\mathbb{R}^n$ by $\bar{g}_R = R^2\bar{g}$. Why did we not bother?

(*Used on p. 185.*)

3-6. Show that two Riemannian metrics g_1 and g_2 are conformal if and only if they define the same angles but not necessarily the same lengths, and that a diffeomorphism is a conformal equivalence if and only if it preserves angles. [Hint: Let (E_i) be a local orthonormal frame for g_1, and consider the g_2-angle between E_i and $(\cos\theta)E_i + (\sin\theta)E_j$.] (*Used on p. 59.*)

3-7. Let $\mathbb{U}^2$ denote the upper half-plane model of the hyperbolic plane (of radius 1), with the metric $\overset{\smile}{g} = (dx^2 + dy^2)/y^2$. Let SL(2, $\mathbb{R}$) denote the group of 2×2 real matrices of determinant 1. Regard $\mathbb{U}^2$ as a subset of the complex plane with coordinate $z = x + iy$, and let

$$A \cdot z = \frac{az+b}{cz+d}, \quad A = \begin{pmatrix} a & b \\ c & d \end{pmatrix} \in \text{SL}(2, \mathbb{R}).$$

Show that this defines a smooth, transitive, orientation-preserving, and isometric action of SL(2, $\mathbb{R}$) on $(\mathbb{U}^2, \overset{\smile}{g})$. Is the induced action transitive on O($\mathbb{U}^2$)?

3-8. Let $\mathbb{B}^2$ denote the Poincaré disk model of the hyperbolic plane (of radius 1), with the metric $\overset{\smile}{g} = (du^2 + dv^2)/(1 - u^2 - v^2)^2$, and let $G \subseteq \text{GL}(2, \mathbb{C})$ be the subgroup defined by (3.18). Regarding $\mathbb{B}^2$ as a subset of the complex plane with coordinate $w = u + iv$, let G act on $\mathbb{B}^2$ by

$$\begin{pmatrix} \alpha & \beta \\ \bar{\beta} & \bar{\alpha} \end{pmatrix} \cdot w = \frac{\alpha z + \beta}{\bar{\beta}z + \bar{\alpha}}.$$

Show that this defines a smooth, transitive, orientation-preserving, and isometric action of G on $(\mathbb{B}^2, \breve{g})$. [Hint: One way to proceed is to define an
action of G on the upper half-plane by $A \cdot z = \kappa^{-1} \circ A \circ \kappa(z)$, where κ is the
Cayley transform defined by (3.14) in the case $R = 1$, and use the result of
Problem 3-7.] (*Used on pp. 73, 185.*)

3-9. Suppose G is a compact Lie group with a left-invariant metric g and a left-
invariant orientation. Show that the Riemannian volume form dV_g is bi-
invariant. [Hint: Show that dV_g is equal to the Riemannian volume form for
a bi-invariant metric.]

3-10. Consider the basis

$$X = \begin{pmatrix} 0 & 1 \\ -1 & 0 \end{pmatrix}, \quad Y = \begin{pmatrix} 0 & i \\ i & 0 \end{pmatrix}, \quad Z = \begin{pmatrix} i & 0 \\ 0 & -i \end{pmatrix}$$

for the Lie algebra $\mathfrak{su}(2)$. For each positive real number a, define a left-
invariant metric g_a on the group $SU(2)$ by declaring X, Y, aZ to be an
orthonormal frame.

(a) Show that g_a is bi-invariant if and only if $a = 1$.
(b) Show that the map defined by (3.16) is an isometry between $(\mathbb{S}^3, \breve{g})$
   and $(SU(2), g_1)$. [Remark: $SU(2)$ with any of these metrics is called a
   Berger sphere, named after Marcel Berger.]

(*Used on pp. 56, 71, 259.*)

3-11. Prove that the formula $\langle A, B \rangle = \operatorname{tr}(A^T B)$ defines a bi-invariant Riemannian
metric on $O(n)$. (See Example 3.16(e).)

3-12. Regard the upper half-space $\mathbb{U}^n$ as a Lie group as described in Example
3.16(f).

(a) Show that for each $R > 0$, the hyperbolic metric $\breve{g}_R^4$ on $\mathbb{U}^n$ is left-
   invariant.
(b) Show that $\mathbb{U}^n$ does not admit any bi-invariant metrics.

(*Used on pp. 68, 72.*)

3-13. Write down an explicit formula for an arbitrary left-invariant metric on the
Heisenberg group H_n of Example 3.16(g) in terms of global coordinates
$(x^1, \ldots, x^n, y^1, \ldots, y^n, z)$, and show that the group has no bi-invariant metrics. (*Used on pp. 68, 72.*)

3-14. Repeat Problem 3-13 for the group Sol of Example 3.16(h). (*Used on p. 72.*)

3-15. Let $\mathbb{R}^{n,1}$ be the $(n + 1)$-dimensional Minkowski space with coordinates
$(\xi, \tau) = (\xi^1, \ldots, \xi^n, \tau)$ and with the Minkowski metric $\bar{q}^{(n,1)} = \sum_i d(\xi^i)^2 -
d\tau^2$. Let $S \subseteq \mathbb{R}^{n,1}$ be the set

$$S = \{(\xi, \tau) : (\xi^1)^2 + \cdots + (\xi^n)^2 = \tau = 1\}.$$

(a) Prove that S is a smooth submanifold diffeomorphic to $\mathbb{S}^{n-1}$, and with the induced metric $g = \iota_S^* \bar{q}^{(n,1)}$ it is isometric to the round unit $(n-1)$-sphere.

(b) Define an action of the orthochronous Lorentz group $O^+(n,1)$ on S as follows: For every $p \in S$, let $\langle p \rangle$ denote the 1-dimensional subspace of $\mathbb{R}^{n,1}$ spanned by p. Given $A \in O^+(n,1)$, show that the image set $A(\langle p \rangle)$ is a 1-dimensional subspace that intersects S in exactly one point, so we can define $A \cdot p$ to be the unique point in $S \cap A(\langle p \rangle)$. Prove that this is a smooth transitive action on S.

(c) Prove that $O^+(n,1)$ acts by conformal diffeomorphisms of (S,g).

3-16. Prove that there is no Riemannian metric on the sphere that is invariant under the group action described in Problem 3-15.

3-17. Given a Lie group G, define an action of the product group $G \times G$ on G by $(\varphi_1, \varphi_2) \cdot \psi = \varphi_1 \psi \varphi_2^{-1}$. Show that this action is transitive, and that the isotropy group of the identity is the diagonal subgroup $\Delta = \{(\varphi, \varphi) : \varphi \in G\}$. Then show that the following diagram commutes:

$$
\begin{array}{ccc}
\Delta & \cong & G \\
\downarrow{\scriptstyle I_e} & & \downarrow{\scriptstyle \mathrm{Ad}} \\
\mathrm{GL}(T_e G) & \cong & \mathrm{GL}(\mathfrak{g}),
\end{array}
$$

where I_e is the isotropy representation of Δ and $\mathfrak{g}$ is the Lie algebra of G, and use this to give an alternative proof of Theorem 3.14.

3-18. Let $\Gamma \subseteq E(2)$ be the subgroup defined by (3.20). Prove that Γ acts freely and properly on $\mathbb{R}^2$ and the orbit space is homeomorphic to the Klein bottle, and conclude that the Klein bottle has a flat metric and a Riemannian covering by the Euclidean plane.

3-19. Show that the Fubini–Study metric on $\mathbb{CP}^n$ (Example 2.30) is homogeneous, isotropic, and symmetric. (*Used on p. 78.*)

3-20. Show that the metric on the Grassmannian $G_k(\mathbb{R}^n)$ defined in Problem 2-7 is homogeneous, isotropic, and symmetric. (*Used on p. 78.*)

3-21. Let $(\widetilde{M}, \widetilde{g})$ be a simply connected Riemannian manifold, and suppose Γ_1 and Γ_2 are countable subgroups of $\mathrm{Iso}(\widetilde{M}, \widetilde{g})$ acting smoothly, freely, and properly on M (when endowed with the discrete topology). For $i = 1, 2$, let $M_i = \widetilde{M}/\Gamma_i$, and let g_i be the Riemannian metric on M_i that makes the quotient map $\pi_i : \widetilde{M} \to M_i$ a Riemannian covering (see Prop. 2.32). Prove that the Riemannian manifolds (M_1, g_1) and (M_2, g_2) are isometric if and only if Γ_1 and Γ_2 are conjugate subgroups of $\mathrm{Iso}(\widetilde{M}, \widetilde{g})$.

3-22. Prove Theorem 3.25 (showing that pseudospheres and pseudohyperbolic spaces are pseudo-Riemannian manifolds). [Hint: Mimic the argument in the proof of Theorem 3.7 that $\mathbb{H}^n(R)$ is Riemannian.]

3-23. Prove Theorem 3.26 (pseudo-Euclidean spaces, pseudospheres, and pseudo-hyperbolic spaces are frame-homogeneous).

3-24. Prove that for all positive integers r and s and every real number $R > 0$, both the pseudohyperbolic space $\mathbb{H}^{r,s}(R)$ and the pseudosphere $\mathbb{S}^{s,r}(R)$ are diffeomorphic to $\mathbb{R}^r \times \mathbb{S}^s$. [Hint: Consider the maps $\varphi, \psi : \mathbb{R}^r \times \mathbb{S}^s \to \mathbb{R}^{r+s+1}$ given by

$$\varphi(x,y) = \left(Rx, \left(\sqrt{1+|x|^2}\right)Ry\right), \quad \psi(x,y) = \left(\left(\sqrt{1+|x|^2}\right)Ry, Rx\right).]$$

3-25. Let $\left(\mathbb{K}^r(R), \breve{g}_R^2\right)$ be the r-dimensional ball of radius R with the Beltrami–Klein metric (3.9), and let $\widetilde{\mathbb{H}}^{r,1}(R)$ be the product manifold $\mathbb{K}^r(R) \times \mathbb{R}$ with the pseudo-Riemannian warped product metric $q = \breve{g}_R^2 \oplus (-f^2 dt^2)$, where $f : \mathbb{K}^r(R) \to \mathbb{R}^+$ is given by

$$f(w) = \frac{R^2}{\sqrt{R^2 - |w|^2}}.$$

Define $F : \widetilde{\mathbb{H}}^{r,1}(R) \to \mathbb{R}^{r,2}$ by

$$F(w,t) = \frac{(Rw, R^2 \cos t, R^2 \sin t)}{\sqrt{R^2 - |w|^2}}.$$

Prove that the image of F is the anti-de Sitter space $\mathbb{H}^{r,1}(R)$, and F defines a pseudo-Riemannian covering of $\left(\mathbb{H}^{r,1}(R), q_R^{(r,1)}\right)$ by $\left(\widetilde{\mathbb{H}}^{r,1}(R), q\right)$. [Remark: We are tacitly extending the notions of warped product metric and Riemannian coverings to the pseudo-Riemannian case in the obvious ways. It follows from the result of Problem 3-24 that $\mathbb{H}^{r,s}(R)$ is simply connected when $s \geq 2$ but $\mathbb{H}^{r,1}(R)$ is not. This shows that $\left(\widetilde{\mathbb{H}}^{r,1}(R), q\right)$, called **universal anti-de Sitter space of radius R**, is the universal pseudo-Riemannian covering manifold of $\left(\mathbb{H}^{r,1}(R), q_R^{(r,1)}\right).]$

Chapter 4
Connections

Our ultimate goal is to define a notion of curvature that makes sense on arbitrary Riemannian manifolds, and to relate it to other geometric and topological properties. Before we can do so, however, we need to study *geodesics*, the generalizations to Riemannian manifolds of straight lines in Euclidean space. There are two key properties satisfied by straight lines in $\mathbb{R}^n$, either of which serves to characterize them uniquely: first, every segment of a straight line is the unique shortest path between its endpoints; and second, straight lines are the only curves that have parametrizations with zero acceleration.

The first of these characterizations—as shortest paths—is probably the most "geometric," so it is tempting to try to use it as a definition of geodesics in Riemannian manifolds. However, this property turns out to be technically difficult to work with as a definition, so instead we will use "zero acceleration" as the defining property and generalize that.

To make sense of acceleration on a manifold, we have to introduce a new object called a *connection*—essentially a coordinate-independent set of rules for taking directional derivatives of vector fields.

We begin this chapter by examining more closely the problem of finding an invariant interpretation for the acceleration of a curve, as a way to motivate the definitions that follow. We then give a rather general definition of a connection, in terms of directional derivatives of sections of vector bundles. After deriving some basic properties of connections, we show how to use them to differentiate vector fields along curves, to define geodesics, and to define "parallel transport" of vectors along curves.

The Problem of Differentiating Vector Fields

To see why we need a new kind of differentiation operator, let us begin by thinking informally about curves in $\mathbb{R}^n$. Let $I \subseteq \mathbb{R}$ be an interval and $\gamma \colon I \to \mathbb{R}^n$ a smooth curve, written in standard coordinates as $\gamma(t) = \big(\gamma^1(t), \ldots, \gamma^n(t)\big)$. Such a curve

© Springer International Publishing AG 2018
J. M. Lee, *Introduction to Riemannian Manifolds*, Graduate Texts
in Mathematics 176, https://doi.org/10.1007/978-3-319-91755-9_4

has a well-defined *velocity* $\gamma'(t)$ and *acceleration* $\gamma''(t)$ at each $t \in I$, computed by differentiating the components:

$$\gamma'(t) = \dot{\gamma}^1(t) \left.\frac{\partial}{\partial x^1}\right|_{\gamma(t)} + \cdots + \dot{\gamma}^n(t) \left.\frac{\partial}{\partial x^n}\right|_{\gamma(t)}, \tag{4.1}$$

$$\gamma''(t) = \ddot{\gamma}^1(t) \left.\frac{\partial}{\partial x^1}\right|_{\gamma(t)} + \cdots + \ddot{\gamma}^n(t) \left.\frac{\partial}{\partial x^n}\right|_{\gamma(t)}. \tag{4.2}$$

(Here and throughout the book, we use dots to denote ordinary derivatives with respect to t when there are superscripts that would make primes hard to read.) A curve γ in $\mathbb{R}^n$ is a straight line if and only if it has a parametrization for which $\gamma''(t) \equiv 0$.

We can also make sense of directional derivatives of vector fields on $\mathbb{R}^n$, just by computing ordinary directional derivatives of the component functions in standard coordinates: given a vector field $Y \in \mathfrak{X}(\mathbb{R}^n)$ and a vector $v \in T_p\mathbb{R}^n$, we define the *Euclidean directional derivative of Y in the direction v* by the formula

$$\overline{\nabla}_v Y = v(Y^1) \left.\frac{\partial}{\partial x^1}\right|_p + \cdots + v(Y^n) \left.\frac{\partial}{\partial x^n}\right|_p,$$

where for each i, $v(Y^i)$ is the result of applying the vector v to the function Y^i:

$$v(Y^i) = v^1 \frac{\partial Y^i(p)}{\partial x^1} + \cdots + v^n \frac{\partial Y^i(p)}{\partial x^n}.$$

If X is another vector field on $\mathbb{R}^n$, we obtain a new vector field $\overline{\nabla}_X Y$ by evaluating $\overline{\nabla}_{X_p} Y$ at each point:

$$\overline{\nabla}_X Y = X(Y^1) \frac{\partial}{\partial x^1} + \cdots + X(Y^n) \frac{\partial}{\partial x^n}. \tag{4.3}$$

More generally, we can play the same game with curves and vector fields on a submanifold of $\mathbb{R}^n$. Suppose $M \subseteq \mathbb{R}^n$ is an embedded submanifold, and consider a smooth curve $\gamma: I \to M$. We want to think of a *geodesic* in M as a curve in M that is "as straight as possible." Of course, if M itself is curved, then $\gamma'(t)$ (thought of as a vector in $\mathbb{R}^n$) will probably have to vary, or else the curve will leave M. But we can try to insist that the velocity not change any more than necessary for the curve to stay in M. One way to do this is to compute the Euclidean acceleration $\gamma''(t)$ as above, and then apply the tangential projection $\pi^\top: T_{\gamma(t)}\mathbb{R}^n \to T_{\gamma(t)}M$ (see Prop. 2.16). This yields a vector $\gamma''(t)^\top = \pi^\top(\gamma''(t))$ tangent to M, which we call the *tangential acceleration of γ*. It is reasonable to say that γ is as straight as it is possible for a curve in M to be if its tangential acceleration is zero.

Similarly, suppose Y is a smooth vector field on (an open subset of) M, and we wish to ask how much Y is varying *in M* in the direction of a vector $v \in T_pM$. Just as in the case of velocity vectors, if we look at it from the point of view of $\mathbb{R}^n$, the vector field Y might be forced to vary just so that it can remain tangent to M. But

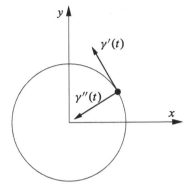

Fig. 4.1: Cartesian coordinates

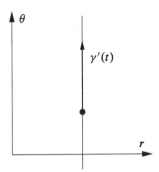

Fig. 4.2: Polar coordinates

one plausible way to answer the question is to extend Y to a smooth vector field $\widetilde{Y}$ on an open subset of $\mathbb{R}^n$, compute the Euclidean directional derivative of $\widetilde{Y}$ in the direction v, and then project orthogonally onto $T_p M$. Let us define the **tangential directional derivative of** Y **in the direction** v to be

$$\nabla_v^\top Y = \pi^\top \left(\overline{\nabla}_v \widetilde{Y} \right). \tag{4.4}$$

Problem 4-1 shows that the tangential directional derivative is well defined and preserved by rigid motions of $\mathbb{R}^n$. However, at this point there is little reason to believe that the tangential directional derivative is an intrinsic invariant of M (one that depends only on the Riemannian geometry of M with its induced metric).

On an abstract Riemannian manifold, for which there is no "ambient Euclidean space" in which to differentiate, this technique is not available. Thus we have to find some way to make sense of the acceleration of a smooth curve in an abstract manifold. Let $\gamma \colon I \to M$ be such a curve. As you know from your study of smooth manifold theory, at each time $t \in I$, the velocity of γ is a well-defined vector $\gamma'(t) \in T_{\gamma(t)} M$ (see Appendix A), whose representation in any coordinates is given by (4.1), just as in Euclidean space.

However, unlike velocity, acceleration has no such coordinate-independent interpretation. For example, consider the parametrized circle in the plane given in Cartesian coordinates by $\gamma(t) = (x(t), y(t)) = (\cos t, \sin t)$ (Fig. 4.1). As a smooth curve in $\mathbb{R}^2$, it has an acceleration vector at time t given by

$$\gamma''(t) = x''(t) \left. \frac{\partial}{\partial x} \right|_{\gamma(t)} + y''(t) \left. \frac{\partial}{\partial x} \right|_{\gamma(t)}$$
$$= -\cos t \left. \frac{\partial}{\partial x} \right|_{\gamma(t)} - \sin t \left. \frac{\partial}{\partial x} \right|_{\gamma(t)}.$$

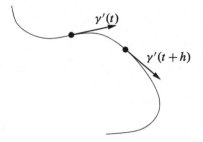

Fig. 4.3: $\gamma'(t)$ and $\gamma'(t+h)$ lie in different vector spaces

But in polar coordinates, the same curve is described by $(r(t), \theta(t)) = (1, t)$ (Fig. 4.2). In these coordinates, if we try to compute the acceleration vector by the analogous formula, we get

$$\gamma''(t) = r''(t) \left. \frac{\partial}{\partial r} \right|_{\gamma(t)} + \theta''(t) \left. \frac{\partial}{\partial \theta} \right|_{\gamma(t)} = 0.$$

The problem is this: to define $\gamma''(t)$ by differentiating $\gamma'(t)$ with respect to t, we have to take a limit of a difference quotient involving the vectors $\gamma'(t+h)$ and $\gamma'(t)$; but these live in different vector spaces ($T_{\gamma(t+h)}M$ and $T_{\gamma(t)}M$ respectively), so it does not make sense to subtract them (Fig. 4.3). The definition of acceleration works in the special case of smooth curves in $\mathbb{R}^n$ expressed in standard coordinates (or more generally, curves in any finite-dimensional vector space expressed in linear coordinates) because each tangent space can be naturally identified with the vector space itself. On a general smooth manifold, there is no such natural identification.

The velocity vector $\gamma'(t)$ is an example of a *vector field along a curve*, a concept for which we will give a rigorous definition presently. To interpret the acceleration of a curve in a manifold, what we need is some coordinate-independent way to differentiate vector fields along curves. To do so, we need a way to compare values of the vector field at different points, or intuitively, to "connect" nearby tangent spaces. This is where a connection comes in: it will be an additional piece of data on a manifold, a rule for computing directional derivatives of vector fields.

Connections

It turns out to be easiest to define a connection first as a way of differentiating sections of vector bundles. The definition is meant to capture the essential properties of the Euclidean and tangential directional derivative operators ($\overline{\nabla}$ and $\nabla^\top$) that we defined above. (We will verify later that those operators actually are connections;

see Examples 4.8 and 4.9.) After defining connections in this general setting, we will adapt the definition to the case of vector fields along curves.

Let $\pi\colon E \to M$ be a smooth vector bundle over a smooth manifold M with or without boundary, and let $\Gamma(E)$ denote the space of smooth sections of E. A **connection in E** is a map

$$\nabla\colon \mathfrak{X}(M) \times \Gamma(E) \to \Gamma(E),$$

written $(X, Y) \mapsto \nabla_X Y$, satisfying the following properties:

(i) $\nabla_X Y$ is linear over $C^\infty(M)$ in X: for $f_1, f_2 \in C^\infty(M)$ and $X_1, X_2 \in \mathfrak{X}(M)$,

$$\nabla_{f_1 X_1 + f_2 X_2} Y = f_1 \nabla_{X_1} Y + f_2 \nabla_{X_2} Y.$$

(ii) $\nabla_X Y$ is linear over $\mathbb{R}$ in Y: for $a_1, a_2 \in \mathbb{R}$ and $Y_1, Y_2 \in \Gamma(E)$,

$$\nabla_X (a_1 Y_1 + a_2 Y_2) = a_1 \nabla_X Y_1 + a_2 \nabla_X Y_2.$$

(iii) ∇ satisfies the following product rule: for $f \in C^\infty(M)$,

$$\nabla_X (f Y) = f \nabla_X Y + (Xf) Y.$$

The symbol ∇ is read "del" or "nabla," and $\nabla_X Y$ is called the **covariant derivative of Y in the direction X**. (This use of the word "covariant" has nothing to do with covariant functors in category theory. It is related, albeit indirectly, to the use of the word in the context of covariant and contravariant tensors, in that it reflects the fact that the components of the covariant derivative have a transformation law that "varies correctly" to give a well-defined meaning independent of coordinates. From the modern coordinate-free point of view, "invariant derivative" would probably be a better term.)

There is a variety of types of connections that are useful in different circumstances. The type of connection we have defined here is sometimes called a **Koszul connection** to distinguish it from other types. Since we have no need to consider other types of connections in this book, we refer to Koszul connections simply as *connections*.

Although a connection is defined by its action on global sections, it follows from the definitions that it is actually a *local operator*, as the next lemma shows.

Lemma 4.1 (Locality). *Suppose ∇ is a connection in a smooth vector bundle $E \to M$. For every $X \in \mathfrak{X}(M)$, $Y \in \Gamma(E)$, and $p \in M$, the covariant derivative $\nabla_X Y|_p$ depends only on the values of X and Y in an arbitrarily small neighborhood of p. More precisely, if $X = \tilde{X}$ and $Y = \tilde{Y}$ on a neighborhood of p, then $\nabla_X Y|_p = \nabla_{\tilde{X}} \tilde{Y}|_p$.*

Proof. First consider Y. Replacing Y by $Y - \tilde{Y}$ shows that it suffices to prove $\nabla_X Y|_p = 0$ if Y vanishes on a neighborhood of p.

Thus suppose Y is a smooth section of E that is identically zero on a neighborhood U of p. Choose a bump function $\varphi \in C^\infty(M)$ with support in U such that

$\varphi(p) = 1$. The hypothesis that Y vanishes on U implies that $\varphi Y \equiv 0$ on all of M, so for every $X \in \mathfrak{X}(M)$, we have $\nabla_X (\varphi Y) = \nabla_X (0 \cdot \varphi Y) = 0 \nabla_X (\varphi Y) = 0$. Thus the product rule gives

$$0 = \nabla_X (\varphi Y) = (X\varphi) Y + \varphi(\nabla_X Y). \tag{4.5}$$

Now $Y \equiv 0$ on the support of φ, so the first term on the right is identically zero. Evaluating (4.5) at p shows that $\nabla_X Y|_p = 0$. The argument for X is similar but easier. □

▶ **Exercise 4.2.** Complete the proof of Lemma 4.1 by showing that $\nabla_X Y$ and $\nabla_{\tilde{X}} Y$ agree at p if $X = \tilde{X}$ on a neighborhood of p.

Proposition 4.3 (Restriction of a Connection). *Suppose ∇ is a connection in a smooth vector bundle $E \to M$. For every open subset $U \subseteq M$, there is a unique connection ∇^U on the restricted bundle $E|_U$ that satisfies the following relation for every $X \in \mathfrak{X}(M)$ and $Y \in \Gamma(E)$:*

$$\nabla^U_{(X|_U)}(Y|_U) = (\nabla_X Y)\big|_U. \tag{4.6}$$

Proof. First we prove uniqueness. Suppose ∇^U is any such connection and $X \in \mathfrak{X}(U)$ and $Y \in \Gamma(E|_U)$ are arbitrary. Given $p \in U$, we can use a bump function to construct a smooth vector field $\tilde{X} \in \mathfrak{X}(M)$ and a smooth section $\tilde{Y} \in \Gamma(E)$ such that $\tilde{X}|_U$ agrees with X and $\tilde{Y}|_U$ with Y on some neighborhood of p, and then Lemma 4.1 together with (4.6) implies

$$\nabla^U_X Y\big|_p = \nabla^U_{(\tilde{X}|_U)}\big(\tilde{Y}|_U\big)\big|_p = \big(\nabla_{\tilde{X}} \tilde{Y}\big)\big|_p. \tag{4.7}$$

Since the right-hand side is completely determined by ∇, this shows that ∇^U is uniquely defined if it exists.

To prove existence, given $X \in \mathfrak{X}(U)$ and $Y \in \Gamma(E|_U)$, for every $p \in U$ we just construct $\tilde{X}$ and $\tilde{Y}$ as above, and define $\nabla^U_X Y|_p$ by (4.7). This is independent of the choices of $\tilde{X}$ and $\tilde{Y}$ by Lemma 4.1, and it is smooth because the same formula holds on some neighborhood of p. The fact that it satisfies the properties of a connection is an easy exercise. □

▶ **Exercise 4.4.** Complete the proof of the preceding proposition by showing that ∇^U is a connection.

In the situation of this proposition, we typically just refer to the restricted connection as ∇ instead of ∇^U; the proposition guarantees that there is no ambiguity in doing so.

Lemma 4.1 tells us that we can compute the value of $\nabla_X Y$ at p knowing only the values of X and Y in a neighborhood of p. In fact, as the next proposition shows, we need only know the value of X at p itself.

Proposition 4.5. *Under the hypotheses of Lemma 4.1, $\nabla_X Y|_p$ depends only on the values of Y in a neighborhood of p and the value of X at p.*

Proof. The claim about Y was proved in Lemma 4.1. To prove the claim about X, it suffices by linearity to assume that $X_p = 0$ and show that $\nabla_X Y|_p = 0$. Choose a coordinate neighborhood U of p, and write $X = X^i \partial_i$ in coordinates on U, with $X^i(p) = 0$. Thanks to Proposition 4.3, it suffices to work with the restricted connection on U, which we also denote by ∇. For every $Y \in \Gamma(E|_U)$, we have

$$\nabla_X Y|_p = \nabla_{X^i \partial_i} Y|_p = X^i(p) \nabla_{\partial_i} Y|_p = 0. \qquad \square$$

Thanks to Propositions 4.3 and 4.5, we can make sense of the expression $\nabla_v Y$ when v is some element of $T_p M$ and Y is a smooth local section of E defined only on some neighborhood of p. To evaluate it, let X be a vector field on a neighborhood of p whose value at p is v, and set $\nabla_v Y = \nabla_X Y|_p$. Proposition 4.5 shows that the result does not depend on the extension chosen. Henceforth, we will interpret covariant derivatives of local sections of bundles in this way without further comment.

Connections in the Tangent Bundle

For Riemannian or pseudo-Riemannian geometry, our primary concern is with connections in the tangent bundle, so for the rest of the chapter we focus primarily on that case. A connection in the tangent bundle is often called simply a **connection on** M. (The terms **affine connection** and **linear connection** are also sometimes used in this context, but there is little agreement on the precise definitions of these terms, so we avoid them.)

Suppose M is a smooth manifold with or without boundary. By the definition we just gave, a connection in TM is a map

$$\nabla: \mathfrak{X}(M) \times \mathfrak{X}(M) \to \mathfrak{X}(M)$$

satisfying properties (i)–(iii) above. Although the definition of a connection resembles the characterization of $(1,2)$-tensor fields given by the tensor characterization lemma (Lemma B.6), a connection in TM is not a tensor field because it is not linear over $C^\infty(M)$ in its second argument, but instead satisfies the product rule.

For computations, we need to examine how a connection appears in terms of a local frame. Let (E_i) be a smooth local frame for TM on an open subset $U \subseteq M$. For every choice of the indices i and j, we can expand the vector field $\nabla_{E_i} E_j$ in terms of this same frame:

$$\nabla_{E_i} E_j = \Gamma_{ij}^k E_k. \tag{4.8}$$

As i, j, and k range from 1 to $n = \dim M$, this defines n^3 smooth functions $\Gamma_{ij}^k: U \to \mathbb{R}$, called the **connection coefficients of** ∇ with respect to the given frame. The following proposition shows that the connection is completely determined in U by its connection coefficients.

Proposition 4.6. *Let M be a smooth manifold with or without boundary, and let ∇ be a connection in TM. Suppose (E_i) is a smooth local frame over an open sub-*

set $U \subseteq M$, and let $\{\Gamma_{ij}^k\}$ be the connection coefficients of ∇ with respect to this frame. For smooth vector fields $X, Y \in \mathfrak{X}(U)$, written in terms of the frame as $X = X^i E_i$, $Y = Y^j E_j$, one has

$$\nabla_X Y = \left(X \left(Y^k \right) + X^i Y^j \Gamma_{ij}^k \right) E_k. \tag{4.9}$$

Proof. Just use the defining properties of a connection and compute:

$$
\begin{aligned}
\nabla_X Y &= \nabla_X \left(Y^j E_j \right) \\
&= X \left(Y^j \right) E_j + Y^j \nabla_{X^i E_i} E_j \\
&= X \left(Y^j \right) E_j + X^i Y^j \nabla_{E_i} E_j \\
&= X \left(Y^j \right) E_j + X^i Y^j \Gamma_{ij}^k E_k.
\end{aligned}
$$

Renaming the dummy index in the first term yields (4.9). $\qquad\square$

Once the connection coefficients (and thus the connection) have been determined in some local frame, they can be determined in any other local frame on the same open set by the result of the following proposition.

Proposition 4.7 (Transformation Law for Connection Coefficients). *Let M be a smooth manifold with or without boundary, and let ∇ be a connection in TM. Suppose we are given two smooth local frames (E_i) and $(\tilde{E}_j)$ for TM on an open subset $U \subseteq M$, related by $\tilde{E}_i = A_i^j E_j$ for some matrix of functions $\left(A_i^j \right)$. Let Γ_{ij}^k and $\tilde{\Gamma}_{ij}^k$ denote the connection coefficients of ∇ with respect to these two frames. Then*

$$\tilde{\Gamma}_{ij}^k = \left(A^{-1} \right)_p^k A_i^q A_j^r \Gamma_{qr}^p + \left(A^{-1} \right)_p^k A_i^q E_q \left(A_j^p \right). \tag{4.10}$$

Proof. Problem 4-3. $\qquad\square$

Observe that the first term above is exactly what the transformation law would be if Γ_{ij}^k were the components of a $(1,2)$-tensor field; but the second term is of a different character, because it involves derivatives of the transition matrix $\left(A_j^p \right)$.

Existence of Connections

So far, we have studied properties of connections but have not produced any, so you might be wondering whether they are plentiful or rare. In fact, they are quite plentiful, as we will show shortly. Let us begin with the simplest example.

Example 4.8 (The Euclidean Connection). In $T\mathbb{R}^n$, define the *Euclidean connection* $\overline{\nabla}$ by formula (4.3). It is easy to check that this satisfies the required properties for a connection, and that its connection coefficients in the standard coordinate frame are all zero. $\qquad/\!/$

Here is a way to construct a large class of examples.

Example 4.9 (The Tangential Connection on a Submanifold of $\mathbb{R}^n$). Let $M \subseteq \mathbb{R}^n$ be an embedded submanifold. Define a connection $\nabla^\top$ on TM, called the *tangential connection*, by setting

$$\nabla_X^\top Y = \pi^\top \left(\bar{\nabla}_{\tilde{X}} \tilde{Y} \big|_M \right),$$

where $\pi^\top$ is the orthogonal projection onto TM (Prop. 2.16), $\bar{\nabla}$ is the Euclidean connection on $\mathbb{R}^n$ (Example 4.8), and $\tilde{X}$ and $\tilde{Y}$ are smooth extensions of X and Y to an open set in $\mathbb{R}^n$. (Such extensions exist by the result of Exercise A.23.) Since the value of $\bar{\nabla}_{\tilde{X}} \tilde{Y}$ at a point $p \in M$ depends only on $\tilde{X}_p = X_p$, this just boils down to defining $\left(\nabla_X^\top Y \right)_p$ to be equal to the tangential directional derivative $\nabla_{X_p}^\top Y$ that we defined in (4.4) above. Problem 4-1 shows that this value is independent of the choice of extension $\tilde{Y}$, so $\nabla^\top$ is well defined. Smoothness is easily verified by expressing $\bar{\nabla}_{\tilde{X}} \tilde{Y}$ in terms of an adapted orthonormal frame (see Prop. 2.14).

It is immediate from the definition that $\nabla_X^\top Y$ is linear over $C^\infty(M)$ in X and over $\mathbb{R}$ in Y, so to show that $\nabla^\top$ is a connection, only the product rule needs to be checked. Let $f \in C^\infty(M)$, and let $\tilde{f}$ be an extension of f to a neighborhood of M in $\mathbb{R}^n$. Then $\tilde{f}\tilde{Y}$ is a smooth extension of fY to a neighborhood of M, so

$$\begin{aligned} \nabla_X^\top(fY) &= \pi^\top \left(\bar{\nabla}_{\tilde{X}} (\tilde{f}\tilde{Y}) \big|_M \right) \\ &= \pi^\top \left((\tilde{X}\tilde{f})\tilde{Y} \big|_M \right) + \pi^\top \left(\tilde{f} \bar{\nabla}_{\tilde{X}} \tilde{Y} \big|_M \right) \\ &= (Xf)Y + f\nabla_X^\top Y. \end{aligned}$$

Thus $\nabla^\top$ is a connection.   //

In fact, there are many connections on $\mathbb{R}^n$, or indeed on every smooth manifold that admits a global frame (for example, every manifold covered by a single smooth coordinate chart). The following lemma shows how to construct all of them explicitly.

Lemma 4.10. *Suppose M is a smooth n-manifold with or without boundary, and M admits a global frame (E_i). Formula (4.9) gives a one-to-one correspondence between connections in TM and choices of n^3 smooth real-valued functions $\{\Gamma_{ij}^k\}$ on M.*

Proof. Every connection determines functions $\{\Gamma_{ij}^k\}$ by (4.8), and Proposition 4.6 shows that those functions satisfy (4.9). On the other hand, given $\{\Gamma_{ij}^k\}$, we can define $\nabla_X Y$ by (4.9); it is easy to see that the resulting expression is smooth if X and Y are smooth, linear over $\mathbb{R}$ in Y, and linear over $C^\infty(M)$ in X. To prove that it is a connection, only the product rule requires checking; this is a straightforward computation left as an exercise. $\qquad\square$

▶ **Exercise 4.11.** Complete the proof of Lemma 4.10.

Proposition 4.12. *The tangent bundle of every smooth manifold with or without boundary admits a connection.*

Proof. Let M be a smooth manifold with or without boundary, and cover M with coordinate charts $\{U_\alpha\}$; the preceding lemma guarantees the existence of a connection ∇^α on each U_α. Choose a partition of unity $\{\varphi_\alpha\}$ subordinate to $\{U_\alpha\}$. We would like to patch the various ∇^α's together by the formula

$$\nabla_X Y = \sum_\alpha \varphi_\alpha \nabla_X^\alpha Y. \tag{4.11}$$

Because the set of supports of the φ_α's is locally finite, the sum on the right-hand side has only finitely many nonzero terms in a neighborhood of each point, so it defines a smooth vector field on M. It is immediate from this definition that $\nabla_X Y$ is linear over $\mathbb{R}$ in Y and linear over $C^\infty(M)$ in X. We have to be a bit careful with the product rule, though, since a linear combination of connections is not necessarily a connection. (You can check, for example, that if ∇^0 and ∇^1 are connections, then neither $2\nabla^0$ nor $\nabla^0 + \nabla^1$ satisfies the product rule.) By direct computation,

$$
\begin{aligned}
\nabla_X(fY) &= \sum_\alpha \varphi_\alpha \nabla_X^\alpha(fY) \\
&= \sum_\alpha \varphi_\alpha \big((Xf)Y + f\nabla_X^\alpha Y\big) \\
&= (Xf)Y \sum_\alpha \varphi_\alpha + f \sum_\alpha \varphi_\alpha \nabla_X^\alpha Y \\
&= (Xf)Y + f\nabla_X Y. \qquad \square
\end{aligned}
$$

Although a connection is not a tensor field, the next proposition shows that the *difference* between two connections is.

Proposition 4.13 (The Difference Tensor). *Let M be a smooth manifold with or without boundary. For any two connections ∇^0 and ∇^1 in TM, define a map $D: \mathfrak{X}(M) \times \mathfrak{X}(M) \to \mathfrak{X}(M)$ by*

$$D(X, Y) = \nabla_X^1 Y - \nabla_X^0 Y.$$

Then D is bilinear over $C^\infty(M)$, and thus defines a $(1,2)$-tensor field called the **difference tensor between ∇^0 and ∇^1.**

Proof. It is immediate from the definition that D is linear over $C^\infty(M)$ in its first argument, because both ∇^0 and ∇^1 are. To show that it is linear over $C^\infty(M)$ in the second argument, expand $D(X, fY)$ using the product rule, and note that the two terms in which f is differentiated cancel each other. $\qquad \square$

Now that we know there is always one connection in TM, we can use the result of the preceding proposition to say exactly how many there are.

Theorem 4.14. *Let M be a smooth manifold with or without boundary, and let ∇^0 be any connection in TM. Then the set $\mathcal{A}(TM)$ of all connections in TM is equal to the following affine space:*

$$\mathcal{A}(TM) = \{\nabla^0 + D : D \in \Gamma(T^{(1,2)}TM)\},$$

where $D \in \Gamma(T^{(1,2)}TM)$ is interpreted as a map from $\mathfrak{X}(M) \times \mathfrak{X}(M)$ to $\mathfrak{X}(M)$ as in Proposition B.1, and $\nabla^0 + D : \mathfrak{X}(M) \times \mathfrak{X}(M) \to \mathfrak{X}(M)$ is defined by

$$(\nabla^0 + D)_X Y = \nabla^0_X Y + D(X,Y).$$

Proof. Problem 4-4. $\qquad\qquad\qquad\qquad\qquad\qquad\qquad\qquad\qquad\qquad\qquad\qquad$ $\square$

Covariant Derivatives of Tensor Fields

By definition, a connection in TM is a rule for computing covariant derivatives of vector fields. We show in this section that every connection in TM automatically induces connections in all tensor bundles over M, and thus gives us a way to compute covariant derivatives of tensor fields of any type.

Proposition 4.15. *Let M be a smooth manifold with or without boundary, and let ∇ be a connection in TM. Then ∇ uniquely determines a connection in each tensor bundle $T^{(k,l)}TM$, also denoted by ∇, such that the following four conditions are satisfied.*

(i) In $T^{(1,0)}TM = TM$, ∇ agrees with the given connection.
(ii) In $T^{(0,0)}TM = M \times \mathbb{R}$, ∇ is given by ordinary differentiation of functions:

$$\nabla_X f = Xf.$$

(iii) ∇ obeys the following product rule with respect to tensor products:

$$\nabla_X(F \otimes G) = (\nabla_X F) \otimes G + F \otimes (\nabla_X G).$$

(iv) ∇ commutes with all contractions: if "tr" denotes a trace on any pair of indices, one covariant and one contravariant, then

$$\nabla_X(\operatorname{tr} F) = \operatorname{tr}(\nabla_X F).$$

This connection also satisfies the following additional properties:

(a) ∇ obeys the following product rule with respect to the natural pairing between a covector field ω and a vector field Y:

$$\nabla_X \langle \omega, Y \rangle = \langle \nabla_X \omega, Y \rangle + \langle \omega, \nabla_X Y \rangle.$$

(b) For all $F \in \Gamma(T^{(k,l)}TM)$, smooth 1-forms $\omega^1, \ldots, \omega^k$, and smooth vector fields $Y_1, \ldots, Y_l$,

$$(\nabla_X F)\big(\omega^1,\dots,\omega^k,Y_1,\dots,Y_l\big) = X\big(F\big(\omega^1,\dots,\omega^k,Y_1,\dots,Y_l\big)\big)$$

$$-\sum_{i=1}^{k} F\big(\omega^1,\dots,\nabla_X\omega^i,\dots,\omega^k,Y_1,\dots,Y_l\big) \qquad (4.12)$$

$$-\sum_{j=1}^{l} F\big(\omega^1,\dots,\omega^k,Y_1,\dots,\nabla_X Y_j,\dots,Y_l\big).$$

Proof. First we show that every family of connections on all tensor bundles satisfying (i)–(iv) also satisfies (a) and (b). Suppose we are given such a family of connections, all denoted by ∇. To prove (a), note that $\langle\omega,Y\rangle = \operatorname{tr}(\omega\otimes Y)$, as can be seen by evaluating both sides in coordinates, where they both reduce to $\omega_i Y^i$. Therefore, (i)–(iv) imply

$$\begin{aligned}
\nabla_X\langle\omega,Y\rangle &= \nabla_X\big(\operatorname{tr}(\omega\otimes Y)\big) \\
&= \operatorname{tr}\big(\nabla_X(\omega\otimes Y)\big) \\
&= \operatorname{tr}\big(\nabla_X\omega\otimes Y + \omega\otimes\nabla_X Y\big) \\
&= \langle\nabla_X\omega,Y\rangle + \langle\omega,\nabla_X Y\rangle.
\end{aligned}$$

Then (b) is proved by induction using a similar computation applied to

$$F\big(\omega^1,\dots,\omega^k,Y_1,\dots,Y_l\big) = \underbrace{\operatorname{tr}\circ\cdots\circ\operatorname{tr}}_{k+l}\big(F\otimes\omega^1\otimes\cdots\otimes\omega^k\otimes Y_1\otimes\cdots\otimes Y_l\big),$$

where each trace operator acts on an upper index of F and the lower index of the corresponding 1-form, or a lower index of F and the upper index of the corresponding vector field.

Next we address uniqueness. Assume again that ∇ represents a family of connections satisfying (i)–(iv), and hence also (a) and (b). Observe that (ii) and (a) imply that the covariant derivative of every 1-form ω can be computed by

$$(\nabla_X\omega)(Y) = X(\omega(Y)) - \omega(\nabla_X Y). \qquad (4.13)$$

It follows that the connection on 1-forms is uniquely determined by the original connection in TM. Similarly, (b) gives a formula that determines the covariant derivative of every tensor field F in terms of covariant derivatives of vector fields and 1-forms, so the connection in every tensor bundle is uniquely determined.

Now to prove existence, we first define covariant derivatives of 1-forms by (4.13), and then we use (4.12) to define ∇ on all other tensor bundles. The first thing that needs to be checked is that the resulting expression is multilinear over $C^\infty(M)$ in each ω^i and Y_j, and therefore defines a smooth tensor field. This is done by inserting $f\omega^i$ in place of ω^i, or fY_j in place of Y_j, and expanding the right-hand side, noting that the two terms in which f is differentiated cancel each other out. Once we know that $\nabla_X F$ is a smooth tensor field, we need to check that it satisfies the defining properties of a connection. Linearity over $C^\infty(M)$ in X and linearity over $\mathbb{R}$ in F

are both evident from (4.12) and (4.13), and the product rule in F follows easily from the fact that differentiation of functions by X satisfies the product rule. $\qquad\square$

While (4.12) and (4.13) are useful for proving the existence and uniqueness of the connections in tensor bundles, they are not very practical for computation, because computing the value of $\nabla_X F$ at a point requires extending all of its arguments to vector fields and covector fields in an open set, and computing a great number of derivatives. For computing the components of a covariant derivative in terms of a local frame, the formulas in the following proposition are far more useful.

Proposition 4.16. *Let M be a smooth manifold with or without boundary, and let ∇ be a connection in TM. Suppose (E_i) is a local frame for M, (ε^j) is its dual coframe, and $\{\Gamma_{ij}^k\}$ are the connection coefficients of ∇ with respect to this frame. Let X be a smooth vector field, and let $X^i E_i$ be its local expression in terms of this frame.*

(a) The covariant derivative of a 1-form $\omega = \omega_i \varepsilon^i$ is given locally by

$$\nabla_X(\omega) = \left(X(\omega_k) - X^j \omega_i \Gamma_{jk}^i\right)\varepsilon^k.$$

(b) If $F \in \Gamma\left(T^{(k,l)}TM\right)$ is a smooth mixed tensor field of any rank, expressed locally as

$$F = F_{j_1 \ldots j_l}^{i_1 \ldots i_k} E_{i_1} \otimes \cdots \otimes E_{i_k} \otimes \varepsilon^{j_1} \otimes \cdots \otimes \varepsilon^{j_l},$$

then the covariant derivative of F is given locally by

$$\nabla_X F = \left(X\left(F_{j_1 \ldots j_l}^{i_1 \ldots i_k}\right) + \sum_{s=1}^{k} X^m F_{j_1 \ldots j_l}^{i_1 \ldots p \ldots i_k} \Gamma_{mp}^{i_s} - \sum_{s=1}^{l} X^m F_{j_1 \ldots p \ldots j_l}^{i_1 \ldots i_k} \Gamma_{mj_s}^{p}\right) \times$$
$$E_{i_1} \otimes \cdots \otimes E_{i_k} \otimes \varepsilon^{j_1} \otimes \cdots \otimes \varepsilon^{j_l}.$$

Proof. Problem 4-5. $\qquad\square$

Because the covariant derivative $\nabla_X F$ of a tensor field (or, as a special case, a vector field) is linear over $C^\infty(M)$ in X, the covariant derivatives of F in all directions can be handily encoded in a single tensor field whose rank is one more than the rank of F, as follows.

Proposition 4.17 (The Total Covariant Derivative). *Let M be a smooth manifold with or without boundary and let ∇ be a connection in TM. For every $F \in \Gamma\left(T^{(k,l)}TM\right)$, the map*

$$\nabla F: \underbrace{\Omega^1(M) \times \cdots \times \Omega^1(M)}_{k \text{ copies}} \times \underbrace{\mathfrak{X}(M) \times \cdots \times \mathfrak{X}(M)}_{l+1 \text{ copies}} \to C^\infty(M)$$

given by

$$(\nabla F)(\omega^1, \ldots, \omega^k, Y_1, \ldots, Y_l, X) = (\nabla_X F)(\omega^1, \ldots, \omega^k, Y_1, \ldots, Y_l) \qquad (4.14)$$

defines a smooth $(k, l + 1)$-*tensor field on* M *called the **total covariant derivative** of* F.

Proof. This follows immediately from the tensor characterization lemma (Lemma B.6): $\nabla_X F$ is a tensor field, so it is multilinear over $C^\infty(M)$ in its $k + l$ arguments; and it is linear over $C^\infty(M)$ in X by definition of a connection. $\square$

When we write the components of a total covariant derivative in terms of a local frame, it is standard practice to use a semicolon to separate indices resulting from differentiation from the preceding indices. Thus, for example, if Y is a vector field written in coordinates as $Y = Y^i E_i$, the components of the $(1, 1)$-tensor field ∇Y are written $Y^i{}_{;j}$, so that

$$\nabla Y = Y^i{}_{;j} E_i \otimes \varepsilon^j,$$

with

$$Y^i{}_{;j} = E_j Y^i + Y^k \Gamma^i_{jk}.$$

For a 1-form ω, the formulas read

$$\nabla \omega = \omega_{i;j} \varepsilon^i \otimes \varepsilon^j, \qquad \text{with } \omega_{i;j} = E_j \omega_i - \omega_k \Gamma^k_{ji}.$$

More generally, the next lemma gives a formula for the components of total covariant derivatives of arbitrary tensor fields.

Proposition 4.18. *Let* M *be a smooth manifold with or without boundary and let* ∇ *be a connection in* TM; *and let* (E_i) *be a smooth local frame for* TM *and* $\{\Gamma^k_{ij}\}$ *the corresponding connection coefficients. The components of the total covariant derivative of a* (k, l)-*tensor field* F *with respect to this frame are given by*

$$F^{i_1 \ldots i_k}_{j_1 \ldots j_l ;m} = E_m \left(F^{i_1 \ldots i_k}_{j_1 \ldots j_l} \right) + \sum_{s=1}^{k} F^{i_1 \ldots p \ldots i_k}_{j_1 \ldots j_l} \Gamma^{i_s}_{mp} - \sum_{s=1}^{l} F^{i_1 \ldots i_k}_{j_1 \ldots p \ldots j_l} \Gamma^p_{mj_s}.$$

▶ **Exercise 4.19.** Prove Proposition 4.18.

▶ **Exercise 4.20.** Suppose F is a smooth (k, l)-tensor field and G is a smooth (r, s)-tensor field. Show that the components of the total covariant derivative of $F \otimes G$ are given by

$$\left(\nabla(F \otimes G) \right)^{i_1 \ldots i_k p_1 \ldots p_r}_{j_1 \ldots j_l q_1 \ldots q_s ;m} = F^{i_1 \ldots i_k}_{j_1 \ldots j_l ;m} G^{p_1 \ldots p_r}_{q_1 \ldots q_s} + F^{i_1 \ldots i_k}_{j_1 \ldots j_l} G^{p_1 \ldots p_r}_{q_1 \ldots q_s ;m}.$$

[Remark: This formula is often written in the following way, more suggestive of the product rule for ordinary derivatives:

$$\left(F^{i_1 \ldots i_k}_{j_1 \ldots j_l} G^{p_1 \ldots p_r}_{q_1 \ldots q_s} \right)_{;m} = F^{i_1 \ldots i_k}_{j_1 \ldots j_l ;m} G^{p_1 \ldots p_r}_{q_1 \ldots q_s} + F^{i_1 \ldots i_k}_{j_1 \ldots j_l} G^{p_1 \ldots p_r}_{q_1 \ldots q_s ;m}.$$

Notice that this does not say that $\nabla(F \otimes G) = (\nabla F) \otimes G + F \otimes (\nabla G)$, because in the first term on the right-hand side of this latter formula, the index resulting from differentiation is not the last lower index.]

Second Covariant Derivatives

Having defined the tensor field ∇F for a (k,l)-tensor field F, we can in turn take its total covariant derivative and obtain a $(k, l+2)$-tensor field $\nabla^2 F = \nabla(\nabla F)$. Given vector fields $X, Y \in \mathfrak{X}(M)$, let us introduce the notation $\nabla^2_{X,Y} F$ for the (k,l)-tensor field obtained by inserting X, Y in the last two slots of $\nabla^2 F$:

$$\nabla^2_{X,Y} F(\ldots) = \nabla^2 F(\ldots, Y, X).$$

Note the reversal of order of X and Y: this is necessitated by our convention that the *last* index position in ∇F is the one resulting from differentiation, while it is conventional to let $\nabla^2_{X,Y}$ stand for differentiating first in the Y direction, then in the X direction. (For this reason, some authors adopt the convention that the new index position introduced by differentiation is the first instead of the last. As usual, be sure to check each author's conventions when you read.)

It is important to be aware that $\nabla^2_{X,Y} F$ is not the same as $\nabla_X(\nabla_Y F)$. The main reason is that the former is linear over $C^\infty(M)$ in Y, while the latter is not. The relationship between the two expressions is given in the following proposition.

Proposition 4.21. *Let M be a smooth manifold with or without boundary and let ∇ be a connection in TM. For every smooth vector field or tensor field F,*

$$\nabla^2_{X,Y} F = \nabla_X(\nabla_Y F) - \nabla_{(\nabla_X Y)} F.$$

Proof. A covariant derivative $\nabla_Y F$ can be expressed as the trace of $\nabla F \otimes Y$ on its last two indices:

$$\nabla_Y F = \mathrm{tr}(\nabla F \otimes Y),$$

as you can verify by noting that both expressions have the same component formula, $F^{i_1 \cdots i_k}_{j_1 \cdots j_l; m} Y^m$. Similarly, $\nabla^2_{X,Y} F$ can be expressed as an iterated trace:

$$\nabla^2_{X,Y} F = \mathrm{tr}\big(\mathrm{tr}(\nabla^2 F \otimes X) \otimes Y\big).$$

(First trace the last index of $\nabla^2 F$ with that of X, and then trace the last remaining free index—originally the second-to-last in $\nabla^2 F$—with that of Y.)

Therefore, since ∇_X commutes with contraction and satisfies the product rule with respect to tensor products (Prop. 4.15), we have

$$
\begin{aligned}
\nabla_X(\nabla_Y F) &= \nabla_X\big(\mathrm{tr}(\nabla F \otimes Y)\big) \\
&= \mathrm{tr}\big(\nabla_X(\nabla F \otimes Y)\big) \\
&= \mathrm{tr}\big(\nabla_X(\nabla F) \otimes Y + \nabla F \otimes \nabla_X Y\big) \\
&= \mathrm{tr}\big(\mathrm{tr}(\nabla^2 F \otimes X) \otimes Y\big) + \mathrm{tr}\big(\nabla F \otimes \nabla_X Y\big) \\
&= \nabla^2_{X,Y} F + \nabla_{(\nabla_X Y)} F. \qquad \square
\end{aligned}
$$

Example 4.22 (The Covariant Hessian). Let u be a smooth function on M. Then $\nabla u \in \Gamma\big(T^{(0,1)}TM\big) = \Omega^1(M)$ is just the 1-form du, because both tensors have the same action on vectors: $\nabla u(X) = \nabla_X u = Xu = du(X)$. The 2-tensor $\nabla^2 u = \nabla(du)$ is called the **covariant Hessian of u**. Proposition 4.21 shows that its action on smooth vector fields X, Y can be computed by the following formula:

$$\nabla^2 u(Y, X) = \nabla^2_{X,Y} u = \nabla_X(\nabla_Y u) - \nabla_{(\nabla_X Y)} u = Y(Xu) - (\nabla_Y X)u.$$

In any local coordinates, it is

$$\nabla^2 u = u_{;ij}\, dx^i \otimes dx^j, \qquad \text{with } u_{;ij} = \partial_j \partial_i u - \Gamma^k_{ji}\partial_k u. \qquad\qquad /\!/$$

Vector and Tensor Fields Along Curves

Now we can address the question that originally motivated the definition of connections: How can we make sense of the derivative of a vector field along a curve?

Let M be a smooth manifold with or without boundary. Given a smooth curve $\gamma: I \to M$, a **vector field along γ** is a continuous map $V: I \to TM$ such that $V(t) \in T_{\gamma(t)}M$ for every $t \in I$; it is a **smooth vector field along γ** if it is smooth as a map from I to TM. We let $\mathfrak{X}(\gamma)$ denote the set of all smooth vector fields along γ. It is a real vector space under pointwise vector addition and multiplication by constants, and it is a module over $C^\infty(I)$ with multiplication defined pointwise:

$$(fX)(t) = f(t)X(t).$$

The most obvious example of a vector field along a smooth curve γ is the curve's velocity: $\gamma'(t) \in T_{\gamma(t)}M$ for each t, and its coordinate expression (4.1) shows that it is smooth. Here is another example: if γ is a curve in $\mathbb{R}^2$, let $N(t) = R\gamma'(t)$, where R is counterclockwise rotation by $\pi/2$, so $N(t)$ is normal to $\gamma'(t)$. In standard coordinates, $N(t) = \big(-\dot\gamma^2(t), \dot\gamma^1(t)\big)$, so N is a smooth vector field along γ.

A large supply of examples is provided by the following construction: suppose $\gamma: I \to M$ is a smooth curve and $\widetilde{V}$ is a smooth vector field on an open subset of M containing the image of γ. Define $V: I \to TM$ by setting $V(t) = \widetilde{V}_{\gamma(t)}$ for each $t \in I$. Since V is equal to the composition $\widetilde{V} \circ \gamma$, it is smooth. A smooth vector field along γ is said to be **extendible** if there exists a smooth vector field $\widetilde{V}$ on a neighborhood of the image of γ that is related to V in this way (Fig. 4.4). Not every vector field along a curve need be extendible; for example, if $\gamma(t_1) = \gamma(t_2)$ but $\gamma'(t_1) \neq \gamma'(t_2)$ (Fig. 4.5), then γ' is not extendible. Even if γ is injective, its velocity need not be extendible, as the next example shows.

Example 4.23. Consider the **figure eight curve** $\gamma: (-\pi, \pi) \to \mathbb{R}^2$ defined by

$$\gamma(t) = (\sin 2t, \sin t).$$

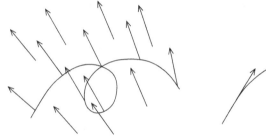

| Fig. 4.4: Extendible vector field | Fig. 4.5: Nonextendible vector field |

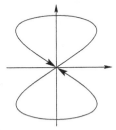

Fig. 4.6: The image of the figure eight curve of Example 4.23

Its image is a set that looks like a figure eight in the plane (Fig. 4.6). Problem 4-7 asks you to show that γ is an injective smooth immersion, but its velocity vector field is not extendible.   //

More generally, a **tensor field along γ** is a continuous map σ from I to some tensor bundle $T^{(k,l)}TM$ such that $\sigma(t) \in T^{(k,l)}\big(T_{\gamma(t)}M\big)$ for each $t \in I$. It is a **smooth tensor field along γ** if it is smooth as a map from I to $T^{(k,l)}TM$, and it is **extendible** if there is a smooth tensor field $\tilde{\sigma}$ on a neighborhood of $\gamma(I)$ such that $\sigma = \tilde{\sigma} \circ \gamma$.

Covariant Derivatives Along Curves

Here is the promised interpretation of a connection as a way to take derivatives of vector fields along curves.

Theorem 4.24 (Covariant Derivative Along a Curve). *Let M be a smooth manifold with or without boundary and let ∇ be a connection in TM. For each smooth curve $\gamma: I \to M$, the connection determines a unique operator*

$$D_t : \mathfrak{X}(\gamma) \to \mathfrak{X}(\gamma),$$

*called the **covariant derivative along** γ, satisfying the following properties:*

(*i*) LINEARITY OVER $\mathbb{R}$:

$$D_t(aV + bW) = aD_tV + bD_tW \qquad for\ a,b \in \mathbb{R}.$$

(*ii*) PRODUCT RULE:

$$D_t(fV) = f'V + fD_tV \qquad for\ f \in C^\infty(I).$$

(*iii*) *If* $V \in \mathfrak{X}(\gamma)$ *is extendible, then for every extension* $\tilde{V}$ *of* V,

$$D_tV(t) = \nabla_{\gamma'(t)}\tilde{V}.$$

There is an analogous operator on the space of smooth tensor fields of any type along γ.

Proof. For simplicity, we prove the theorem for the case of vector fields along γ; the proof for arbitrary tensor fields is essentially identical except for notation.

First we show uniqueness. Suppose D_t is such an operator, and let $t_0 \in I$ be arbitrary. An argument similar to that of Lemma 4.1 shows that the value of D_tV at t_0 depends only on the values of V in any interval $(t_0 - \varepsilon, t_0 + \varepsilon)$ containing t_0. (If t_0 is an endpoint of I, extend a coordinate representation of γ to a slightly bigger open interval, prove the lemma there, and then restrict back to I.)

Choose smooth coordinates (x^i) for M in a neighborhood of $\gamma(t_0)$, and write

$$V(t) = V^j(t)\partial_j\big|_{\gamma(t)}$$

for t near t_0, where $V^1, \ldots, V^n$ are smooth real-valued functions defined on some neighborhood of t_0 in I. By the properties of D_t, since each ∂_j is extendible,

$$
\begin{aligned}
D_tV(t) &= \dot{V}^j(t)\partial_j\big|_{\gamma(t)} + V^j(t)\nabla_{\gamma'(t)}\partial_j\big|_{\gamma(t)} \\
&= \left(\dot{V}^k(t) + \dot{\gamma}^i(t)V^j(t)\Gamma_{ij}^k(\gamma(t))\right)\partial_k\big|_{\gamma(t)}.
\end{aligned}
\tag{4.15}
$$

This shows that such an operator is unique if it exists.

For existence, if $\gamma(I)$ is contained in a single chart, we can *define* D_tV by (4.15); the easy verification that it satisfies the requisite properties is left as an exercise. In the general case, we can cover $\gamma(I)$ with coordinate charts and define D_tV by this formula in each chart, and uniqueness implies that the various definitions agree whenever two or more charts overlap. $\square$

(It is worth noting that in the physics literature, the covariant derivative along a curve is sometimes called the *absolute derivative*.)

▶ **Exercise 4.25.** Complete the proof of Theorem 4.24 by showing that the operator D_t defined in coordinates by (4.15) satisfies properties (i)–(iii).

Apart from its use in proving existence of the covariant derivative along a curve, (4.15) also gives a practical formula for computing such covariant derivatives in coordinates.

Now we can improve Proposition 4.5 by showing that $\nabla_v Y$ actually depends only on the values of Y along any curve through p whose velocity is v.

Proposition 4.26. *Let M be a smooth manifold with or without boundary, let ∇ be a connection in TM, and let $p \in M$ and $v \in T_p M$. Suppose Y and $\widetilde{Y}$ are two smooth vector fields that agree at points in the image of some smooth curve $\gamma : I \to M$ such that $\gamma(t_0) = p$ and $\gamma'(t_0) = v$. Then $\nabla_v Y = \nabla_v \widetilde{Y}$.*

Proof. We can define a smooth vector field Z along γ by $Z(t) = Y_{\gamma(t)} = \widetilde{Y}_{\gamma(t)}$. Since both Y and $\widetilde{Y}$ are extensions of Z, it follows from condition (iii) in Theorem 4.24 that both $\nabla_v Y$ and $\nabla_v \widetilde{Y}$ are equal to $D_t Z(t_0)$. $\qquad\square$

Geodesics

Armed with the notion of covariant differentiation along curves, we can now define acceleration and geodesics.

Let M be a smooth manifold with or without boundary and let ∇ be a connection in TM. For every smooth curve $\gamma : I \to M$, we define the **acceleration of γ** to be the vector field $D_t \gamma'$ along γ. A smooth curve γ is called a **geodesic** (with respect to ∇) if its acceleration is zero: $D_t \gamma' \equiv 0$. In terms of smooth coordinates (x^i), if we write the component functions of γ as $\gamma(t) = (x^1(t), \ldots, x^n(t))$, then it follows from (4.15) that γ is a geodesic if and only if its component functions satisfy the following **geodesic equation**:

$$\ddot{x}^k(t) + \dot{x}^i(t)\dot{x}^j(t)\Gamma_{ij}^k(x(t)) = 0, \tag{4.16}$$

where we use $x(t)$ as an abbreviation for the n-tuple of component functions $(x^1(t), \ldots, x^n(t))$. This is a system of second-order ordinary differential equations (ODEs) for the real-valued functions $x^1, \ldots, x^n$. The next theorem uses ODE theory to prove existence and uniqueness of geodesics with suitable initial conditions. (Because difficulties can arise when a geodesic starts on the boundary or later hits the boundary, we state and prove this theorem only for manifolds without boundary.)

Theorem 4.27 (Existence and Uniqueness of Geodesics). *Let M be a smooth manifold and ∇ a connection in TM. For every $p \in M$, $w \in T_p M$, and $t_0 \in \mathbb{R}$, there exist an open interval $I \subseteq \mathbb{R}$ containing t_0 and a geodesic $\gamma : I \to M$ satisfying $\gamma(t_0) = p$ and $\gamma'(t_0) = w$. Any two such geodesics agree on their common domain.*

Proof. Let (x^i) be smooth coordinates on some neighborhood U of p. A smooth curve in U, written as $\gamma(t) = (x^1(t), \ldots, x^n(t))$, is a geodesic if and only if its

Fig. 4.7: Uniqueness of geodesics

component functions satisfy (4.16). The standard trick for proving existence and uniqueness for such a second-order system is to introduce auxiliary variables $v^i = \dot{x}^i$ to convert it to the following equivalent first-order system in twice the number of variables:

$$\dot{x}^k(t) = v^k(t),$$
$$\dot{v}^k(t) = -v^i(t)v^j(t)\Gamma_{ij}^k(x(t)). \tag{4.17}$$

Treating $(x^1, \ldots, x^n, v^1, \ldots, v^n)$ as coordinates on $U \times \mathbb{R}^n$, we can recognize (4.17) as the equations for the flow of the vector field $G \in \mathfrak{X}(U \times \mathbb{R}^n)$ given by

$$G_{(x,v)} = v^k \frac{\partial}{\partial x^k}\bigg|_{(x,v)} - v^i v^j \Gamma_{ij}^k(x) \frac{\partial}{\partial v^k}\bigg|_{(x,v)}. \tag{4.18}$$

By the fundamental theorem on flows (Thm. A.42), for each $(p, w) \in U \times \mathbb{R}^n$ and $t_0 \in \mathbb{R}$, there exist an open interval I_0 containing t_0 and a unique smooth solution $\zeta \colon I_0 \to U \times \mathbb{R}^n$ to this system satisfying the initial condition $\zeta(t_0) = (p, w)$. If we write the component functions of ζ as $\zeta(t) = (x^i(t), v^i(t))$, then we can easily check that the curve $\gamma(t) = (x^1(t), \ldots, x^n(t))$ in U satisfies the existence claim of the theorem.

To prove the uniqueness claim, suppose $\gamma, \tilde{\gamma} \colon I \to M$ are both geodesics defined on some open interval with $\gamma(t_0) = \tilde{\gamma}(t_0)$ and $\gamma'(t_0) = \tilde{\gamma}'(t_0)$. In any local coordinates around $\gamma(t_0)$, we can define smooth curves $\zeta, \tilde{\zeta} \colon (t_0 - \varepsilon, t_0 + \varepsilon) \to U \times \mathbb{R}^n$ as above. These curves both satisfy the same initial value problem for the system (4.17), so by the uniqueness of ODE solutions, they agree on $(t_0 - \varepsilon, t_0 + \varepsilon)$ for some $\varepsilon > 0$. Suppose for the sake of contradiction that $\gamma(b) \neq \tilde{\gamma}(b)$ for some $b \in I$. First suppose $b > t_0$, and let β be the infimum of numbers $b \in I$ such that $b > t_0$ and $\gamma(b) \neq \tilde{\gamma}(b)$ (Fig. 4.7). Then $\beta \in I$, and by continuity, $\gamma(\beta) = \tilde{\gamma}(\beta)$ and $\gamma'(\beta) = \tilde{\gamma}'(\beta)$. Applying local uniqueness in a neighborhood of β, we conclude that γ and $\tilde{\gamma}$ agree on a neighborhood of β, which contradicts our choice of β. Arguing similarly to the left of t_0, we conclude that $\gamma \equiv \tilde{\gamma}$ on all of I. $\qquad \square$

A geodesic $\gamma \colon I \to M$ is said to be **maximal** if it cannot be extended to a geodesic on a larger interval, that is, if there does not exist a geodesic $\tilde{\gamma} \colon \tilde{I} \to M$ defined on an interval $\tilde{I}$ properly containing I and satisfying $\tilde{\gamma}|_I = \gamma$. A **geodesic segment** is a geodesic whose domain is a compact interval.

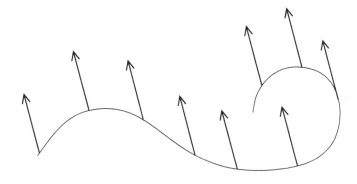

Fig. 4.8: A parallel vector field along a curve

Corollary 4.28. *Let M be a smooth manifold and let ∇ be a connection in TM. For each $p \in M$ and $v \in T_pM$, there is a unique maximal geodesic $\gamma: I \to M$ with $\gamma(0) = p$ and $\gamma'(0) = v$, defined on some open interval I containing 0.*

Proof. Given $p \in M$ and $v \in T_pM$, let I be the union of all open intervals containing 0 on which there is a geodesic with the given initial conditions. By Theorem 4.27, all such geodesics agree where they overlap, so they define a geodesic $\gamma: I \to M$, which is obviously the unique maximal geodesic with the given initial conditions. $\qquad\square$

▶ **Exercise 4.29.** Show that the maximal geodesics on $\mathbb{R}^n$ with respect to the Euclidean connection (4.3) are exactly the constant curves and the straight lines with constant-speed parametrizations.

The unique maximal geodesic γ with $\gamma(0) = p$ and $\gamma'(0) = v$ is often called simply the **geodesic with initial point p and initial velocity v**, and is denoted by γ_v. (For simplicity, we do not specify the initial point p in the notation; it can implicitly be recovered from v by $p = \pi(v)$, where $\pi: TM \to M$ is the natural projection.)

Parallel Transport

Another construction involving covariant differentiation along curves that will be useful later is called *parallel transport*. As we did with geodesics, we restrict attention here to manifolds without boundary.

Let M be a smooth manifold and let ∇ be a connection in TM. A smooth vector or tensor field V along a smooth curve γ is said to be **parallel along γ** (with respect to ∇) if $D_t V \equiv 0$ (Fig. 4.8). Thus a geodesic can be characterized as a curve whose velocity vector field is parallel along the curve.

▶ **Exercise 4.30.** Let $\gamma: I \to \mathbb{R}^n$ be a smooth curve, and let V be a smooth vector field along γ. Show that V is parallel along γ with respect to the Euclidean connection if and only if its component functions (with respect to the standard basis) are constants.

The fundamental fact about parallel vector and tensor fields along curves is that every tangent vector or tensor at any point on a curve can be uniquely extended to a parallel field along the entire curve. Before we prove this claim, let us examine what the equation of parallelism looks like in coordinates. Given a smooth curve γ with a local coordinate representation $\gamma(t) = \left(\gamma^1(t), \ldots, \gamma^n(t)\right)$, formula (4.15) shows that a vector field V is parallel along γ if and only if

$$\dot{V}^k(t) = -V^j(t)\dot{\gamma}^i(t)\Gamma_{ij}^k(\gamma(t)), \qquad k = 1, \ldots, n, \tag{4.19}$$

with analogous expressions based on Proposition 4.18 for tensor fields of other types. In each case, this is a system of first-order linear ordinary differential equations for the unknown coefficients of the vector or tensor field—in the vector case, the functions $\left(V^1(t), \ldots, V^n(t)\right)$. The usual ODE theorem guarantees the existence and uniqueness of a solution for a short time, given any initial values at $t = t_0$; but since the equation is linear, we can actually show much more: there exists a unique solution on the entire parameter interval.

Theorem 4.31 (Existence, Uniqueness, and Smoothness for Linear ODEs). *Let $I \subseteq \mathbb{R}$ be an open interval, and for $1 \le j, k \le n$, let $A_j^k: I \to \mathbb{R}$ be smooth functions. For all $t_0 \in I$ and every initial vector $\left(c^1, \ldots, c^n\right) \in \mathbb{R}^n$, the linear initial value problem*

$$\begin{aligned} \dot{V}^k(t) &= A_j^k(t)V^j(t), \\ V^k(t_0) &= c^k, \end{aligned} \tag{4.20}$$

has a unique smooth solution on all of I, and the solution depends smoothly on $(t, c) \in I \times \mathbb{R}^n$.

Proof. First assume $t_0 = 0$. Let $\left(x^0, x^1, \ldots, x^n\right)$ denote standard coordinates on the manifold $I \times \mathbb{R}^n \subseteq \mathbb{R}^{n+1}$, and consider the vector field $Y \in \mathfrak{X}\left(I \times \mathbb{R}^n\right)$ defined by

$$Y = \frac{\partial}{\partial x^0} + A_j^1\left(x^0\right)x^j \frac{\partial}{\partial x^1} + \cdots + A_j^n\left(x^0\right)x^j \frac{\partial}{\partial x^n}.$$

If $V(t) = \left(V^1(t), \ldots, V^n(t)\right)$ is a solution to (4.20) with $t_0 = 0$ defined on some interval $I_0 \subseteq I$, then the curve $\eta(t) = \left(t, V^1(t), \ldots, V^n(t)\right)$ is an integral curve of Y defined on I_0 satisfying the initial condition

$$\eta(0) = \left(0, c^1, \ldots, c^n\right). \tag{4.21}$$

Conversely, for each $\left(c^1, \ldots, c^n\right) \in \mathbb{R}^n$, there is an integral curve η of Y defined on some open interval $I_0 \subseteq I$ containing 0 and satisfying (4.21). If we write the component functions of η as $\eta(t) = \left(\eta^0(t), \eta^1(t), \ldots, \eta^n(t)\right)$, then $\dot{\eta}^0(t) \equiv 1$ and $\eta^0(0) = 0$, so $\eta(t) = t$ for all t. It then follows that $V(t) = \left(\eta^1(t), \ldots, \eta^n(t)\right)$ solves

(4.20) with $t_0 = 0$. Thus there is a one-to-one correspondence between solutions to (4.20) and integral curves of Y satisfying (4.21).

The fundamental theorem on flows of vector fields (Thm. A.42) guarantees that for each $(c^1,\dots,c^n) \in \mathbb{R}^n$, there exists a maximal integral curve η of Y defined on some open interval containing 0 and satisfying the initial condition (4.21), and the solutions depend smoothly on (t,c). Therefore, there is a solution to (4.20) for t in some maximal interval $I_0 \subseteq I$, and we need only show that $I_0 = I$. Write $I = (a,b)$ and $I_0 = (a_0, b_0)$ (where a, b, a_0, b_0 can be finite or infinite), and assume for the sake of contradiction that $b_0 < b$.

Let us use the differential equation to estimate the derivative of $|V(t)|^2$:

$$\frac{d}{dt}|V(t)|^2 = 2\sum_k \dot{V}^k(t)V^k(t)$$

$$= 2\sum_{j,k} A_j^k(t)V^j(t)V^k(t)$$

$$= 2V(t)^T A(t)V(t) \le 2|A(t)|\,|V(t)|^2.$$

Here $|\cdot|$ denotes the **Frobenius norm** of a vector or matrix, obtained by summing the squares of all the components and taking the square root. (It is just the Euclidean norm of the components.) On the compact interval $[0, b_0] \subseteq I$, the functions A_j^k are all bounded, so there is a constant M such that $|A(t)| \le M$ there. It then follows that

$$\frac{d}{dt}\left(e^{-2Mt}|V(t)|^2\right) = e^{-2Mt}\left(\frac{d}{dt}|V(t)|^2 - 2M|V(t)|^2\right) \le 0,$$

so the expression $e^{-2Mt}|V(t)|^2$ is a nonincreasing function of t, and thus is bounded for all $t \in [0, b_0)$ by its initial value $|V(0)|^2$. This implies $|V(t)|^2 \le e^{2Mt}|V(0)|^2$ for $t \in [0, b_0)$, which in turn implies that the corresponding integral curve of Y stays in the compact set $[0, b_0] \times \bar{B}_R(0) \subseteq I \times \mathbb{R}^n$, where $R = e^{Mb_0}|V(0)|$. This contradicts the escape lemma (Lemma A.43), and shows that $b_0 = b$. The possibility that $a < a_0$ can be ruled out by applying the same reasoning to the vector field $-Y$.

Finally, the case of general $t_0 \in I$ can be reduced to the previous case by making the substitutions $V^k(t) = \tilde{V}^k(t - t_0)$ and $A_j^k(t) = \tilde{A}_j^k(t - t_0)$. $\qquad\square$

Theorem 4.32 (Existence and Uniqueness of Parallel Transport). *Suppose M is a smooth manifold with or without boundary, and ∇ is a connection in TM. Given a smooth curve $\gamma \colon I \to M$, $t_0 \in I$, and a vector $v \in T_{\gamma(t_0)}M$ or tensor $v \in T^{(k,l)}(T_{\gamma(t)}M)$, there exists a unique parallel vector or tensor field V along γ such that $V(t_0) = v$.*

Proof. As in the proof of Theorem 4.24, we carry out the proof for vector fields. The case of tensor fields differs only in notation.

First suppose $\gamma(I)$ is contained in a single coordinate chart. Then V is parallel along γ if and only if its components satisfy the linear system of ODEs (4.19). Theorem 4.31 guarantees the existence and uniqueness of a solution on all of I with any initial condition $V(t_0) = v$.

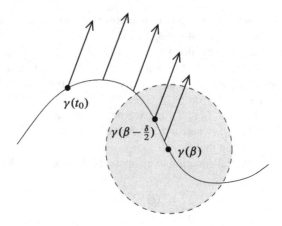

Fig. 4.9: Existence and uniqueness of parallel transports

Now suppose $\gamma(I)$ is not covered by a single chart. Let β denote the supremum of all $b > t_0$ for which a unique parallel transport exists on $[t_0, b]$. (The argument for $t < t_0$ is similar.) We know that $\beta > t_0$, since for b close enough to t_0, $\gamma([t_0, b])$ is contained in a single chart and the above argument applies. Then a unique parallel transport V exists on $[t_0, \beta)$ (Fig. 4.9). If β is equal to $\sup I$, we are done. If not, choose smooth coordinates on an open set containing $\gamma(\beta - \delta, \beta + \delta)$ for some positive δ. Then there exists a unique parallel vector field $\widetilde{V}$ on $(\beta - \delta, \beta + \delta)$ satisfying the initial condition $\widetilde{V}(\beta - \delta/2) = V(\beta - \delta/2)$. By uniqueness, $V = \widetilde{V}$ on their common domain, and therefore $\widetilde{V}$ is a parallel extension of V past β, which is a contradiction.   □

The vector or tensor field whose existence and uniqueness are proved in Theorem 4.32 is called the ***parallel transport of v along γ***. For each $t_0, t_1 \in I$, we define a map

$$P^{\gamma}_{t_0 t_1} : T_{\gamma(t_0)}M \to T_{\gamma(t_1)}M, \tag{4.22}$$

called the ***parallel transport map***, by setting $P^{\gamma}_{t_0 t_1}(v) = V(t_1)$ for each $v \in T_{\gamma(t_0)}M$, where V is the parallel transport of v along γ. This map is linear, because the equation of parallelism is linear. It is in fact an isomorphism, because $P^{\gamma}_{t_1 t_0}$ is an inverse for it.

It is also useful to extend the parallel transport operation to curves that are merely piecewise smooth. Given an admissible curve $\gamma : [a, b] \to M$, a map $V : [a, b] \to TM$ such that $V(t) \in T_{\gamma(t)}M$ for each t is called a ***piecewise smooth vector field along γ*** if V is continuous and there is an admissible partition $(a_0, \ldots, a_k)$ for γ such that V is smooth on each subinterval $[a_{i-1}, a_i]$. We will call any such partition an ***admissible partition for V***. A piecewise smooth vector field V along γ is said to be ***parallel along γ*** if $D_t V = 0$ wherever V is smooth.

Corollary 4.33 (Parallel Transport Along Piecewise Smooth Curves). *Suppose M is a smooth manifold with or without boundary, and ∇ is a connection in TM. Given an admissible curve $\gamma\colon [a,b] \to M$ and a vector $v \in T_{\gamma(t_0)}M$ or tensor $v \in T^{(k,l)}(T_{\gamma(t)}M)$, there exists a unique piecewise smooth parallel vector or tensor field V along γ such that $V(a) = v$, and V is smooth wherever γ is.*

Proof. Let $(a_0,\dots,a_k)$ be an admissible partition for γ. First define $V|_{[a_0,a_1]}$ to be the parallel transport of v along the first smooth segment $\gamma|_{[a_0,a_1]}$; then define $V|_{[a_1,a_2]}$ to be the parallel transport of $V(a_1)$ along the next smooth segment $\gamma|_{[a_1,a_2]}$; and continue by induction. $\square$

Here is an extremely useful tool for working with parallel transport. Given any basis $(b_1,\dots,b_n)$ for $T_{\gamma(t_0)}M$, we can parallel transport the vectors b_i along γ, thus obtaining an n-tuple of parallel vector fields $(E_1,\dots,E_n)$ along γ. Because each parallel transport map is an isomorphism, the vectors $(E_i(t))$ form a basis for $T_{\gamma(t)}M$ at each point $\gamma(t)$. Such an n-tuple of vector fields along γ is called a **parallel frame along γ**. Every smooth (or piecewise smooth) vector field along γ can be expressed in terms of such a frame as $V(t) = V^i(t)E_i(t)$, and then the properties of covariant derivatives along curves, together with the fact that the E_i's are parallel, imply

$$D_t V(t) = \dot{V}^i(t)E_i(t) \tag{4.23}$$

wherever V and γ are smooth. This means that a vector field is parallel along γ if and only if its component functions with respect to the frame (E_i) are constants.

The parallel transport map is the means by which a connection "connects" nearby tangent spaces. The next theorem and its corollary show that parallel transport determines covariant differentiation along curves, and thereby the connection itself.

Theorem 4.34 (Parallel Transport Determines Covariant Differentiation). *Let M be a smooth manifold with or without boundary, and let ∇ be a connection in TM. Suppose $\gamma\colon I \to M$ is a smooth curve and V is a smooth vector field along γ. For each $t_0 \in I$,*

$$D_t V(t_0) = \lim_{t_1 \to t_0} \frac{P^\gamma_{t_1 t_0} V(t_1) - V(t_0)}{t_1 - t_0}. \tag{4.24}$$

Proof. Let (E_i) be a parallel frame along γ, and write $V(t) = V^i(t)E_i(t)$ for $t \in I$. On the one hand, (4.23) shows that $D_t V(t_0) = \dot{V}^i(t_0)E_i(t_0)$.

On the other hand, for every fixed $t_1 \in I$, the parallel transport of the vector $V(t_1)$ along γ is the constant-coefficient vector field $W(t) = V^i(t_1)E_i(t)$ along γ, so $P^\gamma_{t_1 t_0} V(t_1) = V^i(t_1)E_i(t_0)$. Inserting these formulas into (4.24) and taking the limit as $t_1 \to t_0$, we conclude that the right-hand side is also equal to $\dot{V}^i(t_0)E_i(t_0)$. $\square$

Corollary 4.35 (Parallel Transport Determines the Connection). *Let M be a smooth manifold with or without boundary, and let ∇ be a connection in TM. Suppose X and Y are smooth vector fields on M. For every $p \in M$,*

$$\nabla_X Y \big|_p = \lim_{h \to 0} \frac{P^\gamma_{h0} Y_{\gamma(h)} - Y_p}{h}, \tag{4.25}$$

where $\gamma \colon I \to M$ is any smooth curve such that $\gamma(0) = p$ and $\gamma'(0) = X_p$.

Proof. Given $p \in M$ and a smooth curve γ such that $\gamma(0) = p$ and $\gamma'(0) = X_p$, let $V(t)$ denote the vector field along γ determined by Y, so $V(t) = Y_{\gamma(t)}$. By property (iii) of Theorem 4.24, $\nabla_X Y|_p$ is equal to $D_t V(0)$, so the result follows from Theorem 4.34. $\square$

A smooth vector or tensor field on M is said to be **parallel** (with respect to ∇) if it is parallel along every smooth curve in M. For example, Exercise 4.30 shows that every constant-coefficient vector field on $\mathbb{R}^n$ is parallel.

Proposition 4.36. *Suppose M is a smooth manifold with or without boundary, ∇ is a connection in TM, and A is a smooth vector or tensor field on M. Then A is parallel on M if and only if $\nabla A \equiv 0$.*

Proof. Problem 4-12. $\square$

Although Theorem 4.32 showed that it is always possible to extend a vector at a point to a parallel vector field along any given curve, it may not be possible in general to extend it to a *parallel vector field* on an open subset of the manifold. The impossibility of finding such extensions is intimately connected with the phenomenon of curvature, which will occupy a major portion of our attention in the second half of the book.

Pullback Connections

Like vector fields, connections in the tangent bundle cannot be either pushed forward or pulled back by arbitrary smooth maps. However, there is a natural way to pull back such connections by means of a *diffeomorphism*. In this section we define this operation and enumerate some of its most important properties.

Suppose M and $\widetilde{M}$ are smooth manifolds and $\varphi \colon M \to \widetilde{M}$ is a diffeomorphism. For a smooth vector field $X \in \mathfrak{X}(M)$, recall that the *pushforward of X* is the unique vector field $\varphi_* X \in \mathfrak{X}(\widetilde{M})$ that satisfies $d\varphi_p(X_p) = (\varphi_* X)_{\varphi(p)}$ for all $p \in M$. (See Lemma A.36.)

Lemma 4.37 (Pullback Connections). *Suppose M and $\widetilde{M}$ are smooth manifolds with or without boundary. If $\widetilde{\nabla}$ is a connection in $T\widetilde{M}$ and $\varphi \colon M \to \widetilde{M}$ is a diffeomorphism, then the map $\varphi^* \widetilde{\nabla} \colon \mathfrak{X}(M) \times \mathfrak{X}(M) \to \mathfrak{X}(M)$ defined by*

$$\left(\varphi^* \widetilde{\nabla}\right)_X Y = \left(\varphi^{-1}\right)_* \left(\widetilde{\nabla}_{\varphi_* X}(\varphi_* Y)\right) \tag{4.26}$$

*is a connection in TM, called the **pullback of $\widetilde{\nabla}$ by φ**.*

Proof. It is immediate from the definition that $\left(\varphi^* \widetilde{\nabla}\right)_X Y$ is linear over $\mathbb{R}$ in Y. To see that it is linear over $C^\infty(M)$ in X, let $f \in C^\infty(M)$, and let $\widetilde{f} = f \circ \varphi^{-1}$, so $\varphi_*(fX) = \widetilde{f}\varphi_* X$. Then

$$\left(\varphi^*\tilde{\nabla}\right)_{fX}Y = \left(\varphi^{-1}\right)_*\left(\tilde{\nabla}_{\tilde{f}\varphi_*X}(\varphi_*Y)\right)$$
$$= \left(\varphi^{-1}\right)_*\left(\tilde{f}\tilde{\nabla}_{\varphi_*X}(\varphi_*Y)\right)$$
$$= f\left(\varphi^*\tilde{\nabla}\right)_XY.$$

Finally, to prove the product rule in Y, let f and $\tilde{f}$ be as above, and note that (A.7) implies $(\varphi_*X)(\tilde{f}) = (Xf)\circ\varphi^{-1}$. Thus

$$\left(\varphi^*\tilde{\nabla}\right)_X(fY) = \left(\varphi^{-1}\right)_*\left(\tilde{\nabla}_{\varphi_*X}(\tilde{f}\varphi_*Y)\right)$$
$$= \left(\varphi^{-1}\right)_*\left(\tilde{f}\tilde{\nabla}_{\varphi_*X}(\varphi_*Y) + (\varphi_*X)(\tilde{f})\varphi_*Y\right)$$
$$= f\left(\varphi^*\tilde{\nabla}\right)_XY + (Xf)Y. \qquad\square$$

The next proposition shows that various important concepts defined in terms of connections—covariant derivatives along curves, parallel transport, and geodesics—all behave as expected with respect to pullback connections.

Proposition 4.38 (Properties of Pullback Connections). *Suppose M and $\widetilde{M}$ are smooth manifolds with or without boundary, and $\varphi\colon M \to \widetilde{M}$ is a diffeomorphism. Let $\tilde{\nabla}$ be a connection in $T\widetilde{M}$ and let $\nabla = \varphi^*\tilde{\nabla}$ be the pullback connection in TM. Suppose $\gamma\colon I \to M$ is a smooth curve.*

(a) *φ takes covariant derivatives along curves to covariant derivatives along curves: if V is a smooth vector field along γ, then*

$$d\varphi \circ D_t V = \tilde{D}_t(d\varphi \circ V),$$

where D_t is covariant differentiation along γ with respect to ∇, and $\tilde{D}_t$ is covariant differentiation along $\varphi \circ \gamma$ with respect to $\tilde{\nabla}$.
(b) *φ takes geodesics to geodesics: if γ is a ∇-geodesic in M, then $\varphi \circ \gamma$ is a $\tilde{\nabla}$-geodesic in $\widetilde{M}$.*
(c) *φ takes parallel transport to parallel transport: for every $t_0, t_1 \in I$,*

$$d\varphi_{\gamma(t_1)} \circ P_{t_0t_1}^{\gamma} = P_{t_0t_1}^{\varphi\circ\gamma} \circ d\varphi_{\gamma(t_0)}.$$

Proof. Problem 4-13. $\qquad\square$

Problems

4-1. Let $M \subseteq \mathbb{R}^n$ be an embedded submanifold and $Y \in \mathfrak{X}(M)$. For every point $p \in M$ and vector $v \in T_pM$, define $\nabla_v^{\top}Y$ by (4.4).

(a) Show that $\nabla_v^{\top}Y$ does not depend on the choice of extension $\tilde{Y}$ of Y. [Hint: Use Prop. A.28.]

(b) Show that $\nabla_v^\top Y$ is invariant under rigid motions of $\mathbb{R}^n$, in the following sense: if $F \in E(n)$ and $\tilde{M} = F(M)$, then $dF_p(\nabla_v^\top Y) = \nabla_{dF_p(v)}^\top(F_*Y)$.

(Used on pp. 87, 93.)

4-2. In your study of smooth manifolds, you have already seen another way of taking "directional derivatives of vector fields," the Lie derivative $\mathcal{L}_X Y$ (which is equal to the Lie bracket $[X, Y]$; see Prop. A.46). Suppose M is a smooth manifold of positive dimension.

(a) Show that the map $\mathcal{L}: \mathfrak{X}(M) \times \mathfrak{X}(M) \to \mathfrak{X}(M)$ is not a connection.
(b) Show that there are smooth vector fields X and Y on $\mathbb{R}^2$ such that $X = Y = \partial_1$ along the x^1-axis, but the Lie derivatives $\mathcal{L}_X(\partial_2)$ and $\mathcal{L}_Y(\partial_2)$ are not equal on the x^1-axis.

4-3. Prove Proposition 4.7 (the transformation law for the connection coefficients).

4-4. Prove Theorem 4.14 (characterizing the space of connections).

4-5. Prove Proposition 4.16 (local formulas for covariant derivatives of tensor fields).

4-6. Let M be a smooth manifold and let ∇ be a connection in TM. Define a map $\tau: \mathfrak{X}(M) \times \mathfrak{X}(M) \to \mathfrak{X}(M)$ by

$$\tau(X, Y) = \nabla_X Y - \nabla_Y X - [X, Y].$$

(a) Show that τ is a $(1, 2)$-tensor field, called the ***torsion tensor of*** ∇.
(b) We say that ∇ is ***symmetric*** if its torsion vanishes identically. Show that ∇ is symmetric if and only if its connection coefficients with respect to every coordinate frame are symmetric: $\Gamma_{ij}^k = \Gamma_{ji}^k$. [Warning: They might not be symmetric with respect to other frames.]
(c) Show that ∇ is symmetric if and only if the covariant Hessian $\nabla^2 u$ of every smooth function $u \in C^\infty(M)$ is a symmetric 2-tensor field. (See Example 4.22.)
(d) Show that the Euclidean connection $\bar{\nabla}$ on $\mathbb{R}^n$ is symmetric.

(Used on pp. 113, 121, 123.)

4-7. Let $\gamma: (-\pi, \pi) \to \mathbb{R}^2$ be the figure eight curve defined in Example 4.23. Prove that γ is an injective smooth immersion, but its velocity vector field is not extendible.

4-8. Suppose M is a smooth manifold (without boundary), $I \subseteq \mathbb{R}$ is an interval (bounded or not, with or without endpoints), and $\gamma: I \to M$ is a smooth curve.

(a) Show that for every $t_0 \in I$ such that $\gamma'(t_0) \neq 0$, there is a connected neighborhood J of t_0 in I such that every smooth vector field along $\gamma|_J$ is extendible.

(b) Show that if I is an open interval or a compact interval and γ is a smooth embedding, then every smooth vector field along γ is extendible.

4-9. Let M be a smooth manifold, and let ∇^0 and ∇^1 be two connections on TM.

(a) Show that ∇^0 and ∇^1 have the same torsion (Problem 4-6) if and only if their difference tensor is symmetric, i.e., $D(X,Y) = D(Y,X)$ for all X and Y.

(b) Show that ∇^0 and ∇^1 determine the same geodesics if and only if their difference tensor is antisymmetric, i.e., $D(X,Y) = -D(Y,X)$ for all X and Y.

(Used on p. 145.)

4-10. Suppose M is a smooth manifold endowed with a connection, $\gamma: I \to M$ is a smooth curve, and $Y \in \mathfrak{X}(\gamma)$. Prove that if Y is parallel along γ, then it is parallel along every reparametrization of γ.

4-11. Suppose G is a Lie group.

(a) Show that there is a unique connection ∇ in TG with the property that every left-invariant vector field is parallel.

(b) Show that the torsion tensor of ∇ (Problem 4-6) is zero if and only if G is abelian.

4-12. Prove Proposition 4.36 (a vector or tensor field A is parallel if and only if $\nabla A \equiv 0$).

4-13. Prove Proposition 4.38 (properties of pullback connections).

4-14. Let M be a smooth n-manifold and ∇ a connection in TM, let (E_i) be a local frame on some open subset $U \subseteq M$, and let (ε^i) be the dual coframe.

(a) Show that there is a uniquely determined $n \times n$ matrix of smooth 1-forms $(\omega_i{}^j)$ on U, called the **connection 1-forms** for this frame, such that

$$\nabla_X E_i = \omega_i{}^j(X)E_j$$

for all $X \in \mathfrak{X}(U)$.

(b) CARTAN'S FIRST STRUCTURE EQUATION: Prove that these forms satisfy the following equation, due to Élie Cartan:

$$d\varepsilon^j = \varepsilon^i \wedge \omega_i{}^j + \tau^j,$$

where $\tau^1,\ldots,\tau^n \in \Omega^2(M)$ are the **torsion 2-forms**, defined in terms of the torsion tensor τ (Problem 4-6) and the frame (E_i) by

$$\tau(X,Y) = \tau^j(X,Y)E_j.$$

(Used on pp. 145, 222.)

Chapter 5
The Levi-Civita Connection

If we are to use geodesics and covariant derivatives as tools for studying Riemannian geometry, it is evident that we need a way to single out a particular connection on a Riemannian manifold that reflects the properties of the metric. In this chapter, guided by the example of the tangential connection on a submanifold of $\mathbb{R}^n$, we describe two properties that determine a unique connection on every Riemannian manifold. The first property, *compatibility with the metric*, is easy to motivate and understand. The second, *symmetry*, is a bit more mysterious; but it is motivated by the fact that it is invariantly defined, and is always satisfied by the tangential connection. It turns out that these two conditions are enough to determine a unique connection associated with any Riemannian or pseudo-Riemannian metric, called the *Levi-Civita connection* after the early twentieth-century Italian differential geometer Tullio Levi-Civita.

After defining the Levi-Civita connection, we investigate the *exponential map*, which conveniently encodes the collective behavior of geodesics and allows us to study how they change as the initial point and initial velocity vary. Having established the properties of this map, we introduce *normal neighborhoods* and *normal coordinates*, which are essential computational and theoretical tools for studying local geometric properties near a point. Then we introduce the analogous notion for studying properties near a submanifold: *tubular neighborhoods* and *Fermi coordinates*. Finally, we return to our three main model Riemannian manifolds and determine their geodesics.

Except where noted otherwise, the results and proofs of this chapter do not use positivity of the metric, so they apply equally well to Riemannian and pseudo-Riemannian manifolds.

The Tangential Connection Revisited

We are eventually going to show that on each Riemannian manifold there is a natural connection that is particularly well suited to computations in Riemannian geome-

© Springer International Publishing AG 2018
J. M. Lee, *Introduction to Riemannian Manifolds*, Graduate Texts
in Mathematics 176, https://doi.org/10.1007/978-3-319-91755-9_5

try. Since we get most of our intuition about Riemannian manifolds from studying submanifolds of $\mathbb{R}^n$ with the induced metric, let us start by examining that case.

Let $M \subseteq \mathbb{R}^n$ be an embedded submanifold. As a guiding principle, consider the idea mentioned at the beginning of Chapter 4: a geodesic in M should be "as straight as possible." A reasonable way to make this rigorous is to require that the geodesic have no acceleration in directions tangent to the manifold, or in other words that its acceleration vector have zero orthogonal projection onto TM.

The tangential connection defined in Example 4.9 is perfectly suited to this task, because it computes covariant derivatives on M by taking ordinary derivatives in $\mathbb{R}^n$ and projecting them orthogonally to TM.

It is easy to compute covariant derivatives along curves in M with respect to the tangential connection. Suppose $\gamma: I \to M$ is a smooth curve. Then γ can be regarded as either a smooth curve in M or a smooth curve in $\mathbb{R}^n$, and a smooth vector field V along γ that takes its values in TM can be regarded as either a vector field along γ in M or a vector field along γ in $\mathbb{R}^n$. Let $\bar{D}_t V$ denote the covariant derivative of V along γ (as a curve in $\mathbb{R}^n$) with respect to the Euclidean connection $\bar{\nabla}$, and let $D_t^{\top} V$ denote its covariant derivative along γ (as a curve in M) with respect to the tangential connection $\nabla^{\top}$. The next proposition shows that the two covariant derivatives along γ have a simple relationship to each other.

Proposition 5.1. Let $M \subseteq \mathbb{R}^n$ be an embedded submanifold, $\gamma: I \to M$ a smooth curve in M, and V a smooth vector field along γ that takes its values in TM. Then for each $t \in I$,

$$D_t^{\top} V(t) = \pi^{\top}(\bar{D}_t V(t)).$$

Proof. Let $t_0 \in I$ be arbitrary. By Proposition 2.14, on some neighborhood U of $\gamma(t_0)$ in $\mathbb{R}^n$ there is an adapted orthonormal frame for TM, that is, a local orthonormal frame $(E_1, \ldots, E_n)$ for $T\mathbb{R}^n$ such that $(E_1, \ldots, E_k)$ restricts to an orthonormal frame for TM at points of $M \cap U$ (where $k = \dim M$). If $\varepsilon > 0$ is small enough that $\gamma((t_0 - \varepsilon, t_0 + \varepsilon)) \subseteq U$, then for $t \in (t_0 - \varepsilon, t_0 + \varepsilon)$ we can write

$$V(t) = V^1(t) E_1\big|_{\gamma(t)} + \cdots + V^k(t) E_k\big|_{\gamma(t)},$$

for some smooth functions $V^1, \ldots, V^k: (t_0 - \varepsilon, t_0 + \varepsilon) \to \mathbb{R}$. Formula (4.15) yields

$$\pi^{\top}(\bar{D}_t V(t)) = \pi^{\top}\left(\sum_{i=1}^{k}\left(\dot{V}^i(t) E_i\big|_{\gamma(t)} + V^i(t)\bar{\nabla}_{\gamma'(t)} E_i\big|_{\gamma(t)}\right)\right)$$

$$= \sum_{i=1}^{k}\left(\dot{V}^i(t) E_i\big|_{\gamma(t)} + V^i(t)\pi^{\top}\left(\bar{\nabla}_{\gamma'(t)} E_i\big|_{\gamma(t)}\right)\right)$$

$$= \sum_{i=1}^{k}\left(\dot{V}^i(t) E_i\big|_{\gamma(t)} + V^i(t)\nabla_{\gamma'(t)}^{\top} E_i\big|_{\gamma(t)}\right)$$

$$= D_t^{\top} V(t). \qquad \square$$

Corollary 5.2. *Suppose* $M \subseteq \mathbb{R}^n$ *is an embedded submanifold. A smooth curve* $\gamma \colon I \to M$ *is a geodesic with respect to the tangential connection on* M *if and only if its ordinary acceleration* $\gamma''(t)$ *is orthogonal to* $T_{\gamma(t)}M$ *for all* $t \in I$.

Proof. As noted in Example 4.8, the connection coefficients of the Euclidean connection on $\mathbb{R}^n$ are all zero. Thus it follows from (4.15) that the Euclidean covariant derivative of γ' along γ is just its ordinary acceleration: $\bar{D}_t \gamma'(t) = \gamma''(t)$. The corollary then follows from Proposition 5.1. $\qquad\square$

These considerations can be extended to pseudo-Riemannian manifolds as well. Let $\left(\mathbb{R}^{r,s}, \bar{q}^{(r,s)}\right)$ be the pseudo-Euclidean space of signature (r,s). If $M \subseteq \mathbb{R}^{r,s}$ is an embedded Riemannian or pseudo-Riemannian submanifold, then for each $p \in M$, the tangent space $T_p\mathbb{R}^{r,s}$ decomposes as a direct sum $T_pM \oplus N_pM$, where $N_pM = (T_pM)^\perp$ is the orthogonal complement of T_pM with respect to $\bar{q}^{(r,s)}$. We let $\pi^\top \colon T_p\mathbb{R}^{r,s} \to T_pM$ be the $\bar{q}^{(r,s)}$-orthogonal projection, and define the **tangential connection** $\nabla^\top$ on M by

$$\nabla^\top_X Y = \pi^\top\left(\bar{\nabla}_{\widetilde{X}}\,\widetilde{Y}\right),$$

where $\widetilde{X}$ and $\widetilde{Y}$ are smooth extensions of X and Y to a neighborhood of M, and $\bar{\nabla}$ is the ordinary Euclidean connection on $\mathbb{R}^{r,s}$. This is a well-defined connection on M by the same argument as in the Euclidean case, and the next proposition is proved in exactly the same way as Corollary 5.2.

Proposition 5.3. *Suppose* M *is an embedded Riemannian or pseudo-Riemannian submanifold of the pseudo-Euclidean space* $\mathbb{R}^{r,s}$. *A smooth curve* $\gamma \colon I \to M$ *is a geodesic with respect to* $\nabla^\top$ *if and only if* $\gamma''(t)$ *is* $\bar{q}^{(r,s)}$*-orthogonal to* $T_{\gamma(t)}M$ *for all* $t \in I$.

▶ **Exercise 5.4.** Prove the preceding proposition.

Connections on Abstract Riemannian Manifolds

There is a celebrated (and hard) theorem of John Nash [Nas56] that says that every Riemannian metric on a smooth manifold can be realized as the induced metric of some embedding in a Euclidean space. That theorem was later generalized independently by Robert Greene [Gre70] and Chris J. S. Clarke [Cla70] to pseudo-Riemannian metrics. Thus, in a certain sense, we would lose no generality by studying only submanifolds of Euclidean and pseudo-Euclidean spaces with their induced metrics, for which the tangential connection would suffice. However, when we are trying to understand *intrinsic* properties of a Riemannian manifold, an embedding introduces a great deal of extraneous information, and in some cases actually makes it harder to discern which geometric properties depend only on the metric. Our task in this chapter is to distinguish some important properties of the tangential connection that make sense for connections on an abstract Riemannian or pseudo-Riemannian manifold, and to use them to single out a unique connection in the abstract case.

Metric Connections

The Euclidean connection on $\mathbb{R}^n$ has one very nice property with respect to the Euclidean metric: it satisfies the product rule

$$\bar{\nabla}_X \langle Y, Z \rangle = \langle \bar{\nabla}_X Y, Z \rangle + \langle Y, \bar{\nabla}_X Z \rangle,$$

as you can verify easily by computing in terms of the standard basis. (In this formula, the left-hand side represents the covariant derivative of the real-valued function $\langle Y, Z \rangle$ regarded as a $(0,0)$-tensor field, which is really just $X \langle Y, Z \rangle$ by virtue of property (ii) of Prop. 4.15.) The Euclidean connection has the same property with respect to the pseudo-Euclidean metric on $\mathbb{R}^{r,s}$. It is almost immediate that the tangential connection on a Riemannian or pseudo-Riemannian submanifold satisfies the same product rule, if we now interpret all the vector fields as being tangent to M and interpret the inner products as being taken with respect to the induced metric on M (see Prop. 5.8 below).

This property makes sense on an abstract Riemannian or pseudo-Riemannian manifold. Let g be a Riemannian or pseudo-Riemannian metric on a smooth manifold M (with or without boundary). A connection ∇ on TM is said to be **compatible with g**, or to be a **metric connection**, if it satisfies the following product rule for all $X, Y, Z \in \mathfrak{X}(M)$:

$$\nabla_X \langle Y, Z \rangle = \langle \nabla_X Y, Z \rangle + \langle Y, \nabla_X Z \rangle. \tag{5.1}$$

The next proposition gives several alternative characterizations of compatibility with a metric, any one of which could be used as the definition.

Proposition 5.5 (Characterizations of Metric Connections). *Let (M, g) be a Riemannian or pseudo-Riemannian manifold (with or without boundary), and let ∇ be a connection on TM. The following conditions are equivalent:*

(a) *∇ is compatible with g: $\nabla_X \langle Y, Z \rangle = \langle \nabla_X Y, Z \rangle + \langle Y, \nabla_X Z \rangle$.*

(b) *g is parallel with respect to ∇: $\nabla g \equiv 0$.*

(c) *In terms of any smooth local frame (E_i), the connection coefficients of ∇ satisfy*

$$\Gamma^l_{ki} g_{lj} + \Gamma^l_{kj} g_{il} = E_k(g_{ij}). \tag{5.2}$$

(d) *If V, W are smooth vector fields along any smooth curve γ, then*

$$\frac{d}{dt} \langle V, W \rangle = \langle D_t V, W \rangle + \langle V, D_t W \rangle. \tag{5.3}$$

(e) *If V, W are parallel vector fields along a smooth curve γ in M, then $\langle V, W \rangle$ is constant along γ.*

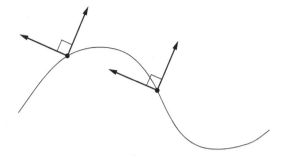

Fig. 5.1: A parallel orthonormal frame

(f) *Given any smooth curve γ in M, every parallel transport map along γ is a linear isometry.*

(g) *Given any smooth curve γ in M, every orthonormal basis at a point of γ can be extended to a parallel orthonormal frame along γ (Fig. 5.1).*

Proof. First we prove (a) $\Leftrightarrow$ (b). By (4.14) and (4.12), the total covariant derivative of the symmetric 2-tensor g is given by

$$(\nabla g)(Y, Z, X) = (\nabla_X g)(Y, Z) = X(g(Y, Z)) - g(\nabla_X Y, Z) - g(Y, \nabla_X Z).$$

This is zero for all X, Y, Z if and only if (5.1) is satisfied for all X, Y, Z.

To prove (b) $\Leftrightarrow$ (c), note that Proposition 4.18 shows that the components of ∇g in terms of a smooth local frame (E_i) are

$$g_{ij;k} = E_k(g_{ij}) - \Gamma_{ki}^l g_{lj} - \Gamma_{kj}^l g_{il}.$$

These are all zero if and only if (5.2) is satisfied.

Next we prove (a) $\Leftrightarrow$ (d). Assume (a), and let V, W be smooth vector fields along a smooth curve $\gamma : I \to M$. Given $t_0 \in I$, in a neighborhood of $\gamma(t_0)$ we may choose coordinates (x^i) and write $V = V^i \partial_i$ and $W = W^j \partial_j$ for some smooth functions $V^i, W^j : I \to \mathbb{R}$. Applying (5.1) to the extendible vector fields ∂_i, ∂_j, we obtain

$$\begin{aligned}
\frac{d}{dt}\langle V, W \rangle &= \frac{d}{dt}\left(V^i W^j \langle \partial_i, \partial_j \rangle \right) \\
&= \left(\dot{V}^i W^j + V^i \dot{W}^j \right)\langle \partial_i, \partial_j \rangle + V^i W^j \left(\langle \nabla_{\gamma'(t)} \partial_i, \partial_j \rangle + \langle \partial_i, \nabla_{\gamma'(t)} \partial_j \rangle \right) \\
&= \langle D_t V, W \rangle + \langle V, D_t W \rangle,
\end{aligned}$$

which proves (d). Conversely, if (d) holds, then in particular it holds for extendible vector fields along γ, and then (a) follows from part (iii) of Theorem 4.24.

Now we will prove (d) $\Rightarrow$ (e) $\Rightarrow$ (f) $\Rightarrow$ (g) $\Rightarrow$ (d). Assume first that (d) holds. If V and W are parallel along γ, then (5.3) shows that $\langle V, W \rangle$ has zero derivative with respect to t, so it is constant along γ.

Now assume (e). Let v_0, w_0 be arbitrary vectors in $T_{\gamma(t_0)}M$, and let V, W be their parallel transports along γ, so that $V(t_0) = v_0$, $W(t_0) = w_0$, $P_{t_0 t_1}^{\gamma} v_0 = V(t_1)$, and $P_{t_0 t_1}^{\gamma} w_0 = W(t_1)$. Because $\langle V, W \rangle$ is constant along γ, it follows that $\langle P_{t_0 t_1}^{\gamma} v_0, P_{t_0 t_1}^{\gamma} w_0 \rangle = \langle V(t_1), W(t_1) \rangle = \langle V(t_0), W(t_0) \rangle = \langle v_0, w_0 \rangle$, so $P_{t_0 t_1}^{\gamma}$ is a linear isometry.

Next, assuming (f), we suppose $\gamma \colon I \to M$ is a smooth curve and (b_i) is an orthonormal basis for $T_{\gamma(t_0)}M$, for some $t_0 \in I$. We can extend each b_i by parallel transport to obtain a smooth parallel vector field E_i along γ, and the assumption that parallel transport is a linear isometry guarantees that the resulting n-tuple (E_i) is an orthonormal frame at all points of γ.

Finally, assume that (g) holds, and let (E_i) be a parallel orthonormal frame along γ. Given smooth vector fields V and W along γ, we can express them in terms of this frame as $V = V^i E_i$ and $W = W^j E_j$. The fact that the frame is orthonormal means that the metric coefficients $g_{ij} = \langle E_i, E_j \rangle$ are constants along γ (± 1 or 0), and the fact that it is parallel means that $D_t V = \dot{V}^i E_i$ and $D_t W = \dot{W}^i E_i$. Thus both sides of (5.3) reduce to the following expression:

$$g_{ij}\left(\dot{V}^i W^j + V^i \dot{W}^j\right). \tag{5.4}$$

This proves (d). $\square$

Corollary 5.6. *Suppose (M, g) is a Riemannian or pseudo-Riemannian manifold with or without boundary, ∇ is a metric connection on M, and $\gamma \colon I \to M$ is a smooth curve.*

(a) *$|\gamma'(t)|$ is constant if and only if $D_t \gamma'(t)$ is orthogonal to $\gamma'(t)$ for all $t \in I$.*
(b) *If γ is a geodesic, then $|\gamma'(t)|$ is constant.*

▶ **Exercise 5.7.** Prove the preceding corollary.

Proposition 5.8. *If M is an embedded Riemannian or pseudo-Riemannian submanifold of $\mathbb{R}^n$ or $\mathbb{R}^{r,s}$, the tangential connection on M is compatible with the induced Riemannian or pseudo-Riemannian metric.*

Proof. We will show that $\nabla^{\top}$ satisfies (5.1). Suppose $X, Y, Z \in \mathfrak{X}(M)$, and let $\widetilde{X}, \widetilde{Y}, \widetilde{Z}$ be smooth extensions of them to an open subset of $\mathbb{R}^n$ or $\mathbb{R}^{r,s}$. At points of M, we have

$$\begin{aligned}
\nabla_X^{\top} \langle Y, Z \rangle &= X \langle Y, Z \rangle = \widetilde{X} \langle \widetilde{Y}, \widetilde{Z} \rangle \\
&= \overline{\nabla}_{\widetilde{X}} \langle \widetilde{Y}, \widetilde{Z} \rangle \\
&= \langle \overline{\nabla}_{\widetilde{X}} \widetilde{Y}, \widetilde{Z} \rangle + \langle \widetilde{Y}, \overline{\nabla}_{\widetilde{X}} \widetilde{Z} \rangle \\
&= \langle \pi^{\top}(\overline{\nabla}_{\widetilde{X}} \widetilde{Y}), \widetilde{Z} \rangle + \langle \widetilde{Y}, \pi^{\top}(\overline{\nabla}_{\widetilde{X}} \widetilde{Z}) \rangle \\
&= \langle \nabla_X^{\top} Y, Z \rangle + \langle Y, \nabla_X^{\top} Z \rangle,
\end{aligned}$$

where the next-to-last equality follows from the fact that $\widetilde{Z}$ and $\widetilde{Y}$ are tangent to M. $\square$

Symmetric Connections

It turns out that every abstract Riemannian or pseudo-Riemannian manifold admits many different metric connections (see Problem 5-1), so requiring compatibility with the metric is not sufficient to pin down a unique connection on such a manifold. To do so, we turn to another key property of the tangential connection. Recall the definition (4.3) of the Euclidean connection. The expression on the right-hand side of that definition is reminiscent of part of the coordinate expression for the Lie bracket:

$$[X,Y] = X(Y^i)\frac{\partial}{\partial x^i} - Y(X^i)\frac{\partial}{\partial x^i}.$$

In fact, the two terms in the Lie bracket formula are exactly the coordinate expressions for $\overline{\nabla}_X Y$ and $\overline{\nabla}_Y X$. Therefore, the Euclidean connection satisfies the following identity for all smooth vector fields X, Y:

$$\overline{\nabla}_X Y - \overline{\nabla}_Y X = [X,Y].$$

This expression has the virtue that it is coordinate-independent and makes sense for every connection on the tangent bundle. We say that a connection ∇ on the tangent bundle of a smooth manifold M is *symmetric* if

$$\nabla_X Y - \nabla_Y X \equiv [X,Y] \text{ for all } X, Y \in \mathfrak{X}(M).$$

The symmetry condition can also be expressed in terms of the *torsion tensor* of the connection, which was introduced in Problem 4-6; this is the smooth $(1,2)$-tensor field $\tau : \mathfrak{X}(M) \times \mathfrak{X}(M) \to \mathfrak{X}(M)$ defined by

$$\tau(X,Y) = \nabla_X Y - \nabla_Y X - [X,Y].$$

Thus a connection ∇ is symmetric if and only if its torsion vanishes identically. It follows from the result of Problem 4-6 that a connection is symmetric if and only if its connection coefficients in every coordinate frame satisfy $\Gamma_{ij}^k = \Gamma_{ji}^k$; this is the origin of the term "symmetric."

Proposition 5.9. *If M is an embedded (pseudo-)Riemannian submanifold of a (pseudo-)Euclidean space, then the tangential connection on M is symmetric.*

Proof. Let M be an embedded Riemannian or pseudo-Riemannian submanifold of $\mathbb{R}^n$, where $\mathbb{R}^n$ is endowed either with the Euclidean metric or with a pseudo-Euclidean metric $\overline{q}^{(r,s)}$, $r + s = n$. Let $X, Y \in \mathfrak{X}(M)$, and let $\widetilde{X}, \widetilde{Y}$ be smooth extensions of them to an open subset of the ambient space. If $\iota : M \hookrightarrow \mathbb{R}^n$ represents the inclusion map, it follows that X and Y are ι-related to $\widetilde{X}$ and $\widetilde{Y}$, respectively, and thus by the naturality of the Lie bracket (Prop. A.39), $[X,Y]$ is ι-related to $[\widetilde{X}, \widetilde{Y}]$. In particular, $[\widetilde{X}, \widetilde{Y}]$ is tangent to M, and its restriction to M is equal to $[X,Y]$. Therefore,

$$\nabla_X^\mathsf{T} Y - \nabla_Y^\mathsf{T} X = \pi^\mathsf{T}\big(\bar{\nabla}_{\tilde{X}}\tilde{Y}\big|_M - \bar{\nabla}_{\tilde{Y}}\tilde{X}\big|_M\big)$$
$$= \pi^\mathsf{T}\big([\tilde{X},\tilde{Y}]\big|_M\big)$$
$$= [\tilde{X},\tilde{Y}]\big|_M$$
$$= [X,Y]. \qquad\qquad \square$$

The last two propositions show that if we wish to single out a connection on each Riemannian or pseudo-Riemannian manifold in such a way that it matches the tangential connection when the manifold is presented as an embedded submanifold of $\mathbb{R}^n$ or $\mathbb{R}^{r,s}$ with the induced metric, then we must require at least that the connection be compatible with the metric and symmetric. It is a pleasant fact that these two conditions are enough to determine a unique connection.

Theorem 5.10 (Fundamental Theorem of Riemannian Geometry). *Let (M,g) be a Riemannian or pseudo-Riemannian manifold (with or without boundary). There exists a unique connection ∇ on TM that is compatible with g and symmetric. It is called the **Levi-Civita connection of g** (or also, when g is positive definite, the **Riemannian connection**).*

Proof. We prove uniqueness first, by deriving a formula for ∇. Suppose, therefore, that ∇ is such a connection, and let $X,Y,Z \in \mathfrak{X}(M)$. Writing the compatibility equation three times with X,Y,Z cyclically permuted, we obtain

$$X\langle Y,Z\rangle = \langle \nabla_X Y,Z\rangle + \langle Y,\nabla_X Z\rangle,$$
$$Y\langle Z,X\rangle = \langle \nabla_Y Z,X\rangle + \langle Z,\nabla_Y X\rangle,$$
$$Z\langle X,Y\rangle = \langle \nabla_Z X,Y\rangle + \langle X,\nabla_Z Y\rangle.$$

Using the symmetry condition on the last term in each line, this can be rewritten as

$$X\langle Y,Z\rangle = \langle \nabla_X Y,Z\rangle + \langle Y,\nabla_Z X\rangle + \langle Y,[X,Z]\rangle,$$
$$Y\langle Z,X\rangle = \langle \nabla_Y Z,X\rangle + \langle Z,\nabla_X Y\rangle + \langle Z,[Y,X]\rangle,$$
$$Z\langle X,Y\rangle = \langle \nabla_Z X,Y\rangle + \langle X,\nabla_Y Z\rangle + \langle X,[Z,Y]\rangle.$$

Adding the first two of these equations and subtracting the third, we obtain

$$X\langle Y,Z\rangle + Y\langle Z,X\rangle - Z\langle X,Y\rangle =$$
$$2\langle \nabla_X Y,Z\rangle + \langle Y,[X,Z]\rangle + \langle Z,[Y,X]\rangle - \langle X,[Z,Y]\rangle.$$

Finally, solving for $\langle \nabla_X Y,Z\rangle$, we get

$$\langle \nabla_X Y,Z\rangle = \tfrac{1}{2}\big(X\langle Y,Z\rangle + Y\langle Z,X\rangle - Z\langle X,Y\rangle$$
$$- \langle Y,[X,Z]\rangle - \langle Z,[Y,X]\rangle + \langle X,[Z,Y]\rangle\big). \quad (5.5)$$

Now suppose ∇^1 and ∇^2 are two connections on TM that are symmetric and compatible with g. Since the right-hand side of (5.5) does not depend on the connection, it follows that $\langle \nabla^1_X Y - \nabla^2_X Y, Z \rangle = 0$ for all X, Y, Z. This can happen only if $\nabla^1_X Y = \nabla^2_X Y$ for all X and Y, so $\nabla^1 = \nabla^2$.

To prove existence, we use (5.5), or rather a coordinate version of it. It suffices to prove that such a connection exists in each coordinate chart, for then uniqueness ensures that the connections in different charts agree where they overlap.

Let $(U, (x^i))$ be any smooth local coordinate chart. Applying (5.5) to the coordinate vector fields, whose Lie brackets are zero, we obtain

$$\langle \nabla_{\partial_i} \partial_j, \partial_l \rangle = \tfrac{1}{2} \left(\partial_i \langle \partial_j, \partial_l \rangle + \partial_j \langle \partial_l, \partial_i \rangle - \partial_l \langle \partial_i, \partial_j \rangle \right). \tag{5.6}$$

Recall the definitions of the metric coefficients and the connection coefficients:

$$g_{ij} = \langle \partial_i, \partial_j \rangle, \qquad \nabla_{\partial_i} \partial_j = \Gamma^m_{ij} \partial_m.$$

Inserting these into (5.6) yields

$$\Gamma^m_{ij} g_{ml} = \tfrac{1}{2} \left(\partial_i g_{jl} + \partial_j g_{il} - \partial_l g_{ij} \right). \tag{5.7}$$

Finally, multiplying both sides by the inverse matrix g^{kl} and noting that $g_{ml} g^{kl} = \delta^k_m$, we get

$$\Gamma^k_{ij} = \tfrac{1}{2} g^{kl} \left(\partial_i g_{jl} + \partial_j g_{il} - \partial_l g_{ij} \right). \tag{5.8}$$

This formula certainly defines a connection in each chart, and it is evident from the formula that $\Gamma^k_{ij} = \Gamma^k_{ji}$, so the connection is symmetric by Problem 4-6(b). Thus only compatibility with the metric needs to be checked. Using (5.7) twice, we get

$$\Gamma^l_{ki} g_{lj} + \Gamma^l_{kj} g_{il} = \tfrac{1}{2} \left(\partial_k g_{ij} + \partial_i g_{kj} - \partial_j g_{ki} \right) + \tfrac{1}{2} \left(\partial_k g_{ji} + \partial_j g_{ki} - \partial_i g_{kj} \right)$$
$$= \partial_k g_{ij}.$$

By Proposition 5.5(c), this shows that ∇ is compatible with g. $\qquad \square$

A bonus of this proof is that it gives us explicit formulas that can be used for computing the Levi-Civita connection in various circumstances.

Corollary 5.11 (Formulas for the Levi-Civita Connection). *Let (M, g) be a Riemannian or pseudo-Riemannian manifold (with or without boundary), and let ∇ be its Levi-Civita connection.*

(a) IN TERMS OF VECTOR FIELDS: *If X, Y, Z are smooth vector fields on M, then*

$$\langle \nabla_X Y, Z \rangle = \tfrac{1}{2} \big(X \langle Y, Z \rangle + Y \langle Z, X \rangle - Z \langle X, Y \rangle$$
$$- \langle Y, [X, Z] \rangle - \langle Z, [Y, X] \rangle + \langle X, [Z, Y] \rangle \big). \tag{5.9}$$

*(This is known as **Koszul's formula**.)*

(b) IN COORDINATES: *In any smooth coordinate chart for M, the coefficients of the Levi-Civita connection are given by*

$$\Gamma^k_{ij} = \tfrac{1}{2} g^{kl} \left(\partial_i g_{jl} + \partial_j g_{il} - \partial_l g_{ij} \right). \tag{5.10}$$

(c) IN A LOCAL FRAME: *Let (E_i) be a smooth local frame on an open subset $U \subseteq M$, and let $c_{ij}^k : U \to \mathbb{R}$ be the n^3 smooth functions defined by*

$$[E_i, E_j] = c_{ij}^k E_k. \tag{5.11}$$

Then the coefficients of the Levi-Civita connection in this frame are

$$\Gamma_{ij}^k = \tfrac{1}{2} g^{kl} \left(E_i g_{jl} + E_j g_{il} - E_l g_{ij} - g_{jm} c_{il}^m - g_{lm} c_{ji}^m + g_{im} c_{lj}^m \right). \tag{5.12}$$

(d) IN A LOCAL ORTHONORMAL FRAME: *If g is Riemannian, (E_i) is a smooth local orthonormal frame, and the functions c_{ij}^k are defined by (5.11), then*

$$\Gamma_{ij}^k = \tfrac{1}{2} \left(c_{ij}^k - c_{ik}^j - c_{jk}^i \right). \tag{5.13}$$

Proof. We derived (5.9) and (5.10) in the proof of Theorem 5.10. To prove (5.12), apply formula (5.9) with $X = E_i$, $Y = E_j$, and $Z = E_l$, to obtain

$$\begin{aligned} \Gamma_{ij}^q g_{ql} &= \langle \nabla_{E_i} E_j, E_l \rangle \\ &= \tfrac{1}{2} \left(E_i g_{jl} + E_j g_{il} - E_l g_{ij} - g_{jm} c_{il}^m - g_{lm} c_{ji}^m + g_{im} c_{lj}^m \right). \end{aligned}$$

Multiplying both sides by g^{kl} and simplifying yields (5.12). Finally, under the hypotheses of (d), we have $g_{ij} = \delta_{ij}$, so (5.12) reduces to (5.13) after rearranging and using the fact that c_{ij}^k is antisymmetric in i, j. $\square$

On every Riemannian or pseudo-Riemannian manifold, we will always use the Levi-Civita connection from now on without further comment. Geodesics with respect to this connection are called **Riemannian** (or **pseudo-Riemannian**) **geodesics**, or simply "geodesics" as long as there is no risk of confusion. The connection coefficients Γ_{ij}^k of the Levi-Civita connection in coordinates, given by (5.10), are called the **Christoffel symbols** of g.

The next proposition shows that these connections are familiar ones in the case of embedded submanifolds of Euclidean or pseudo-Euclidean spaces.

Proposition 5.12.

(a) *The Levi-Civita connection on a (pseudo-)Euclidean space is equal to the Euclidean connection.*

(b) *Suppose M is an embedded (pseudo-)Riemannian submanifold of a (pseudo-)Euclidean space. Then the Levi-Civita connection on M is equal to the tangential connection $\nabla^\top$.*

Proof. We observed earlier in this chapter that the Euclidean connection is symmetric and compatible with both the Euclidean metric $\bar{g}$ and the pseudo-Euclidean metrics $\bar{q}^{(r,s)}$, which implies (a). Part (b) then follows from Propositions 5.8 and 5.9. $\square$

An important consequence of the definition is that because Levi-Civita connections are defined in coordinate-independent terms, they behave well with respect to isometries. Recall the definition of the *pullback of a connection* (see Lemma 4.37).

Proposition 5.13 (Naturality of the Levi-Civita Connection. *Suppose* (M,g) *and* $(\widetilde{M},\widetilde{g})$ *are Riemannian or pseudo-Riemannian manifolds with or without boundary, and let* ∇ *denote the Levi-Civita connection of* g *and* $\widetilde{\nabla}$ *that of* $\widetilde{g}$. *If* $\varphi\colon M \to \widetilde{M}$ *is an isometry, then* $\varphi^*\widetilde{\nabla} = \nabla$.

Proof. By uniqueness of the Levi-Civita connection, it suffices to show that the pullback connection $\varphi^*\widetilde{\nabla}$ is symmetric and compatible with g. The fact that φ is an isometry means that for any $X, Y \in \mathfrak{X}(M)$ and $p \in M$,

$$\langle Y_p, Z_p \rangle = \langle d\varphi_p(Y_p), d\varphi_p(Z_p) \rangle = \langle (\varphi_* Y)_{\varphi(p)}, (\varphi_* Z)_{\varphi(p)} \rangle,$$

or in other words, $\langle Y, Z \rangle = \langle \varphi_* Y, \varphi_* Z \rangle \circ \varphi$. Therefore,

$$
\begin{aligned}
X \langle Y, Z \rangle &= X(\langle \varphi_* Y, \varphi_* Z \rangle \circ \varphi) \\
&= \big((\varphi_* X) \langle \varphi_* Y, \varphi_* Z \rangle \big) \circ \varphi \\
&= \big(\langle \widetilde{\nabla}_{\varphi_* X}(\varphi_* Y), \varphi_* Z \rangle + \langle \varphi_* Y, \widetilde{\nabla}_{\varphi_* X}(\varphi_* Z) \rangle \big) \circ \varphi \\
&= \langle (\varphi^{-1})_* \widetilde{\nabla}_{\varphi_* X}(\varphi_* Y), Z \rangle + \langle Y, (\varphi^{-1})_* \widetilde{\nabla}_{\varphi_* X}(\varphi_* Z) \rangle \\
&= \langle (\varphi^*\widetilde{\nabla})_X Y, Z \rangle + \langle Y, (\varphi^*\widetilde{\nabla})_X Z \rangle,
\end{aligned}
$$

which shows that the pullback connection is compatible with g. Symmetry is proved as follows:

$$
\begin{aligned}
(\varphi^*\widetilde{\nabla})_X Y - (\varphi^*\widetilde{\nabla})_Y X &= (\varphi^{-1})_* \big(\widetilde{\nabla}_{\varphi_* X}(\varphi_* Y) - \widetilde{\nabla}_{\varphi_* Y}(\varphi_* X) \big) \\
&= (\varphi^{-1})_* [\varphi_* X, \varphi_* Y] \\
&= [X, Y]. \qquad \qquad \square
\end{aligned}
$$

Corollary 5.14 (Naturality of Geodesics). *Suppose* (M,g) *and* $(\widetilde{M},\widetilde{g})$ *are Riemannian or pseudo-Riemannian manifolds with or without boundary, and* $\varphi\colon M \to \widetilde{M}$ *is a local isometry. If* γ *is a geodesic in* M, *then* $\varphi \circ \gamma$ *is a geodesic in* $\widetilde{M}$.

Proof. This is an immediate consequence of Proposition 4.38, together with the fact that being a geodesic is a local property. $\qquad \square$

Like every connection on the tangent bundle, the Levi-Civita connection induces connections on all tensor bundles.

Proposition 5.15. *Suppose* (M,g) *is a Riemannian or pseudo-Riemannian manifold. The connection induced on each tensor bundle by the Levi-Civita connection is compatible with the induced inner product on tensors, in the sense that* $X \langle F, G \rangle = \langle \nabla_X F, G \rangle + \langle F, \nabla_X G \rangle$ *for every vector field* X *and every pair of smooth tensor fields* $F, G \in \Gamma(T^{(k,l)}TM)$.

Proof. Since every tensor field can be written as a sum of tensor products of vector and/or covector fields, it suffices to consider the case in which $F = \alpha_1 \otimes \cdots \otimes \alpha_{k+l}$ and $G = \beta_1 \otimes \cdots \otimes \beta_{k+l}$, where α_i and β_i are covariant or contravariant 1-tensor fields, as appropriate. In this case, the formula follows from (2.15) by a routine computation. $\qquad\square$

Proposition 5.16. *Let (M, g) be an oriented Riemannian manifold. The Riemannian volume form of g is parallel with respect to the Levi-Civita connection.*

Proof. Let $p \in M$ and $v \in T_p M$ be arbitrary, and let $\gamma \colon (-\varepsilon, \varepsilon) \to M$ be a smooth curve satisfying $\gamma(0) = p$ and $\gamma'(0) = v$. Let $(E_1, \ldots, E_n)$ be a parallel oriented orthonormal frame along γ. Since $dV_g(E_1, \ldots, E_n) \equiv 1$ and $D_t E_i \equiv 0$ along γ, formula (4.12) shows that $\nabla_v(dV_g) = D_t(dV_g)|_{t=0} = 0$. $\qquad\square$

Proposition 5.17. *The musical isomorphisms commute with the total covariant derivative operator: if F is any smooth tensor field with a contravariant ith index position, and $\flat$ represents the operation of lowering the ith index, then*

$$\nabla\left(F^\flat\right) = (\nabla F)^\flat. \tag{5.14}$$

Similarly, if G has a covariant ith position and $\sharp$ denotes raising the ith index, then

$$\nabla\left(G^\sharp\right) = (\nabla G)^\sharp. \tag{5.15}$$

Proof. The discussion on page 27 shows that $F^\flat = \operatorname{tr}(F \otimes g)$, where the trace is taken on the ith and last indices of $F \otimes g$. Because g is parallel, for every vector field X we have $\nabla_X(F \otimes g) = (\nabla_X F) \otimes g$. Because ∇_X commutes with traces, therefore,

$$\nabla_X\left(F^\flat\right) = \nabla_X\left(\operatorname{tr}(F \otimes g)\right) = \operatorname{tr}\left((\nabla_X F) \otimes g\right) = (\nabla_X F)^\flat.$$

This shows that when X is inserted into the last index position on both sides of (5.14), the results are equal. Since X is arbitrary, this proves (5.14).

Because the sharp and flat operators are inverses of each other when applied to the same index position, (5.15) follows by substituting $F = G^\sharp$ into (5.14) and applying $\sharp$ to both sides. $\qquad\square$

The Exponential Map

Throughout this section, we let (M, g) be a Riemannian or pseudo-Riemannian n-manifold, endowed with its Levi-Civita connection. Corollary 4.28 showed that each initial point $p \in M$ and each initial velocity vector $v \in T_p M$ determine a unique maximal geodesic γ_v. To deepen our understanding of geodesics, we need to study their collective behavior, and in particular, to address the following question: How do geodesics change if we vary the initial point or the initial velocity? The dependence of geodesics on the initial data is encoded in a map from the tangent bundle into the manifold, called the *exponential map*, whose properties are fundamental to the further study of Riemannian geometry.

(It is worth noting that the existence of the exponential map and the basic properties expressed in Proposition 5.19 below hold for every connection in TM, not just for the Levi-Civita connection. For simplicity, we restrict attention here to the latter case, because that is all we need. We also restrict to manifolds without boundary, in order to avoid complications with geodesics running into a boundary.)

The next lemma shows that geodesics with proportional initial velocities are related in a simple way.

Lemma 5.18 (Rescaling Lemma). *For every $p \in M$, $v \in T_pM$, and $c, t \in \mathbb{R}$,*

$$\gamma_{cv}(t) = \gamma_v(ct), \tag{5.16}$$

whenever either side is defined.

Proof. If $c = 0$, then both sides of (5.16) are equal to p for all $t \in \mathbb{R}$, so we may assume that $c \neq 0$. It suffices to show that $\gamma_{cv}(t)$ exists and (5.16) holds whenever the right-hand side is defined. (The same argument with the substitutions $v = c'v'$, $t = c't'$, and $c = 1/c'$ then implies that the conclusion holds when only the left-hand side is known to be defined.)

Suppose the maximal domain of γ_v is the open interval $I \subseteq \mathbb{R}$. For simplicity, write $\gamma = \gamma_v$, and define a new curve $\tilde{\gamma}: c^{-1}I \to M$ by $\tilde{\gamma}(t) = \gamma(ct)$, where $c^{-1}I = \{c^{-1}t : t \in I\}$. We will show that $\tilde{\gamma}$ is a geodesic with initial point p and initial velocity cv; it then follows by uniqueness and maximality that it must be equal to γ_{cv}.

It is immediate from the definition that $\tilde{\gamma}(0) = \gamma(0) = p$. Choose any smooth local coordinates on M and write the coordinate representation of γ as $\gamma(t) = (\gamma^1(t), \ldots, \gamma^n(t))$; then the chain rule gives

$$\dot{\tilde{\gamma}}^i(t) = \frac{d}{dt}\gamma^i(ct)$$
$$= c\dot{\gamma}^i(ct).$$

In particular, it follows that $\tilde{\gamma}'(0) = c\gamma'(0) = cv$.

Now let D_t and $\tilde{D}_t$ denote the covariant differentiation operators along γ and $\tilde{\gamma}$, respectively. Using the chain rule again in coordinates yields

$$\tilde{D}_t\tilde{\gamma}'(t) = \left(\frac{d}{dt}\dot{\tilde{\gamma}}^k(t) + \Gamma_{ij}^k(\tilde{\gamma}(t))\dot{\tilde{\gamma}}^i(t)\dot{\tilde{\gamma}}^j(t)\right)\partial_k$$
$$= \left(c^2\ddot{\gamma}^k(ct) + c^2\Gamma_{ij}^k(\gamma(ct))\dot{\gamma}^i(ct)\dot{\gamma}^j(ct)\right)\partial_k$$
$$= c^2 D_t\gamma'(ct) = 0.$$

Thus $\tilde{\gamma}$ is a geodesic, so $\tilde{\gamma} = \gamma_{cv}$, as claimed. $\qquad\square$

The assignment $v \mapsto \gamma_v$ defines a map from TM to the set of geodesics in M. More importantly, by virtue of the rescaling lemma, it allows us to define a map from (a subset of) the tangent bundle to M itself, which sends each line through the origin in $T_p M$ to a geodesic.

Define a subset $\mathcal{E} \subseteq TM$, the **domain of the exponential map**, by

$$\mathcal{E} = \{v \in TM : \gamma_v \text{ is defined on an interval containing } [0,1]\},$$

and then define the **exponential map** $\exp \colon \mathcal{E} \to M$ by

$$\exp(v) = \gamma_v(1).$$

For each $p \in M$, the **restricted exponential map at p**, denoted by $\exp_p$, is the restriction of exp to the set $\mathcal{E}_p = \mathcal{E} \cap T_p M$.

The exponential map of a Riemannian manifold should not be confused with the exponential map of a Lie group. The two are closely related for bi-invariant metrics (see Problem 5-8), but in general they need not be. To avoid confusion, we always designate the exponential map of a Lie group G by $\exp^G$, and reserve the undecorated notation exp for the Riemannian exponential map.

The next proposition describes some essential features of the exponential map. Recall that a subset of a vector space V is said to be **star-shaped** with respect to a point $x \in S$ if for every $y \in S$, the line segment from x to y is contained in S.

Proposition 5.19 (Properties of the Exponential Map). *Let (M, g) be a Riemannian or pseudo-Riemannian manifold, and let* $\exp \colon \mathcal{E} \to M$ *be its exponential map.*

(a) *$\mathcal{E}$ is an open subset of TM containing the image of the zero section, and each set $\mathcal{E}_p \subseteq T_p M$ is star-shaped with respect to 0.*
(b) *For each $v \in TM$, the geodesic γ_v is given by*

$$\gamma_v(t) = \exp(tv) \tag{5.17}$$

for all t such that either side is defined.
(c) *The exponential map is smooth.*
(d) *For each point $p \in M$, the differential $d\left(\exp_p\right)_0 \colon T_0(T_p M) \cong T_p M \to T_p M$ is the identity map of $T_p M$, under the usual identification of $T_0(T_p M)$ with $T_p M$.*

Proof. Write $n = \dim M$. The rescaling lemma with $t = 1$ says precisely that $\exp(cv) = \gamma_{cv}(1) = \gamma_v(c)$ whenever either side is defined; this is (b). Moreover, if $v \in \mathcal{E}_p$, then by definition γ_v is defined at least on $[0, 1]$. Thus for $0 \le t \le 1$, the rescaling lemma says that

$$\exp_p(tv) = \gamma_{tv}(1) = \gamma_v(t)$$

is defined. This shows that $\mathcal{E}_p$ is star-shaped with respect to 0.

Next we will show that $\mathcal{E}$ is open and exp is smooth. To do so, we revisit the proof of the existence and uniqueness theorem for geodesics (Theorem 4.27) and reformulate it in a more invariant way. Let $\left(x^i\right)$ be any smooth local coordinates on an open set $U \subseteq M$, let $\pi: TM \to M$ be the projection, and let (x^i, v^i) denote the associated natural coordinates for $\pi^{-1}(U) \subseteq TM$ (see p. 384). In terms of these coordinates, formula (4.18) defines a smooth vector field G on $\pi^{-1}(U)$. The integral curves of G are the curves $\eta(t) = (x^1(t), \ldots, x^n(t), v^1(t), \ldots, v^n(t))$ that satisfy the system of ODEs given by (4.17), which is equivalent to the geodesic equation under the substitution $v^k = \dot{x}^k$, as we observed in the proof of Theorem 4.27. Stated somewhat more invariantly, every integral curve of G on $\pi^{-1}(U)$ projects to a geodesic under $\pi: TM \to M$ (which in these coordinates is just $\pi(x, v) = x$); conversely, every geodesic $\gamma(t) = (x^1(t), \ldots, x^n(t))$ in U lifts to an integral curve of G in $\pi^{-1}(U)$ by setting $v^i(t) = \dot{x}^i(t)$.

The importance of G stems from the fact that it actually defines a *global* vector field on the total space of TM, called the **geodesic vector field**. We could verify this by computing the transformation law for the components of G under a change of coordinates and showing that they take the same form in every coordinate chart; but fortunately there is a way to avoid that messy computation. The key observation, to be proved below, is that G acts on a function $f \in C^\infty(TM)$ by

$$Gf(p, v) = \frac{d}{dt}\bigg|_{t=0} f\big(\gamma_v(t), \gamma_v'(t)\big). \tag{5.18}$$

(Here and whenever convenient, we use the notations (p, v) and v interchangeably for an element $v \in T_p M$, depending on whether we wish to emphasize the point at which v is tangent.) Since this formula is independent of coordinates, it shows that the various definitions of G given by (4.18) in different coordinate systems agree.

To prove that G satisfies (5.18), we write the components of the geodesic $\gamma_v(t)$ as $x^i(t)$ and those of its velocity as $v^i(t) = \dot{x}^i(t)$. Using the chain rule and the geodesic equation in the form (4.17), we can write the right-hand side of (5.18) as

$$\left(\frac{\partial f}{\partial x^k}(x(t), v(t))\dot{x}^k(t) + \frac{\partial f}{\partial v^k}(x(t), v(t))\dot{v}^k(t) \right)\bigg|_{t=0}$$

$$= \frac{\partial f}{\partial x^k}(p, v)v^k - \frac{\partial f}{\partial v^k}(p, v)v^i v^j \Gamma_{ij}^k(p)$$

$$= Gf(p, v).$$

The fundamental theorem on flows (Thm. A.42) shows that there exist an open set $\mathcal{D} \subseteq \mathbb{R} \times TM$ containing $\{0\} \times TM$ and a smooth map $\theta: \mathcal{D} \to TM$, such that each curve $\theta^{(p,v)}(t) = \theta(t, (p, v))$ is the unique maximal integral curve of G starting at (p, v), defined on an open interval containing 0.

Now suppose $(p, v) \in \mathcal{E}$. This means that the geodesic γ_v is defined at least on the interval $[0, 1]$, and therefore so is the integral curve of G starting at $(p, v) \in TM$. Since $(1, (p, v)) \in \mathcal{D}$, there is a neighborhood of $(1, (p, v))$ in $\mathbb{R} \times TM$ on which the flow of G is defined (Fig. 5.2). In particular, this means that there is a neighborhood of (p, v) on which the flow exists for $t \in [0, 1]$, and therefore on which the exponential map is defined. This shows that $\mathcal{E}$ is open.

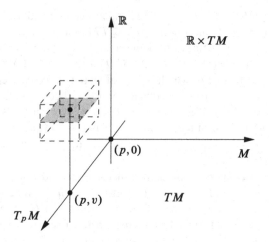

Fig. 5.2: $\mathcal{E}$ is open

Since geodesics are projections of integral curves of G, it follows that the exponential map can be expressed as

$$\exp_p(v) = \gamma_v(1) = \pi \circ \theta(1, (p, v))$$

wherever it is defined, and therefore $\exp_p(v)$ is a smooth function of (p, v).

To compute $d\left(\exp_p\right)_0(v)$ for an arbitrary vector $v \in T_pM$, we just need to choose a curve τ in T_pM starting at 0 whose initial velocity is v, and compute the initial velocity of $\exp_p \circ \tau$. A convenient curve is $\tau(t) = tv$, which yields

$$d\left(\exp_p\right)_0(v) = \left.\frac{d}{dt}\right|_{t=0} (\exp_p \circ \tau)(t) = \left.\frac{d}{dt}\right|_{t=0} \exp_p(tv) = \left.\frac{d}{dt}\right|_{t=0} \gamma_v(t) = v.$$

Thus $d\left(\exp_p\right)_0$ is the identity map.   □

Corollary 5.14 on the naturality of geodesics translates into the following important property of the exponential map.

Proposition 5.20 (Naturality of the Exponential Map). *Suppose (M, g) and $\left(\widetilde{M}, \widetilde{g}\right)$ are Riemannian or pseudo-Riemannian manifolds and $\varphi \colon M \to \widetilde{M}$ is a local isometry. Then for every $p \in M$, the following diagram commutes:*

$$
\begin{array}{ccc}
\mathcal{E}_p & \xrightarrow{\ d\varphi_p\ } & \widetilde{\mathcal{E}}_{\varphi(p)} \\
{\scriptstyle \exp_p} \downarrow & & \downarrow {\scriptstyle \exp_{\varphi(p)}} \\
M & \xrightarrow{\ \varphi\ } & \widetilde{M},
\end{array}
$$

where $\mathcal{E}_p \subseteq T_p M$ and $\widetilde{\mathcal{E}}_{\varphi(p)} \subseteq T_{\varphi(p)}\widetilde{M}$ are the domains of the restricted exponential maps $\exp_p$ (with respect to g) and $\exp_{\varphi(p)}$ (with respect to $\widetilde{g}$), respectively.

▶ **Exercise 5.21.** Prove Proposition 5.20.

An important consequence of the naturality of the exponential map is the following proposition, which says that local isometries of connected manifolds are completely determined by their values and differentials at a single point.

Proposition 5.22. Let (M, g) and $(\widetilde{M}, \widetilde{g})$ be Riemannian or pseudo-Riemannian manifolds, with M connected. Suppose $\varphi, \psi \colon M \to \widetilde{M}$ are local isometries such that for some point $p \in M$, we have $\varphi(p) = \psi(p)$ and $d\varphi_p = d\psi_p$. Then $\varphi \equiv \psi$.

Proof. Problem 5-10.   □

A Riemannian or pseudo-Riemannian manifold (M, g) is said to be **geodesically complete** if every maximal geodesic is defined for all $t \in \mathbb{R}$, or equivalently if the domain of the exponential map is all of TM. It is easy to construct examples of manifolds that are not geodesically complete; for example, in every proper open subset of $\mathbb{R}^n$ with its Euclidean metric or with a pseudo-Euclidean metric, there are geodesics that reach the boundary in finite time. Similarly, on $\mathbb{R}^n$ with the metric $(\sigma^{-1})^* \mathring{g}$ obtained from the sphere by stereographic projection, there are geodesics that escape to infinity in finite time. Geodesically complete manifolds are the natural setting for global questions in Riemannian or pseudo-Riemannian geometry; beginning with Chapter 6, most of our attention will be focused on them.

Normal Neighborhoods and Normal Coordinates

We continue to let (M, g) be a Riemannian or pseudo-Riemannian manifold of dimension n (without boundary). Recall that for every $p \in M$, the restricted exponential map $\exp_p$ maps the open subset $\mathcal{E}_p \subseteq T_p M$ smoothly into M. Because $d(\exp_p)_0$ is invertible, the inverse function theorem guarantees that there exist a neighborhood V of the origin in $T_p M$ and a neighborhood U of p in M such that $\exp_p \colon V \to U$ is a diffeomorphism. A neighborhood U of $p \in M$ that is the diffeomorphic image under $\exp_p$ of a *star-shaped* neighborhood of $0 \in T_p M$ is called a **normal neighborhood of** p.

Every orthonormal basis (b_i) for $T_p M$ determines a basis isomorphism $B \colon \mathbb{R}^n \to T_p M$ by $B(x^1, \ldots, x^n) = x^i b_i$. If $U = \exp_p(V)$ is a normal neighborhood of p, we can combine this isomorphism with the exponential map to get a smooth coordinate map $\varphi = B^{-1} \circ (\exp_p|_V)^{-1} \colon U \to \mathbb{R}^n$:

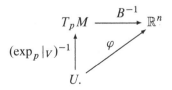

Such coordinates are called (*Riemannian or pseudo-Riemannian*) *normal coordinates centered at* p.

Proposition 5.23 (Uniqueness of Normal Coordinates). *Let* (M, g) *be a Riemannian or pseudo-Riemannian* n-*manifold,* p *a point of* M, *and* U *a normal neighborhood of* p. *For every normal coordinate chart on* U *centered at* p, *the coordinate basis is orthonormal at* p; *and for every orthonormal basis* (b_i) *for* $T_p M$, *there is a unique normal coordinate chart* (x^i) *on* U *such that* $\partial_i|_p = b_i$ *for* $i = 1, \ldots, n$. *In the Riemannian case, any two normal coordinate charts* (x^i) *and* $(\tilde{x}^j)$ *are related by*

$$\tilde{x}^j = A_i^j x^i \tag{5.19}$$

for some (constant) matrix $(A_i^j) \in O(n)$.

Proof. Let φ be a normal coordinate chart on U centered at p, with coordinate functions (x^i). By definition, this means that $\varphi = B^{-1} \circ \exp_p^{-1}$, where $B \colon \mathbb{R}^n \to T_p M$ is the basis isomorphism determined by some orthonormal basis (b_i) for $T_p M$. Note that $d\varphi_p^{-1} = d(\exp_p)_0 \circ dB_0 = B$ because $d(\exp_p)_0$ is the identity and B is linear. Thus $\partial_i|_p = d\varphi_p^{-1}(\partial_i|_0) = B(\partial_i|_0) = b_i$, which shows that the coordinate basis is orthonormal at p. Conversely, every orthonormal basis (b_i) for $T_p M$ yields a basis isomorphism B and thus a normal coordinate chart $\varphi = B^{-1} \circ \exp_p^{-1}$, which satisfies $\partial_i|_p = b_i$ by the computation above.

If $\tilde{\varphi} = \tilde{B}^{-1} \circ \exp_p^{-1}$ is another such chart, then

$$\tilde{\varphi} \circ \varphi^{-1} = \tilde{B}^{-1} \circ \exp_p^{-1} \circ \exp_p \circ B = \tilde{B}^{-1} \circ B,$$

which is a linear isometry of $\mathbb{R}^n$ and therefore has the form (5.19) in terms of standard coordinates on $\mathbb{R}^n$. Since $(\tilde{x}^j)$ and (x^i) are the same coordinates if and only if (A_i^j) is the identity matrix, this shows that the normal coordinate chart associated with a given orthonormal basis is unique.   □

Proposition 5.24 (Properties of Normal Coordinates). *Let* (M, g) *be a Riemannian or pseudo-Riemannian* n-*manifold, and let* $(U, (x^i))$ *be any normal coordinate chart centered at* $p \in M$.

(a) *The coordinates of* p *are* $(0, \ldots, 0)$.

(b) *The components of the metric at* p *are* $g_{ij} = \delta_{ij}$ *if* g *is Riemannian, and* $g_{ij} = \pm\delta_{ij}$ *otherwise.*

(c) *For every* $v = v^i \partial_i|_p \in T_p M$, *the geodesic* γ_v *starting at* p *with initial velocity* v *is represented in normal coordinates by the line*

$$\gamma_v(t) = (tv^1, \ldots, tv^n), \tag{5.20}$$

as long as t *is in some interval* I *containing* 0 *such that* $\gamma_v(I) \subseteq U$.

(d) *The Christoffel symbols in these coordinates vanish at* p.

(e) *All of the first partial derivatives of* g_{ij} *in these coordinates vanish at* p.

Proof. Part (a) follows directly from the definition of normal coordinates, and parts (b) and (c) follow from Propositions 5.23 and 5.19(b), respectively.

To prove (d), let $v = v^i \partial_i|_p \in T_p M$ be arbitrary. The geodesic equation (4.16) for $\gamma_v(t) = (tv^1, \ldots, tv^n)$ simplifies to

$$\Gamma_{ij}^k(tv) v^i v^j = 0.$$

Evaluating this expression at $t = 0$ shows that $\Gamma_{ij}^k(0) v^i v^j = 0$ for every index k and every vector v. In particular, with $v = \partial_a$ for some fixed a, this shows that $\Gamma_{aa}^k = 0$ for each a and k (no summation). Substituting $v = \partial_a + \partial_b$ and $v = \partial_b - \partial_a$ for any fixed pair of indices a and b and subtracting, we conclude also that $\Gamma_{ab}^k = 0$ at p for all a, b, k. Finally, (e) follows from (d) together with (5.2) in the case $E_k = \partial_k$. $\square$

Because they are given by the simple formula (5.20), the geodesics starting at p and lying in a normal neighborhood of p are called **radial geodesics**. (But be warned that geodesics that do not pass through p do not in general have a simple form in normal coordinates.)

Tubular Neighborhoods and Fermi Coordinates

The exponential map and normal coordinates give us a good understanding of the behavior of geodesics starting a point. In this section, we generalize those constructions to geodesics starting on any embedded submanifold. We restrict attention to the Riemannian case, because we will be using the Riemannian distance function.

Suppose (M, g) is a Riemannian manifold, $P \subseteq M$ is an embedded submanifold, and $\pi \colon NP \to P$ is the normal bundle of P in M. Let $\mathcal{E} \subseteq TM$ denote the domain of the exponential map of M, and let $\mathcal{E}_P = \mathcal{E} \cap NP$. Let $E \colon \mathcal{E}_P \to M$ denote the restriction of exp (the exponential map of M) to $\mathcal{E}_P$. We call E the **normal exponential map of P in M**.

A **normal neighborhood of P in M** is an open subset $U \subseteq M$ that is the diffeomorphic image under E of an open subset $V \subseteq \mathcal{E}_P$ whose intersection with each fiber $N_x P$ is star-shaped with respect to 0. We will be primarily interested in normal neighborhoods of the following type: a normal neighborhood of P in M is called a **tubular neighborhood** if it is the diffeomorphic image under E of a subset $V \subseteq \mathcal{E}_P$ of the form

$$V = \{(x, v) \in NP : |v|_g < \delta(x)\}, \tag{5.21}$$

for some positive continuous function $\delta \colon P \to \mathbb{R}$ (Fig. 5.3). If U is the diffeomorphic image of such a set V for a constant function $\delta(x) \equiv \varepsilon$, then it is called a **uniform tubular neighborhood** of radius ε, or an **ε-tubular neighborhood**.

Theorem 5.25 (Tubular Neighborhood Theorem). *Let (M, g) be a Riemannian manifold. Every embedded submanifold of M has a tubular neighborhood in M, and every compact submanifold has a uniform tubular neighborhood.*

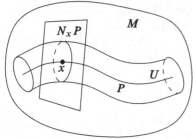
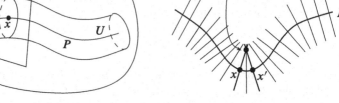

Fig. 5.3: A tubular neighborhood   Fig. 5.4: Injectivity of E

Proof. Let $P \subseteq M$ be an embedded submanifold, and let $P_0 \subseteq NP$ be the subset $\{(x,0) : x \in P\}$ (the image of the zero section of NP). We begin by showing that the normal exponential map E is a local diffeomorphism on a neighborhood of P_0. By the inverse function theorem, it suffices to show that the differential $dE_{(x,0)}$ is bijective at each point $(x,0) \in P_0$. The restriction of E to P_0 is just the diffeomorphism $P_0 \to P$ followed by the embedding $P \hookrightarrow M$, so $dE_{(x,0)}$ maps the subspace $T_{(x,0)}P_0 \subseteq T_{(x,0)}NP$ isomorphically onto $T_x P$. On the other hand, on the fiber $N_x P$, E agrees with the restricted exponential map $\exp_x$, which is a diffeomorphism near 0, so $dE_{(x,0)}$ maps $T_{(x,0)}(N_x P) \subseteq T_{(x,0)}NP$ isomorphically onto $N_x P$. Since $T_x M = T_x P \oplus N_x P$, this shows that $dE_{(x,0)}$ is surjective, and hence it is bijective for dimensional reasons. Thus E is a diffeomorphism on a neighborhood of $(x,0)$ in NP, which we can take to be of the form

$$V_\delta(x) = \left\{(x',v') \in NP : d_g(x,x') < \delta, \ |v'|_g < \delta\right\} \qquad (5.22)$$

for some $\delta > 0$. (Here we are using the fact that P is embedded in M, so it has the subspace topology.)

To complete the proof, we need to show that there is a set $V \subseteq \mathcal{E}_P$ of the form (5.21) on which E is a diffeomorphism onto its image. For each point $x \in P$, define

$$\Delta(x) = \sup\left\{\delta \le 1 : E \text{ is a diffeomorphism from } V_\delta(x) \text{ to its image}\right\}. \qquad (5.23)$$

The argument in the preceding paragraph implies that $\Delta(x)$ is positive for each x. Note that E is injective on the entire set $V_{\Delta(x)}(x)$, because any two points $(x_1, v_1), (x_2, v_2)$ in this set are in $V_\delta(x)$ for some $\delta < \Delta(x)$. Because it is an injective local diffeomorphism, E is actually a diffeomorphism from $V_{\Delta(x)}(x)$ onto its image.

Next we show that the function $\Delta: P \to \mathbb{R}$ is continuous. For any $x, x' \in P$, if $d_g(x,x') < \Delta(x)$, then the triangle inequality shows that $V_\delta(x')$ is contained in $V_{\Delta(x)}(x)$ for $\delta = \Delta(x) - d_g(x,x')$, which implies that $\Delta(x') \ge \Delta(x) - d_g(x,x')$, or

$$\Delta(x) - \Delta(x') \le d_g(x,x'). \qquad (5.24)$$

If $d_g(x, x') \geq \Delta(x)$, then (5.24) holds for trivial reasons. Reversing the roles of x and x' yields an analogous inequality, which shows that $|\Delta(x) - \Delta(x')| \leq d_g(x, x')$, so Δ is continuous.

Let $V = \{(x, v) \in NP : |v|_g < \frac{1}{2}\Delta(x)\}$, which is an open subset of NP containing P_0. We show that E is injective on V. Suppose (x, v) and (x', v') are points in V such that $E(x, v) = E(x', v')$ (Fig. 5.4). Assume without loss of generality that $\Delta(x') \leq \Delta(x)$. Because $\exp_x(v) = \exp_{x'}(v')$, there is an admissible curve from x to x' of length $|v|_g + |v'|_g$, and thus

$$d_g(x, x') \leq |v|_g + |v'|_g < \tfrac{1}{2}\Delta(x) + \tfrac{1}{2}\Delta(x') \leq \Delta(x).$$

Therefore, both (x, v) and (x', v') are in $V_{\Delta(x)}(x)$. Since E is injective on this set, this implies $(x, v) = (x', v')$.

The set $U = E(V)$ is open in M because $E|_V$ is a local diffeomorphism and thus an open map, and $E : V \to U$ is a diffeomorphism. Therefore, U is a tubular neighborhood of P.

Finally, if P is compact, then the continuous function $\frac{1}{2}\Delta$ achieves a minimum value $\varepsilon > 0$ on P, so U contains a uniform tubular neighborhood of radius ε. $\square$

Fermi Coordinates

Now we will construct coordinates on a tubular neighborhood that are analogous to Riemannian normal coordinates around a point. Let P be an embedded p-dimensional submanifold of a Riemannian n-manifold (M, g), and let $U \subseteq M$ be a normal neighborhood of P, with $U = E(V)$ for some appropriate open subset $V \subseteq NP$.

Let (W_0, ψ) be a smooth coordinate chart for P, and let $(E_1, \dots, E_{n-p})$ be a local orthonormal frame for the normal bundle NP; by shrinking W_0 if necessary, we can assume that the coordinates and the local frame are defined on the same open subset $W_0 \subseteq P$. Let $\widehat{W}_0 = \psi(W_0) \subseteq \mathbb{R}^p$, and let $NP|_{W_0}$ be the portion of the normal bundle over W_0. The coordinate map ψ and frame (E_j) yield a diffeomorphism $B : \widehat{W}_0 \times \mathbb{R}^{n-p} \to NP|_{W_0}$ defined by

$$B\left(x^1, \dots, x^p, v^1, \dots, v^{n-p}\right) = \left(q, \; v^1 E_1|_q + \cdots + v^{n-p} E_{n-p}|_q\right),$$

where $q = \psi^{-1}(x^1, \dots, x^p)$. Let $V_0 = V \cap NP|_{W_0} \subseteq NP$ and $U_0 = E(V_0) \subseteq M$, and define a smooth coordinate map $\varphi : U_0 \to \mathbb{R}^n$ by $\varphi = B^{-1} \circ \left(E|_{V_0}\right)^{-1}$:

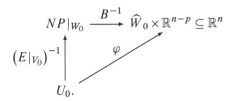

The coordinate map can also be written

$$\varphi: E\left(q, v^1 E_1|_q + \cdots + v^{n-p} E_{n-p}|_q\right) \mapsto \left(x^1(q), \ldots, x^p(q), v^1, \ldots, v^{n-p}\right). \quad (5.25)$$

Coordinates of this form are called **Fermi coordinates**, after the Italian physicist Enrico Fermi (1901–1954), who first introduced them in the special case in which P is the image of a geodesic in M. The generalization to arbitrary submanifolds was first introduced by Alfred Gray [Gra82]. (See also [Gra04] for a detailed study of the geometry of tubular neighborhoods.)

Here is the analogue of Proposition 5.24 for Fermi coordinates.

Proposition 5.26 (Properties of Fermi Coordinates). *Let P be an embedded p-dimensional submanifold of a Riemannian n-manifold (M, g), let U be a normal neighborhood of P in M, and let $(x^1, \ldots, x^p, v^1, \ldots, v^{n-p})$ be Fermi coordinates on an open subset $U_0 \subseteq U$. For convenience, we also write $x^{p+j} = v^j$ for $j = 1, \ldots, n - p$.*

(a) *$P \cap U_0$ is the set of points where $x^{p+1} = \cdots = x^n = 0$.*
(b) *At each point $q \in P \cap U_0$, the metric components satisfy the following:*

$$g_{ij} = g_{ji} = \begin{cases} 0, & 1 \leq i \leq p \text{ and } p+1 \leq j \leq n, \\ \delta_{ij}, & p+1 \leq i, j \leq n. \end{cases}$$

(c) *For every $q \in P \cap U_0$ and $v = v^1 E_1|_q + \cdots + v^{n-p} E_{n-p}|_q \in N_q P$, the geodesic γ_v starting at q with initial velocity v is the curve with coordinate expression $\gamma_v(t) = \left(x^1(q), \ldots, x^p(q), t v^1, \ldots, t v^{n-p}\right)$.*
(d) *At each $q \in P \cap U_0$, the Christoffel symbols in these coordinates satisfy $\Gamma^k_{ij} = 0$, provided $p + 1 \leq i, j \leq n$.*
(e) *At each $q \in P \cap U_0$, the partial derivatives $\partial_i g_{jk}(q)$ vanish for $p + 1 \leq i, j, k \leq n$.*

Proof. Problem 5-18. $\qquad\qquad\qquad\qquad\qquad\qquad\qquad\qquad\qquad\qquad\qquad\qquad\qquad$ □

Geodesics of the Model Spaces

In this section we determine the geodesics of the three types of frame-homogeneous Riemannian manifolds defined in Chapter 3. We could, of course, compute the Christoffel symbols of these metrics in suitable coordinates, and try to find the geodesics by solving the appropriate differential equations; but for these spaces, much easier methods are available based on symmetry and other geometric considerations.

Euclidean Space

On $\mathbb{R}^n$ with the Euclidean metric, Proposition 5.12 shows that the Levi-Civita connection is the Euclidean connection. Therefore, as one would expect, constant-coefficient vector fields are parallel, and the Euclidean geodesics are straight lines with constant-speed parametrizations (Exercises 4.29 and 4.30). Every Euclidean space is geodesically complete.

Spheres

Because the round metric on the sphere $\mathbb{S}^n(R)$ is induced by the Euclidean metric on $\mathbb{R}^{n+1}$, it is easy to determine the geodesics on a sphere using Corollary 5.2. Define a *great circle* on $\mathbb{S}^n(R)$ to be any subset of the form $\mathbb{S}^n(R) \cap \Pi$, where $\Pi \subseteq \mathbb{R}^{n+1}$ is a 2-dimensional linear subspace.

Proposition 5.27. *A nonconstant curve on $\mathbb{S}^n(R)$ is a maximal geodesic if and only if it is a periodic constant-speed curve whose image is a great circle. Thus every sphere is geodesically complete.*

Proof. Let $p \in \mathbb{S}^n(R)$ be arbitrary. Because $f(x) = |x|^2$ is a defining function for $\mathbb{S}^n(R)$, a vector $v \in T_p\mathbb{R}^{n+1}$ is tangent to $\mathbb{S}^n(R)$ if and only if $df_p(v) = 2\langle v, p \rangle = 0$, where we think of p as a vector by means of the usual identification of $\mathbb{R}^{n+1}$ with $T_p\mathbb{R}^{n+1}$. Thus $T_p\mathbb{S}^n(R)$ is exactly the set of vectors orthogonal to p.

Suppose v is an arbitrary nonzero vector in $T_p\mathbb{S}^n(R)$. Let $a = |v|/R$ and $\hat{v} = v/a$ (so $|\hat{v}| = R$), and consider the smooth curve $\gamma \colon \mathbb{R} \to \mathbb{R}^{n+1}$ given by

$$\gamma(t) = (\cos at)p + (\sin at)\hat{v}.$$

By direct computation, $|\gamma(t)|^2 = R^2$, so $\gamma(t) \in \mathbb{S}^n(R)$ for all t. Moreover,

$$\gamma'(t) = -a(\sin at)p + a(\cos at)\hat{v},$$
$$\gamma''(t) = -a^2(\cos at)p - a^2(\sin at)\hat{v}.$$

Because $\gamma''(t)$ is proportional to $\gamma(t)$ (thinking of both as vectors in $\mathbb{R}^{n+1}$), it follows that $\gamma''(t)$ is $\bar{g}$-orthogonal to $T_{\gamma(t)}\mathbb{S}^n(R)$, so γ is a geodesic in $\mathbb{S}^n(R)$ by Corollary 5.2. Since $\gamma(0) = p$ and $\gamma'(0) = a\hat{v} = v$, it follows that $\gamma = \gamma_v$.

Each γ_v is periodic of period $2\pi/a$, and has constant speed by Corollary 5.6 (or by direct computation). The image of γ_v is the great circle formed by the intersection of $\mathbb{S}^n(R)$ with the linear subspace spanned by $\{p, \hat{v}\}$, as you can check.

Conversely, suppose C is a great circle formed by intersecting $\mathbb{S}^n(R)$ with a 2-dimensional subspace Π, and let $\{v, w\}$ be an orthonormal basis for Π. Then C is the image of the geodesic with initial point $p = Rw$ and initial velocity v. $\square$

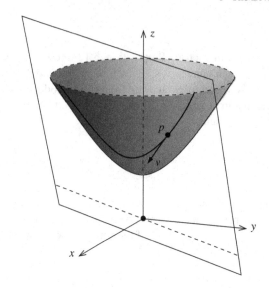

Fig. 5.5: A great hyperbola

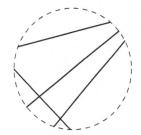

Fig. 5.6: Geodesics of $\mathbb{K}^n(R)$ Fig. 5.7: Geodesics of $\mathbb{B}^n(R)$

Hyperbolic Spaces

The geodesics of hyperbolic spaces can be determined by an analogous procedure using the hyperboloid model.

Proposition 5.28. *A nonconstant curve in a hyperbolic space is a maximal geodesic if and only if it is a constant-speed embedding of* $\mathbb{R}$ *whose image is one of the following:*

(a) HYPERBOLOID MODEL: *The intersection of* $\mathbb{H}^n(R)$ *with a 2-dimensional lin-ear subspace of* $\mathbb{R}^{n,1}$, *called a **great hyperbola** (Fig. 5.5).*

(b) BELTRAMI–KLEIN MODEL: *The interior of a line segment whose endpoints both lie on* $\partial\mathbb{K}^n(R)$ *(Fig. 5.6).*

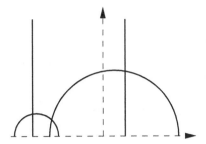

Fig. 5.8: Geodesics of $\mathbb{U}^n(R)$

(c) BALL MODEL: *The interior of a diameter of* $\mathbb{B}^n(R)$, *or the intersection of* $\mathbb{B}^n(R)$ *with a Euclidean circle that intersects* $\partial\mathbb{B}^n(R)$ *orthogonally (Fig. 5.7).*

(d) HALF-SPACE MODEL: *The intersection of* $\mathbb{U}^n(R)$ *with one of the following: a line parallel to the* y*-axis or a Euclidean circle with center on* $\partial\mathbb{U}^n(R)$ *(Fig. 5.8).*

Every hyperbolic space is geodesically complete.

Proof. We begin with the hyperboloid model, for which the proof is formally quite similar to what we just did for the sphere. Since the Riemannian connection on $\mathbb{H}^n(R)$ is equal to the tangential connection by Proposition 5.12, it follows from Corollary 5.2 that a smooth curve $\gamma\colon I \to \mathbb{H}^n(R)$ is a geodesic if and only if its acceleration $\gamma''(t)$ is everywhere $\bar{q}$-orthogonal to $T_{\gamma(t)}\mathbb{H}^n(R)$ (where $\bar{q} = \bar{q}^{(n,1)}$ is the Minkowski metric).

Let $p \in \mathbb{H}^n(R)$ be arbitrary. Note that $f(x) = \bar{q}(x,x)$ is a defining function for $\mathbb{H}^n(R)$, and (3.10) shows that the gradient of f at p is equal to $2p$ (where we regard p as a vector in $T_p\mathbb{R}^{n,1}$ as before). It follows that a vector $v \in T_p\mathbb{R}^{n,1}$ is tangent to $\mathbb{H}^n(R)$ if and only if $\bar{q}(p,v) = 0$. Let $v \in T_p\mathbb{H}^n(R)$ be an arbitrary nonzero vector. Put $a = |v|_{\bar{q}}/R = \bar{q}(v,v)^{1/2}/R$ and $\hat{v} = v/a$, and define $\gamma\colon \mathbb{R} \to \mathbb{R}^{n,1}$ by

$$\gamma(t) = (\cosh at)p + (\sinh at)\hat{v}.$$

Direct computation shows that γ takes its values in $\mathbb{H}^n(R)$ and that its acceleration vector is everywhere proportional to $\gamma(t)$. Thus $\gamma''(t)$ is $\bar{q}$-orthogonal to $T_{\gamma(t)}\mathbb{H}^n(R)$, so γ is a geodesic in $\mathbb{H}^n(R)$ and therefore has constant speed. Because it satisfies the initial conditions $\gamma(0) = p$ and $\gamma'(0) = v$, it is equal to γ_v. Note that γ_v is a smooth embedding of $\mathbb{R}$ into $\mathbb{H}^n(R)$ whose image is the great hyperbola formed by the intersection between $\mathbb{H}^n(R)$ and the plane spanned by $\{p, \hat{v}\}$.

Conversely, suppose Π is any 2-dimensional linear subspace of $\mathbb{R}^{n,1}$ that has nontrivial intersection with $\mathbb{H}^n(R)$. Choose $p \in \Pi \cap \mathbb{H}^n(R)$, and let v be another nonzero vector in Π that is $\bar{q}$-orthogonal to p, which implies $v \in T_p\mathbb{H}^n(R)$. Using the computation above, we see that the image of the geodesic γ_v is the great hyperbola formed by the intersection of Π with $\mathbb{H}^n(R)$.

Before considering the other three models, note that since maximal geodesics in $\mathbb{H}^n(R)$ are constant-speed embeddings of $\mathbb{R}$, it follows from naturality that maximal geodesics in each of the other models are also constant-speed embeddings of $\mathbb{R}$. Thus each model is geodesically complete, and to determine the geodesics in the other models we need only determine their images.

Consider the Beltrami–Klein model. Recall the isometry $c \colon \mathbb{H}^n(R) \to \mathbb{K}^n(R)$ given by $c(\xi, \tau) = R\xi/\tau$ (see (3.11)). The image of a maximal geodesic in $\mathbb{H}^n(R)$ is a great hyperbola, which is the set of points $(\xi, \tau) \in \mathbb{H}^n(R)$ that solve a system of $n - 1$ independent linear equations. Simple algebra shows that (ξ, τ) satisfies a linear equation $\alpha_i \xi^i + \beta\tau = 0$ if and only if $w = c(\xi, \tau) = R\xi/\tau$ satisfies the affine equation $\alpha_i w^i = -\beta R$. Thus c maps each great hyperbola onto the intersection of $\mathbb{K}^n(R)$ with an affine subspace of $\mathbb{R}^n$, and since it is the image of a smooth curve, it must be the intersection of $\mathbb{K}^n(R)$ with a straight line.

Next consider the Poincaré ball model. First consider the 2-dimensional case, and recall the inverse hyperbolic stereographic projection $\pi^{-1} \colon \mathbb{B}^2(R) \to \mathbb{H}^2(R)$ constructed in Chapter 3:

$$\pi^{-1}(u) = (\xi, \tau) = \left(\frac{2R^2 u}{R^2 - |u|^2}, R\frac{R^2 + |u|^2}{R^2 - |u|^2} \right).$$

In this case, a great hyperbola is the set of points on $\mathbb{H}^2(R)$ that satisfy a single linear equation $\alpha_i \xi^i + \beta\tau = 0$. In the special case $\beta = 0$, this hyperbola is mapped by π to a straight line segment through the origin, as can easily be seen from the geometric definition of π. If $\beta \neq 0$, we can assume (after multiplying through by a constant if necessary) that $\beta = -1$, and write the linear equation as $\tau = \alpha_i \xi^i = \alpha \cdot \xi$ (where the dot represents the Euclidean dot product between elements of $\mathbb{R}^2$). Under π^{-1}, this pulls back to the equation

$$R\frac{R^2 + |u|^2}{R^2 - |u|^2} = \frac{2R^2 \alpha \cdot u}{R^2 - |u|^2}$$

on the disk, which simplifies to

$$|u|^2 - 2R\alpha \cdot u + R^2 = 0.$$

Completing the square, we can write this as

$$|u - R\alpha|^2 = R^2(|\alpha|^2 - 1). \tag{5.26}$$

If $|\alpha|^2 \leq 1$, this locus is either empty or a point on $\partial\mathbb{B}^2(R)$, so it contains no points in $\mathbb{B}^2(R)$. Since we are assuming that it is the image of a maximal geodesic, we must therefore have $|\alpha|^2 > 1$. In that case, (5.26) is the equation of a circle with center $R\alpha$ and radius $R\sqrt{|\alpha|^2 - 1}$. At a point u_0 where the circle intersects $\partial\mathbb{B}^2(R)$, the three points 0, u_0, and $R\alpha$ form a triangle with sides $|u_0| = R$, $|R\alpha|$, and $|u_0 - R\alpha|$ (Fig. 5.9), which satisfy the Pythagorean identity by (5.26); therefore the circle meets $\partial\mathbb{B}^2(R)$ in a right angle.

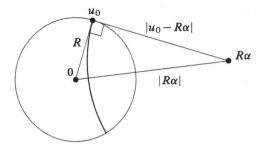

Fig. 5.9: Geodesics are arcs of circles orthogonal to the boundary of $\mathbb{H}^2(R)$

In the higher-dimensional case, a geodesic on $\mathbb{H}^n(R)$ is determined by a 2-plane. If the 2-plane contains the point $(0,\ldots,0,R)$, then the corresponding geodesic on $\mathbb{B}^n(R)$ is a line through the origin as before. Otherwise, we can use an orthogonal transformation in the $(\xi^1,\ldots,\xi^n)$ variables (which preserves $\breve{g}_R$) to move this 2-plane so that it lies in the (ξ^1,ξ^2,τ) subspace, and then we are in the same situation as in the 2-dimensional case.

Finally, consider the upper half-space model. The 2-dimensional case is easiest to analyze using complex notation. Recall the complex formula for the Cayley transform $\kappa\colon \mathbb{U}^2(R) \to \mathbb{B}^2(R)$ given in Chapter 3:

$$\kappa(z) = w = iR\frac{z-iR}{z+iR}.$$

Substituting this into equation (5.26) and writing $w = u+iv$ and $\alpha = a+ib$ in place of $u = (u^1,u^2)$, $\alpha = (\alpha^1,\alpha^2)$, we get

$$R^2\frac{|z-iR|^2}{|z+iR|^2} - iR^2\bar{\alpha}\frac{z-iR}{z+iR} + iR^2\alpha\frac{\bar{z}+iR}{\bar{z}-iR} + R^2|\alpha|^2 = R^2(|\alpha|^2 - 1).$$

Multiplying through by $(z+iR)(\bar{z}-iR)/2R^2$ and simplifying yields

$$(1-b)|z|^2 - 2aRx + (b+1)R^2 = 0.$$

This is the equation of a circle with center on the x-axis, unless $b=1$, in which case the condition $|\alpha|^2 > 1$ forces $a \neq 0$, and then it is a straight line $x = $ constant. The other class of geodesics on the ball, line segments through the origin, can be handled similarly.

In the higher-dimensional case, suppose first that $\gamma\colon \mathbb{R} \to \mathbb{U}^n(R)$ is a maximal geodesic such that $\gamma(0)$ lies on the y-axis and $\gamma'(0)$ is in the span of $\{\partial/\partial x^1, \partial/\partial y\}$. From the explicit formula (3.15) for κ, it follows that $\kappa \circ \gamma(0)$ lies on the v-axis in the ball, and $(\kappa \circ \gamma)'(0)$ is in the span of $\{\partial/\partial u^1, \partial/\partial v\}$. The image of the geodesic $\kappa \circ \gamma$ is either part of a line through the origin or an arc of a circle perpendicular to $\partial\mathbb{B}^n(R)$, both of which are contained in the (u^1,v)-plane. By the argument in the preceding paragraph, it then follows that the image of γ is contained in the (x^1, y)-plane and is either a vertical half-line or a semicircle centered on the $y = 0$ hyperplane. For the general case, note that translations and orthogonal transformations in the x-variables preserve vertical half-lines and circles centered on the $y = 0$ hyperplane in $\mathbb{U}^n(R)$, and they also preserve the metric $\breve{g}_R^3$. Given an arbitrary maximal geodesic $\gamma \colon \mathbb{R} \to \mathbb{U}^n(R)$, after applying an x-translation we may assume that $\gamma(0)$ lies on the y-axis, and after an orthogonal transformation in the x variables, we may assume that $\gamma'(0)$ is in the span of $\{\partial/\partial x^1, \partial/\partial y\}$; then the argument above shows that the image of γ is either a vertical half-line or a semicircle centered on the $y = 0$ hyperplane. $\square$

Euclidean and Non-Euclidean Geometries

In two dimensions, our model spaces can be interpreted as models of classical Euclidean and non-Euclidean plane geometries.

Euclidean Plane Geometry

Euclid's axioms for plane and spatial geometry, written around 300 BCE, became a model for axiomatic treatments of geometry, and indeed for all of mathematics. As standards of rigor evolved, mathematicians revised and added to Euclid's axioms in various ways. One axiom system that meets modern standards of rigor was created by David Hilbert [Hil71]. Here (in somewhat simplified form) are his axioms for plane geometry. (See [Hil71, Gan73, Gre93] for more complete treatments of Hilbert's axioms, and see [LeeAG] for a different axiomatic approach based on the real number system.)

Hilbert's Axioms For Euclidean Plane Geometry. The terms *point*, *line*, *lies on*, *between*, and *congruent* are primitive terms, and are thus left undefined. We make the following definitions:

- Given a line l and a point P, we say that l *contains* P if P lies on l.
- A set of points is said to be *collinear* if there is a line that contains them all.
- Given two distinct points A, B, the *segment* $\overline{AB}$ is the set consisting of A, B, and all points C such that C is between A and B.
- The notation $\overline{AB} \cong \overline{A'B'}$ means that $\overline{AB}$ is congruent to $\overline{A'B'}$.
- Given two distinct points A, B, the *ray* $\overrightarrow{AB}$ is the set consisting of A, B, and all points C such that either C is between A and B or B is between A and C.
- An *interior point* of the ray $\overrightarrow{AB}$ is a point that lies on $\overrightarrow{AB}$ and is not equal to A.
- Given three noncollinear points A, O, B, the *angle* $\angle AOB$ is the union of the rays $\overrightarrow{OA}$ and $\overrightarrow{OB}$.

- The notation $\angle ABC \cong \angle A'B'C'$ means that $\angle ABC$ is congruent to $\angle A'B'C'$.
- Given a line l and two points A, B that do not lie on l, we say that **A and B are on the same side of** l if no point of $\overline{AB}$ lies on l.
- Two lines are said to be **parallel** if there is no point that lies on both of them.

These terms are assumed to satisfy the following postulates, among others:

- **Incidence Postulates:**
  - (a) *For any two distinct points A, B, there exists a unique line that contains both of them.*
  - (b) *There exist at least two points on each line, and there exist at least three noncollinear points.*
- **Order Postulates:**
  - (a) *If a point B lies between a point A and a point C, then A, B, C are three distinct points of a line, and B also lies between C and A.*
  - (b) *Given two distinct points A and C, there always exists at least one point B such that C lies between A and B.*
  - (c) *Given three distinct points on a line, no more than one of them lies between the other two.*
  - (d) *Let A, B, C be three noncollinear points, and let l be a line that does not contain any of them. If l contains a point of $\overline{AB}$, then it also contains a point of $\overline{AC}$ or $\overline{BC}$.*
- **Congruence Postulates:**
  - (a) *If A, B are two points on a line l, and A' is a point on a line l', then it is always possible to find a point B' on a given ray of l' starting at A' such that $\overline{AB} \cong \overline{A'B'}$.*
  - (b) *If segments $\overline{A'B'}$ and $\overline{A''B''}$ are congruent to the same segment $\overline{AB}$, then $\overline{A'B'}$ and $\overline{A''B''}$ are congruent to each other.*
  - (c) *On a line l, let $\overline{AB}$ and $\overline{BC}$ be two segments that, except for B, have no points in common. Furthermore, on the same or another line l', let $\overline{A'B'}$ and $\overline{B'C'}$ be two segments that, except for B', have no points in common. In that case, if $\overline{AB} \cong \overline{A'B'}$ and $\overline{BC} \cong \overline{B'C'}$, then $\overline{AC} \cong \overline{A'C'}$.*
  - (d) *Let $\angle rs$ be an angle and l' a line, and let a definite side of l' be given. Let $\overrightarrow{r'}$ be a ray on l' starting at a point O'. Then there exists one and only one ray $\overrightarrow{s'}$ such that $\angle r's' \cong \angle rs$ and at the same time all the interior points of $\overrightarrow{s'}$ lie on the given side of l'.*
  - (e) *If for two triangles $\triangle ABC$ and $\triangle A'B'C'$ the congruences $\overline{AB} \cong \overline{A'B'}$, $\overline{AC} \cong \overline{A'C'}$, and $\angle BAC \cong \angle B'A'C'$ hold, then $\angle ABC \cong \angle A'B'C'$ and $\angle ACB \cong \angle A'C'B'$ as well.*
- **Euclidean Parallel Postulate:** *Given a line l and a point A that does not lie on l, there exists a unique line that contains A and is parallel to l.*

Given an axiomatic system such as this one, an **interpretation** of the system is simply an assignment of a definition for each of the primitive terms. An interpretation is called a **model** of the axiomatic system provided that each of the axioms becomes a theorem when the primitive terms are given the assigned meanings.

We are all familiar with the Cartesian plane as an interpretation of Euclidean plane geometry. Formally, in this interpretation, we make the following definitions:

- A **point** is an element of $\mathbb{R}^2$.
- A **line** is the image of a maximal geodesic with respect to the Euclidean metric.
- Given a point A and a line l, we say that A **lies on** l if $A \in l$.
- Given three distinct points A, B, C, we say that B **is between** A **and** C if B is on the geodesic segment joining A to C.
- Given two sets of points S and S', we say that S **is congruent to** S' if there is a Euclidean isometry $\varphi \colon \mathbb{R}^2 \to \mathbb{R}^2$ such that $\varphi(S) = S'$.

With this interpretation, it will come as no surprise that Hilbert's postulates are all theorems; proving them is just a standard exercise in plane analytic geometry.

▶ **Exercise 5.29.** Verify that all of Hilbert's axioms are theorems when the primitive terms are given the interpretations listed above.

More interesting is the application of Riemannian geometry to non-Euclidean geometry. Hilbert's axioms can be easily modified to yield axioms for plane hyperbolic geometry, simply by replacing the Euclidean parallel postulate by the following:

- **Hyperbolic Parallel Postulate:** *Given a line l and a point A that does not lie on l, there exist at least two distinct lines that contain A and are parallel to l.*

We obtain an interpretation of this new axiomatic system by giving definitions to the primitive terms just as we did above, but now with $\mathbb{R}^2$ replaced by $\mathbb{H}^2$ and $\bar{g}$ replaced by any hyperbolic metric $\breve{g}_R$. (The axioms we have listed here do not distinguish among hyperbolic metrics of different radii.) In Problem 5-19, you will be asked to prove that some of Hilbert's axioms are theorems under this interpretation.

In addition to hyperbolic geometry, it is possible to construct another version of non-Euclidean geometry, in which the Euclidean parallel postulate is replaced by the following assertion:

- **Elliptic Parallel Postulate:** *No two lines are parallel.*

Unfortunately, we cannot simply replace the Euclidean parallel postulate with this one and leave the other axioms alone, because it already follows from Hilbert's other axioms that for every line l and every point $A \notin l$ there exists at least one line through A that is parallel to l (for a proof, see [Gre93], for example). Nonetheless, we already know of an interesting geometry that satisfies the elliptic parallel postulate—the sphere $\mathbb{S}^2$ with the round metric $\mathring{g}$. To construct a consistent axiomatic system including the elliptic parallel postulate, some of the other axioms need to be modified.

If we take the sphere as a guide, with images of maximal geodesics as lines, then we can see already that the first incidence postulate needs to be abandoned, because if $A, B \in \mathbb{S}^2$ are **antipodal points** (meaning that $B = -A$), then there are infinitely many lines containing A and B. Any axiomatic system for which $(\mathbb{S}^2, \mathring{g})$ is a model is called **double elliptic geometry**, because every pair of distinct lines intersects in exactly two points.

It is also possible to construct an elliptic geometry in which the incidence postulates hold, as in the following example.

Example 5.30. The real projective plane $\mathbb{R}\mathbb{P}^2$ has a frame-homogeneous Riemannian metric g that is locally isometric to a round metric on $\mathbb{S}^n$ (see Example 2.34 and Problem 3-2). As Problem 5-20 shows, single elliptic geometry satisfies Hilbert's incidence postulates as well as the elliptic parallel postulate. This interpretation is called *single elliptic geometry*. //

Problems

5-1. Let (M, g) be a Riemannian or pseudo-Riemannian manifold, and let ∇ be its Levi-Civita connection. Suppose $\widetilde{\nabla}$ is another connection on TM, and D is the difference tensor between ∇ and $\widetilde{\nabla}$ (Prop. 4.13). Let $D^\flat$ denote the covariant 3-tensor field defined by $D^\flat(X, Y, Z) = \langle D(X, Y), Z \rangle$. Show that $\widetilde{\nabla}$ is compatible with g if and only if $D^\flat$ is antisymmetric in its last two arguments: $D^\flat(X, Y, Z) = -D^\flat(X, Z, Y)$ for all $X, Y, Z \in \mathfrak{X}(M)$. Conclude that on every Riemannian or pseudo-Riemannian manifold of dimension at least 2, the space of metric connections is an infinite-dimensional affine space. (*Used on p. 121.*)

5-2. Let ∇ be a connection on the tangent bundle of a Riemannian manifold (M, g). Show that ∇ is compatible with g if and only if the connection 1-forms $\omega_i{}^j$ (Problem 4-14) with respect to each local frame (E_i) satisfy

$$g_{jk}\omega_i{}^k + g_{ik}\omega_j{}^k = dg_{ij}.$$

Show that this implies that with respect to every local orthonormal frame, the matrix $(\omega_i{}^j)$ is skew-symmetric.

5-3. Define a connection on $\mathbb{R}^3$ by setting (in standard coordinates)

$$\Gamma_{12}^3 = \Gamma_{23}^1 = \Gamma_{31}^2 = 1,$$
$$\Gamma_{21}^3 = \Gamma_{32}^1 = \Gamma_{13}^2 = -1,$$

with all other connection coefficients equal to zero. Show that this connection is compatible with the Euclidean metric and has the same geodesics as the Euclidean connection, but is not symmetric. (See Problem 4-9.)

5-4. Let C be an embedded smooth curve in the half-plane $H = \{(r, z) : r > 0\}$, and let $S_C \subseteq \mathbb{R}^3$ be the surface of revolution determined by C as in Example 2.20. Let $\gamma(t) = (a(t), b(t))$ be a unit-speed local parametrization of C, and let X be the parametrization of S_C given by (2.11).

    (a) Compute the Christoffel symbols of the induced metric on S_C in (t, θ) coordinates.

(b) Show that each meridian is the image of a geodesic on S_C.

(c) Determine necessary and sufficient conditions for a latitude circle to be the image of a geodesic.

5-5. Recall that a vector field Y defined on (an open subset of) a Riemannian manifold is said to be *parallel* if $\nabla Y \equiv 0$.

(a) Let $p \in \mathbb{R}^n$ and $v \in T_p\mathbb{R}^n$. Show that there is a unique parallel vector field Y on $\mathbb{R}^n$ such that $Y_p = v$.

(b) Let $X(\varphi, \theta) = (\sin\varphi \cos\theta, \sin\varphi \sin\theta, \cos\varphi)$ be the spherical coordinate parametrization of an open subset U of the unit sphere $\mathbb{S}^2$ (see Example 2.20 and Problem 5-4), and let $X_\theta = X_*(\partial_\theta)$, $X_\varphi = X_*(\partial_\varphi)$ denote the coordinate vector fields associated with this parametrization. Compute $\nabla_{X_\theta}(X_\varphi)$ and $\nabla_{X_\varphi}(X_\varphi)$, and conclude that X_φ is parallel along the equator and along each meridian $\theta = \theta_0$.

(c) Let $p = (1,0,0) \in \mathbb{S}^2$. Show that there is no parallel vector field W on any neighborhood of p in $\mathbb{S}^2$ such that $W_p = X_\varphi|_p$.

(d) Use (a) and (c) to show that no neighborhood of p in $(\mathbb{S}^2, \mathring{g})$ is isometric to an open subset of $(\mathbb{R}^2, \bar{g})$.

(Used on p. 194.)

5-6. Suppose $\pi : (\widetilde{M}, \widetilde{g}) \to (M, g)$ is a Riemannian submersion. If Z is any vector field on M, we let $\widetilde{Z}$ denote its horizontal lift to $\widetilde{M}$ (see Prop. 2.25).

(a) Show that for every pair of vector fields $X, Y \in \mathfrak{X}(M)$, we have

$$\langle \widetilde{X}, \widetilde{Y} \rangle = \langle X, Y \rangle \circ \pi;$$
$$[\widetilde{X}, \widetilde{Y}]^H = \widetilde{[X,Y]};$$
$$[\widetilde{X}, W] \text{ is vertical if } W \in \mathfrak{X}(\widetilde{M}) \text{ is vertical.}$$

(b) Let $\widetilde{\nabla}$ and ∇ denote the Levi-Civita connections of $\widetilde{g}$ and g, respectively. Show that for every pair of vector fields $X, Y \in \mathfrak{X}(M)$, we have

$$\widetilde{\nabla}_{\widetilde{X}}\widetilde{Y} = \widetilde{\nabla_X Y} + \tfrac{1}{2}[\widetilde{X}, \widetilde{Y}]^V. \tag{5.27}$$

[Hint: Let $\widetilde{Z}$ be a horizontal lift and W a vertical vector field on $\widetilde{M}$, and compute $\langle \widetilde{\nabla}_{\widetilde{X}}\widetilde{Y}, \widetilde{Z} \rangle$ and $\langle \widetilde{\nabla}_{\widetilde{X}}\widetilde{Y}, W \rangle$ using (5.9).]

(Used on p. 224.)

5-7. Suppose (M_1, g_1) and (M_2, g_2) are Riemannian manifolds.

(a) Prove that if $M_1 \times M_2$ is endowed with the product metric, then a curve $\gamma : I \to M_1 \times M_2$ of the form $\gamma(t) = (\gamma_1(t), \gamma_2(t))$ is a geodesic if and only if γ_i is a geodesic in (M_i, g_i) for $i = 1, 2$.

(b) Now suppose $f : M_1 \to \mathbb{R}^+$ is a strictly positive smooth function, and $M_1 \times_f M_2$ is the resulting warped product (see Example 2.24). Let $\gamma_1 : I \to M_1$ be a smooth curve and q_0 a point in M_2, and define

$\gamma: I \to M_1 \times_f M_2$ by $\gamma(t) = (\gamma_1(t), q_0)$. Prove that γ is a geodesic with respect to the warped product metric if and only if γ_1 is a geodesic with respect to g_1.

(*Used on p. 316.*)

5-8. Let G be a Lie group and $\mathfrak{g}$ its Lie algebra. Suppose g is a bi-invariant Riemannian metric on G, and $\langle \cdot, \cdot \rangle$ is the corresponding inner product on $\mathfrak{g}$ (see Prop. 3.12). Let $\mathrm{ad}: \mathfrak{g} \to \mathfrak{gl}(\mathfrak{g})$ denote the adjoint representation of $\mathfrak{g}$ (see Appendix C).

(a) Show that $\mathrm{ad}(X)$ is a skew-adjoint endomorphism of $\mathfrak{g}$ for every $X \in \mathfrak{g}$:

$$\langle \mathrm{ad}(X)Y, Z \rangle = -\langle Y, \mathrm{ad}(X)Z \rangle.$$

[Hint: Take the derivative of $\langle \mathrm{Ad}(\exp^G tX)Y, \mathrm{Ad}(\exp^G tX)Z \rangle$ with respect to t at $t = 0$, where $\exp^G$ is the Lie group exponential map of G, and use the fact that $\mathrm{Ad}_* = \mathrm{ad}$.]

(b) Show that $\nabla_X Y = \frac{1}{2}[X, Y]$ whenever X and Y are left-invariant vector fields on G.

(c) Show that the geodesics of g starting at the identity are exactly the one-parameter subgroups. Conclude that under the canonical isomorphism of $\mathfrak{g} \cong T_e G$ described in Proposition C.3, the restricted Riemannian exponential map at the identity coincides with the Lie group exponential map $\exp^G: \mathfrak{g} \to G$. (See Prop. C.7.)

(d) Let $\mathbb{R}^+$ be the set of positive real numbers, regarded as a Lie group under multiplication. Show that $g = t^{-2} dt^2$ is a bi-invariant metric on $\mathbb{R}^+$, and the restricted Riemannian exponential map at 1 is given by $c\partial/\partial t \mapsto e^c$.

(*Used on pp. 128, 224.*)

5-9. Suppose (M, g) is a Riemannian manifold and (U, φ) is a smooth coordinate chart on a neighborhood of $p \in M$ such that $\varphi(p) = 0$ and $\varphi(U)$ is star-shaped with respect to 0. Prove that this chart is a normal coordinate chart for g if and only if $g_{ij}(p) = \delta_{ij}$ and the following identity is satisfied on U:

$$x^i x^j \Gamma_{ij}^k(x) \equiv 0.$$

5-10. Prove Proposition 5.22 (a local isometry is determined by its value and differential at one point).

5-11. Recall the groups $\mathrm{E}(n)$, $\mathrm{O}(n+1)$, and $\mathrm{O}^+(n, 1)$ defined in Chapter 3, which act isometrically on the model Riemannian manifolds $(\mathbb{R}^n, \bar{g})$, $(\mathbb{S}^n(R), \overset{\circ}{g}_R)$, and $(\mathbb{H}^n(R), \breve{g}_R)$, respectively.

(a) Show that

$$\mathrm{Iso}\left(\mathbb{R}^n,\bar{g}\right) = \mathrm{E}(n),$$
$$\mathrm{Iso}\left(\mathbb{S}^n(R),\mathring{g}_R\right) = \mathrm{O}(n+1),$$
$$\mathrm{Iso}\left(\mathbb{H}^n(R),\breve{g}_R\right) = \mathrm{O}^+(n,1).$$

(b) Show that in each case, for each point p in $\mathbb{R}^n$, $\mathbb{S}^n(R)$, or $\mathbb{H}^n(R)$, the isotropy group at p is a subgroup isomorphic to $\mathrm{O}(n)$.

(c) Strengthen the result above by showing that if (M,g) is one of the Riemannian manifolds $\left(\mathbb{R}^n,\bar{g}\right)$, $\left(\mathbb{S}^n(R),\mathring{g}_R\right)$, or $\left(\mathbb{H}^n(R),\breve{g}_R\right)$, U is a connected open subset of M, and $\varphi\colon U \to M$ is a local isometry, then φ is the restriction to U of an element of $\mathrm{Iso}(M,g)$.

(Used on pp. 57, 58, 67, 348, 349.)

5-12. Suppose M is a connected n-dimensional Riemannian manifold, and G is a Lie group acting isometrically and effectively on M. Show that $\dim G \leq n(n+1)/2$. *(Used on p. 261.)*

5-13. Let (M,g) be a Riemannian manifold.

(a) Show that the following formula holds for every smooth 1-form $\eta \in \Omega^1(M)$:
$$d\eta(X,Y) = (\nabla_X \eta)(Y) - (\nabla_Y \eta)(X).$$

(b) Generalize this to an arbitrary k-form $\eta \in \Omega^k(M)$ as follows:

$$d\eta = (-1)^k (k+1)\,\mathrm{Alt}(\nabla\eta),$$

where Alt denotes the alternation operator defined in (B.9). [Hint: For each $p \in M$, do the computation in normal coordinates centered at p, and note that both sides of the equation are well defined, independently of the choice of coordinates.]

(Used on p. 209.)

5-14. Suppose (M,g) is a Riemannian manifold, and let div and Δ be the divergence and Laplace operators defined on pages 32–33.

(a) Show that for every vector field $X \in \mathfrak{X}(M)$, $\mathrm{div}\,X$ can be written in terms of the total covariant derivative as $\mathrm{div}\,X = \mathrm{tr}(\nabla X)$, and that if $X = X^i E_i$ in terms of some local frame, then $\mathrm{div}\,X = X^i{}_{;i}$. [Hint: Show that it suffices to prove the formulas at the origin in normal coordinates.]

(b) Show that the Laplace operator acting on a smooth function u can be expressed as
$$\Delta u = \mathrm{tr}_g\left(\nabla^2 u\right), \tag{5.28}$$

and in terms of any local frame,

$$\Delta u = g^{ij} u_{;ij} = u_{;i}{}^i. \tag{5.29}$$

(Used on pp. 218, 256, 333.)

5-15. Suppose (M, g) is a compact Riemannian n-manifold (without boundary) and $u \in C^\infty(M)$ is an eigenfunction of M, meaning that $-\Delta u = \lambda u$ for some constant λ (see Prob. 2-24). Prove that

$$\lambda \int_M |\operatorname{grad} u|^2 \, dV_g \le n \int_M \left|\nabla^2 u\right|^2 \, dV_g.$$

[Hint: Consider the 2-tensor field $\nabla^2 u - \frac{1}{n}(\Delta u)g$, and use one of Green's identities (Prob. 2-23).] (*Used on p. 223.*)

5-16. By analogy with the formula $\operatorname{div} X = \operatorname{tr}(\nabla X)$ developed in Problem 5-14, we can define a divergence operator on tensor fields of any rank on a Riemannian manifold. If F is any smooth k-tensor field (covariant, contravariant, or mixed), we define the *divergence of F* by

$$\operatorname{div} F = \operatorname{tr}_g (\nabla F),$$

where the trace is taken on the last two indices of the $(k+1)$-tensor field ∇F. (If F is purely contravariant, then tr_g can be replaced with tr, because the next-to-last index of ∇F is already an upper index.) Extend the integration by parts formula of Problem 2-22 as follows: if F is a smooth covariant k-tensor field and G is a smooth covariant $(k+1)$-tensor field on a compact smooth Riemannian manifold (M, g) with boundary, then

$$\int_M \langle \nabla F, G \rangle \, dV_g = \int_{\partial M} \langle F \otimes N^\flat, G \rangle \, dV_{\hat{g}} - \int_M \langle F, \operatorname{div} G \rangle \, dV_g,$$

where $\hat{g}$ is the induced metric on ∂M. This is often written more suggestively as

$$\int_M F_{i_1 \dots i_k ; j} G^{i_1 \dots i_k j} \, dV_g$$
$$= \int_{\partial M} F_{i_1 \dots i_k} G^{i_1 \dots i_k j} N_j \, dV_{\hat{g}} - \int_M F_{i_1 \dots i_k} G^{i_1 \dots i_k j}{}_{; j} \, dV_g.$$

5-17. Suppose (M, g) is a Riemannian manifold and $P \subseteq M$ is an embedded submanifold. Show that P has a tubular neighborhood that is diffeomorphic to the total space of the normal bundle NP, by a diffeomorphism that sends the zero section of NP to P. [Hint: First show that the function δ in (5.21) can be chosen to be smooth.]

5-18. Prove Proposition 5.26 (properties of Fermi coordinates).

5-19. Use $\mathbb{H}^2$ with the metric $\breve{g}$ to construct an interpretation of Hilbert's axioms with the hyperbolic parallel postulate substituted for the Euclidean one, and prove that the incidence postulates, congruence postulate (e), and the hyperbolic parallel postulate are theorems in this geometry. (*Used on p. 144.*)

5-20. Show that single elliptic geometry (Example 5.30) satisfies Hilbert's incidence postulates and the elliptic parallel postulate if **points** are defined as elements of $\mathbb{RP}^2$ and **lines** are defined as images of maximal geodesics.

5-21. Let (M, g) be a Riemannian or pseudo-Riemannian manifold and $p \in M$. Show that for every orthonormal basis $(b_1, \ldots, b_n)$ for $T_p M$, there is a smooth orthonormal frame (E_i) on a neighborhood of p such that $E_i|_p = b_i$ and $(\nabla E_i)_p = 0$ for each i.

5-22. A smooth vector field X on a Riemannian manifold is called a **Killing vector field** if the Lie derivative of the metric with respect to X vanishes. By Proposition B.10, this is equivalent to the requirement that the metric be invariant under the flow of X. Prove that X is a Killing vector field if and only if the covariant 2-tensor field $(\nabla X)^\flat$ is antisymmetric. [Hint: Use Prop. B.9.] (*Used on pp. 190, 315.*)

5-23. Let (M, g) be a connected Riemannian manifold and $p \in M$. An **admissible loop based at p** is an admissible curve $\gamma: [a, b] \to M$ such that $\gamma(a) = \gamma(b) = p$. For each such loop γ, let P^γ denote the parallel transport operator $P^\gamma_{ab}: T_p M \to T_p M$ along γ, and let $\mathrm{Hol}(p) \subseteq \mathrm{GL}\left(T_p M\right)$ denote the set of all automorphisms of $T_p M$ obtained in this way:

$$\mathrm{Hol}(p) = \left\{ P^\gamma : \gamma \text{ is an admissible loop based at } p \right\}.$$

(a) Show that $\mathrm{Hol}(p)$ is a subgroup of $\mathrm{O}\left(T_p M\right)$ (the set of all linear isometries of $T_p M$), called the **holonomy group at p**.

(b) Let $\mathrm{Hol}^0(p) \subseteq \mathrm{Hol}(p)$ denote the subset obtained by restricting to loops γ that are path-homotopic to the constant loop. Show that $\mathrm{Hol}^0(p)$ is a normal subgroup of $\mathrm{Hol}(p)$, called the **restricted holonomy group at p**.

(c) Given $p, q \in M$, show that there is an isomorphism of $\mathrm{GL}\left(T_p M\right)$ with $\mathrm{GL}\left(T_q M\right)$ that takes $\mathrm{Hol}(p)$ to $\mathrm{Hol}(q)$.

(d) Show that M is orientable if and only if $\mathrm{Hol}(p) \subseteq \mathrm{SO}\left(T_p M\right)$ (the set of linear isometries with determinant $+1$) for some $p \in M$.

(e) Show that g is flat if and only if $\mathrm{Hol}^0(p)$ is the trivial group for some $p \in M$.

Chapter 6
Geodesics and Distance

In this chapter, we explore the relationships among geodesics, lengths, and distances on a Riemannian manifold. A primary goal is to show that all length-minimizing curves are geodesics, and that all geodesics are locally length-minimizing. Later, we prove the *Hopf–Rinow theorem*, which states that a connected Riemannian manifold is geodesically complete if and only if it is complete as a metric space. At the end of the chapter, we study *distance functions* (which express the distance to a point or other subset) and show that they can be used to construct coordinates, called *semigeodesic coordinates*, that put a metric in a particularly simple form.

Most of the results of this chapter do not apply to general pseudo-Riemannian metrics, at least not without substantial modification. For this reason, we restrict our focus here to the Riemannian case. (For a treatment of lengths of curves in the pseudo-Riemannian setting, see [O'N83].) Also, the theory of minimizing curves becomes considerably more complicated in the presence of a nonempty boundary; thus, unless otherwise stated, throughout this chapter we assume that (M, g) is a Riemannian manifold without boundary. And because we will be using the Riemannian distance function, we assume for most results that M is connected.

Geodesics and Minimizing Curves

Let (M, g) be a Riemannian manifold. An admissible curve γ in M is said to be a ***minimizing curve*** if $L_g(\gamma) \le L_g(\tilde{\gamma})$ for every admissible curve $\tilde{\gamma}$ with the same endpoints. When M is connected, it follows from the definition of the Riemannian distance that γ is minimizing if and only if $L_g(\gamma)$ is equal to the distance between its endpoints.

Our first goal in this section is to show that all minimizing curves are geodesics. To do so, we will think of the length function L_g as a functional on the set of all admissible curves in M with fixed starting and ending points. (Real-valued functions whose domains are themselves sets of functions are typically called ***functionals***.) Our project is to search for minima of this functional.

© Springer International Publishing AG 2018

J. M. Lee, *Introduction to Riemannian Manifolds*, Graduate Texts in Mathematics 176, https://doi.org/10.1007/978-3-319-91755-9_6

From calculus, we might expect that a necessary condition for a curve γ to be minimizing would be that the "derivative" of L_g vanish at γ, in some sense. This brings us to the brink of the subject known as the *calculus of variations*: the use of calculus to identify and analyze extrema of functionals defined on spaces of functions or maps. In its fully developed state, the calculus of variations allows one to apply all the usual tools of multivariable calculus in the infinite-dimensional setting of function spaces—tools such as directional derivatives, gradients, critical points, local extrema, saddle points, and Hessians. For our purposes, however, we do not need to formalize the theory of calculus in the infinite-dimensional setting. It suffices to note that if γ is a minimizing curve, and $\{\Gamma_s : s \in (-\varepsilon, \varepsilon)\}$ is a one-parameter family of admissible curves with the same endpoints such that $L_g(\Gamma_s)$ is a differentiable function of s and $\Gamma_0 = \gamma$, then by elementary calculus, the s-derivative of $L_g(\Gamma_s)$ must vanish at $s = 0$ because $L_g(\Gamma_s)$ attains a minimum there.

Families of Curves

To make this rigorous, we introduce some more definitions. Let (M, g) be a Riemannian manifold.

Given intervals $I, J \subseteq \mathbb{R}$, a continuous map $\Gamma : J \times I \to M$ is called a **one-parameter family of curves**. Such a family defines two collections of curves in M: the **main curves** $\Gamma_s(t) = \Gamma(s, t)$ defined for $t \in I$ by holding s constant, and the **transverse curves** $\Gamma^{(t)}(s) = \Gamma(s, t)$ defined for $s \in J$ by holding t constant.

If such a family Γ is smooth (or at least continuously differentiable), we denote the velocity vectors of the main and transverse curves by

$$\partial_t \Gamma(s, t) = (\Gamma_s)'(t) \in T_{\Gamma(s,t)} M; \qquad \partial_s \Gamma(s, t) = \Gamma^{(t)\prime}(s) \in T_{\Gamma(s,t)} M.$$

Each of these is an example of a **vector field along Γ**, which is a continuous map $V : J \times I \to TM$ such that $V(s, t) \in T_{\Gamma(s,t)} M$ for each (s, t).

The families of curves that will interest us most in this chapter are of the following type. A one-parameter family Γ is called an **admissible family of curves** if (i) its domain is of the form $J \times [a, b]$ for some open interval J; (ii) there is a partition $(a_0, \ldots, a_k)$ of $[a, b]$ such that Γ is smooth on each rectangle of the form $J \times [a_{i-1}, a_i]$; and (iii) $\Gamma_s(t) = \Gamma(s, t)$ is an admissible curve for each $s \in J$ (Fig. 6.1). Every such partition is called an **admissible partition for the family**.

If $\gamma : [a, b] \to M$ is a given admissible curve, a **variation of γ** is an admissible family of curves $\Gamma : J \times [a, b] \to M$ such that J is an open interval containing 0 and $\Gamma_0 = \gamma$. It is called a **proper variation** if in addition, all of the main curves have the same starting and ending points: $\Gamma_s(a) = \gamma(a)$ and $\Gamma_s(b) = \gamma(b)$ for all $s \in J$.

In the case of an admissible family, the transverse curves are smooth on J for each t, but the main curves are in general only piecewise regular. Thus the velocity vector fields $\partial_s \Gamma$ and $\partial_t \Gamma$ are smooth on each rectangle $J \times [a_{i-1}, a_i]$, but not generally on the whole domain.

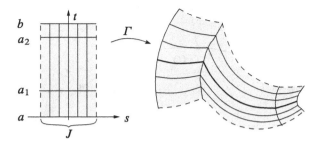

Fig. 6.1: An admissible family

We can say a bit more about $\partial_s \Gamma$, though. If Γ is an admissible family, a *piecewise smooth vector field along Γ* is a (continuous) vector field along Γ whose restriction to each rectangle $J \times [a_{i-1}, a_i]$ is smooth for some admissible partition $(a_0, \ldots, a_k)$ for Γ. In fact, $\partial_s \Gamma$ is always such a vector field. To see that it is continuous on the whole domain $J \times [a, b]$, note on the one hand that for each $i = 1, \ldots, k-1$, the values of $\partial_s \Gamma$ along the set $J \times \{a_i\}$ depend only on the values of Γ on that set, since the derivative is taken only with respect to the s variable; on the other hand, $\partial_s \Gamma$ is continuous (in fact smooth) on each subrectangle $J \times [a_{i-1}, a_i]$ and $J \times [a_i, a_{i+1}]$, so the right-hand and left-hand limits at $t = a_i$ must be equal. Therefore $\partial_s \Gamma$ is always a piecewise smooth vector field along Γ. (However, $\partial_t \Gamma$ is typically not continuous at $t = a_i$.)

If Γ is a variation of γ, the *variation field of Γ* is the piecewise smooth vector field $V(t) = \partial_s \Gamma(0, t)$ along γ. We say that a vector field V along γ is *proper* if $V(a) = 0$ and $V(b) = 0$; it follows easily from the definitions that the variation field of every proper variation is itself proper.

Lemma 6.1. *If γ is an admissible curve and V is a piecewise smooth vector field along γ, then V is the variation field of some variation of γ. If V is proper, the variation can be taken to be proper as well.*

Proof. Suppose γ and V satisfy the hypotheses, and set $\Gamma(s, t) = \exp_{\gamma(t)}(s V(t))$ (Fig. 6.2). By compactness of $[a, b]$, there is some positive ε such that Γ is defined on $(-\varepsilon, \varepsilon) \times [a, b]$. By composition, Γ is smooth on $(-\varepsilon, \varepsilon) \times [a_{i-1}, a_i]$ for each subinterval $[a_{i-1}, a_i]$ on which V is smooth, and it is continuous on its whole domain. By the properties of the exponential map, the variation field of Γ is V. Moreover, if $V(a) = 0$ and $V(b) = 0$, the definition gives $\Gamma(s, a) \equiv \gamma(a)$ and $\Gamma(s, b) \equiv \gamma(b)$, so Γ is proper. $\square$

If V is a piecewise smooth vector field along Γ, we can compute the covariant derivative of V either along the main curves (at points where V is smooth) or along the transverse curves; the resulting vector fields along Γ are denoted by $D_t V$ and $D_s V$ respectively.

A key ingredient in the proof that minimizing curves are geodesics is the symmetry of the Levi-Civita connection. It enters into our proofs in the form of the

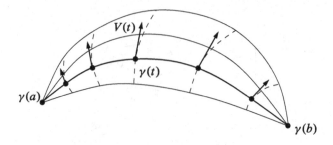

Fig. 6.2: Every vector field along γ is a variation field

following lemma. (Although we state and use this lemma only for the Levi-Civita connection, the proof shows that it is actually true for every symmetric connection in TM.)

Lemma 6.2 (Symmetry Lemma). *Let $\Gamma: J \times [a,b] \to M$ be an admissible family of curves in a Riemannian manifold. On every rectangle $J \times [a_{i-1}, a_i]$ where Γ is smooth,*

$$D_s \partial_t \Gamma = D_t \partial_s \Gamma.$$

Proof. This is a local question, so we may compute in local coordinates (x^i) around a point $\Gamma(s_0, t_0)$. Writing the components of Γ as $\Gamma(s,t) = (x^1(s,t), \dots, x^n(s,t))$, we have

$$\partial_t \Gamma = \frac{\partial x^k}{\partial t} \partial_k; \qquad \partial_s \Gamma = \frac{\partial x^k}{\partial s} \partial_k.$$

Then, using the coordinate formula (4.15) for covariant derivatives along curves, we obtain

$$D_s \partial_t \Gamma = \left(\frac{\partial^2 x^k}{\partial s \partial t} + \frac{\partial x^i}{\partial t} \frac{\partial x^j}{\partial s} \Gamma_{ji}^k \right) \partial_k;$$

$$D_t \partial_s \Gamma = \left(\frac{\partial^2 x^k}{\partial t \partial s} + \frac{\partial x^i}{\partial s} \frac{\partial x^j}{\partial t} \Gamma_{ji}^k \right) \partial_k.$$

Reversing the roles of i and j in the second line above and using the symmetry condition $\Gamma_{ji}^k = \Gamma_{ij}^k$, we conclude that these two expressions are equal. $\square$

Minimizing Curves Are Geodesics

We can now compute an expression for the derivative of the length functional along a variation of a curve. Traditionally, the derivative of a functional on a space of maps is called its *first variation*.

Theorem 6.3 (First Variation Formula). *Let (M, g) be a Riemannian manifold. Suppose $\gamma: [a,b] \to M$ is a unit-speed admissible curve, $\Gamma: J \times [a,b] \to M$ is a*

Fig. 6.3: $\Delta_i \gamma'$ is the "jump" in γ' at a_i

variation of γ, and V is its variation field. Then $L_g(\Gamma_s)$ is a smooth function of s, and

$$
\frac{d}{ds}\Big|_{s=0} L_g(\Gamma_s) = -\int_a^b \langle V, D_t \gamma' \rangle dt - \sum_{i=1}^{k-1} \langle V(a_i), \Delta_i \gamma' \rangle
$$
$$
+ \langle V(b), \gamma'(b) \rangle - \langle V(a), \gamma'(a) \rangle, \quad (6.1)
$$

where $(a_0, \ldots, a_k)$ is an admissible partition for V, and for each $i = 1, \ldots, k-1$, $\Delta_i \gamma' = \gamma'(a_i^+) - \gamma'(a_i^-)$ is the "jump" in the velocity vector field γ' at a_i (Fig. 6.3). In particular, if Γ is a proper variation, then

$$
\frac{d}{ds}\Big|_{s=0} L_g(\Gamma_s) = -\int_a^b \langle V, D_t \gamma' \rangle dt - \sum_{i=1}^{k-1} \langle V(a_i), \Delta_i \gamma' \rangle. \quad (6.2)
$$

Proof. On every rectangle $J \times [a_{i-1}, a_i]$ where Γ is smooth, since the integrand in $L_g(\Gamma_s)$ is smooth and the domain of integration is compact, we can differentiate under the integral sign as many times as we wish. Because $L_g(\Gamma_s)$ is a finite sum of such integrals, it follows that it is a smooth function of s.

For brevity, let us introduce the notations

$$
T(s,t) = \partial_t \Gamma(s,t), \qquad S(s,t) = \partial_s \Gamma(s,t).
$$

Differentiating on the interval $[a_{i-1}, a_i]$ yields

$$
\frac{d}{ds} L_g \left(\Gamma_s \big|_{[a_{i-1}, a_i]} \right) = \int_{a_{i-1}}^{a_i} \frac{\partial}{\partial s} \langle T, T \rangle^{1/2} dt
$$
$$
= \int_{a_{i-1}}^{a_i} \frac{1}{2} \langle T, T \rangle^{-1/2} \, 2 \langle D_s T, T \rangle \, dt \quad (6.3)
$$
$$
= \int_{a_{i-1}}^{a_i} \frac{1}{|T|} \langle D_t S, T \rangle \, dt,
$$

where we have used the symmetry lemma in the last line. Setting $s = 0$ and noting that $S(0,t) = V(t)$ and $T(0,t) = \gamma'(t)$ (which has length 1), we get

$$\frac{d}{ds}\bigg|_{s=0} L_g\big(\Gamma_s|_{[a_{i-1},a_i]}\big) = \int_{a_{i-1}}^{a_i} \langle D_t V, \gamma' \rangle \, dt$$

$$= \int_{a_{i-1}}^{a_i} \left(\frac{d}{dt} \langle V, \gamma' \rangle - \langle V, D_t \gamma' \rangle \right) dt$$

$$= \langle V(a_i), \gamma'(a_i^-) \rangle - \langle V(a_{i-1}), \gamma'(a_{i-1}^+) \rangle$$

$$- \int_{a_{i-1}}^{a_i} \langle V, D_t \gamma' \rangle \, dt.$$

(The second equality follows from (5.3), and the third from the fundamental theorem of calculus.) Finally, summing over i, we obtain (6.1).   □

Because every admissible curve has a unit-speed parametrization and length is independent of parametrization, the requirement in the above proposition that γ be of unit speed is not a real restriction, but rather just a computational convenience.

Theorem 6.4. *In a Riemannian manifold, every minimizing curve is a geodesic when it is given a unit-speed parametrization.*

Proof. Suppose $\gamma \colon [a,b] \to M$ is minimizing and of unit speed, and $(a_0, \dots, a_k)$ is an admissible partition for γ. If Γ is any proper variation of γ, then $L_g(\Gamma_s)$ is a smooth function of s that achieves its minimum at $s = 0$, so it follows from elementary calculus that $d\big(L_g(\Gamma_s)\big)/ds = 0$ when $s = 0$. Since every proper vector field along γ is the variation field of some proper variation, the right-hand side of (6.2) must vanish for every such V.

First we show that $D_t \gamma' = 0$ on each subinterval $[a_{i-1}, a_i]$, so γ is a "broken geodesic." Choose one such interval, and let $\varphi \in C^\infty(\mathbb{R})$ be a bump function such that $\varphi > 0$ on (a_{i-1}, a_i) and $\varphi = 0$ elsewhere. Then (6.2) with $V = \varphi D_t \gamma'$ becomes

$$0 = - \int_{a_{i-1}}^{a_i} \varphi \left| D_t \gamma' \right|^2 dt.$$

Since the integrand is nonnegative and $\varphi > 0$ on (a_{i-1}, a_i), this shows that $D_t \gamma' = 0$ on each such subinterval.

Next we need to show that $\Delta_i \gamma' = 0$ for each i between 0 and k, which is to say that γ has no corners. For each such i, we can use a bump function in a coordinate chart to construct a piecewise smooth vector field V along γ such that $V(a_i) = \Delta_i \gamma'$ and $V(a_j) = 0$ for $j \neq i$. Then (6.2) reduces to $-|\Delta_i \gamma'|^2 = 0$, so $\Delta_i \gamma' = 0$ for each i.

Finally, since the two one-sided velocity vectors of γ match up at each a_i, it follows from uniqueness of geodesics that $\gamma|_{[a_i, a_{i+1}]}$ is the continuation of the geodesic $\gamma|_{[a_{i-1}, a_i]}$, and therefore γ is smooth.   □

The preceding proof has an enlightening geometric interpretation. Under the assumption that $D_t \gamma' \neq 0$, the first variation with $V = \varphi D_t \gamma'$ is negative, which shows that deforming γ in the direction of its acceleration vector field decreases its length (Fig. 6.4). Similarly, the length of a broken geodesic γ is decreased by deforming it

Fig. 6.4: Deforming in the direction of the acceleration

Fig. 6.5: Rounding the corner

in the direction of a vector field V such that $V(a_i) = \Delta_i \gamma'$ (Fig. 6.5). Geometrically, this corresponds to "rounding the corner."

The first variation formula actually tells us a bit more than is claimed in Theorem 6.4. In proving that γ is a geodesic, we did not use the full strength of the assumption that the length of Γ_s takes a *minimum* when $s = 0$; we only used the fact that its derivative is zero. We say that an admissible curve γ is a ***critical point of*** L_g if for every proper variation Γ_s of γ, the derivative of $L_g(\Gamma_s)$ with respect to s is zero at $s = 0$. Therefore we can strengthen Theorem 6.4 in the following way.

Corollary 6.5. *A unit-speed admissible curve γ is a critical point for L_g if and only if it is a geodesic.*

Proof. If γ is a critical point, the proof of Theorem 6.4 goes through without modification to show that γ is a geodesic. Conversely, if γ is a geodesic, then the first term on the right-hand side of (6.2) vanishes by the geodesic equation, and the second term vanishes because γ' has no jumps. □

The geodesic equation $D_t \gamma' = 0$ thus characterizes the critical points of the length functional. In general, the equation that characterizes critical points of a functional on a space of maps is called the ***variational equation*** or the ***Euler–Lagrange equation*** of the functional. Many interesting equations in differential geometry arise as variational equations. We touch briefly on three others in this book: the Einstein equation (7.34), the Yamabe equation (7.59), and the minimal hypersurface equation $H \equiv 0$ (Thm. 8.18).

Geodesics Are Locally Minimizing

Next we turn to the converse of Theorem 6.4. It is easy to see that the literal converse is not true, because not every geodesic segment is minimizing. For example, every geodesic segment on $\mathbb{S}^n$ that goes more than halfway around the sphere is not minimizing, because the other portion of the same great circle is a shorter curve segment between the same two points. For that reason, we concentrate initially on *local* minimization properties of geodesics.

As usual, let (M, g) be a Riemannian manifold. A regular (or piecewise regular) curve $\gamma : I \to M$ is said to be ***locally minimizing*** if every $t_0 \in I$ has a neighborhood $I_0 \subseteq I$ such that whenever $a, b \in I_0$ with $a < b$, the restriction of γ to $[a, b]$ is minimizing.

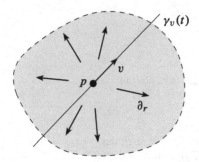

Fig. 6.6: The radial vector field in a normal neighborhood

Lemma 6.6. *Every minimizing admissible curve segment is locally minimizing.*

▶ **Exercise 6.7.** Prove the preceding lemma.

Our goal in this section is to show that geodesics are locally minimizing. The proof will be based on a careful analysis of the geodesic equation in Riemannian normal coordinates.

If ε is a positive number such that $\exp_p$ is a diffeomorphism from the ball $B_\varepsilon(0) \subseteq T_p M$ to its image (where the radius of the ball is measured with respect to the norm defined by g_p), then the image set $\exp_p\big(B_\varepsilon(0)\big)$ is a normal neighborhood of p, called a **geodesic ball in M**, or sometimes an **open geodesic ball** for clarity.

Also, if the closed ball $\bar{B}_\varepsilon(0)$ is contained in an open set $V \subseteq T_p M$ on which $\exp_p$ is a diffeomorphism onto its image, then $\exp_p\big(\bar{B}_\varepsilon(0)\big)$ is called a **closed geodesic ball**, and $\exp_p\big(\partial B_\varepsilon(0)\big)$ is called a **geodesic sphere**. Given such a V, by compactness there exists $\varepsilon' > \varepsilon$ such that $B_{\varepsilon'}(0) \subseteq V$, so every closed geodesic ball is contained in an open geodesic ball of larger radius. In Riemannian normal coordinates centered at p, the open and closed geodesic balls and geodesic spheres centered at p are just the coordinate balls and spheres.

Suppose U is a normal neighborhood of $p \in M$. Given any normal coordinates (x^i) on U centered at p, define the **radial distance function** $r : U \to \mathbb{R}$ by

$$r(x) = \sqrt{(x^1)^2 + \cdots + (x^n)^2}, \tag{6.4}$$

and the **radial vector field** on $U \smallsetminus \{p\}$ (see Fig. 6.6), denoted by ∂_r, by

$$\partial_r = \frac{x^i}{r(x)} \frac{\partial}{\partial x^i}. \tag{6.5}$$

In Euclidean space, $r(x)$ is the distance to the origin, and ∂_r is the unit vector field pointing radially outward from the origin. (The notation is suggested by the fact that ∂_r is a coordinate derivative in polar or spherical coordinates.)

Lemma 6.8. *In every normal neighborhood U of $p \in M$, the radial distance function and the radial vector field are well defined, independently of the choice of normal coordinates. Both r and ∂_r are smooth on $U \smallsetminus \{p\}$, and r^2 is smooth on all of U.*

Proof. Proposition 5.23 shows that any two normal coordinate charts on U are related by $\tilde{x}^i = A^i_j x^j$ for some orthogonal matrix (A^i_j), and a straightforward computation shows that both r and ∂_r are invariant under such coordinate changes. The smoothness statements follow directly from the coordinate formulas. $\square$

The crux of the proof that geodesics are locally minimizing is the following deceptively simple geometric lemma.

Theorem 6.9 (The Gauss Lemma). *Let (M, g) be a Riemannian manifold, let U be a geodesic ball centered at $p \in M$, and let ∂_r denote the radial vector field on $U \smallsetminus \{p\}$. Then ∂_r is a unit vector field orthogonal to the geodesic spheres in $U \smallsetminus \{p\}$.*

Proof. We will work entirely in normal coordinates (x^i) on U centered at p, using the properties of normal coordinates described in Proposition 5.24.

Let $q \in U \smallsetminus \{p\}$ be arbitrary, with coordinate representation $(q^1, \dots, q^n)$, and let $b = r(q) = \sqrt{(q^1)^2 + \cdots + (q^n)^2}$, where r is the radial distance function defined by (6.4). It follows that $\partial_r|_q$ has the coordinate representation

$$\partial_r\big|_q = \frac{q^i}{b} \frac{\partial}{\partial x^i}\bigg|_q.$$

Let $v = v^i \partial_i|_p \in T_p M$ be the tangent vector at p with components $v^i = q^i/b$. By Proposition 5.24(c), the radial geodesic with initial velocity v is given in these coordinates by

$$\gamma_v(t) = (tv^1, \dots, tv^n).$$

It satisfies $\gamma_v(0) = p$, $\gamma_v(b) = q$, and $\gamma_v'(b) = v^i \partial_i|_q = \partial_r|_q$. Because g_p is equal to the Euclidean metric in these coordinates, we have

$$\big|\gamma_v'(0)\big|_g = |v|_g = \sqrt{(v^1)^2 + \cdots + (v^n)^2} = \frac{1}{b}\sqrt{(q^1)^2 + \cdots + (q^n)^2} = 1,$$

so v is a unit vector, and thus γ_v is a unit-speed geodesic. It follows that $\partial_r|_q = \gamma_v'(b)$ is also a unit vector.

To prove that ∂_r is orthogonal to the geodesic spheres let q, b, and v be as above, and let $\Sigma_b = \exp_p(\partial B_b(0))$ be the geodesic sphere containing q. In these coordinates, Σ_b is the set of points satisfying the equation $(x^1)^2 + \cdots + (x^n)^2 = b^2$. Let $w \in T_q M$ be any vector tangent to Σ_b at q. We need to show that $\langle w, \partial_r|_q \rangle_g = 0$.

Choose a smooth curve $\sigma : (-\varepsilon, \varepsilon) \to \Sigma_b$ satisfying $\sigma(0) = q$ and $\sigma'(0) = w$, and write its coordinate representation in (x^i)-coordinates as $\sigma(s) = (\sigma^1(s), \dots, \sigma^n(s))$. The fact that $\sigma(s)$ lies in Σ_b for all s means that

$$\big(\sigma^1(s)\big)^2 + \cdots + \big(\sigma^n(s)\big)^2 = b^2. \tag{6.6}$$

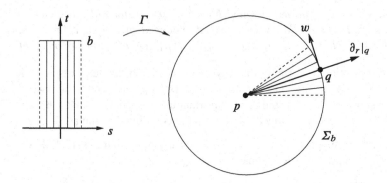

Fig. 6.7: Proof of the Gauss lemma

Define a smooth map $\Gamma : (-\varepsilon, \varepsilon) \times [0, b] \to U$ (Fig. 6.7) by

$$\Gamma(s, t) = \left(\frac{t}{b}\sigma^1(s), \ldots, \frac{t}{b}\sigma^n(s)\right).$$

For each $s \in (-\varepsilon, \varepsilon)$, Γ_s is a geodesic by Proposition 5.24(c). Its initial velocity is $\Gamma_s'(0) = (1/b)\sigma^i(s)\partial_i|_p$, which is a unit vector by (6.6) and the fact that g_p is the Euclidean metric in coordinates; thus each Γ_s is a unit-speed geodesic.

As before, let $S = \partial_s \Gamma$ and $T = \partial_t \Gamma$. It follows from the definitions that

$$S(0, 0) = \frac{d}{ds}\bigg|_{s=0} \Gamma_s(0) = 0;$$

$$T(0, 0) = \frac{d}{dt}\bigg|_{t=0} \gamma_v(t) = v;$$

$$S(0, b) = \frac{d}{ds}\bigg|_{s=0} \sigma(s) = w;$$

$$T(0, b) = \frac{d}{dt}\bigg|_{t=b} \gamma_v(t) = \gamma_v'(b) = \partial_r|_q.$$

Therefore $\langle S, T \rangle$ is zero when $(s, t) = (0, 0)$ and equal to $\langle w, \partial_r|_q \rangle$ when $(s, t) = (0, b)$, so to prove the theorem it suffices to show that $\langle S, T \rangle$ is independent of t.

We compute

$$
\begin{aligned}
\frac{\partial}{\partial t}\langle S, T \rangle &= \langle D_t S, T \rangle + \langle S, D_t T \rangle && \text{(compatibility with the metric)} \\
&= \langle D_s T, T \rangle + \langle S, D_t T \rangle && \text{(symmetry lemma)} \\
&= \langle D_s T, T \rangle + 0 && \text{(each } \Gamma_s \text{ is a geodesic)} \\
&= \frac{1}{2}\frac{\partial}{\partial s}|T|^2 = 0 && (|T| = |\Gamma_s'| \equiv 1 \text{ for all } (s, t)).
\end{aligned}
$$

(6.7)

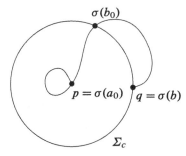

Fig. 6.8: Radial geodesics are minimizing

This proves the theorem.   □

We will use the Gauss lemma primarily in the form of the next corollary.

Corollary 6.10. *Let U be a geodesic ball centered at $p \in M$, and let r and ∂_r be the radial distance and radial vector field as defined by (6.4) and (6.5). Then* grad $r = \partial_r$ *on* $U \smallsetminus \{p\}$.

Proof. By the result of Problem 2-10, it suffices to show that ∂_r is orthogonal to the level sets of r and $\partial_r(r) \equiv |\partial_r|_g^2$. The first claim follows directly from the Gauss lemma, and the second from the fact that $\partial_r(r) \equiv 1$ by direct computation in normal coordinates, which in turn is equal to $|\partial_r|_g^2$ by the Gauss lemma.   □

Here is the payoff: our first step toward proving that geodesics are locally minimizing. Note that this is not yet the full strength of the theorem we are aiming for, because it shows only that for each point on a geodesic, sufficiently small segments of the geodesic *starting at that point* are minimizing. We will remove this restriction after a little more work below.

Proposition 6.11. *Let (M, g) be a Riemannian manifold. Suppose $p \in M$ and q is contained in a geodesic ball around p. Then (up to reparametrization) the radial geodesic from p to q is the unique minimizing curve in M from p to q.*

Proof. Choose $\varepsilon > 0$ such that $\exp_p\big(B_\varepsilon(0)\big)$ is a geodesic ball containing q. Let $\gamma : [0, c] \to M$ be the radial geodesic from p to q parametrized by arc length, and write $\gamma(t) = \exp_p(tv)$ for some unit vector $v \in T_pM$. Then $L_g(\gamma) = c$, since γ has unit speed.

To show that γ is minimizing, we need to show that every other admissible curve from p to q has length at least c. Let $\sigma : [0, b] \to M$ be an arbitrary admissible curve from p to q, which we may assume to be parametrized by arc length as well. Let $a_0 \in [0, b]$ denote the last time that $\sigma(t) = p$, and $b_0 \in [0, b]$ the first time after a_0 that $\sigma(t)$ meets the geodesic sphere Σ_c of radius c around p (Fig. 6.8). Then the composite function $r \circ \sigma$ is continuous on $[a_0, b_0]$ and piecewise smooth in (a_0, b_0), so we can apply the fundamental theorem of calculus to conclude that

$$r(\sigma(b_0)) - r(\sigma(a_0)) = \int_{a_0}^{b_0} \frac{d}{dt} r(\sigma(t)) \, dt$$

$$= \int_{a_0}^{b_0} dr(\sigma'(t)) \, dt$$

$$= \int_{a_0}^{b_0} \langle \operatorname{grad} r|_{\sigma(t)}, \sigma'(t) \rangle \, dt \qquad (6.8)$$

$$\leq \int_{a_0}^{b_0} |\operatorname{grad} r|_{\sigma(t)}| \, |\sigma'(t)| \, dt$$

$$= \int_{a_0}^{b_0} |\sigma'(t)| \, dt$$

$$= L_g\big(\sigma|_{[a_0,b_0]}\big) \leq L_g(\sigma).$$

Thus $L_g(\sigma) \geq r(\sigma(b_0)) - r(\sigma(a_0)) = c$, so γ is minimizing.

Now suppose $L_g(\sigma) = c$. Then $b = c$, and both inequalities in (6.8) are equalities. Because we assume that σ is a unit-speed curve, the second of these equalities implies that $a_0 = 0$ and $b_0 = b = c$, since otherwise the segments of σ before $t = a_0$ and after $t = b_0$ would contribute positive lengths. The first equality then implies that the nonnegative expression $|\operatorname{grad} r|_{\sigma(t)}| \, |\sigma'(t)| - \langle \operatorname{grad} r|_{\sigma(t)}, \sigma'(t) \rangle$ is identically zero on $[0,b]$, which is possible only if $\sigma'(t)$ is a positive multiple of $\operatorname{grad} r|_{\sigma(t)}$ for each t. Since we assume that σ has unit speed, we must have $\sigma'(t) = \operatorname{grad} r|_{\sigma(t)} = \partial_r|_{\sigma(t)}$. Thus σ and γ are both integral curves of ∂_r passing through q at time $t = c$, so $\sigma = \gamma$.   □

The next two corollaries show how radial distance functions, balls, and spheres in normal coordinates are related to their global metric counterparts.

Corollary 6.12. *Let (M,g) be a connected Riemannian manifold and $p \in M$. Within every open or closed geodesic ball around p, the radial distance function $r(x)$ defined by (6.4) is equal to the Riemannian distance from p to x in M.*

Proof. Since every closed geodesic ball is contained in an open geodesic ball of larger radius, we need only consider the open case. If x is in the open geodesic ball $\exp_p\big(B_c(0)\big)$, the radial geodesic γ from p to x is minimizing by Proposition 6.11. Since its velocity is equal to ∂_r, which is a unit vector in both the g-norm and the Euclidean norm in normal coordinates, the g-length of γ is equal to its Euclidean length, which is $r(x)$.   □

Corollary 6.13. *In a connected Riemannian manifold, every open or closed geodesic ball is also an open or closed metric ball of the same radius, and every geodesic sphere is a metric sphere of the same radius.*

Proof. Let (M,g) be a Riemannian manifold, and let $p \in M$ be arbitrary. First, let $V = \exp_p\big(\bar{B}_c(0)\big) \subseteq M$ be a closed geodesic ball of radius $c > 0$ around p. Suppose q is an arbitrary point of M. If $q \in V$, then Corollary 6.12 shows that q

is also in the closed metric ball of radius c. Conversely, suppose $q \notin V$. Let S be the geodesic sphere $\exp_p\left(\partial B_c(0)\right)$. The complement of S is the disjoint union of the open sets $\exp_p\left(B_c(0)\right)$ and $M \smallsetminus \exp_p\left(\overline{B}_c(0)\right)$, and hence disconnected. Thus if $\gamma : [a,b] \to M$ is any admissible curve from p to q, there must be a time $t_0 \in (a,b)$ when $\gamma(t_0) \in S$, and then Corollary 6.12 shows that the length of $\gamma|_{[a,t_0]}$ must be at least c. Since $\gamma|_{[t_0,b]}$ must have positive length, it follows that $d_g(p,q) > c$, so q is not in the closed metric ball of radius c around p.

Next, let $W = \exp_p\left(B_c(0)\right)$ be an open geodesic ball of radius c. Since W is the union of all closed geodesic balls around p of radius less than c, and the open metric ball of radius c is similarly the union of all closed metric metric balls of smaller radii, the result of the preceding paragraph shows that W is equal to the open metric ball of radius c.

Finally, if $S = \exp_p\left(\partial B_c(0)\right)$ is a geodesic sphere of radius c, the arguments above show that S is equal to the closed metric ball of radius c minus the open metric ball of radius c, which is exactly the metric sphere of radius c.   □

The last corollary suggests a simplified notation for geodesic balls and spheres in M. From now on, we will use the notations $B_c(p) = \exp_p\left(B_c(0)\right)$, $\overline{B}_c(p) = \exp_p\left(\overline{B}_c(0)\right)$, and $S_c(p) = \exp_p\left(\partial B_c(0)\right)$ for open and closed geodesic balls and geodesic spheres, which we now know are also open and closed metric balls and spheres. (To avoid confusion, we refrain from using this notation for other metric balls and spheres unless explicitly stated.)

Uniformly Normal Neighborhoods

We continue to let (M,g) be a Riemannian manifold. In order to prove that geodesics in M are locally minimizing, we need the following refinement of the concept of normal neighborhoods. A subset $W \subseteq M$ is called **uniformly normal** if there exists some $\delta > 0$ such that W is contained in a geodesic ball of radius δ around each of its points (Fig. 6.9). If δ is any such constant, we will also say that W is **uniformly δ-normal**. Clearly every subset of a uniformly δ-normal set is itself uniformly δ-normal.

Lemma 6.14 (Uniformly Normal Neighborhood Lemma). *Given $p \in M$ and any neighborhood U of p, there exists a uniformly normal neighborhood of p contained in U.*

Proof. Choose a normal coordinate chart $\left(U_0, (x^i)\right)$ centered at p and contained in U, and let (x^i, v^i) be the corresponding natural coordinates for $\pi^{-1}(U_0) \subseteq TM$. Because this is a local question, we might as well identify U_0 with an open subset of $\mathbb{R}^n$, and identify TM with $U_0 \times \mathbb{R}^n$. The exponential map for the Riemannian manifold (U_0, g) is defined on an open subset $\mathcal{E} \subseteq U_0 \times \mathbb{R}^n$. Consider the map $E : \mathcal{E} \to U_0 \times U_0$ defined by

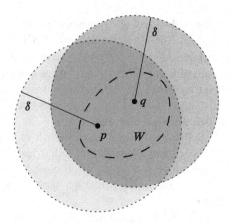

Fig. 6.9: Uniformly normal neighborhood

$$E(x,v) = (x, \exp_x v).$$

The differential of E at $(p,0)$ is represented by the matrix

$$dE_{(p,0)} = \begin{pmatrix} \dfrac{\partial x^i}{\partial x^j} & \dfrac{\partial x^i}{\partial v^j} \\[2mm] \dfrac{\partial \exp^i}{\partial x^j} & \dfrac{\partial \exp^i}{\partial v^j} \end{pmatrix} = \begin{pmatrix} \mathrm{Id} & 0 \\ * & \mathrm{Id} \end{pmatrix},$$

which is invertible. By the inverse function theorem, therefore, there are neighborhoods $\mathcal{U} \subseteq U_0 \times \mathbb{R}^n$ of $(p,0)$ and $\mathcal{V} \subseteq U_0 \times U_0$ of (p,p) such that E restricts to a diffeomorphism from $\mathcal{U}$ to $\mathcal{V}$. Shrinking both neighborhoods if necessary, we may assume that $\mathcal{U}$ is a product set of the form $W \times B_\varepsilon(0)$, where W is a neighborhood of p and $B_\varepsilon(0)$ is a Euclidean ball in v-coordinates. It follows that for each $x \in W$, $\exp_x$ maps $B_\varepsilon(0)$ smoothly onto the open set $\mathcal{V}_x = \{y : (x,y) \in \mathcal{V}\}$, and it is a diffeomorphism because its inverse is given explicitly by $\exp_x^{-1}(y) = \pi_2 \circ E^{-1}(x,y)$, where $\pi_2 : U_0 \times \mathbb{R}^n \to \mathbb{R}^n$ is the projection. Shrinking W still further if necessary, we may assume that the metric g satisfies an estimate of the form (2.21) for all $x \in W$. This means that for each $x \in W$, the coordinate ball $B_\varepsilon(0) \subseteq T_x M$ contains a g_x-ball of radius at least ε/c. Put $\delta = \varepsilon/c$; we have shown that for each $x \in W$, there is a g-geodesic ball of radius δ in M centered at x.

Now, shrinking W once more, we may assume that its diameter (with respect to the metric d_g) is less than δ. It follows that for each $x \in W$, the entire set W is contained in the metric ball of radius δ around x, and Corollary 6.13 shows that this metric ball is also a geodesic ball of radius δ. Thus W is the required uniformly normal neighborhood of p. □

We are now ready to prove the main result of this section.

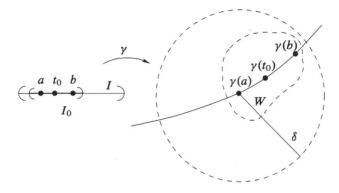

Fig. 6.10: Geodesics are locally minimizing

Theorem 6.15. *Every Riemannian geodesic is locally minimizing.*

Proof. Let (M, g) be a Riemannian manifold. Suppose $\gamma: I \to M$ is a geodesic, which we may assume to be defined on an open interval, and let $t_0 \in I$. Let W be a uniformly normal neighborhood of $\gamma(t_0)$, and let $I_0 \subseteq I$ be the connected component of $\gamma^{-1}(W)$ containing t_0. If $a, b \in I_0$ with $a < b$, then the definition of uniformly normal neighborhood implies that $\gamma(b)$ is contained in a geodesic ball centered at $\gamma(a)$ (Fig. 6.10). Therefore, by Proposition 6.11, the radial geodesic segment from $\gamma(a)$ to $\gamma(b)$ is the unique minimizing curve segment between these points. However, the restriction of γ to $[a, b]$ is also a geodesic segment from $\gamma(a)$ to $\gamma(b)$ lying in the same geodesic ball, and thus $\gamma|_{[a,b]}$ must coincide with this minimizing geodesic. $\qquad\square$

It is interesting to note that the Gauss lemma and the uniformly normal neighborhood lemma also yield another proof that minimizing curves are geodesics, without using the first variation formula. On the principle that knowing more than one proof of an important fact always deepens our understanding of it, we present this proof for good measure.

Another proof of Theorem 6.4. Suppose $\gamma: [a, b] \to M$ is a minimizing admissible curve. Just as in the preceding proof, for every $t_0 \in [a, b]$ we can find a connected neighborhood I_0 of t_0 such that $\gamma(I_0)$ is contained in a uniformly normal neighborhood W. Then for every $a_0, b_0 \in I_0$, the same argument as above shows that the unique minimizing curve segment from $\gamma(a_0)$ to $\gamma(b_0)$ is a radial geodesic. Since the restriction of γ to $[a_0, b_0]$ is such a minimizing curve segment, it must coincide with this radial geodesic. Therefore γ solves the geodesic equation in a neighborhood of t_0. Since t_0 was arbitrary, γ is a geodesic. $\qquad\square$

Given a Riemannian manifold (M, g) (without boundary), for each point $p \in M$ we define the **injectivity radius of M at p**, denoted by $\mathrm{inj}(p)$, to be the supremum of all $a > 0$ such that $\exp_p$ is a diffeomorphism from $B_a(0) \subseteq T_p M$ onto its image.

If there is no upper bound to the radii of such balls (as is the case, for example, on $\mathbb{R}^n$), then we set $\mathrm{inj}(p) = \infty$. Then we define the ***injectivity radius of*** M, denoted by $\mathrm{inj}(M)$, to be the infimum of $\mathrm{inj}(p)$ as p ranges over points of M. It can be zero, positive, or infinite. (The terminology is explained by Problem 10-24.)

Lemma 6.16. *If (M, g) is a compact Riemannian manifold, then $\mathrm{inj}(M)$ is positive.*

Proof. For each $x \in M$, there is a positive number $\delta(x)$ such that x is contained in a uniformly $\delta(x)$-normal neighborhood W_x, and $\mathrm{inj}(x') \geq \delta(x)$ for each $x' \in W$. Since M is compact, it is covered by finitely many such neighborhoods $W_{x_1}, \dots, W_{x_k}$. Therefore, $\mathrm{inj}(M)$ is at least equal to the minimum of $\delta(x_1), \dots, \delta(x_k)$. It cannot be infinite, because a compact metric space is bounded, and a geodesic ball of radius c contains points whose distances from the center are arbitrarily close to c. □

In addition to uniformly normal neighborhoods, there is another, more specialized, kind of normal neighborhood that is frequently useful. Let (M, g) be a Riemannian manifold. A subset $U \subseteq M$ is said to be ***geodesically convex*** if for each $p, q \in U$, there is a unique minimizing geodesic segment from p to q in M, and the image of this geodesic segment lies entirely in U.

The next theorem says that every sufficiently small geodesic ball is geodesically convex.

Theorem 6.17. *Let (M, g) be a Riemannian manifold. For each $p \in M$, there exists $\varepsilon_0 > 0$ such that every geodesic ball centered at p of radius less than or equal to ε_0 is geodesically convex.*

Proof. Problem 6-5. □

Completeness

Suppose (M, g) is a connected Riemannian manifold. Now that we can view M as a metric space, it is time to address one of the most important questions one can ask about a metric space: Is it complete? In general, the answer is no: for example, if M is an open ball in $\mathbb{R}^n$ with its Euclidean metric, then every sequence in M that converges in $\mathbb{R}^n$ to a point in ∂M is Cauchy, but not convergent in M.

In Chapter 5, we introduced another notion of completeness for Riemannian and pseudo-Riemannian manifolds: recall that such a manifold is said to be ***geodesically complete*** if every maximal geodesic is defined for all $t \in \mathbb{R}$. For clarity, we will use the phrase ***metrically complete*** for a connected Riemannian manifold that is complete as a metric space with the Riemannian distance function, in the sense that every Cauchy sequence converges.

The Hopf–Rinow theorem, which we will state and prove below, shows that these two notions of completeness are equivalent for connected Riemannian manifolds. Before we prove it, let us establish a preliminary result, which will have other important consequences besides the Hopf–Rinow theorem itself.

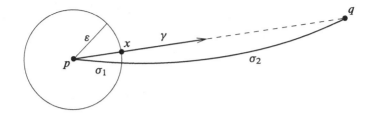

Fig. 6.11: Proof that $\gamma|_{[0,\varepsilon]}$ aims at q

Lemma 6.18. *Suppose (M,g) is a connected Riemannian manifold, and there is a point $p \in M$ such that $\exp_p$ is defined on the whole tangent space T_pM.*

(a) Given any other $q \in M$, there is a minimizing geodesic segment from p to q.
(b) M is metrically complete.

Proof. Let $q \in M$ be arbitrary. If $\gamma: [a,b] \to M$ is a geodesic segment starting at p, let us say that $\boldsymbol{\gamma}$ *aims at* q if γ is minimizing and

$$d_g(p,q) = d_g(p,\gamma(b)) + d_g(\gamma(b),q). \tag{6.9}$$

(This would be the case, for example, if γ were an initial segment of a minimizing geodesic from p to q; but we are not assuming that.) To prove (a), it suffices to show that there is a geodesic segment $\gamma: [a,b] \to M$ that begins at p, aims at q, and has length equal to $d_g(p,q)$, for then the fact that γ is minimizing means that $d_g(p,\gamma(b)) = L_g(\gamma) = d_g(p,q)$, and (6.9) becomes

$$d_g(p,q) = d_g(p,q) + d_g(\gamma(b),q),$$

which implies $\gamma(b) = q$. Since γ is a segment from p to q of length $d_g(p,q)$, it is the desired minimizing geodesic segment.

Choose $\varepsilon > 0$ such that there is a closed geodesic ball $\bar{B}_\varepsilon(p)$ around p that does not contain q. Since the distance function on a metric space is continuous, there is a point x in the geodesic sphere $S_\varepsilon(p)$ where $d_g(x,q)$ attains its minimum on the compact set $S_\varepsilon(p)$. Let γ be the maximal unit-speed geodesic whose restriction to $[0,\varepsilon]$ is the radial geodesic segment from p to x (Fig. 6.11); by assumption, γ is defined for all $t \in \mathbb{R}$.

We begin by showing that $\gamma|_{[0,\varepsilon]}$ aims at q. Since it is minimizing by Proposition 6.11 (noting that every closed geodesic ball is contained in a larger open one), we need only show that (6.9) holds with $b = \varepsilon$, or

$$d_g(p,q) = d_g(p,x) + d_g(x,q). \tag{6.10}$$

To this end, let $\sigma: [a_0,b_0] \to M$ be any admissible curve from p to q. Let t_0 be the first time σ hits $S_\varepsilon(p)$, and let σ_1 and σ_2 denote the restrictions of σ_1 to $[a_0,t_0]$ and $[t_0,b_0]$, respectively (Fig. 6.11). Since every point in $S_\varepsilon(p)$ is at a distance ε from

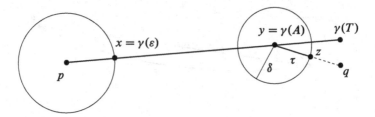

Fig. 6.12: Proof that $A = T$

p, we have $L_g(\sigma_1) \geq d_g(p,\sigma(t_0)) = d_g(p,x)$; and by our choice of x we have $L_g(\sigma_2) \geq d_g(\sigma(t_0),q) \geq d_g(x,q)$. Putting these two inequalities together yields

$$L_g(\sigma) = L_g(\sigma_1) + L_g(\sigma_2) \geq d_g(p,x) + d_g(x,q).$$

Taking the infimum over all such σ, we find that $d_g(p,q) \geq d_g(p,x) + d_g(x,q)$. The opposite inequality is just the triangle inequality, so (6.10) holds.

Now let $T = d_g(p,q)$ and

$$\mathcal{A} = \left\{ b \in [0,T] : \gamma|_{[0,b]} \text{ aims at } q \right\}.$$

We have just shown that $\varepsilon \in \mathcal{A}$. Let $A = \sup \mathcal{A} \geq \varepsilon$. By continuity of the distance function, it is easy to see that $\mathcal{A}$ is closed in $[0,T]$, and therefore $A \in \mathcal{A}$. If $A = T$, then $\gamma|_{[0,T]}$ is a geodesic of length $T = d_g(p,q)$ that aims at q, and by the remark above we are done. So we assume $A < T$ and derive a contradiction.

Let $y = \gamma(A)$, and choose $\delta > 0$ such that there is a closed geodesic ball $\bar{B}_\delta(y)$ around y, small enough that it does not contain q (Fig. 6.12). The fact that $A \in \mathcal{A}$ means that

$$d_g(y,q) = d_g(p,q) - d_g(p,y) = T - A.$$

Let $z \in S_\delta(y)$ be a point where $d_g(z,q)$ attains its minimum, and let $\tau : [0,\delta] \to M$ be the unit-speed radial geodesic from y to z. By exactly the same argument as before, τ aims at q, so

$$d_g(z,q) = d_g(y,q) - d_g(y,z) = (T - A) - \delta. \tag{6.11}$$

By the triangle inequality and (6.11),

$$\begin{aligned} d_g(p,z) &\geq d_g(p,q) - d_g(z,q) \\ &= T - (T - A - \delta) = A + \delta. \end{aligned}$$

Therefore, the admissible curve consisting of $\gamma|_{[0,A]}$ (of length A) followed by τ (of length δ) is a minimizing curve from p to z. This means that it has no corners, so z must lie on γ, and in fact, $z = \gamma(A + \delta)$. But then (6.11) says that

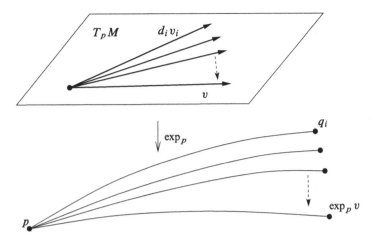

Fig. 6.13: Cauchy sequences converge

$$d_g(p,q) = T = (A+\delta) + d_g(z,q) = d_g(p,z) + d_g(z,q),$$

so $\gamma|_{[0,A+\delta]}$ aims at q and $A + \delta \in \mathcal{A}$, which is a contradiction. This completes the proof of (a).

To prove (b), we need to show that every Cauchy sequence in M converges. Let (q_i) be a Cauchy sequence in M. For each i, let $\gamma_i(t) = \exp_p(tv_i)$ be a unit-speed minimizing geodesic from p to q_i, and let $d_i = d_g(p,q_i)$, so that $q_i = \exp_p(d_i v_i)$ (Fig. 6.13). The sequence (d_i) is bounded in $\mathbb{R}$ (because Cauchy sequences in a metric space are bounded), and the sequence (v_i) consists of unit vectors in T_pM, so the sequence of vectors $(d_i v_i)$ in T_pM is bounded. Therefore a subsequence $(d_{i_k} v_{i_k})$ converges to some $v \in T_pM$. By continuity of the exponential map, $q_{i_k} = \exp_p(d_{i_k} v_{i_k}) \to \exp_p v$, and since the original sequence (q_i) is Cauchy, it converges to the same limit. $\square$

The next theorem is the main result of this section.

Theorem 6.19 (Hopf–Rinow). *A connected Riemannian manifold is metrically complete if and only if it is geodesically complete.*

Proof. Let (M,g) be a connected Riemannian manifold. Suppose first that M is geodesically complete. Then in particular, it satisfies the hypothesis of Lemma 6.18, so it is metrically complete.

Conversely, suppose M is metrically complete, and assume for the sake of contradiction that it is not geodesically complete. Then there is some unit-speed geodesic $\gamma \colon [0,b) \to M$ that has no extension to a geodesic on any interval $[0,b')$ for $b' > b$. Let (t_i) be any increasing sequence in $[0,b)$ that approaches b, and set $q_i = \gamma(t_i)$. Since γ is parametrized by arc length, the length of $\gamma|_{[t_i,t_j]}$ is exactly $|t_j - t_i|$, so $d_g(q_i, q_j) \le |t_j - t_i|$ and (q_i) is a Cauchy sequence in M. By completeness, (q_i) converges to some point $q \in M$.

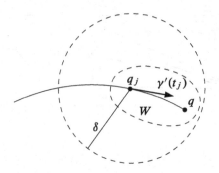

Fig. 6.14: γ extends past q

Let W be a uniformly δ-normal neighborhood of q for some $\delta > 0$. Choose j large enough that $t_j > b - \delta$ and $q_j \in W$ (Fig. 6.14). The fact that $B_\delta(q_j)$ is a geodesic ball means that every unit-speed geodesic starting at q_j exists at least for $t \in [0, \delta)$. In particular, this is true of the geodesic σ with $\sigma(0) = q_j$ and $\sigma'(0) = \gamma'(t_j)$. Define $\widetilde{\gamma} : [0, t_j + \delta) \to M$ by

$$\widetilde{\gamma}(t) = \begin{cases} \gamma(t), & t \in [0, b), \\ \sigma(t - t_j), & t \in (t_j - \delta, t_j + \delta). \end{cases}$$

Note that both expressions on the right-hand side are geodesics, and they have the same position and velocity when $t = t_j$. Therefore, by uniqueness of geodesics, the two definitions agree where they overlap. Since $t_j + \delta > b$, $\widetilde{\gamma}$ is an extension of γ past b, which is a contradiction. $\qquad\square$

A connected Riemannian manifold is simply said to be **complete** if it is either geodesically complete or metrically complete; the Hopf–Rinow theorem then implies that it is both. For disconnected manifolds, we interpret "complete" to mean *geodesically complete*, which is equivalent to the requirement that each component be a complete metric space. As mentioned in the previous chapter, complete manifolds are the natural setting for global questions in Riemannian geometry.

We conclude this section by stating three important corollaries, whose proofs are easy applications of Lemma 6.18 and the Hopf–Rinow theorem.

Corollary 6.20. *If M is a connected Riemannian manifold and there exists a point $p \in M$ such that the restricted exponential map $\exp_p$ is defined on all of T_pM, then M is complete.* $\qquad\square$

Corollary 6.21. *If M is a complete, connected Riemannian manifold, then any two points in M can be joined by a minimizing geodesic segment.* $\qquad\square$

Corollary 6.22. *If M is a compact Riemannian manifold, then every maximal geodesic in M is defined for all time.* $\qquad\square$

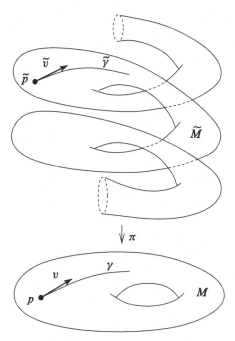

Fig. 6.15: Lifting geodesics

The Hopf–Rinow theorem and Corollary 6.20 are key ingredients in the following theorem about Riemannian covering maps. This theorem will play a key role in the proofs of some of the local-to-global theorems in Chapter 12.

Theorem 6.23. *Suppose $\left(\widetilde{M},\widetilde{g}\right)$ and (M,g) are connected Riemannian manifolds with $\widetilde{M}$ complete, and $\pi\colon \widetilde{M} \to M$ is a local isometry. Then M is complete and π is a Riemannian covering map.*

Proof. A fundamental property of covering maps is the *path-lifting property* (Prop. A.54(b)): if π is a covering map, then every continuous path $\gamma\colon I \to M$ lifts to a path $\widetilde{\gamma}$ in $\widetilde{M}$ such that $\pi \circ \widetilde{\gamma} = \gamma$. We begin by proving that π possesses the path-lifting property for geodesics (Fig. 6.15): if $p \in M$ is a point in the image of π, $\gamma\colon I \to M$ is any geodesic starting at p, and $\widetilde{p}$ is any point in $\pi^{-1}(p)$, then γ has a unique lift starting at $\widetilde{p}$. The lifted curve is necessarily also a geodesic because π is a local isometry.

To prove the path-lifting property for geodesics, suppose $p \in \pi(M)$ and $\widetilde{p} \in \pi^{-1}(p)$, and let $\gamma\colon I \to M$ be any geodesic with $p = \gamma(0)$. Let $v = \gamma'(0)$ and $\widetilde{v} = (d\pi_{\widetilde{p}})^{-1}(v) \in T_{\widetilde{p}}\widetilde{M}$ (which is well defined because $d\pi_{\widetilde{p}}$ is an isomorphism), and let $\widetilde{\gamma}$ be the geodesic in $\widetilde{M}$ with initial point $\widetilde{p}$ and initial velocity $\widetilde{v}$. Because $\widetilde{M}$ is complete, $\widetilde{\gamma}$ is defined on all of $\mathbb{R}$. Since π is a local isometry, it takes geodesics to geodesics; and since by construction $\pi\left(\widetilde{\gamma}(0)\right) = \gamma(0)$ and $d\pi_{\widetilde{p}}\left(\widetilde{\gamma}'(0)\right) = \gamma'(0)$, we must have $\pi \circ \widetilde{\gamma} = \gamma$ on I, so $\widetilde{\gamma}|_I$ is a lift of γ starting at $\widetilde{p}$.

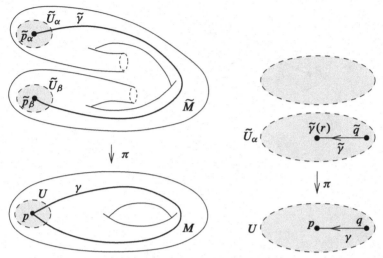

Fig. 6.16: Proof that $\widetilde{U}_\alpha \cap \widetilde{U}_\beta = \varnothing$ Fig. 6.17: Proof that $\pi^{-1}(U) \subseteq \bigcup_\alpha \widetilde{U}_\alpha$

To show that M is complete, let p be any point in the image of π. If $\gamma : I \to M$ is any geodesic starting at p, then γ has a lift $\widetilde{\gamma} : I \to \widetilde{M}$. Because $\widetilde{M}$ is complete, $\pi \circ \widetilde{\gamma}$ is a geodesic defined for all t that coincides with γ on I, so γ extends to all of $\mathbb{R}$. Thus M is complete by Corollary 6.20.

Next we show that π is surjective. Choose some point $\widetilde{p} \in \widetilde{M}$, write $p = \pi(\widetilde{p})$, and let $q \in M$ be arbitrary. Because M is connected and complete, there is a minimizing unit-speed geodesic segment γ from p to q. Letting $\widetilde{\gamma}$ be the lift of γ starting at $\widetilde{p}$ and $r = d_g(p, q)$, we have $\pi\big(\widetilde{\gamma}(r)\big) = \gamma(r) = q$, so q is in the image of π.

To show that π is a smooth covering map, we need to show that every point of M has a neighborhood U that is evenly covered, which means that $\pi^{-1}(U)$ is a disjoint union of connected open sets $\widetilde{U}_\alpha$ such that $\pi|_{\widetilde{U}_\alpha} : \widetilde{U}_\alpha \to U$ is a diffeomorphism. We will show, in fact, that every geodesic ball is evenly covered.

Let $p \in M$, and let $U = B_\varepsilon(p)$ be a geodesic ball centered at p. Write $\pi^{-1}(p) = \{\widetilde{p}_\alpha\}_{\alpha \in A}$, and for each α let $\widetilde{U}_\alpha$ denote the *metric* ball of radius ε around $\widetilde{p}_\alpha$ (we are not claiming that $\widetilde{U}_\alpha$ is a geodesic ball). The first step is to show that the various sets $\widetilde{U}_\alpha$ are disjoint. For every $\alpha \neq \beta$, there is a minimizing geodesic segment $\widetilde{\gamma}$ from $\widetilde{p}_\alpha$ to $\widetilde{p}_\beta$ because $\widetilde{M}$ is complete. The projected curve $\gamma = \pi \circ \widetilde{\gamma}$ is a geodesic segment that starts and ends at p (Fig. 6.16), whose length is the same as that of $\widetilde{\gamma}$. Such a geodesic must leave U and reenter it (since all geodesics passing through p and lying in U are radial line segments), and thus must have length at least 2ε. This means that $d_{\widetilde{g}}(\widetilde{p}_\alpha, \widetilde{p}_\beta) \geq 2\varepsilon$, and thus by the triangle inequality, $\widetilde{U}_\alpha \cap \widetilde{U}_\beta = \varnothing$.

The next step is to show that $\pi^{-1}(U) = \bigcup_\alpha \widetilde{U}_\alpha$. If $\widetilde{q}$ is any point in $\widetilde{U}_\alpha$, then there is a geodesic $\widetilde{\gamma}$ of length less than ε from $\widetilde{p}_\alpha$ to $\widetilde{q}$, and then $\pi \circ \widetilde{\gamma}$ is a geodesic of the same length from p to $\pi(\widetilde{q})$, showing that $\pi(\widetilde{q}) \in U$. It follows that $\bigcup_\alpha \widetilde{U}_\alpha \subseteq \pi^{-1}(U)$.

Conversely, suppose $\widetilde{q} \in \pi^{-1}(U)$, and set $q = \pi(\widetilde{q})$. This means that $q \in U$, so there is a minimizing radial geodesic γ in U from q to p, and $r = d_g(q, p) < \varepsilon$. Let $\widetilde{\gamma}$ be the lift of γ starting at $\widetilde{q}$ (Fig. 6.17). It follows that $\pi\big(\widetilde{\gamma}(r)\big) = \gamma(r) = p$. Therefore $\widetilde{\gamma}(r) = \widetilde{p}_\alpha$ for some α, and $d_{\widetilde{g}}(\widetilde{q}, \widetilde{p}_\alpha) \le L_g(\widetilde{\gamma}) = r < \varepsilon$, so $\widetilde{q} \in \widetilde{U}_\alpha$.

It remains only to show that $\pi \colon \widetilde{U}_\alpha \to U$ is a diffeomorphism for each α. It is certainly a local diffeomorphism (because π is). It is bijective because its inverse can be constructed explicitly: it is the map sending each radial geodesic starting at p to its lift starting at $\widetilde{p}_\alpha$. This completes the proof. $\square$

Corollary 6.24. *Suppose $\widetilde{M}$ and M are connected Riemannian manifolds, and $\pi \colon \widetilde{M} \to M$ is a Riemannian covering map. Then M is complete if and only if $\widetilde{M}$ is complete.*

Proof. A Riemannian covering map is, in particular, a local isometry. Thus if $\widetilde{M}$ is complete, π satisfies the hypotheses of Theorem 6.23, which implies that M is also complete.

Conversely, suppose M is complete. Let $\widetilde{p} \in \widetilde{M}$ and $\widetilde{v} \in T_p \widetilde{M}$ be arbitrary, and let $p = \pi(\widetilde{p})$ and $v = d\pi_{\widetilde{p}}(\widetilde{v})$. Completeness of M implies that the geodesic γ with $\gamma(0) = p$ and $\gamma'(0) = v$ is defined for all $t \in \mathbb{R}$, and then its lift $\widetilde{\gamma} \colon \mathbb{R} \to \widetilde{M}$ starting at $\widetilde{p}$ is a geodesic in $\widetilde{M}$ with initial velocity $\widetilde{v}$, also defined for all t. $\square$

Corollary 6.21 to the Hopf–Rinow theorem shows that any two points in a complete, connected Riemannian manifold can be joined by a minimizing geodesic segment. The next proposition gives a refinement of that statement.

Proposition 6.25. *Suppose (M, g) is a complete, connected Riemannian manifold, and $p, q \in M$. Every path-homotopy class of paths from p to q contains a geodesic segment γ that minimizes length among all admissible curves in the same path-homotopy class.*

Proof. Let $\pi \colon \widetilde{M} \to M$ be the universal covering manifold of M, endowed with the pullback metric $\widetilde{g} = \pi^* g$. Given $p, q \in M$ and a path $\sigma \colon [0, 1] \to M$ from p to q, choose a point $\widetilde{p} \in \pi^{-1}(p)$, and let $\widetilde{\sigma} \colon [0, 1] \to \widetilde{M}$ be the lift of σ starting at $\widetilde{p}$, and set $\widetilde{q} = \widetilde{\sigma}(1)$. By Corollary 6.21, there is a minimizing $\widetilde{g}$-geodesic segment $\widetilde{\gamma}$ from $\widetilde{p}$ to $\widetilde{q}$, and because π is a local isometry, $\gamma = \pi \circ \widetilde{\gamma}$ is a geodesic in M from p to q. If γ_1 is any other admissible curve from p to q in the same path-homotopy class, then by the monodromy theorem (Prop. A.54(c)), its lift $\widetilde{\gamma}_1$ starting at $\widetilde{p}$ also ends at $\widetilde{q}$. Thus $\widetilde{\gamma}_1$ is no longer than $\widetilde{\gamma}$, which implies γ_1 is no longer than γ. $\square$

Closed Geodesics

Suppose (M, g) is a connected Riemannian manifold. A ***closed geodesic*** in M is a nonconstant geodesic segment $\gamma \colon [a, b] \to M$ such that $\gamma(a) = \gamma(b)$ and $\gamma'(a) = \gamma'(b)$.

▶ **Exercise 6.26.** Show that a geodesic segment is closed if and only if it extends to a periodic geodesic defined on all of $\mathbb{R}$.

Round spheres have the remarkable property that all of their geodesics are closed when restricted to appropriate intervals. Of course, this is not typically the case, even for compact Riemannian manifolds; but it is natural to wonder whether closed geodesics exist in more general manifolds. Much work has been done in Riemannian geometry to determine how many closed geodesics exist in various situations. Here we can only touch on the simplest case; these results will be useful in some of the proofs of local-to-global theorems in Chapter 12.

A continuous path $\sigma : [0,1] \to M$ is called a *loop* if $\sigma(0) = \sigma(1)$. Two loops $\sigma_0, \sigma_1 : [0,1] \to M$ are said to be *freely homotopic* if they are homotopic through closed paths (but not necessarily preserving the base point), that is, if there exists a homotopy $H : [0,1] \times [0,1] \to M$ satisfying

$$H(s,0) = \sigma_0(s) \text{ and } H(s,1) = \sigma_1(s) \text{ for all } s \in [0,1],$$
$$H(0,t) = H(1,t) \text{ for all } t \in [0,1]. \tag{6.12}$$

This is an equivalence relation on the set of all loops in M, and an equivalence class is called a *free homotopy class*. The *trivial free homotopy class* is the equivalence class of any constant path.

▶ **Exercise 6.27.** Given a connected manifold M and a point $x \in M$, show that a loop based at x represents the trivial free homotopy class if and only if it represents the identity element of $\pi_1(M,x)$.

The next proposition shows that closed geodesics are easy to find on compact Riemannian manifolds that are not simply connected.

Proposition 6.28 (Existence of Closed Geodesics). *Suppose (M,g) is a compact, connected Riemannian manifold. Every nontrivial free homotopy class in M is represented by a closed geodesic that has minimum length among all admissible loops in the given free homotopy class.*

Proof. Problem 6-17.   □

The previous proposition guarantees the existence of at least one closed geodesic on every non-simply-connected compact Riemannian manifold. In fact, it was proved in 1951 by the Russian mathematicians Lazar Lyusternik and Abram Fet that closed geodesics exist on *every* compact Riemannian manifold, but the proof in the simply connected case is considerably harder. Proofs can be found in [Jos17] and [Kli95].

Distance Functions

Suppose (M,g) is a connected Riemannian manifold and $S \subseteq M$ is any subset. For each point $x \in M$, we define the *distance from x to S* to be

$$d_g(x,S) = \inf\{d_g(x,p) : p \in S\}.$$

Lemma 6.29. *Suppose (M, g) is a connected Riemannian manifold and $S \subseteq M$ is any subset.*

(a) $d_g(x, S) \le d_g(x, y) + d_g(y, S)$ *for all $x, y \in M$.*
(b) $x \mapsto d_g(x, S)$ *is a continuous function on M.*

▶ **Exercise 6.30.** Prove the preceding lemma.

The simplest example of a distance function occurs when the set S is just a singleton, $S = \{p\}$. Inside a geodesic ball around p, Corollary 6.12 shows that $d_g(x, S) = r(x)$, the radial distance function, and Corollary 6.10 shows that it has unit gradient where it it smooth (everywhere inside the geodesic ball except at p itself). The next theorem is a far-reaching generalization of that result.

Theorem 6.31. *Suppose (M, g) is a connected Riemannian manifold, $S \subseteq M$ is arbitrary, and $f : M \to [0, \infty)$ is the distance to S, that is, $f(x) = d_g(x, S)$ for all $x \in M$. If f is continuously differentiable on some open subset $U \subseteq M \smallsetminus S$, then $|\operatorname{grad} f| \equiv 1$ on U.*

Proof. Suppose $U \subseteq M \smallsetminus S$ is an open subset on which f is continuously differentiable, and $x \in U$. We will show first that $\left|\operatorname{grad} f|_x\right| \le 1$, and then that $\left|\operatorname{grad} f|_x\right| \ge 1$.

To prove the first inequality, we may assume $\operatorname{grad} f|_x \neq 0$, for otherwise the inequality is trivial. Let $v \in T_x M$ be any unit vector, and let γ be the unit-speed geodesic with $\gamma(0) = x$ and $\gamma'(0) = v$. Then for every positive t sufficiently small that $\gamma|_{[0,t]}$ is minimizing, Lemma 6.29 gives $d_g(\gamma(t), S) \le d_g(\gamma(t), \gamma(0)) + d_g(\gamma(0), S)$, or equivalently $f(\gamma(t)) \le t + f(\gamma(0))$. Therefore, since f is differentiable at x,

$$\left.\frac{d}{dt}\right|_{t=0} f(\gamma(t)) = \lim_{t \searrow 0} \frac{f(\gamma(t)) - f(\gamma(0))}{t} \le 1.$$

In particular, taking $v = (\operatorname{grad} f|_x)/\left|\operatorname{grad} f|_x\right|$ (the unit vector in the direction of $\operatorname{grad} f|_x$), we obtain

$$\left.\frac{d}{dt}\right|_{t=0} f(\gamma(t)) = df_x(\gamma'(0)) = \left\langle \operatorname{grad} f|_x, v\right\rangle = \left|\operatorname{grad} f|_x\right|.$$

This proves that $\left|\operatorname{grad} f|_x\right| \le 1$.

To prove the reverse inequality, assume for the sake of contradiction that $\left|\operatorname{grad} f|_x\right| < 1$. Since we are assuming that $\operatorname{grad} f$ is continuous on U, there exist $\delta, \varepsilon > 0$ such that $\bar{B}_\varepsilon(x)$ is a closed geodesic ball contained in U and $|\operatorname{grad} f| \le 1 - \delta$ on $\bar{B}_\varepsilon(x)$. Let c be a positive constant less than $\varepsilon\delta$. By definition of $d_g(x, S)$, there is an admissible curve $\alpha : [0, b] \to M$ (which we may assume to be parametrized by arc length) such that $\alpha(0) = x$, $\alpha(b) \in S$, and $b = L_g(\alpha) < d_g(x, S) + c$.

Since we are assuming $\bar{B}_\varepsilon(x) \subseteq U \subseteq M \smallsetminus S$, we have that $b > \varepsilon$, so $\alpha|_{[\varepsilon,b]}$ is an admissible curve from $\alpha(\varepsilon)$ to S. On the one hand,

$$d_g(\alpha(\varepsilon), S) \le L_g(\alpha|_{[\varepsilon,b]}) = b - \varepsilon < d_g(x, S) + c - \varepsilon. \tag{6.13}$$

On the other hand, for $0 \leq t \leq \varepsilon$, the fact that $\alpha(t) \in \bar{B}_\varepsilon(x)$ implies

$$\left| \frac{d}{dt} f(\alpha(t)) \right| = \left| \langle \operatorname{grad} f |_{\alpha(t)}, \alpha'(t) \rangle \right| \leq \left| \operatorname{grad} f |_{\alpha(t)} \right| |\alpha'(t)| \leq 1 - \delta.$$

Thus $f(\alpha(t)) \geq f(x) - (1-\delta)t$ for all such t. In particular, for $t = \varepsilon$, this means that

$$d_g(\alpha(\varepsilon), S) \geq d_g(x, S) - (1-\delta)\varepsilon. \tag{6.14}$$

Combining (6.13) and (6.14) yields $c > \varepsilon\delta$, contradicting our choice of c. $\square$

Motivated by the previous theorem, if (M, g) is a Riemannian manifold and $U \subseteq M$ is an open subset, we define a **local distance function on** U to be a continuously differentiable function $f \colon U \to \mathbb{R}$ such that $|\operatorname{grad} f|_g \equiv 1$ in U. Theorem 6.34 and Corollary 6.35 below will justify this terminology. But first, we develop some important general properties of local distance functions.

Theorem 6.32. *Suppose (M, g) is a Riemannian manifold and f is a smooth local distance function on an open subset $U \subseteq M$. Then $\nabla_{\operatorname{grad} f} (\operatorname{grad} f) \equiv 0$, and each integral curve of $\operatorname{grad} f$ is a unit-speed geodesic.*

Proof. Let $F \in \mathfrak{X}(U)$ denote the unit vector field $\operatorname{grad} f$. The definition of the gradient shows that for every vector field W, we have

$$Wf = df(W) = \langle F, W \rangle, \tag{6.15}$$

and therefore

$$Ff = \langle F, F \rangle = |\operatorname{grad} f|^2 = 1. \tag{6.16}$$

For every smooth vector field W on U, we have

$$
\begin{aligned}
\langle W, \nabla_F F \rangle &= F \langle W, F \rangle - \langle \nabla_F W, F \rangle && \text{(compatibility with } g\text{)} \\
&= F W f - \langle [F, W], F \rangle - \langle \nabla_W F, F \rangle && \text{((6.15), symmetry of } \nabla\text{)} \\
&= F W f - [F, W] f - \tfrac{1}{2} W |F|^2 && \text{((6.15), compatibility with } g\text{)} \\
&= W F f - \tfrac{1}{2} W |F|^2 && \text{(definition of } [F, W]\text{)} \\
&= 0 && \text{(since } Ff \equiv |F|^2 \equiv 1\text{)}.
\end{aligned}
$$

Since W is arbitrary, this proves that $\nabla_F F \equiv 0$.

If $\gamma \colon I \to U$ is an integral curve of F, then the fact that γ' is extendible implies

$$D_t \gamma'(t) = \nabla_F F |_{\gamma(t)} = 0,$$

so γ is a geodesic. $\square$

Lemma 6.33. *Suppose (M, g) is a Riemannian manifold, $K \subseteq M$, and $f \colon K \to \mathbb{R}$ is a continuous function whose restriction to some open set $W \subseteq K$ is a smooth local distance function. For every admissible curve $\sigma \colon [a_0, b_0] \to K$ such that $\sigma\big((a_0, b_0)\big) \subseteq W$, we have*

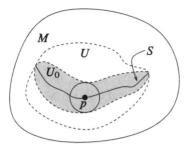

Fig. 6.18: Proving that $f(x) = d_g(x, S)$ in a neighborhood of S

$$L_g(\sigma) \geq |f(\sigma(b_0)) - f(\sigma(a_0))|.$$

Proof. This is proved exactly as in (6.8), noting that the only properties of r we used in that computation were that it is continuous on the image of σ and continuously differentiable on $\sigma((a_0, b_0))$, and its gradient has unit length there. $\square$

The next theorem and its corollary explain why the name "local distance function" is justified. Its proof is an adaptation of the proof of Proposition 6.11.

Theorem 6.34. *Suppose* (M, g) *is a Riemannian manifold,* $U \subseteq M$ *is an open subset,* $S \subseteq U$, *and* $f : U \to [0, \infty)$ *is a continuous function such that* $f^{-1}(0) = S$ *and* f *is a smooth local distance function on* $U \smallsetminus S$. *Then there is a neighborhood* $U_0 \subseteq U$ *of* S *in which* $f(x)$ *is equal to the distance in* M *from* x *to* S.

Proof. For each $p \in S$, there are positive numbers ε_p, δ_p such that $B_{\varepsilon_p}(p)$ is a uniformly δ_p-normal geodesic ball and $B_{2\varepsilon_p}(p) \subseteq U$. This means that $B_{\varepsilon_p}(p)$ is contained in the open geodesic ball of radius δ_p around each of its points. In particular, $B_{\varepsilon_p}(p) \subseteq B_{\delta_p}(p)$, which means that $\delta_p \geq \varepsilon_p$, and thus every geodesic starting at a point of $B_{\varepsilon_p}(p)$ is defined at least for $t \in (-\varepsilon_p, \varepsilon_p)$. Let U_0 be the union of all of the geodesic balls $B_{\varepsilon_p}(p)$ for $p \in S$, which is a neighborhood of S contained in U (Fig. 6.18).

Let $x \in U_0$ be arbitrary, and let $c = f(x)$. We will show that $d_g(x, S) = c$. If $x \in S$, then $d_g(x, S) = 0 = c$, so we may as well assume $x \notin S$.

There is some $p \in S$ such that $x \in B_{\varepsilon_p}(p)$, which means that $d_g(x, S) < \varepsilon_p$ and geodesics starting at x are defined at least on $(-\varepsilon_p, \varepsilon_p)$. Also, if $\alpha : [0, b] \to B_{\varepsilon}(p)$ is the radial geodesic segment from p to x, it follows from Lemma 6.33 that $L_g(\alpha) \geq |f(x) - f(p)| = c$, and we conclude that $c \leq L_g(\alpha) < \varepsilon_p$ as well.

Let $\gamma : (-\varepsilon_p, \varepsilon_p) \to U$ be the unit-speed geodesic starting at x with initial velocity equal to $-\operatorname{grad} f|_x$. By Theorem 6.32 and uniqueness of geodesics, γ coincides with an integral curve of $-\operatorname{grad} f$ as long as $\gamma(t) \in U \smallsetminus S$, which is to say as long as $f(\gamma(t)) \neq 0$. For all such t we have

$$\frac{d}{dt} f(\gamma(t)) = \langle \operatorname{grad} f|_{\gamma(t)}, \gamma'(t) \rangle = -|\operatorname{grad} f|_{\gamma(t)}|^2 = -1,$$

so $f(\gamma(t)) = c - t$ as long as $t < c$, and by continuity, $f(\gamma(c)) = 0$. This means that $\gamma(c) \in S$, and $\gamma|_{[0,c]}$ is a curve segment of length c connecting x with S, so $d_g(x, S) \le c$.

To prove the reverse inequality, suppose $\alpha \colon [a, b] \to M$ is any admissible curve starting at x and ending at a point of S. Assume first that $\alpha(t) \in U$ for all $t \in [a, b]$, and let $b_0 \in [a, b]$ be the first time that $\alpha(b_0) \in S$. Then Lemma 6.33 shows that $L_g(\alpha) \ge L_g(\alpha|_{[a, b_0]}) \ge |f(\alpha(b_0)) - f(\alpha(a))| = c$. On the other hand, suppose $\alpha(t) \in M \smallsetminus U$ for some t. The triangle inequality implies $B_{\varepsilon_p}(x) \subseteq B_{2\varepsilon_p}(p) \subseteq U$, so there is a first time $b_0 \in [a, b]$ such that $d_g(x, \alpha(b_0)) \ge \varepsilon_p$. Then $L_g(\alpha) \ge L_g(\alpha|_{[a, b_0]}) \ge \varepsilon_p > c$. Taken together, these two inequalities show that $L_g(\alpha) \ge c$ for every such α, which implies $d_g(x, S) \ge c$. $\qquad \square$

See Problem 6-27 for a global version of the preceding theorem.

Corollary 6.35. *Let (M, g) be a Riemannian manifold, and let f be a smooth local distance function on an open subset $U \subseteq M$. If c is a real number such that $S = f^{-1}(c)$ is nonempty, then there is a neighborhood U_0 of S in U on which $|f(x) - c|$ is equal to the distance in M from x to S.*

▶ **Exercise 6.36.** Prove the preceding corollary.

Distance Functions for Embedded Submanifolds

The most important local distance functions are those associated with embedded submanifolds. As we will see in this section, such distance functions are always smooth near the manifold.

Suppose (M, g) is a Riemannian n-manifold (without boundary) and $P \subseteq M$ is an embedded k-dimensional submanifold. Let NP denote the normal bundle of P in M, and let $U \subseteq M$ be a normal neighborhood of P in M, which is the diffeomorphic image of a certain open subset $V \subseteq NP$ under the normal exponential map. (Such a neighborhood always exists by Thm. 5.25.) We begin by constructing generalizations of the radial distance function and radial vector field (see (6.4) and (6.5)). Recall the definition of Fermi coordinates from Chapter 5 (see Prop. 5.26).

Proposition 6.37. *Let P be an embedded submanifold of a Riemannian manifold (M, g) and let U be any normal neighborhood of P in M. There exist a unique continuous function $r \colon U \to [0, \infty)$ and smooth vector field ∂_r on $U \smallsetminus P$ that have the following coordinate representations in terms of any Fermi coordinates $\left(x^1, \ldots, x^k, v^1, \ldots, v^{n-k}\right)$ for P on a subset $U_0 \subseteq U$:*

$$r\left(x^1, \ldots, x^k, v^1, \ldots, v^{n-k}\right) = \sqrt{(v^1)^2 + \cdots + (v^{n-k})^2}, \tag{6.17}$$

$$\partial_r = \frac{v^1}{r(x, v)} \frac{\partial}{\partial v^1} + \cdots + \frac{v^{n-k}}{r(x, v)} \frac{\partial}{\partial v^{n-k}}. \tag{6.18}$$

The function r is smooth on $U \smallsetminus P$, and r^2 is smooth on all of U.

Proof. The uniqueness, continuity, and smoothness claims follow immediately from the coordinate expressions (6.17) and (6.18), so we need only prove that r and ∂_r can be globally defined so as to have the indicated coordinate expressions in any Fermi coordinates.

Let $V \subseteq NS$ be the subset that is mapped diffeomorphically onto U by the normal exponential map E. Define a function $\rho \colon V \to [0, \infty)$ by $\rho(p, v) = |v|_g$, and define $r \colon U \to [0, \infty)$ by $r = \rho \circ E^{-1}$. Any Fermi coordinates for P are defined by choosing local coordinates $(x^1, \dots, x^k)$ for P and a local orthonormal frame (E_α) for NP, and assigning the coordinates $(x^1(p), \dots, x^k(p), v^1, \dots, v^{n-k})$ to the point $E(p, v^\alpha E_\alpha|_p)$. (Here we are using the summation convention with Greek indices running from 1 to $n - k$.) Because the frame is orthonormal, for each $(p, v) = (p, v^\alpha E_\alpha|_p) \in V$ we have $r(E(p, v))^2 = \rho(p, v)^2 = (v^1)^2 + \cdots + (v^{n-k})^2$, which shows that r has the coordinate representation (6.17).

To define ∂_r, let q be an arbitrary point in $U \smallsetminus P$. Then $q = \exp_p(v)$ for a unique $(p, v) \in V$, and the curve $\gamma \colon [0, 1] \to U$ given by $\gamma(t) = \exp_p(tv)$ is a geodesic from p to q. Define

$$\partial_r|_q = \frac{1}{r(q)} \gamma'(1), \tag{6.19}$$

which is independent of the choice of coordinates. Proposition 5.26 shows that in any Fermi coordinates, if we write $v = v^\alpha E_\alpha|_p$, then γ has the coordinate formula $\gamma(t) = (x^1, \dots, x^k, tv^1, \dots, tv^{n-k})$, and therefore $\gamma'(t) = v^\alpha \partial/\partial v^\alpha|_{\gamma(t)}$. It follows that ∂_r has the coordinate formula (6.18). $\square$

By analogy with the special case in which P is a point, we call r the **radial distance function for** P and ∂_r the **radial vector field for** P.

Theorem 6.38 (Gauss Lemma for Submanifolds). *Let P be an embedded submanifold of a Riemannian manifold (M, g), let U be a normal neighborhood of P in M, and let r and ∂_r be defined as in Proposition 6.37. On $U \smallsetminus P$, ∂_r is a unit vector field orthogonal to the level sets of r.*

Proof. The proof is a dressed-up version of the proof of the ordinary Gauss lemma. Let $q \in U \smallsetminus P$ be arbitrary, and let $(x^1, \dots, x^k, v^1, \dots, v^{n-k})$ be the coordinate representation of q in some choice of Fermi coordinates associated with a local orthonormal frame (E_α) for NP. As in the proof of Proposition 6.37, $q = \gamma(1)$, where γ is the geodesic $\exp_p(tv)$ for some $p \in P$ and $v = v^\alpha E_\alpha|_p \in N_pM$. Since the frame (E_α) is orthonormal, we have

$$|\gamma'(0)|_g = |v|_g = \sqrt{(v^1)^2 + \cdots + (v^{n-k})^2} = r(q).$$

Since geodesics have constant speed, it follows that $|\gamma'(1)|_g = r(q)$ as well, and then (6.19) shows that $\partial_r|_q$ is a unit vector.

Next we show that ∂_r is orthogonal to the level sets of r. Suppose $q \in U \smallsetminus P$, and write $q = \exp_{p_0}(v_0)$ for some $p_0 \in P$ and $v_0 \in N_pP$ with $v_0 \neq 0$. Let $b = r(q) = |v_0|_g$, so q lies in the level set $r^{-1}(b)$. The coordinate representation (6.17) shows that this is a regular level set, and hence an embedded submanifold of U.

Let $w \in T_q M$ be an arbitrary vector tangent to this level set, and let $\sigma : (-\varepsilon, \varepsilon) \to U$ be a smooth curve lying in the same level set, with $\sigma(0) = q$ and $\sigma'(0) = w$. We can write $\sigma(s) = \exp_{x(s)}(v(s))$, where $x(s) \in P$ and $v(s) \in N_{x(s)} P$ with $|v(s)|_g = b$. The initial condition $\sigma(0) = q$ translates to $x(0) = p_0$ and $v(0) = v_0$. Define a smooth one-parameter family of curves $\Gamma : (-\varepsilon, \varepsilon) \times [0, b] \to M$ by

$$\Gamma(s, t) = \exp_{x(s)} \left(\frac{t}{b} v^1(s), \dots, \frac{t}{b} v^n(s) \right).$$

Since $|v(s)|_g / b \equiv 1$, each Γ_s is a unit-speed geodesic.

Write $T(s, t) = \partial_t \Gamma(s, t)$ and $S(s, t) = \partial_s \Gamma(s, t)$ as in the proof of Theorem 6.9. We have the following endpoint conditions:

$$S(0, 0) = \frac{d}{ds}\bigg|_{s=0} x(s) = x'(0);$$

$$T(0, 0) = \frac{d}{dt}\bigg|_{t=0} \exp_{p_0}\left(\frac{t}{b} v_0\right) = \frac{1}{b} v_0;$$

$$S(0, b) = \frac{d}{ds}\bigg|_{s=0} \sigma(s) = w;$$

$$T(0, b) = \frac{d}{dt}\bigg|_{t=b} \exp_{p_0}\left(\frac{t}{b} v_0\right) = \partial_r \big|_q.$$

Then the same computation as in (6.7) shows that $(\partial / \partial t)\langle S, T \rangle_g \equiv 0$, and therefore $\langle w, \partial_r|_q \rangle_g = \langle S(0, b), T(0, b) \rangle_g = \langle S(0, 0), T(0, 0) \rangle_g = (1/b)\langle x'(0), v_0 \rangle_g$, which is zero because $x'(0)$ is tangent to P and v_0 is normal to it. This proves that ∂_r is orthogonal to the level sets of r.   □

Corollary 6.39. *Assume the hypotheses of Theorem 6.38.*

(a) *∂_r is equal to the gradient of r on $U \smallsetminus P$.*
(b) *r is a local distance function.*
(c) *Each unit-speed geodesic $\gamma : [a, b) \to U$ with $\gamma'(a)$ normal to P coincides with an integral curve of ∂_r on (a, b).*
(d) *P has a tubular neighborhood in which the distance in M to P is equal to r.*

Proof. By direct computation in Fermi coordinates using formulas (6.17) and (6.18), $\partial_r(r) \equiv 1$, which is equal to $|\partial_r|_g^2$ by the previous theorem. Thus $\partial_r = \operatorname{grad} r$ on $U \smallsetminus P$ by Problem 2-10. Because $\operatorname{grad} r$ is a unit vector field, r is a local distance function. By Proposition 5.26, the geodesics in U that start normal to P are represented in any Fermi coordinates by $t \mapsto (x^1, \dots, x^k, tv^1, \dots, tv^{n-k})$, and such a geodesic has unit speed if and only if $(v^1)^2 + \cdots + (v^{n-k})^2 = 1$. Another direct computation shows that each such curve is an integral curve of ∂_r wherever $r \neq 0$.

Finally, to prove (d), note that Theorem 6.34 shows that there is *some* neighborhood $\tilde{U}_0$ of P in M on which $r(x) = d_g(x, P)$; if we take U_0 to be a tubular neighborhood of P in $\tilde{U}_0$, then U_0 satisfies the conclusion.   □

When P is compact, we can say more.

Theorem 6.40. *Suppose (M, g) is a connected Riemannian manifold, $P \subseteq M$ is a compact submanifold, and U_ε is an ε-tubular neighborhood of P. Then U_ε is also an ε-neighborhood in the metric space sense, and inside U_ε, the distance in M to P is equal to the function r defined in Proposition 6.37.*

Proof. First we show that r can be extended continuously to $\bar{U}_\varepsilon$ by setting $r(q) = \varepsilon$ for $q \in \partial U_\varepsilon$. Indeed, suppose $q \in \partial U_\varepsilon$ and q_i is any sequence of points in U_ε converging to q. Then $\limsup_i r(q_i) \le \varepsilon$ because $r(q_i) < \varepsilon$ for each i. Let $c = \liminf_i r(q_i)$; we will prove the result by showing that $c = \varepsilon$. Suppose for the sake of contradiction that $c < \varepsilon$. By passing to a subsequence, we may assume that $r(q_i) \to c$. We can write $q_i = \exp_{p_i}(v_i)$ for $p_i \in P$ and $v_i \in N_{p_i} P$, and because P is compact and $\lim_i |v_i|_g = \lim_i r(q_i) = c$, we can pass to a further subsequence and assume that $(p_i, v_i) \to (p, v) \in NP$ with $|v|_g = c < \varepsilon$. Then we have $q = \lim_i q_i = \lim_i \exp_{p_i}(v_i) = \exp_p v$, which lies in the open set U_ε, contradicting our assumption that $q \in \partial U_\varepsilon$. Henceforth, we regard r as a continuous function on $\bar{U}_\varepsilon$.

Now to prove the theorem, let W_ε denote the ε-neighborhood of P in the metric space sense. Suppose first that $q \in M \smallsetminus U_\varepsilon$, and suppose $\alpha \colon [a, b] \to M$ is any admissible curve from a point of P to q. There is a first time $b_0 \in [a, b]$ that $\alpha(b_0) \in \partial U_\varepsilon$, and then Lemma 6.33 shows that

$$L_g(\alpha) \ge L_g\big(\alpha|_{[a, b_0]}\big) \ge |r(\alpha(b_0)) - r(\alpha(a))| = \varepsilon.$$

Thus $q \notin U_\varepsilon \Rightarrow q \notin W_\varepsilon$, or equivalently $W_\varepsilon \subseteq U_\varepsilon$.

Conversely, suppose $q \in U_\varepsilon$. Then q is connected to P by a geodesic segment of length $r(q)$, so $d_g(q, P) \le r(q)$. To prove the reverse inequality, suppose $\alpha \colon [a, b] \to M$ is any admissible curve starting at a point of P and ending at q. If $\alpha(t)$ remains in U for all $t \in [a, b]$, then Lemma 6.33 shows that

$$L_g(\alpha) \ge L_g\big(\alpha|_{[a_0, b]}\big) \ge |r(\gamma(b)) - r(\gamma(a_0))| = r(q),$$

where a_0 is the last time that $\alpha(a_0) \in P$. On the other hand, if $\alpha(t)$ does not remain in U, then there is a first time b_0 such that $\alpha(b_0) \in \partial U_\varepsilon$, and the argument in the preceding paragraph shows that $L_g(\alpha) \ge \varepsilon > r(q)$. Thus $d_g(q, P) = r(q)$ for all $q \in U_\varepsilon$. Since $r(q) < \varepsilon$ for all such q, it follows also that $U_\varepsilon \subseteq W_\varepsilon$. $\square$

Semigeodesic Coordinates

Local distance functions can be used to build coordinate charts near submanifolds in which the metric has a particularly simple form. We begin by describing the kind of coordinates we are looking for.

Let (M, g) be an n-dimensional Riemannian manifold. Smooth local coordinates $(x^1, \dots, x^n)$ on an open subset $U \subseteq M$ are called **semigeodesic coordinates** if each x^n-coordinate curve $t \mapsto (x^1, \dots, x^{n-1}, t)$ is a unit-speed geodesic that meets each level set of x^n orthogonally.

Because of the distinguished role played by the last coordinate function, through-out the rest of this section we will use the summation convention with Latin indices running from 1 to n and Greek indices running from 1 to $n-1$.

We will see below that semigeodesic coordinates are easy to construct. But first, let us develop some alternative characterizations of them.

Proposition 6.41 (Characterizations of Semigeodesic Coordinates). *Let (M, g) be a Riemannian n-manifold, and let $(x^1, \ldots, x^n)$ be smooth coordinates on an open subset of M. The following are equivalent:*

(a) (x^i) *are semigeodesic coordinates.*
(b) $|\partial_n|_g \equiv 1$ *and* $\langle \partial_\alpha, \partial_n \rangle_g \equiv 0$ *for* $\alpha = 1, \ldots, n-1$.
(c) $|dx^n|_g \equiv 1$ *and* $\langle dx^\alpha, dx^n \rangle_g \equiv 0$ *for* $\alpha = 1, \ldots, n-1$.
(d) $|\operatorname{grad} x^n|_g \equiv 1$ *and* $\langle \operatorname{grad} x^\alpha, \operatorname{grad} x^n \rangle_g \equiv 0$ *for* $\alpha = 1, \ldots, n-1$.
(e) x^n *is a local distance function and* $x^1, \ldots, x^{n-1}$ *are constant along the integral curves of* $\operatorname{grad} x^n$.
(f) $\operatorname{grad} x^n \equiv \partial_n$.

Proof. We begin by showing that (b) $\Leftrightarrow$ (c) $\Leftrightarrow$ (d) $\Leftrightarrow$ (e) and (c) $\Leftrightarrow$ (f). Note that (b) is equivalent to the coordinate matrix of g having the block form $\begin{pmatrix} * & 0 \\ 0 & 1 \end{pmatrix}$, where the asterisk represents an arbitrary $(n-1) \times (n-1)$ positive definite symmetric matrix, while (c) is equivalent to the inverse matrix having the same form. It follows from Cramer's rule that the matrix of g has this form if and only if its inverse does, and thus (b) is equivalent to (c).

The equivalence of (c) and (d) follows from the definitions of the gradient and of the inner product on 1-forms: for all $i, j = 1, \ldots, n$,

$$\langle dx^i, dx^j \rangle_g = \langle (dx^i)^\#, (dx^j)^\# \rangle_g = \langle \operatorname{grad} x^i, \operatorname{grad} x^j \rangle_g.$$

The equivalence of (d) and (e) also follows from the definition of the gradient: $\langle \operatorname{grad} x^\alpha, \operatorname{grad} x^n \rangle_g = dx^\alpha(\operatorname{grad} x^n) = (\operatorname{grad} x^n)(x^\alpha)$ for each α, which means that x^α is constant along the grad x^n integral curves if and only if $\langle \operatorname{grad} x^\alpha, \operatorname{grad} x^n \rangle_g = 0$. Finally, by examining the individual components of the coordinate formula $\operatorname{grad} x^n = g^{nj} \partial_j$, we see that (c) is also equivalent to (f).

To complete the proof, we show that (a) $\Leftrightarrow$ (b). Assume first that (a) holds. Because the x^n-coordinate curves have unit speed, it follows that $|\partial_n|_g \equiv 1$. The tangent space to any x^n-level set is spanned at each point by $\partial_1, \ldots, \partial_{n-1}$, and (a) guarantees that ∂_n is orthogonal to each of these, showing that (b) holds. Conversely, if we assume (b), the first part of the proof shows that (f) holds as well, so $|\operatorname{grad} x^n|_g \equiv |\partial_n|_g \equiv 1$, showing that x^n is a local distance function. Thus the x^n-coordinate curves are also integral curves of grad x^n and hence are unit-speed geodesics by Theorem 6.32. The fact that $\langle \partial_\alpha, \partial_n \rangle = 0$ for $\alpha = 2, \ldots, n$ implies that these geodesics are orthogonal to the level sets of x^n, thus proving (a). $\square$

Part (b) of this proposition leads to the following simplified coordinate represen-tations for the metric and Christoffel symbols in semigeodesic coordinates. Recall that implied summations with Greek indices run from 1 to $n-1$.

Corollary 6.42. *Let (x^i) be semigeodesic coordinates on an open subset of a Riemannian n-manifold (M, g).*

(a) The metric has the following coordinate expression:

$$g = (dx^n)^2 + g_{\alpha\beta}(x^1, \ldots, x^n)dx^\alpha dx^\beta.$$

(b) The Christoffel symbols of g have the following coordinate expressions:

$$\begin{aligned}
\Gamma^n_{nn} &= \Gamma^\alpha_{nn} = \Gamma^n_{\alpha n} = \Gamma^n_{n\alpha} = 0, \\
\Gamma^n_{\alpha\beta} &= -\tfrac{1}{2}\partial_n g_{\alpha\beta}, \\
\Gamma^\beta_{n\alpha} &= \Gamma^\beta_{\alpha n} = \tfrac{1}{2}g^{\beta\gamma}\partial_n g_{\alpha\gamma}, \\
\Gamma^\gamma_{\alpha\beta} &= \hat{\Gamma}^\gamma_{\alpha\beta},
\end{aligned} \tag{6.20}$$

where for each fixed value of x^n, the quantities $\hat{\Gamma}^\gamma_{\alpha\beta}$ are the Christoffel symbols in (x^α) coordinates for the induced metric $\hat{g}$ on the level set $x^n = $ constant.

Proof. Part (a) follows immediately from part (b) of Proposition 6.41, and (b) is proved by inserting $g_{nn} = 1$ and $g_{\alpha n} = g_{n\alpha} = 0$ into formula (5.8) for the Christoffel symbols. □

Proposition 6.41(e) gives us an effective way to construct semigeodesic coordinates: if r is any smooth local distance function (for example, the distance from a point or a smooth submanifold), just set $x^n = r$, choose any local coordinates $x^1, \ldots, x^{n-1}$ for a level set of r, and then extend them to be constant along the integral curves of grad r. Here are some explicit examples.

Example 6.43 (Fermi Coordinates for a Hypersurface). Suppose P is an embedded hypersurface in a Riemannian manifold (M, g), and let $(x^1, \ldots, x^{n-1}, v)$ be any Fermi coordinates for P on an open subset $U \subseteq M$ (see (5.25)). In this case, the function r defined by (6.17) is just $r(x^1, \ldots, x^{n-1}, v) = (v^2)^{1/2} = |v|$, so v is a local distance function on $U \smallsetminus P$. It follows from Corollary 6.39 that there is a neighborhood U_0 of P on which $|v|$ is equal to the distance from P. Moreover, (6.18) reduces to $\partial_r = \pm \partial/\partial_v$, which is equal to grad $|v|$ by Corollary 6.39, so it follows from Proposition 6.41(f) that Fermi coordinates for a hypersurface are automatically semigeodesic coordinates. //

Example 6.44 (Boundary Normal Coordinates). Suppose (M, g) is a smooth Riemannian manifold with nonempty boundary. The results in this chapter do not apply directly to manifolds with boundary, but we can embed M in its double $\widetilde{M}$ (Prop. A.31), extend the metric smoothly to $\widetilde{M}$, and construct Fermi coordinates $(x^1, \ldots, x^{n-1}, v)$ for ∂M in $\widetilde{M}$. By replacing v with $-v$ if necessary, we can arrange that $v > 0$ in Int M, and then these Fermi coordinates restrict to smooth boundary coordinates for M that are also semigeodesic coordinates. Such coordinates are called **boundary normal coordinates** for M. //

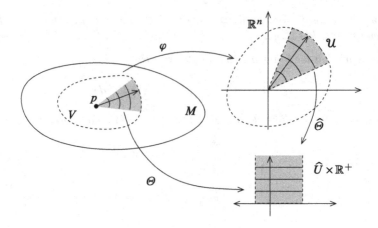

Fig. 6.19: Polar normal coordinates

Example 6.45 (Polar Normal Coordinates). *Polar coordinates* for $\mathbb{R}^n$ are con-
structed by choosing a smooth local parametrization $\widehat{\psi} \colon \widehat{U} \to U \subseteq \mathbb{S}^{n-1}$ for an
open subset U of $\mathbb{S}^{n-1}$, and defining $\widehat{\Psi} \colon \widehat{U} \times \mathbb{R}^+ \to \mathbb{R}^n$ by $\widehat{\Psi}(\theta^1, \ldots, \theta^{n-1}, r) =$
$r\widehat{\psi}(\theta^1, \ldots, \theta^{n-1})$. It is straightforward to show that the differential of $\widehat{\Psi}$ van-
ishes nowhere, so $\widehat{\Theta} = \widehat{\Psi}^{-1}$ is a smooth coordinate map on the open subset
$\mathcal{U} = \widehat{\Psi}(\widehat{U} \times \mathbb{R}^+) \subseteq \mathbb{R}^n \smallsetminus \{0\}$. Familiar examples are ordinary polar coordinates in
the plane and spherical coordinates in $\mathbb{R}^3$. Such coordinates have the property
that the last coordinate function is $r(x) = |x|$.

Now let (M, g) be a Riemannian n-manifold, p a point in M, and φ any normal
coordinate chart defined on a normal neighborhood V of p. For every choice of polar
coordinates $(\mathcal{U}, \widehat{\Theta})$ for $\mathbb{R}^n \smallsetminus \{0\}$, we obtain a smooth coordinate map $\Theta = \widehat{\Theta} \circ \varphi$ on
an open subset of $V \smallsetminus \{p\}$ (see Fig. 6.19). Such coordinates are called *polar normal
coordinates*. They have the property that the last coordinate function r is the radial
distance function on V, and the other coordinates are constant along the integral
curves of grad r, so they are semigeodesic coordinates.   //

Example 6.46 (Polar Fermi Coordinates). Now let P be an embedded subman-
ifold of (M, g), and let $\varphi = (x^1, \ldots, x^k, v^1, \ldots, v^{n-k})$ be Fermi coordinates on a
neighborhood U_0 of a point $p \in P$. Then any polar coordinate map $\widehat{\Theta}$ for $\mathbb{R}^{n-k}$
can be applied to the variables $(v^1, \ldots, v^{n-k})$ to yield a coordinate chart $\Theta =$
$(\mathrm{Id}_{\mathbb{R}^k} \times \widehat{\Theta}) \circ \varphi$ on an open subset of $U_0 \smallsetminus P$, taking values in $\mathbb{R}^k \times \mathbb{R}^{n-k-1} \times \mathbb{R}^+$.
If we write the coordinate functions as $(x^1, \ldots, x^k, \theta^1, \ldots, \theta^{n-k}, r)$, it follows from
Proposition 5.26 that each coordinate curve $t \mapsto (x^1, \ldots, x^k, \theta^1, \ldots, \theta^{n-k}, t)$ is a
unit-speed geodesic. Thus these are semigeodesic coordinates, called *polar Fermi
coordinates*. The polar normal coordinates described above are just the special case
$P = \{p\}$.   //

Problems

6-1. Suppose M is a nonempty connected Riemannian 1-manifold. Show that if M is noncompact, then it is isometric to an open interval in $\mathbb{R}$ with the Euclidean metric, while if it is compact, it is isometric to a circle $\mathbb{S}^1(R) = \{x \in \mathbb{R}^2 : |x| = R\}$ with its induced metric for some $R > 0$, using the following steps.

(a) Let $\gamma: I \to M$ be any maximal unit-speed geodesic. Show that its image is open and closed, and therefore γ is surjective.

(b) Show that if γ is injective, then it is an isometry between I with its Euclidean metric and M.

(c) Now suppose $\gamma(t_1) = \gamma(t_2)$ for some $t_1 \neq t_2$. In case $\gamma'(t_1) = \gamma'(t_2)$, show that γ is periodic, and descends to a global isometry from an appropriate circle to M.

(d) It remains only to rule out the case $\gamma(t_1) = \gamma(t_2)$ and $\gamma'(t_1) = -\gamma'(t_2)$. If this occurs, let $t_0 = (t_1 + t_2)/2$, and define geodesics α and β by

$$\alpha(t) = \gamma(t_0 + t), \qquad \beta(t) = \gamma(t_0 - t).$$

Use uniqueness of geodesics to conclude that $\alpha \equiv \beta$ on their common domain, and show that this contradicts the fact that γ is injective on some neighborhood of t_0.

6-2. Let n be a positive integer and R a positive real number.

(a) Prove that the Riemannian distance between any two points p, q in $\mathbb{S}^n(R)$ with the round metric is given by

$$d_{\mathring{g}_R}(p,q) = R \arccos \frac{\langle p, q \rangle}{R^2},$$

where $\langle \cdot, \cdot \rangle$ is the Euclidean inner product on $\mathbb{R}^{n+1}$.

(b) Prove that the metric space $\left(\mathbb{S}^n(R), d_{\mathring{g}_R} \right)$ has diameter πR.

(Used on pp. 39, 359.)

6-3. Let n be a positive integer and R a positive real number. Prove that the Riemannian distance between any two points in the Poincaré ball model $\left(\mathbb{B}^n(R), \breve{g}_R \right)$ of hyperbolic space of radius R is given by

$$d_{\breve{g}_R}(p,q) = R \operatorname{arccosh} \left(1 + \frac{2R^2 |p - q|^2}{\left(R^2 - |p|^2 \right)\left(R^2 - |q|^2 \right)} \right),$$

where $|\cdot|$ represents the Euclidean norm in $\mathbb{R}^n$. [Hint: First use the result of Problem 3-5 to show that it suffices to consider the case $R = 1$. Then use a rotation to reduce to the case $n = 2$, and use the group action of Problem 3-8 to show that it suffices to consider the case in which p is the origin.]

6-4. In Chapter 2, we started with a Riemannian metric and used it to define the Riemannian distance function. This problem shows how to go back the other way: the distance function determines the Riemannian metric. Let (M, g) be a connected Riemannian manifold.

(a) Show that if $\gamma: (-\varepsilon, \varepsilon) \to M$ is any smooth curve, then

$$|\gamma'(0)|_g = \lim_{t \searrow 0} \frac{d_g(\gamma(0), \gamma(t))}{t}.$$

(b) Show that if g and $\widetilde{g}$ are two Riemannian metrics on M such that $d_g(p, q) = d_{\widetilde{g}}(p, q)$ for all $p, q \in M$, then $g = \widetilde{g}$.

6-5. Prove Theorem 6.17 (sufficiently small geodesic balls are geodesically convex) as follows.

(a) Let (M, g) be a Riemannian manifold, let $p \in M$ be fixed, and let W be a uniformly normal neighborhood of p. For $\varepsilon > 0$ small enough that $B_{3\varepsilon}(p) \subseteq W$, define a subset $W_\varepsilon \subseteq TM \times \mathbb{R}$ by

$$W_\varepsilon = \{(q, v, t) \in TM \times \mathbb{R} : q \in B_\varepsilon(p), \ v \in T_q M, \ |v| = 1, \ |t| < 2\varepsilon\}.$$

Define $f: W_\varepsilon \to \mathbb{R}$ by

$$f(q, v, t) = d_g(p, \exp_q(tv))^2.$$

Show that f is smooth. [Hint: Use normal coordinates centered at p.]

(b) Show that if ε is chosen small enough, then $\partial^2 f / \partial t^2 > 0$ on W_ε. [Hint: Compute $f(p, v, t)$ explicitly and use continuity. Be careful to verify that ε can be chosen independently of v.]

(c) Suppose ε is chosen as in (b). Show that if $q_1, q_2 \in B_\varepsilon(p)$ and γ is a minimizing geodesic segment from q_1 to q_2, then $d_g(p, \gamma(t))$ attains its maximum at one of the endpoints of γ.

(d) Show that $B_\varepsilon(p)$ is geodesically convex.

6-6. Suppose (M, g) is a Riemannian manifold. For each $x \in M$, define the **convexity radius of M at x**, denoted by $\mathrm{conv}(x)$, to be the supremum of all $\varepsilon > 0$ such that there is a geodesically convex geodesic ball of radius ε centered at x. Show that $\mathrm{conv}(x)$ is a continuous function of x.

6-7. We now have two kinds of "metrics" on a connected Riemannian manifold: the Riemannian metric and the distance function. Correspondingly, there are two definitions of "isometry" between connected Riemannian manifolds: a **Riemannian isometry** is a diffeomorphism that pulls one Riemannian metric back to the other, and a **metric isometry** is a homeomorphism that preserves distances. Proposition 2.51 shows that every Riemannian isometry is a metric isometry. This problem outlines a proof of the converse. Suppose (M, g) and $(\widetilde{M}, \widetilde{g})$ are connected Riemannian manifolds, and $\varphi: M \to \widetilde{M}$ is a metric isometry.

(a) Show that for every $p \in M$ and $v, w \in T_p M$, we have

$$\lim_{t \to 0} \frac{d_g(\exp_p tv, \exp_p tw)}{t} = |v - w|_g.$$

[Hint: Use the Taylor series for g in Riemannian normal coordinates on a convex geodesic ball centered at p.]

(b) Show that φ takes geodesics to geodesics.

(c) For each $p \in M$, show that there exist an open ball $B_\varepsilon(0) \subseteq T_p M$ and a continuous map $\psi: B_\varepsilon(0) \to T_{\varphi(p)} \widetilde{M}$ satisfying $\psi(0) = 0$ and $\exp_{\varphi(p)} \psi(v) = \varphi(\exp_p v)$ for all $v \in B_\varepsilon(0)$.

(d) With ψ as above, show that $|\psi(v) - \psi(w)|_{\widetilde{g}} = |v - w|_g$ for all $v, w \in B_\varepsilon(0)$, and conclude from Problem 2-2 that ψ is the restriction of a linear isometry.

(e) With p and ψ as above, show that φ is smooth on a neighborhood of p and $d\varphi_p = \psi$.

(f) Conclude that φ is a Riemannian isometry.

6-8. Suppose (M, g) and $(\widetilde{M}, \widetilde{g})$ are connected Riemannian manifolds (not necessarily complete), and for each $i \in \mathbb{Z}^+$, $\varphi_i: M \to \widetilde{M}$ is a Riemannian isometry such that φ_i converges pointwise to a map $\varphi: M \to \widetilde{M}$. Show that φ is a Riemannian isometry. [Hint: Once you have shown that φ is a local isometry, to show that φ is surjective, suppose y is a limit point of $\varphi(M)$. Choose $x \in M$ such that $\varphi(x)$ lies in a uniformly normal neighborhood of y, and show that there exists a convergent sequence of points $(x_i, v_i) \in TM$ such that $\varphi_i(x_i) = \varphi(x)$ and $\varphi_i(\exp_{x_i} v_i) = y$.]

6-9. Suppose (M, g) is a (not necessarily connected) Riemannian manifold. In Problem 2-30, you were asked to show that there is a distance function $d: M \times M \to \mathbb{R}$ that induces the given topology and restricts to the Riemannian distance on each connected component of M. Show that if each component of M is geodesically complete, then d can be chosen so that (M, d) is a complete metric space. Show also that if M has infinitely many components, then d can be chosen so that (M, d) is not complete.

6-10. A curve $\gamma: [0, b) \to M$ (with $0 < b \le \infty$) is said to **diverge to infinity** if for every compact set $K \subseteq M$, there is a time $T \in [0, b)$ such that $\gamma(t) \notin K$ for $t > T$. (For those who are familiar with one-point compactifications, this means that $\gamma(t)$ converges to the "point at infinity" in the one-point compactification of M as $t \nearrow b$.) Prove that a connected Riemannian manifold is complete if and only if every regular curve that diverges to infinity has infinite length. (The length of a curve whose domain is not compact is just the supremum of the lengths of its restrictions to compact subintervals.)

6-11. Suppose (M, g) is a connected Riemannian manifold, $P \subseteq M$ is a connected embedded submanifold, and $\widehat{g}$ is the induced Riemannian metric on P.

(a) Show that $d_{\widehat{g}}(p, q) \ge d_g(p, q)$ for $p, q \in P$.

(b) Prove that if (M,g) is complete and P is closed in M, then $(P,\hat{g})$ is complete.

(c) Give an example of a complete Riemannian manifold (M,g) and a connected embedded submanifold $P \subseteq M$ that is complete but not closed in M.

6-12. Let (M,g) be a connected Riemannian manifold.

(a) Suppose there exists $\delta > 0$ such that for each $p \in M$, every maximal unit-speed geodesic starting at p is defined at least on an interval of the form $(-\delta,\delta)$. Prove that M is complete.

(b) Prove that if M has positive or infinite injectivity radius, then it is complete.

(c) Prove that if M is homogeneous, then it is complete.

(d) Give an example of a complete, connected Riemannian manifold that has zero injectivity radius.

6-13. Let G be a connected compact Lie group. Show that the Lie group exponential map of G is surjective. [Hint: Use Problem 5-8.]

6-14. Let (M,g) be a connected Riemannian manifold.

(a) Show that M is complete if and only if the compact subsets of M are exactly the closed and bounded ones.

(b) Show that M is compact if and only if it is complete and bounded.

6-15. Let S be the unit 2-sphere minus its north and south poles, and let $M = (0,\pi) \times \mathbb{R}$. Define $q: M \to S$ by $q(\varphi,\theta) = (\sin\varphi\cos\theta,\ \sin\varphi\sin\theta,\ \cos\varphi)$, and let g be the metric on M given by pulling back the round metric: $g = q^*\mathring{g}$. (Think of M as an infinitely long onion skin wrapping infinitely many times around an onion.) Prove that the Riemannian manifold (M,g) has diameter π, but contains infinitely long properly embedded geodesics.

6-16. Suppose (M,g) is a complete, connected Riemannian manifold with positive or infinite injectivity radius.

(a) Let $\rho \in (0,\infty]$ denote the injectivity radius of M, and define $T^\rho M$ to be the subset of TM consisting of vectors of length less than ρ, and D^ρ to be the subset $\{(p,q) : d_g(p,q) < \rho\} \subseteq M \times M$. Define $E: T^\rho M \to D^\rho$ by $E(x,v) = (x,\exp_x v)$. Prove that E is a diffeomorphism.

(b) Use part (a) to prove that if B is a topological space and $F,G: B \to M$ are continuous maps such that $d_g(F(x),G(x)) < \mathrm{inj}(M)$ for all $x \in B$, then F and G are homotopic.

6-17. Prove Proposition 6.28 (existence of a closed geodesic in a free homotopy class). [Hint: Use Prop. 6.25 to show that the given free homotopy class is represented by a *geodesic loop*, i.e., a geodesic whose starting and ending points are the same. Show that the lengths of such loops have a positive greatest lower bound; then choose a sequence of geodesic loops whose lengths approach that lower bound, and show that a subsequence converges uniformly to a geodesic loop whose length is equal to the lower bound. Use Problem 6-16 to show that the limiting curve is in the given free homotopy class, and apply the first variation formula to show that the limiting curve is in fact a closed geodesic.]

6-18. A connected Riemannian manifold (M, g) is said to be **k-point homogeneous** if for any two ordered k-tuples $(p_1, \ldots, p_k)$ and $(q_1, \ldots, q_k)$ of points in M such that $d_g(p_i, p_j) = d_g(q_i, q_j)$ for all i, j, there is an isometry $\varphi \colon M \to M$ such that $\varphi(p_i) = q_i$ for $i = 1, \ldots, k$. Show that (M, g) is 2-point homogeneous if and only if it is isotropic. [Hint: Assuming that M is isotropic, first show that it is homogeneous by considering the midpoint of a geodesic segment joining sufficiently nearby points $p, q \in M$, and then use the result of Problem 6-12(c) to show that it is complete.] (*Used on pp. 56, 261.*)

6-19. Prove that every Riemannian symmetric space is homogeneous. [Hint: Proceed as in Problem 6-18.] (*Used on p. 78.*)

6-20. A connected Riemannian manifold is said to be **extendible** if it is isometric to a proper open subset of a larger connected Riemannian manifold.

(a) Show that every complete, connected Riemannian manifold is nonextendible.

(b) Show that the converse is false by giving an example of a nonextendible connected Riemannian manifold that is not complete.

6-21. Let (M, g) be a complete, connected, noncompact Riemannian manifold. Define a **ray** in M to be a geodesic γ whose domain is $[0, \infty)$, and such that the restriction of γ to $[0, b]$ is minimizing for every $b > 0$. Prove that for each $p \in M$ there is a ray in M starting at p.

6-22. Let (M, g) be a connected Riemannian manifold with boundary. Prove that (M, d_g) is a complete metric space if and only if the following condition holds: for every geodesic $\gamma \colon [0, b) \to M$ that cannot be extended to a geodesic on any interval $[0, b')$ with $b' > b$, $\gamma(t)$ converges to a point of ∂M as $t \nearrow b$.

6-23. In some treatments of Riemannian geometry, instead of minimizing the length functional, one considers the following **energy functional** for an admissible curve $\gamma \colon [a, b] \to M$:

$$E(\gamma) = \tfrac{1}{2} \int_a^b |\gamma'(t)|^2 \, dt.$$

(Note that $E(\gamma)$ is *not* independent of parametrization.)

(a) Prove that an admissible curve is a critical point for E (with respect to proper variations) if and only if it is a geodesic (which means, in particular, that it has constant speed).

(b) Prove that if γ is an admissible curve that minimizes energy among admissible curves with the same endpoints, then it also minimizes length.

(c) Prove that if γ is an admissible curve that minimizes length among admissible curves with the same endpoints, then it minimizes energy if and only if it has constant speed.

[Remark: For our limited purposes, it is easier and more straightforward to use the length functional. But because the energy functional does not involve the square root function, and its critical points automatically have constant-speed parametrizations, it is sometimes more useful for proving the existence of geodesics with certain properties.]

6-24. Let (M,g) be a Riemannian manifold. Recall that a vector field $X \in \mathfrak{X}(M)$ is called a **Killing vector field** if $\mathcal{L}_X g = 0$ (see Problem 5-22).

(a) Prove that a Killing vector field that is normal to a geodesic at one point is normal everywhere along the geodesic.
(b) Prove that if a Killing vector field vanishes at a point p, then it is tangent to geodesic spheres centered at p.
(c) Prove that a Killing vector field on an odd-dimensional manifold cannot have an isolated zero.

(Used on p. 315.)

6-25. Suppose (M,g) is a Riemannian manifold and $f : M \to \mathbb{R}$ is a smooth local distance function. Prove that f is a Riemannian submersion.

6-26. Suppose (M,g) is a complete, connected Riemannian manifold and $S \subseteq M$ is a closed subset.

(a) Show that for every $p \in M \smallsetminus S$, there is a geodesic segment from p to a point of S whose length is equal to $d_g(p, S)$.
(b) In case S is a properly embedded submanifold of M, show that every geodesic segment satisfying the conclusion of (a) intersects S orthogonally.
(c) Give a counterexample to (a) if S is not closed.

6-27. Suppose (M,g) is a complete, connected Riemannian manifold, $S \subseteq M$ is a closed subset, and $f : M \to \mathbb{R}$ is a continuous function such that $f^{-1}(0) = S$ and f is smooth with unit gradient on $M \smallsetminus S$. Prove that $f(x) = d_g(x, S)$ for all $x \in M$.

6-28. Suppose (M,g) is a complete, connected Riemannian n-manifold whose isometry group is a Lie group acting smoothly on M. (The Myers–Steenrod theorem shows that this is always the case when M is connected, but we have not proved this.) Prove that if G is a closed subgroup of $\mathrm{Iso}(M)$, then G acts properly on M, using the following outline. Suppose (φ_m) is a sequence in G and (p_m) is a sequence in M such that $p_m \to p_0$ and $\varphi_m(p_m) \to q_0$.

(a) Prove that $\varphi_m(p_0) \to q_0$.
(b) Let $B_r(p_0)$ be a geodesic ball centered at p_0; let $0 < \varepsilon < r$; let $(b_1,\ldots,b_n)$ be an orthonormal basis for $T_{p_0}M$; and let $p_i = \exp_{p_0}(\varepsilon b_i)$ for $i = 1,\ldots,n$. Prove that there exist a linear isometry $A : T_{p_0}M \to$

$T_{q_0}M$ and a subsequence (φ_{m_j}) such that $\varphi_{m_j}(p_i) \to \exp_{q_0}(\varepsilon A b_i)$ for $i = 1, \ldots, n$. [Hint: Use Prop. 5.20.]

(c) Prove that there is an isometry $\varphi \in G$ such that $\varphi_{m_j} \to \varphi$ pointwise, meaning that $\varphi_{m_j}(x) \to \varphi(x)$ for every $x \in M$.

(d) Prove that $\varphi_{m_j} \to \varphi$ in the topology of G. [Hint: Define a map $F : G \to M^{n+1}$ by $F(\psi) = (\psi(p_0), \psi(p_1), \ldots, \psi(p_n))$, where $p_1, \ldots, p_n$ are the points introduced in part (b). Show that F is continuous, injective, and closed, and therefore is a homeomorphism onto its image.]

(Used on p. 349.)

6-29. Suppose (M, g) is a Riemannian n-manifold that admits a nonzero parallel vector field X. Show that for each $p \in M$, there exist a neighborhood U of p, a Riemannian $(n-1)$-manifold N, and an open interval $I \subseteq \mathbb{R}$ with its Euclidean metric such that U is isometric to the Riemannian product $N \times I$. [Hint: It might be helpful to prove that the 1-form $X^\flat$ is closed and to compute the Lie derivative $\mathcal{L}_X g$.]

6-30. Suppose (M, g) is a Riemannian n-manifold, $p \in M$, and $B_\varepsilon(p)$ is a geodesic ball centered at p. Prove that for every δ such that $0 < \delta < \varepsilon$,

$$\mathrm{Vol}\left(\bar{B}_\delta(p)\right) = \int_0^\delta \mathrm{Area}(\partial B_r(p))\, dr,$$

where $\mathrm{Area}(\partial B_r(p))$ represents the $(n-1)$-dimensional volume of $\partial B_r(p)$ with its induced Riemannian metric.

6-31. Suppose (M, g) is a Riemannian n-manifold, $P \subseteq M$ is a compact embedded submanifold, and U is an ε-tubular neighborhood of P for some $\varepsilon > 0$. For $0 < \delta < \varepsilon$, define U_δ and P_δ by

$$U_\delta = \{x \in U : d_g(x, P) \le \delta\},$$
$$P_\delta = \{x \in U : d_g(x, P) = \delta\}.$$

(a) Prove that U_δ is a regular domain and P_δ is a compact, embedded submanifold of M for each $\delta \in (0, \varepsilon)$.

(b) Generalize the result of Problem 6-30 by proving that

$$\mathrm{Vol}(U_\delta) = \int_0^\delta \mathrm{Area}(P_r)\, dr,$$

where $\mathrm{Area}(P_r)$ denotes the $(n-1)$-dimensional volume of P_r with the induced Riemannian metric.

Chapter 7
Curvature

In this chapter, we begin our study of the local invariants of Riemannian metrics. Starting with the question whether all Riemannian metrics are locally isometric, we are led to a definition of the Riemannian curvature tensor as a measure of the failure of second covariant derivatives to commute. Then we prove the main result of this chapter: a Riemannian manifold has zero curvature if and only if it is *flat*, or locally isometric to Euclidean space. Next, we derive the basic symmetries of the curvature tensor, and introduce the Ricci, scalar, and Weyl curvature tensors. At the end of the chapter, we explore how the curvature can be used to detect conformal flatness. As you will see, the results of this chapter apply essentially unchanged to pseudo-Riemannian metrics.

Local Invariants

For any geometric structure defined on smooth manifolds, it is of great interest to address the *local equivalence question*: Are all examples of the structure locally equivalent to each other (under an appropriate notion of local equivalence)?

There are some interesting and useful structures in differential geometry that have the property that all such structures on manifolds of the same dimension are locally equivalent to each other. For example:

- NONVANISHING VECTOR FIELDS: Every nonvanishing vector field can be written as $X = \partial/\partial x^1$ in suitable local coordinates, so they are all locally equivalent.
- RIEMANNIAN METRICS ON A 1-MANIFOLD: Problem 2-1 shows that every Riemannian 1-manifold is locally isometric to $\mathbb{R}$ with its Euclidean metric.
- SYMPLECTIC FORMS: A *symplectic form* on a smooth manifold M is a closed 2-form ω that is *nondegenerate* at each $p \in M$, meaning that $\omega_p(v, w) = 0$ for all $w \in T_p M$ only if $v = 0$. By the theorem of Darboux [LeeSM, Thm. 22.13], every symplectic form can be written in suitable coordinates as $\sum_i dx^i \wedge dy^i$. Thus all symplectic forms on $2n$-manifolds are locally equivalent.

© Springer International Publishing AG 2018
J. M. Lee, *Introduction to Riemannian Manifolds*, Graduate Texts in Mathematics 176, https://doi.org/10.1007/978-3-319-91755-9_7

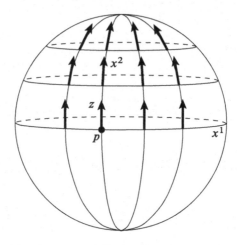

Fig. 7.1: Result of parallel transport along the x^1-axis and the x^2-coordinate lines

On the other hand, Problem 5-5 showed that the round 2-sphere and the Euclidean plane are not locally isometric.

The most important technique for proving that two geometric structures are not locally equivalent is to find *local invariants*, which are quantities that must be preserved by local equivalences. In order to address the general problem of local equivalence of Riemannian or pseudo-Riemannian metrics, we will define a local invariant for all such metrics called *curvature*. Initially, its definition will have nothing to do with the curvature of curves described in Chapter 1, but later we will see that the two concepts are intimately related.

To motivate the definition of curvature, let us look back at the argument outlined in Problem 5-5 for showing that the sphere and the plane are not locally isometric. The key idea is that every tangent vector in the plane can be extended to a parallel vector field, so every Riemannian manifold that is locally isometric to $\mathbb{R}^2$ must have the same property locally.

Given a Riemannian 2-manifold M, here is one way to attempt to construct a parallel extension of a vector $z \in T_p M$: working in any smooth local coordinates (x^1, x^2) centered at p, first parallel transport z along the x^1-axis, and then parallel transport the resulting vectors along the coordinate lines parallel to the x^2-axis (Fig. 7.1). The result is a vector field Z that, by construction, is parallel along every x^2-coordinate line and along the x^1-axis. The question is whether this vector field is parallel along x^1-coordinate lines other than the x^1-axis, or in other words, whether $\nabla_{\partial_1} Z \equiv 0$. Observe that $\nabla_{\partial_1} Z$ vanishes when $x^2 = 0$. If we could show that

$$\nabla_{\partial_2} \nabla_{\partial_1} Z = 0, \tag{7.1}$$

then it would follow that $\nabla_{\partial_1} Z \equiv 0$, because the zero vector field is the unique parallel transport of zero along the x^2-curves. If we knew that

$$\nabla_{\partial_2}\nabla_{\partial_1} Z = \nabla_{\partial_1}\nabla_{\partial_2} Z, \tag{7.2}$$

then (7.1) would follow immediately, because $\nabla_{\partial_2} Z = 0$ everywhere by construction. Indeed, on $\mathbb{R}^2$ with the Euclidean metric, direct computation shows that

$$\overline{\nabla}_{\partial_2}\overline{\nabla}_{\partial_1} Z = \overline{\nabla}_{\partial_2}\left(\left(\partial_1 Z^k\right)\partial_k\right)$$
$$= \left(\partial_2\partial_1 Z^k\right)\partial_k,$$

and $\overline{\nabla}_{\partial_1}\overline{\nabla}_{\partial_2} Z$ is equal to the same thing, because ordinary second partial derivatives commute. However, (7.2) might not hold for an arbitrary Riemannian metric; indeed, it is precisely the noncommutativity of such second covariant derivatives that forces this construction to fail on the sphere. Lurking behind this noncommutativity is the fact that the sphere is "curved."

To express this noncommutativity in a coordinate-independent way, let us look more closely at the quantity $\overline{\nabla}_X\overline{\nabla}_Y Z - \overline{\nabla}_Y\overline{\nabla}_X Z$ when X, Y, and Z are smooth vector fields. On $\mathbb{R}^2$ with the Euclidean connection, we just showed that this always vanishes if $X = \partial_1$ and $Y = \partial_2$; however, for arbitrary vector fields this may no longer be true. In fact, in $\mathbb{R}^n$ with the Euclidean connection we have

$$\overline{\nabla}_X\overline{\nabla}_Y Z = \overline{\nabla}_X\left(Y\left(Z^k\right)\partial_k\right) = XY\left(Z^k\right)\partial_k,$$

and similarly $\overline{\nabla}_Y\overline{\nabla}_X Z = YX\left(Z^k\right)\partial_k$. The difference between these two expressions is $\left(XY\left(Z^k\right) - YX\left(Z^k\right)\right)\partial_k = \overline{\nabla}_{[X,Y]}Z$. Therefore, the following relation holds for all vector fields X, Y, Z defined on an open subset of $\mathbb{R}^n$:

$$\overline{\nabla}_X\overline{\nabla}_Y Z - \overline{\nabla}_Y\overline{\nabla}_X Z = \overline{\nabla}_{[X,Y]}Z.$$

Recall that a Riemannian manifold is said to be *flat* if it is locally isometric to a Euclidean space, that is, if every point has a neighborhood that is isometric to an open set in $\mathbb{R}^n$ with its Euclidean metric. Similarly, a pseudo-Riemannian manifold is flat if it is locally isometric to a pseudo-Euclidean space. The computation above leads to the following simple necessary condition for a Riemannian or pseudo-Riemannian manifold to be flat. We say that a connection ∇ on a smooth manifold M satisfies the *flatness criterion* if whenever X, Y, Z are smooth vector fields defined on an open subset of M, the following identity holds:

$$\nabla_X\nabla_Y Z - \nabla_Y\nabla_X Z = \nabla_{[X,Y]}Z. \tag{7.3}$$

Example 7.1. The metric on the n-torus induced by the embedding in $\mathbb{R}^{2n}$ given in Example 2.21 is flat, because each point has a coordinate neighborhood in which the metric is Euclidean.   //

Proposition 7.2. *If (M, g) is a flat Riemannian or pseudo-Riemannian manifold, then its Levi-Civita connection satisfies the flatness criterion.*

Proof. We just showed that the Euclidean connection on $\mathbb{R}^n$ satisfies (7.3). By naturality, the Levi-Civita connection on every manifold that is locally isometric to a Euclidean or pseudo-Euclidean space must also satisfy the same identity.   □

The Curvature Tensor

Motivated by the computation in the preceding section, we make the following definition. Let (M, g) be a Riemannian or pseudo-Riemannian manifold, and define a map $R\colon \mathfrak{X}(M) \times \mathfrak{X}(M) \times \mathfrak{X}(M) \to \mathfrak{X}(M)$ by

$$R(X, Y)Z = \nabla_X \nabla_Y Z - \nabla_Y \nabla_X Z - \nabla_{[X,Y]} Z.$$

Proposition 7.3. *The map R defined above is multilinear over $C^\infty(M)$, and thus defines a $(1, 3)$-tensor field on M.*

Proof. The map R is obviously multilinear over $\mathbb{R}$. For $f \in C^\infty(M)$,

$$
\begin{aligned}
R(X, fY)Z &= \nabla_X \nabla_{fY} Z - \nabla_{fY} \nabla_X Z - \nabla_{[X, fY]} Z \\
&= \nabla_X (f \nabla_Y Z) - f \nabla_Y \nabla_X Z - \nabla_{f[X,Y]+(Xf)Y} Z \\
&= (Xf) \nabla_Y Z + f \nabla_X \nabla_Y Z - f \nabla_Y \nabla_X Z \\
&\quad - f \nabla_{[X,Y]} Z - (Xf) \nabla_Y Z \\
&= f R(X, Y)Z.
\end{aligned}
$$

The same proof shows that R is linear over $C^\infty(M)$ in X, because $R(X,Y)Z = -R(Y,X)Z$ from the definition. The remaining case to be checked is linearity over $C^\infty(M)$ in Z; this is left to Problem 7-1.

By the tensor characterization lemma (Lemma B.6), the fact that R is multilinear over $C^\infty(M)$ implies that it is a $(1, 3)$-tensor field.   □

Thanks to this proposition, for each pair of vector fields $X, Y \in \mathfrak{X}(M)$, the map $R(X, Y)\colon \mathfrak{X}(M) \to \mathfrak{X}(M)$ given by $Z \mapsto R(X, Y)Z$ is a smooth bundle endomorphism of TM, called the **curvature endomorphism determined by X and Y**. The tensor field R itself is called the **(Riemann) curvature endomorphism** or the **(1, 3)-curvature tensor**. (Some authors call it simply the *curvature tensor*, but we reserve that name instead for another closely related tensor field, defined below.)

As a $(1,3)$-tensor field, the curvature endomorphism can be written in terms of any local frame with one upper and three lower indices. We adopt the convention that the *last* index is the contravariant (upper) one. (This is contrary to our default assumption that covector arguments come first.) Thus, for example, the curvature endomorphism can be written in terms of local coordinates (x^i) as

$$R = R_{ijk}{}^l \, dx^i \otimes dx^j \otimes dx^k \otimes \partial_l,$$

where the coefficients $R_{ijk}{}^l$ are defined by

$$R\left(\partial_i, \partial_j\right)\partial_k = R_{ijk}{}^l \partial_l.$$

The next proposition shows how to compute the components of R in coordinates.

Proposition 7.4. *Let* (M, g) *be a Riemannian or pseudo-Riemannian manifold. In terms of any smooth local coordinates, the components of the* $(1, 3)$*-curvature tensor are given by*

$$R_{ijk}{}^l = \partial_i \Gamma_{jk}^l - \partial_j \Gamma_{ik}^l + \Gamma_{jk}^m \Gamma_{im}^l - \Gamma_{ik}^m \Gamma_{jm}^l. \tag{7.4}$$

Proof. Problem 7-2.   □

Importantly for our purposes, the curvature endomorphism also measures the failure of second covariant derivatives *along families of curves* to commute. Given a smooth one-parameter family of curves $\Gamma\colon J \times I \to M$, recall from Chapter 6 that the velocity fields $\partial_t \Gamma(s,t) = (\Gamma_s)'(t)$ and $\partial_s \Gamma(s,t) = \Gamma^{(t)'}(s)$ are smooth vector fields along Γ.

Proposition 7.5. *Suppose* (M, g) *is a smooth Riemannian or pseudo-Riemannian manifold and* $\Gamma\colon J \times I \to M$ *is a smooth one-parameter family of curves in* M. *Then for every smooth vector field* V *along* Γ,

$$D_s D_t V - D_t D_s V = R(\partial_s \Gamma, \partial_t \Gamma)V. \tag{7.5}$$

Proof. This is a local question, so for each $(s,t) \in J \times I$, we can choose smooth coordinates (x^i) defined on a neighborhood of $\Gamma(s,t)$ and write

$$\Gamma(s,t) = \left(\gamma^1(s,t), \ldots, \gamma^n(s,t)\right), \qquad V(s,t) = V^j(s,t)\partial_j\big|_{\Gamma(s,t)}.$$

Formula (4.15) yields

$$D_t V = \frac{\partial V^i}{\partial t}\partial_i + V^i D_t \partial_i.$$

Therefore, applying (4.15) again, we get

$$D_s D_t V = \frac{\partial^2 V^i}{\partial s \partial t}\partial_i + \frac{\partial V^i}{\partial t} D_s \partial_i + \frac{\partial V^i}{\partial s} D_t \partial_i + V^i D_s D_t \partial_i.$$

Interchanging s and t and subtracting, we see that all the terms except the last cancel:

$$D_s D_t V - D_t D_s V = V^i (D_s D_t \partial_i - D_t D_s \partial_i). \tag{7.6}$$

Now we need to compute the commutator in parentheses. For brevity, let us write

$$S = \partial_s \Gamma = \frac{\partial \gamma^k}{\partial s}\partial_k; \qquad T = \partial_t \Gamma = \frac{\partial \gamma^j}{\partial t}\partial_j.$$

Because ∂_i is extendible,

$$D_t \partial_i = \nabla_T \partial_i = \frac{\partial \gamma^j}{\partial t}\nabla_{\partial_j}\partial_i,$$

and therefore, because $\nabla_{\partial_j} \partial_i$ is also extendible,

$$
\begin{aligned}
D_s D_t \partial_i &= D_s \left(\frac{\partial \gamma^j}{\partial t} \nabla_{\partial_j} \partial_i \right) \\
&= \frac{\partial^2 \gamma^j}{\partial s \partial t} \nabla_{\partial_j} \partial_i + \frac{\partial \gamma^j}{\partial t} \nabla_S \left(\nabla_{\partial_j} \partial_i \right) \\
&= \frac{\partial^2 \gamma^j}{\partial s \partial t} \nabla_{\partial_j} \partial_i + \frac{\partial \gamma^j}{\partial t} \frac{\partial \gamma^k}{\partial s} \nabla_{\partial_k} \nabla_{\partial_j} \partial_i .
\end{aligned}
$$

Interchanging $s \leftrightarrow t$ and $j \leftrightarrow k$ and subtracting, we find that the first terms cancel, and we get

$$
\begin{aligned}
D_s D_t \partial_i - D_t D_s \partial_i &= \frac{\partial \gamma^j}{\partial t} \frac{\partial \gamma^k}{\partial s} \left(\nabla_{\partial_k} \nabla_{\partial_j} \partial_i - \nabla_{\partial_j} \nabla_{\partial_k} \partial_i \right) \\
&= \frac{\partial \gamma^j}{\partial t} \frac{\partial \gamma^k}{\partial s} R(\partial_k, \partial_j) \partial_i = R(S, T) \partial_i .
\end{aligned}
$$

Finally, inserting this into (7.6) yields the result. $\square$

For many purposes, the information contained in the curvature endomorphism is much more conveniently encoded in the form of a *covariant* 4-tensor. We define the **(Riemann) curvature tensor** to be the $(0, 4)$-tensor field $Rm = R^\flat$ (also denoted by *Riem* by some authors) obtained from the $(1, 3)$-curvature tensor R by lowering its last index. Its action on vector fields is given by

$$
Rm(X, Y, Z, W) = \langle R(X, Y)Z, W \rangle_g . \tag{7.7}
$$

In terms of any smooth local coordinates it is written

$$
Rm = R_{ijkl} dx^i \otimes dx^j \otimes dx^k \otimes dx^l ,
$$

where $R_{ijkl} = g_{lm} R_{ijk}{}^m$. Thus (7.4) yields

$$
R_{ijkl} = g_{lm} \left(\partial_i \Gamma^m_{jk} - \partial_j \Gamma^m_{ik} + \Gamma^p_{jk} \Gamma^m_{ip} - \Gamma^p_{ik} \Gamma^m_{jp} \right) . \tag{7.8}
$$

It is appropriate to note here that there is much variation in the literature with respect to index positions in the definitions of the curvature endomorphism and curvature tensor. While almost all authors define the $(1, 3)$-curvature tensor as we have, there are a few (notably [dC92, GHL04]) whose definition is the negative of ours. There is much less agreement on the definition of the $(0, 4)$-curvature tensor: whichever definition is chosen for the curvature endomorphism, you will see the curvature tensor defined as in (7.7) but with various permutations of (X, Y, Z, W) on the right-hand side. After applying the symmetries of the curvature tensor that we will prove later in this chapter, however, all of the definitions agree up to sign. There are various arguments to support one choice or another; we have made a choice that makes equation (7.7) easy to remember. You just have to be careful when you begin reading any book or article to determine the author's sign convention.

The next proposition gives one reason for our interest in the curvature tensor.

Proposition 7.6. *The curvature tensor is a local isometry invariant: if (M,g) and $\left(\widetilde{M},\widetilde{g}\right)$ are Riemannian or pseudo-Riemannian manifolds and $\varphi\colon M \to \widetilde{M}$ is a local isometry, then $\varphi^*\widetilde{Rm} = Rm$.*

▶ **Exercise 7.7.** Prove Proposition 7.6.

Flat Manifolds

To give a qualitative geometric interpretation to the curvature tensor, we will show that it is precisely the obstruction to being locally isometric to Euclidean (or pseudo-Euclidean) space. (In Chapter 8, after we have developed more machinery, we will be able to give a far more detailed *quantitative* interpretation.) The crux of the proof is the following lemma.

Lemma 7.8. *Suppose M is a smooth manifold, and ∇ is any connection on M satisfying the flatness criterion. Given $p \in M$ and any vector $v \in T_pM$, there exists a parallel vector field V on a neighborhood of p such that $V_p = v$.*

Proof. Let $p \in M$ and $v \in T_pM$ be arbitrary, and let $\left(x^1,\ldots,x^n\right)$ be any smooth coordinates for M centered at p. By shrinking the coordinate neighborhood if necessary, we may assume that the image of the coordinate map is an open cube $C_\varepsilon = \{x : |x^i| < \varepsilon,\ i = 1,\ldots,n\}$. We use the coordinate map to identify the coordinate domain with C_ε.

Begin by parallel transporting v along the x^1-axis; then from each point on the x^1-axis, parallel transport along the coordinate line parallel to the x^2-axis; then successively parallel transport along coordinate lines parallel to the x^3 through x^n-axes (Fig. 7.2). The result is a vector field V defined in C_ε. The fact that V is smooth follows from an inductive application of the theorem concerning smooth dependence of solutions to linear ODEs on initial conditions (Thm. 4.31); the details are left as an exercise.

Since $\nabla_X V$ is linear over $C^\infty(M)$ in X, to show that V is parallel, it suffices to show that $\nabla_{\partial_i} V = 0$ for each $i = 1,\ldots,n$. By construction, $\nabla_{\partial_1} V = 0$ on the x^1-axis, $\nabla_{\partial_2} V = 0$ on the (x^1,x^2)-plane, and in general $\nabla_{\partial_k} V = 0$ on the slice $M_k \subseteq C_\varepsilon$ defined by $x^{k+1} = \cdots = x^n = 0$. We will prove the following fact by induction on k:

$$\nabla_{\partial_1} V = \cdots = \nabla_{\partial_k} V = 0 \quad \text{on } M_k. \tag{7.9}$$

For $k = 1$, this is true by construction, and for $k = n$, it means that V is parallel on the whole cube C_ε. So assume that (7.9) holds for some k. By construction, $\nabla_{\partial_{k+1}} V = 0$ on all of M_{k+1}, and for $i \leq k$, the inductive hypothesis shows that $\nabla_{\partial_i} V = 0$ on the hyperplane $M_k \subseteq M_{k+1}$.

Since $[\partial_{k+1}, \partial_i] = 0$, the flatness criterion gives

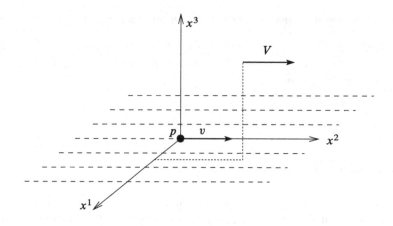

Fig. 7.2: Proof of Lemma 7.8

$$\nabla_{\partial_{k+1}}(\nabla_{\partial_i} V) = \nabla_{\partial_i}(\nabla_{\partial_{k+1}} V) = 0 \quad \text{on } M_{k+1}.$$

This shows that $\nabla_{\partial_i} V$ is parallel along the x^{k+1}-curves starting on M_k. Since $\nabla_{\partial_i} V$ vanishes on M_k and the zero vector field is the unique parallel transport of zero, we conclude that $\nabla_{\partial_i} V$ is zero on each x^{k+1}-curve. Since every point of M_{k+1} is on one of these curves, it follows that $\nabla_{\partial_i} V = 0$ on all of M_{k+1}. This completes the inductive step to show that V is parallel. $\square$

▶ **Exercise 7.9.** Prove that the vector field V constructed in the preceding proof is smooth.

Theorem 7.10. *A Riemannian or pseudo-Riemannian manifold is flat if and only if its curvature tensor vanishes identically.*

Proof. One direction is immediate: Proposition 7.2 showed that the Levi-Civita connection of a flat metric satisfies the flatness criterion, so its curvature endomorphism is identically zero, which implies that the curvature tensor is also zero.

Now suppose (M, g) has vanishing curvature tensor. This means that the curvature endomorphism vanishes as well, so the Levi-Civita connection satisfies the flatness criterion. We begin by showing that g shares one important property with Euclidean and pseudo-Euclidean metrics: it admits a parallel orthonormal frame in a neighborhood of each point.

Let $p \in M$, and choose an orthonormal basis $(b_1, \dots, b_n)$ for $T_p M$. In the pseudo-Riemannian case, we may assume that the basis is in standard order (with positive entries before negative ones in the matrix $g_{ij} = g_p(b_i, b_j)$). Lemma 7.8 shows that there exist parallel vector fields $E_1, \dots, E_n$ on a neighborhood U of p such that $E_i|_p = b_i$ for each $i = 1, \dots, n$. Because parallel transport preserves inner products, the vector fields (E_j) are orthonormal (and hence linearly independent) in all of U.

Because the Levi-Civita connection is symmetric, we have

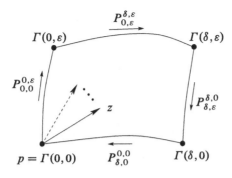

Fig. 7.3: The curvature endomorphism and parallel transport around a closed loop

$$[E_i, E_j] = \nabla_{E_i} E_j - \nabla_{E_j} E_i = 0.$$

Thus the vector fields $(E_1, \ldots, E_n)$ form a commuting orthonormal frame on U. The canonical form theorem for commuting vector fields (Prop. A.48) shows that there are coordinates $(y^1, \ldots, y^n)$ on a (possibly smaller) neighborhood of p such that $E_i = \partial/\partial y^i$ for $i = 1, \ldots, n$. In any such coordinates, $g_{ij} = g(\partial_i, \partial_j) = g(E_i, E_j) = \pm \delta_{ij}$, so the map $y = (y^1, \ldots, y^n)$ is an isometry from a neighborhood of p to an open subset of the appropriate Euclidean or pseudo-Euclidean space. $\qquad\square$

Using similar ideas, we can give a more precise interpretation of the meaning of the curvature tensor: it is a measure of the extent to which parallel transport around a small rectangle fails to be the identity map.

Theorem 7.11. *Let (M, g) be a Riemannian or pseudo-Riemannian manifold; let I be an open interval containing 0; let $\Gamma \colon I \times I \to M$ be a smooth one-parameter family of curves; and let $p = \Gamma(0,0)$, $x = \partial_s \Gamma(0,0)$, and $y = \partial_t \Gamma(0,0)$. For any $s_1, s_2, t_1, t_2 \in I$, let $P^{s_1, t_2}_{s_1, t_1} \colon T_{\Gamma(s_1, t_1)} M \to T_{\Gamma(s_1, t_2)} M$ denote parallel transport along the curve $t \mapsto \Gamma(s_1, t)$ from time t_1 to time t_2, and let $P^{s_2, t_1}_{s_1, t_1} \colon T_{\Gamma(s_1, t_1)} M \to T_{\Gamma(s_2, t_1)} M$ denote parallel transport along the curve $s \mapsto \Gamma(s, t_1)$ from time s_1 to time s_2. (See Fig. 7.3.) Then for every $z \in T_p M$,*

$$R(x, y)z = \lim_{\delta, \varepsilon \to 0} \frac{P^{0,0}_{\delta,0} \circ P^{\delta,0}_{\delta,\varepsilon} \circ P^{\delta,\varepsilon}_{0,\varepsilon} \circ P^{0,\varepsilon}_{0,0}(z) - z}{\delta \varepsilon}. \tag{7.10}$$

Proof. Define a vector field Z along Γ by first parallel transporting z along the curve $t \mapsto \Gamma(0, t)$, and then for each t, parallel transporting $Z(0, t)$ along the curve $s \mapsto \Gamma(s, t)$. The resulting vector field along Γ is smooth by another application of Theorem 4.31; and by construction, it satisfies $D_t Z(0, t) = 0$ for all $t \in I$, and $D_s Z(s, t) = 0$ for all $(s, t) \in I \times I$. Proposition 7.5 shows that

$$R(x, y)z = D_s D_t Z(0,0) - D_t D_s Z(0,0) = D_s D_t Z(0,0).$$

Thus we need only show that $D_s D_t Z(0,0)$ is equal to the limit on the right-hand side of (7.10).

From Theorem 4.34, we have

$$D_t Z(s,0) = \lim_{\varepsilon \to 0} \frac{P_{s,\varepsilon}^{s,0}(Z(s,\varepsilon)) - Z(s,0)}{\varepsilon}, \tag{7.11}$$

$$D_s D_t Z(0,0) = \lim_{\delta \to 0} \frac{P_{\delta,0}^{0,0}(D_t Z(\delta,0)) - D_t Z(0,0)}{\delta}. \tag{7.12}$$

Evaluating (7.11) first at $s = \delta$ and then at $s = 0$, and inserting the resulting expressions into (7.12), we obtain

$$D_s D_t Z(0,0)$$

$$= \lim_{\delta,\varepsilon \to 0} \frac{P_{\delta,0}^{0,0} \circ P_{\delta,\varepsilon}^{\delta,0}(Z(\delta,\varepsilon)) - P_{\delta,0}^{0,0}(Z(\delta,0)) - P_{0,\varepsilon}^{0,0}(Z(0,\varepsilon)) + Z(0,0)}{\delta\varepsilon}. \tag{7.13}$$

Here we have used the fact that parallel transport is linear, so the ε-limit can be pulled past $P_{\delta,0}^{0,0}$.

Now, the fact that Z is parallel along $t \mapsto \Gamma(0,t)$ and along all of the curves $s \mapsto \Gamma(s,t)$ implies

$$P_{\delta,0}^{0,0}(Z(\delta,0)) = P_{0,\varepsilon}^{0,0}(Z(0,\varepsilon)) = Z(0,0) = z,$$

$$Z(\delta,\varepsilon) = P_{0,\varepsilon}^{\delta,\varepsilon}(Z(0,\varepsilon)) = P_{0,\varepsilon}^{\delta,\varepsilon} \circ P_{0,0}^{0,\varepsilon}(z).$$

Inserting these relations into (7.13) yields (7.10).   □

Symmetries of the Curvature Tensor

The curvature tensor on a Riemannian or pseudo-Riemannian manifold has a number of symmetries besides the obvious skew-symmetry in its first two arguments.

Proposition 7.12 (Symmetries of the Curvature Tensor). *Let (M,g) be a Riemannian or pseudo-Riemannian manifold. The $(0,4)$-curvature tensor of g has the following symmetries for all vector fields W, X, Y, Z:*

(a) $Rm(W,X,Y,Z) = -Rm(X,W,Y,Z)$.
(b) $Rm(W,X,Y,Z) = -Rm(W,X,Z,Y)$.
(c) $Rm(W,X,Y,Z) = Rm(Y,Z,W,X)$.
(d) $Rm(W,X,Y,Z) + Rm(X,Y,W,Z) + Rm(Y,W,X,Z) = 0$.

Before we begin the proof, a few remarks are in order. First, as the proof will show, (a) is a trivial consequence of the definition of the curvature endomorphism; (b) follows from the compatibility of the Levi-Civita connection with the metric; (d) follows from the symmetry of the connection; and (c) follows from (a), (b), and (d). The identity in (d) is called the **algebraic Bianchi identity** (or more traditionally but less informatively, the *first Bianchi identity*). It is easy to show using (a)–(d) that a three-term sum obtained by cyclically permuting *any* three arguments of Rm is also zero. Finally, it is useful to record the form of these symmetries in terms of components with respect to any basis:

(a') $R_{ijkl} = -R_{jikl}$.
(b') $R_{ijkl} = -R_{ijlk}$.
(c') $R_{ijkl} = R_{klij}$.
(d') $R_{ijkl} + R_{jkil} + R_{kijl} = 0$.

Proof of Proposition 7.12. Identity (a) is immediate from the definition of the curvature tensor, because $R(W, X)Y = -R(X, W)Y$. To prove (b), it suffices to show that $Rm(W, X, Y, Y) = 0$ for all Y, for then (b) follows from the expansion of $Rm(W, X, Y + Z, Y + Z) = 0$. Using compatibility with the metric, we have

$$WX|Y|^2 = W(2\langle \nabla_X Y, Y \rangle) = 2\langle \nabla_W \nabla_X Y, Y \rangle + 2\langle \nabla_X Y, \nabla_W Y \rangle;$$
$$XW|Y|^2 = X(2\langle \nabla_W Y, Y \rangle) = 2\langle \nabla_X \nabla_W Y, Y \rangle + 2\langle \nabla_W Y, \nabla_X Y \rangle;$$
$$[W, X]|Y|^2 = 2\langle \nabla_{[W,X]} Y, Y \rangle.$$

When we subtract the second and third equations from the first, the left-hand side is zero. The terms $2\langle \nabla_X Y, \nabla_W Y \rangle$ and $2\langle \nabla_W Y, \nabla_X Y \rangle$ cancel on the right-hand side, giving

$$0 = 2\langle \nabla_W \nabla_X Y, Y \rangle - 2\langle \nabla_X \nabla_W Y, Y \rangle - 2\langle \nabla_{[W,X]} Y, Y \rangle$$
$$= 2\langle R(W, X)Y, Y \rangle$$
$$= 2Rm(W, X, Y, Y).$$

Next we prove (d). From the definition of Rm, this will follow immediately from

$$R(W, X)Y + R(X, Y)W + R(Y, W)X = 0.$$

Using the definition of R and the symmetry of the connection, the left-hand side expands to

$$(\nabla_W \nabla_X Y - \nabla_X \nabla_W Y - \nabla_{[W,X]} Y)$$
$$+ (\nabla_X \nabla_Y W - \nabla_Y \nabla_X W - \nabla_{[X,Y]} W)$$
$$+ (\nabla_Y \nabla_W X - \nabla_W \nabla_Y X - \nabla_{[Y,W]} X)$$
$$= \nabla_W (\nabla_X Y - \nabla_Y X) + \nabla_X (\nabla_Y W - \nabla_W Y) + \nabla_Y (\nabla_W X - \nabla_X W)$$
$$- \nabla_{[W,X]} Y - \nabla_{[X,Y]} W - \nabla_{[Y,W]} X$$
$$= \nabla_W [X, Y] + \nabla_X [Y, W] + \nabla_Y [W, X]$$
$$- \nabla_{[W,X]} Y - \nabla_{[X,Y]} W - \nabla_{[Y,W]} X$$
$$= [W, [X, Y]] + [X, [Y, W]] + [Y, [W, X]].$$

This is zero by the Jacobi identity.

Finally, we show that identity (c) follows from the other three. Writing the algebraic Bianchi identity four times with indices cyclically permuted gives

$$Rm(W,X,Y,Z)+Rm(X,Y,W,Z)+Rm(Y,W,X,Z)=0,$$
$$Rm(X,Y,Z,W)+Rm(Y,Z,X,W)+Rm(Z,X,Y,W)=0,$$
$$Rm(Y,Z,W,X)+Rm(Z,W,Y,X)+Rm(W,Y,Z,X)=0,$$
$$Rm(Z,W,X,Y)+Rm(W,X,Z,Y)+Rm(X,Z,W,Y)=0.$$

Now add up all four equations. Applying (b) four times makes all the terms in the first two columns cancel. Then applying (a) and (b) in the last column yields $2Rm(Y,W,X,Z) - 2Rm(X,Z,Y,W) = 0$, which is equivalent to (c). □

There is one more identity that is satisfied by the covariant derivatives of the curvature tensor on every Riemannian manifold. Classically, it was called the *second Bianchi identity*, but modern authors tend to use the more informative name *differential Bianchi identity*.

Proposition 7.13 (Differential Bianchi Identity). *The total covariant derivative of the curvature tensor satisfies the following identity:*

$$\nabla Rm(X,Y,Z,V,W) + \nabla Rm(X,Y,V,W,Z) + \nabla Rm(X,Y,W,Z,V) = 0. \quad (7.14)$$

In components, this is

$$R_{ijkl;m} + R_{ijlm;k} + R_{ijmk;l} = 0. \quad (7.15)$$

Proof. First of all, by the symmetries of Rm, (7.14) is equivalent to

$$\nabla Rm(Z,V,X,Y,W) + \nabla Rm(V,W,X,Y,Z) + \nabla Rm(W,Z,X,Y,V) = 0. \quad (7.16)$$

This can be proved by a long and tedious computation, but there is a standard shortcut for such calculations in Riemannian geometry that makes our task immeasurably easier. To prove that (7.16) holds at a particular point p, it suffices by multilinearity to prove the formula when X, Y, Z, V, W are basis vectors with respect to some frame. The shortcut consists in choosing a special frame for each point p to simplify the computations there.

Let p be an arbitrary point, let (x^i) be normal coordinates centered at p, and let X, Y, Z, V, W be arbitrary coordinate basis vector fields. These vectors satisfy two properties that simplify our computations enormously: (1) their commutators vanish identically, since $[\partial_i, \partial_j] \equiv 0$; and (2) their covariant derivatives vanish at p, since $\Gamma^k_{ij}(p) = 0$ (Prop. 5.24(d)).

Using these facts and the compatibility of the connection with the metric, the first term in (7.16) evaluated at p becomes

$$(\nabla_W Rm)(Z,V,X,Y) = \nabla_W\big(Rm(Z,V,X,Y)\big)$$
$$= \nabla_W\langle R(Z,V)X,\,Y\rangle$$
$$= \nabla_W\langle \nabla_Z\nabla_V X - \nabla_V\nabla_Z X - \nabla_{[Z,V]}X,\,Y\rangle$$
$$= \langle \nabla_W\nabla_Z\nabla_V X - \nabla_W\nabla_V\nabla_Z X,\,Y\rangle.$$

Write this equation three times, with the vector fields W,Z,V cyclically permuted. Summing all three gives

$$\nabla Rm(Z,V,X,Y,W) + \nabla Rm(V,W,X,Y,Z) + \nabla Rm(W,Z,X,Y,V)$$
$$= \langle \nabla_W\nabla_Z\nabla_V X - \nabla_W\nabla_V\nabla_Z X$$
$$+ \nabla_Z\nabla_V\nabla_W X - \nabla_Z\nabla_W\nabla_V X$$
$$+ \nabla_V\nabla_W\nabla_Z X - \nabla_V\nabla_Z\nabla_W X,\,Y\rangle$$
$$= \langle R(W,Z)(\nabla_V X) + R(Z,V)(\nabla_W X) + R(V,W)(\nabla_Z X),\,Y\rangle$$
$$= 0,$$

where the last line follows because $\nabla_V X = \nabla_W X = \nabla_Z X = 0$ at p. □

The Ricci Identities

The curvature endomorphism also appears as the obstruction to commutation of total covariant derivatives. Recall from Chapter 4 that if F is any smooth tensor field of type (k,l), then its second covariant derivative $\nabla^2 F = \nabla(\nabla F)$ is a smooth $(k,l+2)$-tensor field, and for vector fields X and Y, the notation $\nabla^2_{X,Y} F$ denotes $\nabla^2 F(\dots,Y,X)$. Given vector fields X and Y, let $R(X,Y)^*: T^*M \to T^*M$ denote the dual map to $R(X,Y)$, defined by

$$\big(R(X,Y)^*\eta\big)(Z) = \eta\big(R(X,Y)Z\big).$$

Theorem 7.14 (Ricci Identities). *On a Riemannian or pseudo-Riemannian manifold M, the second total covariant derivatives of vector and tensor fields satisfy the following identities. If Z is a smooth vector field,*

$$\nabla^2_{X,Y} Z - \nabla^2_{Y,X} Z = R(X,Y)Z. \tag{7.17}$$

If β is a smooth 1-form,

$$\nabla^2_{X,Y}\beta - \nabla^2_{Y,X}\beta = -R(X,Y)^*\beta. \tag{7.18}$$

And if B is a smooth (k,l)-tensor field,

$$(\nabla^2_{X,Y} B - \nabla^2_{Y,X} B)(\omega^1, \dots, \omega^k, V_1, \dots, V_l)$$
$$= B(R(X,Y)^*\omega^1, \omega^2, \dots, \omega^k, V_1, \dots, V_l) + \cdots$$
$$+ B(\omega^1, \dots, \omega^{k-1}, R(X,Y)^*\omega^k, V_1, \dots, V_l) \qquad (7.19)$$
$$- B(\omega^1, \dots, \omega^k, R(X,Y)V_1, V_2, \dots, V_l) - \cdots$$
$$- B(\omega^1, \dots, \omega^k, V_1, \dots, V_{l-1}, R(X,Y)V_l),$$

for all covector fields ω^i and vector fields V_j. In terms of any local frame, the component versions of these formulas read

$$Z^i{}_{;pq} - Z^i{}_{;qp} = -R_{pqm}{}^i Z^m, \qquad (7.20)$$

$$\beta_{j;pq} - \beta_{j;qp} = R_{pqj}{}^m \beta_m, \qquad (7.21)$$

$$B^{i_1 \dots i_k}_{j_1 \dots j_l;pq} - B^{i_1 \dots i_k}_{j_1 \dots j_l;qp} = -R_{pqm}{}^{i_1} B^{mi_2 \dots i_k}_{j_1 \dots j_l} - \cdots - R_{pqm}{}^{i_k} B^{i_1 \dots i_{k-1}m}_{j_1 \dots j_l}$$
$$+ R_{pqj_1}{}^m B^{i_1 \dots i_k}_{mj_2 \dots j_l} + \cdots + R_{pqj_l}{}^m B^{i_1 \dots i_k}_{j_1 \dots j_{l-1}m}. \qquad (7.22)$$

Proof. For any tensor field B and vector fields X, Y, Proposition 4.21 implies

$$\nabla^2_{X,Y} B - \nabla^2_{Y,X} B = \nabla_X \nabla_Y B - \nabla_{(\nabla_X Y)} B - \nabla_Y \nabla_X B + \nabla_{(\nabla_Y X)} B$$
$$= \nabla_X \nabla_Y B - \nabla_Y \nabla_X B - \nabla_{[X,Y]} B, \qquad (7.23)$$

where the last equality follows from the symmetry of the connection. In particular, this holds when $B = Z$ is a vector field, so (7.17) follows directly from the definition of the curvature endomorphism.

Next we prove (7.18). Using (4.13) repeatedly, we compute

$$(\nabla_X \nabla_Y \beta)(Z) = X((\nabla_Y \beta)(Z)) - (\nabla_Y \beta)(\nabla_X Z)$$
$$= X(Y(\beta(Z)) - \beta(\nabla_Y Z)) - (\nabla_Y \beta)(\nabla_X Z)$$
$$= XY(\beta(Z)) - (\nabla_X \beta)(\nabla_Y Z) - \beta(\nabla_X \nabla_Y Z) - (\nabla_Y \beta)(\nabla_X Z). \qquad (7.24)$$

Reversing the roles of X and Y, we get

$$(\nabla_Y \nabla_X \beta)(Z) = YX(\beta(Z)) - (\nabla_Y \beta)(\nabla_X Z) - \beta(\nabla_Y \nabla_X Z) - (\nabla_X \beta)(\nabla_Y Z), \qquad (7.25)$$

and applying (4.13) one more time yields

$$(\nabla_{[X,Y]} \beta)(Z) = [X,Y](\beta(Z)) - \beta(\nabla_{[X,Y]} Z). \qquad (7.26)$$

Now subtract (7.25) and (7.26) from (7.24): all but three of the terms cancel, yielding

$$(\nabla_X \nabla_Y \beta - \nabla_Y \nabla_X \beta - \nabla_{[X,Y]} \beta)(Z) = -\beta(\nabla_X \nabla_Y Z - \nabla_Y \nabla_X Z - \nabla_{[X,Y]} Z)$$
$$= -\beta(R(X,Y)Z),$$

which is equivalent to (7.18).

Next consider the action of $\nabla^2_{X,Y} - \nabla^2_{Y,X}$ on an arbitrary tensor product:

$$\left(\nabla^2_{X,Y} - \nabla^2_{Y,X}\right)(F \otimes G)$$
$$= \left(\nabla_X \nabla_Y - \nabla_Y \nabla_X - \nabla_{[X,Y]}\right)(F \otimes G)$$
$$= \nabla_X \nabla_Y F \otimes G + \nabla_Y F \otimes \nabla_X G + \nabla_X F \otimes \nabla_Y G + F \otimes \nabla_X \nabla_Y G$$
$$\quad - \nabla_Y \nabla_X F \otimes G - \nabla_X F \otimes \nabla_Y G - \nabla_Y F \otimes \nabla_X G - F \otimes \nabla_Y \nabla_X G$$
$$\quad - \nabla_{[X,Y]} F \otimes G - F \otimes \nabla_{[X,Y]} G$$
$$= \left(\nabla^2_{X,Y} F - \nabla^2_{Y,X} F\right) \otimes G + F \otimes \left(\nabla^2_{X,Y} G - \nabla^2_{Y,X} G\right).$$

A simple induction using this relation together with (7.17) and (7.18) shows that for all smooth vector fields $W_1, \dots, W_k$ and 1-forms $\eta^1, \dots, \eta^l$,

$$\left(\nabla^2_{X,Y} - \nabla^2_{Y,X}\right)\left(W_1 \otimes \cdots \otimes W_k \otimes \eta^1 \otimes \cdots \otimes \eta^l\right)$$
$$= \left(R(X,Y)W_1\right) \otimes W_2 \otimes \cdots \otimes W_k \otimes \eta^1 \otimes \cdots \otimes \eta^l + \cdots$$
$$\quad + W_1 \otimes \cdots \otimes W_{k-1} \otimes \left(R(X,Y)W_k\right) \otimes \eta^1 \otimes \cdots \otimes \eta^l$$
$$\quad + W_1 \otimes \cdots \otimes W_k \otimes \left(-R(X,Y)^*\eta^1\right) \otimes \eta^2 \otimes \cdots \otimes \eta^l + \cdots$$
$$\quad + W_1 \otimes \cdots \otimes W_k \otimes \eta^1 \otimes \cdots \otimes \eta^{l-1} \otimes \left(-R(X,Y)^*\eta^l\right).$$

Since every tensor field can be written as a sum of tensor products of vector fields and 1-forms, this implies (7.19).

Finally, the component formula (7.22) follows by applying (7.19) to

$$\left(\nabla^2_{E_q,E_p} B - \nabla^2_{E_p,E_q} B\right)\left(\varepsilon^{i_1}, \dots, \varepsilon^{i_k}, E_{j_1}, \dots, E_{j_l}\right),$$

where (E_j) and (ε^i) represent a local frame and its dual coframe, respectively, and using

$$R(E_q, E_p)E_j = R_{qpj}{}^m E_m = -R_{pqj}{}^m E_m,$$
$$R(E_q, E_p)^*\varepsilon^i = R_{qpm}{}^i \varepsilon^m = -R_{pqm}{}^i \varepsilon^m.$$

The other two component formulas are special cases of (7.22). $\qquad\square$

Ricci and Scalar Curvatures

Suppose (M, g) is an n-dimensional Riemannian or pseudo-Riemannian manifold. Because 4-tensors are so complicated, it is often useful to construct simpler tensors that summarize some of the information contained in the curvature tensor. The most important such tensor is the ***Ricci curvature*** or ***Ricci tensor***, denoted by Rc (or often *Ric* in the literature), which is the covariant 2-tensor field defined as the trace of the curvature endomorphism on its first and last indices. Thus for vector fields X, Y,

$$Rc(X, Y) = \text{tr}\left(Z \mapsto R(Z, X)Y\right).$$

The components of Rc are usually denoted by R_{ij}, so that

$$R_{ij} = R_{kij}{}^k = g^{km} R_{kijm}.$$

The *scalar curvature* is the function S defined as the trace of the Ricci tensor:

$$S = \operatorname{tr}_g Rc = R_i{}^i = g^{ij} R_{ij}.$$

It is probably not clear at this point why the Ricci tensor or scalar curvature might be interesting, and we do not yet have the tools to give them geometric interpretations. But be assured that there is such an interpretation; see Proposition 8.32.

Lemma 7.15. *The Ricci curvature is a symmetric 2-tensor field. It can be expressed in any of the following ways:*

$$R_{ij} = R_{kij}{}^k = R_{ik}{}^k{}_j = -R_{ki}{}^k{}_j = -R_{ikj}{}^k.$$

▶ **Exercise 7.16.** Prove Lemma 7.15, using the symmetries of the curvature tensor.

It is sometimes useful to decompose the Ricci tensor into a multiple of the metric and a complementary piece with zero trace. Define the *traceless Ricci tensor of g* as the following symmetric 2-tensor:

$$\overset{\circ}{Rc} = Rc - \frac{1}{n} Sg.$$

Proposition 7.17. *Let (M,g) be a Riemannian or pseudo-Riemannian n-manifold. Then $\operatorname{tr}_g \overset{\circ}{Rc} \equiv 0$, and the Ricci tensor decomposes orthogonally as*

$$Rc = \overset{\circ}{Rc} + \frac{1}{n} Sg. \tag{7.27}$$

Therefore, in the Riemannian case,

$$|Rc|_g^2 = \left|\overset{\circ}{Rc}\right|_g^2 + \frac{1}{n} S^2. \tag{7.28}$$

Remark. The statement about norms, and others like it that we will prove below, works only in the Riemannian case because of the additional absolute value signs required to compute norms in the pseudo-Riemannian case. The pseudo-Riemannian analogue would be $\langle Rc, Rc \rangle_g = \left\langle \overset{\circ}{Rc}, \overset{\circ}{Rc} \right\rangle_g + \frac{1}{n} S^2$, but this is not as useful.

Proof. Note that in every local frame, we have

$$\operatorname{tr}_g g = g_{ij} g^{ji} = \delta_i^i = n.$$

It then follows directly from the definition of $\overset{\circ}{Rc}$ that $\operatorname{tr}_g \overset{\circ}{Rc} \equiv 0$ and (7.27) holds. The fact that the decomposition is orthogonal follows easily from the fact that for every symmetric 2-tensor h, we have

$$\langle h, g \rangle = g^{ik} g^{jl} h_{ij} g_{kl} = g^{ij} h_{ij} = \text{tr}_g h,$$

and therefore $\langle \overset{\circ}{Rc}, g \rangle = \text{tr}_g \overset{\circ}{Rc} = 0$. Finally, (7.28) follows from (7.27) and the fact that $\langle g, g \rangle = \text{tr}_g g = n$. $\qquad\qquad\qquad\qquad\qquad\qquad\qquad\qquad\qquad\qquad\square$

The next proposition, which follows directly from the differential Bianchi identity, expresses some important relationships among the covariant derivatives of the various curvature tensors. To express it concisely, it is useful to introduce another operator on tensor fields. If T is a smooth 2-tensor field on a Riemannian or pseudo-Riemannian manifold, we define the *exterior covariant derivative of* T to be the 3-tensor field DT defined by

$$(DT)(X, Y, Z) = -(\nabla T)(X, Y, Z) + (\nabla T)(X, Z, Y). \qquad (7.29)$$

In terms of components, this is

$$(DT)_{ijk} = -T_{ij;k} + T_{ik;j}.$$

(This operator is a generalization of the ordinary exterior derivative of a 1-form, which can be expressed in terms of the total covariant derivative by $(d\eta)(Y, Z) = -(\nabla \eta)(Y, Z) + (\nabla \eta)(Z, Y)$ by the result of Problem 5-13. The exterior covariant derivative can be generalized to other types of tensors as well, but this is the only case we need.)

Proposition 7.18 (Contracted Bianchi Identities). *Let* (M, g) *be a Riemannian or pseudo-Riemannian manifold. The covariant derivatives of the Riemann, Ricci, and scalar curvatures of* g *satisfy the following identities:*

$$\text{tr}_g (\nabla Rm) = -D(Rc), \qquad\qquad\qquad (7.30)$$
$$\text{tr}_g (\nabla Rc) = \tfrac{1}{2} dS, \qquad\qquad\qquad (7.31)$$

where the trace in each case is on the first and last indices. In components, this is

$$R_{ijkl;}{}^{i} = R_{jk;l} - R_{jl;k}, \qquad\qquad\qquad (7.32)$$
$$R_{il;}{}^{i} = \tfrac{1}{2} S_{;l}. \qquad\qquad\qquad (7.33)$$

Proof. Start with the component form (7.15) of the differential Bianchi identity, raise the index m, and then contract on the indices i, m to obtain (7.32). (Note that covariant differentiation commutes with contraction by Proposition 4.15 and with the musical isomorphisms by Proposition 5.17, so it does not matter whether the indices that are raised and contracted come before or after the semicolon.) Then do the same with the indices j, k and simplify to obtain (7.33). The coordinate-free formulas (7.30) and (7.31) follow by expanding everything out in components. $\qquad\square$

It is important to note that if the sign convention chosen for the curvature tensor is the opposite of ours, then the Ricci tensor must be defined as the trace of Rm on the first and third (or second and fourth) indices. (The trace on the first two or last two indices is always zero by antisymmetry.) The definition is chosen so that the Ricci and scalar curvatures have the same meaning for everyone, regardless of the conventions chosen for the full curvature tensor. So, for example, if a manifold is said to have positive scalar curvature, there is no ambiguity as to what is meant.

A Riemannian or pseudo-Riemannian metric is said to be an *Einstein metric* if its Ricci tensor is a constant multiple of the metric—that is,

$$Rc = \lambda g \quad \text{for some constant} \lambda. \tag{7.34}$$

This equation is known as the *Einstein equation*. As the next proposition shows, for connected manifolds of dimension greater than 2, it is not necessary to assume that λ is constant; just assuming that the Ricci tensor is a function times the metric is sufficient.

Proposition 7.19 (Schur's Lemma). *Suppose (M, g) is a connected Riemannian or pseudo-Riemannian manifold of dimension $n \geq 3$ whose Ricci tensor satisfies $Rc = fg$ for some smooth real-valued function f. Then f is constant and g is an Einstein metric.*

Proof. Taking traces of both sides of $Rc = fg$ shows that $f = \frac{1}{n}S$, so the traceless Ricci tensor is identically zero. It follows that $\nabla \mathring{Rc} \equiv 0$. Because the covariant derivative of the metric is zero, this implies the following equation in any coordinate chart:

$$0 = R_{ij;k} - \frac{1}{n}S_{;k}g_{ij}.$$

Tracing this equation on i and k, and comparing with the contracted Bianchi identity (7.33), we conclude that

$$0 = \frac{1}{2}S_{;j} - \frac{1}{n}S_{;j}.$$

Because $n \geq 3$, this implies $S_{;j} = 0$. But $S_{;j}$ is the component of $\nabla S = dS$, so connectedness of M implies that S is constant and thus so is f.   □

Corollary 7.20. *If (M, g) is a connected Riemannian or pseudo-Riemannian manifold of dimension $n \geq 3$, then g is Einstein if and only if $\mathring{Rc} = 0$.*

Proof. Suppose first that g is an Einstein metric with $Rc = \lambda g$. Taking traces of both sides, we find that $\lambda = \frac{1}{n}S$, and therefore $\mathring{Rc} = Rc - \lambda g = 0$. Conversely, if $\mathring{Rc} = 0$, Schur's lemma implies that g is Einstein.   □

By an argument analogous to those of Chapter 6, Hilbert showed (see [Bes87, Thm. 4.21]) that Einstein metrics are critical points for the *total scalar curvature functional* $\mathcal{S}(g) = \int_M S \, dV_g$ on the space of all metrics on M with fixed volume. Thus Einstein metrics can be viewed as "optimal" metrics in a certain sense, and as such they form an appealing higher-dimensional analogue of locally homogeneous metrics on 2-manifolds, with which one might hope to prove some sort of generalization of the uniformization theorem (Thm. 3.22). Although the statement of such a theorem cannot be as elegant as that of its 2-dimensional ancestor because there are known examples of smooth, compact manifolds that admit no Einstein metrics [Bes87, Chap. 6], there is still a reasonable hope that "most" higher-dimensional manifolds (in some sense) admit Einstein metrics. This is an active field of current research; see [Bes87] for a sweeping survey of Einstein metrics.

The term "Einstein metric" originated, as you might guess, in physics: the central assertion of Einstein's general theory of relativity is that physical spacetime is modeled by a 4-manifold that carries a Lorentz metric whose Ricci curvature satisfies the following *Einstein field equation:*

$$Rc - \frac{1}{2}Sg = T, \tag{7.35}$$

where T is a certain symmetric 2-tensor field (the *stress–energy tensor*) that describes the density, momentum, and stress of the matter and energy present at each point in spacetime. It is shown in physics books (e.g., [CB09, pp. 51–53]) that (7.35) is the variational equation for a certain functional, called the *Einstein–Hilbert action*, on the space of all Lorentz metrics on a given 4-manifold. Einstein's theory can then be interpreted as the assertion that a physically realistic spacetime must be a critical point for this functional.

In the special case $T \equiv 0$, (7.35) reduces to the *vacuum Einstein field equation* $Rc = \frac{1}{2}Sg$. Taking traces of both sides and recalling that $\mathrm{tr}_g\, g = \dim M = 4$, we obtain $S = 2S$, which implies $S = 0$. Therefore, the vacuum Einstein equation is equivalent to $Rc = 0$, which means that g is a (pseudo-Riemannian) Einstein metric in the mathematical sense of the word. (At one point in the development of the theory, Einstein considered adding a term Λg to the left-hand side of (7.35), where Λ is a constant that he called the *cosmological constant*. With this modification the vacuum Einstein field equation would be exactly the same as the mathematicians' Einstein equation (7.34). Einstein soon decided that the cosmological constant was a mistake on physical grounds; however, researchers in general relativity have recently begun to believe that a theory with a nonzero cosmological constant might in fact have physical relevance.)

Other than these special cases and the obvious formal similarity between (7.35) and (7.34), there is no direct connection between the physicists' version of the Einstein equation and the mathematicians' version. The mathematical interest in Riemannian Einstein metrics stems more from their potential applications to uniformization in higher dimensions than from their relation to physics.

Another approach to generalizing the uniformization theorem to higher dimensions is to search for metrics of constant *scalar* curvature. These are also critical points of the total scalar curvature functional, but only with respect to variations of the metric with fixed volume within a given conformal equivalence class. Thus it makes sense to ask whether, given a metric g on a manifold M, there exists a metric $\tilde{g}$ conformal to g that has constant scalar curvature. This is called the *Yamabe problem*, because it was first posed in 1960 by Hidehiko Yamabe, who claimed to have proved that the answer is always yes when M is compact. Yamabe's proof was later found to be in error, and it was two dozen years before the proof was finally completed by Richard Schoen; see [LP87] for an expository account of Schoen's solution. When M is noncompact, the issues are much subtler, and much current research is focused on determining exactly which conformal classes contain metrics of constant scalar curvature.

The Weyl Tensor

As noted above, the Ricci and scalar curvatures contain only part of the information encoded into the curvature tensor. In this section, we introduce a tensor field called the *Weyl tensor*, which encodes all the rest.

We begin by considering some linear-algebraic aspects of tensors that have the symmetries of the curvature tensor. Suppose V is an n-dimensional real vector space. Let $\mathcal{R}(V^*) \subseteq T^4(V^*)$ denote the vector space of all covariant 4-tensors T on V that have the symmetries of the $(0,4)$ Riemann curvature tensor:

(a) $T(w,x,y,z) = -T(x,w,y,z)$.
(b) $T(w,x,y,z) = -T(w,x,z,y)$.
(c) $T(w,x,y,z) = T(y,z,w,x)$.
(d) $T(w,x,y,z) + T(x,y,w,z) + T(y,w,x,z) = 0$.

(As the proof of Prop. 7.12 showed, (c) follows from the other three symmetries, so it would suffice to assume only (a), (b), and (d); but it is more convenient to include all four symmetries in the definition.) An element of $\mathcal{R}(V^*)$ is called an **algebraic curvature tensor** on V.

Proposition 7.21. *If the vector space V has dimension n, then*

$$\dim \mathcal{R}(V^*) = \frac{n^2(n^2-1)}{12}. \tag{7.36}$$

Proof. Let $\mathcal{B}(V^*)$ denote the linear subspace of $T^4(V^*)$ consisting of tensors satisfying properties (a)–(c), and let $\Sigma^2(\Lambda^2(V)^*)$ denote the space of symmetric bilinear forms on the vector space $\Lambda^2(V)$ of alternating contravariant 2-tensors on V. Define a map $\Phi: \Sigma^2(\Lambda^2(V)^*) \to \mathcal{B}(V^*)$ as follows:

$$\Phi(B)(w,x,y,z) = B(w \wedge x, \ y \wedge z).$$

It is easy to check that $\Phi(B)$ satisfies (a)–(c), so $\Phi(B) \in \mathcal{B}(V^*)$, and that Φ is a linear map. In fact, it is an isomorphism, which we prove by constructing an inverse for it. Choose a basis $(b_1, \ldots, b_n)$ for V, so the collection $\{b_i \wedge b_j : i < j\}$ is a basis for $\Lambda^2(V)$. Define a map $\Psi: \mathcal{B}(V^*) \to \Sigma^2(\Lambda^2(V)^*)$ by setting

$$\Psi(T)(b_i \wedge b_j, \ b_k \wedge b_l) = T(b_i, b_j, b_k, b_l)$$

when $i < j$ and $k < l$, and extending by bilinearity. A straightforward computation shows that Ψ is an inverse for Φ.

The upshot of the preceding construction is that

$$\dim \mathcal{B}(V^*) = \dim\left(\Sigma^2\left(\Lambda^2(V)^*\right)\right) = \frac{\binom{n}{2}\left(\binom{n}{2}+1\right)}{2} = \frac{n(n-1)(n^2-n+2)}{8},$$

where we have used the facts that $\dim \Lambda^2(V) = \binom{n}{2} = n(n-1)/2$ and the dimension of the space of symmetric bilinear forms on a vector space of dimension m is $m(m+1)/2$.

Now consider the linear map $\pi: \mathcal{B}(V^*) \to T^4(V^*)$ defined by

$$\pi(T)(w,x,y,z) = \tfrac{1}{3}\big(T(w,x,y,z) + T(x,y,w,z) + T(y,w,x,z)\big).$$

By definition, $\mathcal{R}(V^*)$ is equal to the kernel of π. In fact, π is equal to the restriction to $\mathcal{B}(V^*)$ of the operator $\mathrm{Alt}: T^4(V^*) \to \Lambda^4(V^*)$ defined by (B.9): thanks to the symmetries (a)–(c), the 24 terms in the definition of $\mathrm{Alt}\,T$ can be arranged in three groups of eight in such a way that all the terms in each group reduce to one of the terms in the definition of π. Thus the image of π is contained in $\Lambda^4(V^*)$. In fact, the image is all of $\Lambda^4(V^*)$: every $T \in \Lambda^4(V^*)$ satisfies (a)–(c) and thus lies in $\mathcal{B}(V^*)$, and $\pi(T) = \mathrm{Alt}\,T = T$ for each such tensor.

Therefore, using the rank–nullity theorem of linear algebra, we conclude that

$$\dim \mathcal{R}(V^*) = \dim \mathcal{B}(V^*) - \dim \Lambda^4(V^*) = \frac{n(n-1)(n^2-n+2)}{8} - \binom{n}{4},$$

and simplification yields (7.36). $\qquad\qquad\qquad\qquad\qquad\qquad\qquad\qquad\qquad\qquad\square$

Let us now assume that our vector space V is endowed with a (not necessarily positive definite) scalar product $g \in \Sigma^2(V^*)$. Let $\mathrm{tr}_g: \mathcal{R}(V^*) \to \Sigma^2(V^*)$ denote the trace operation (with respect to g) on the first and last indices (so that, for example, $Rc = \mathrm{tr}_g(Rm)$). It is natural to wonder whether this operator is surjective and what its kernel is, as a way of asking how much of the information contained in the Riemann curvature tensor is captured by the Ricci tensor. One way to try to answer the question is to attempt to construct a right inverse for the trace operator—a linear map $G: \Sigma^2(V^*) \to \mathcal{R}(V^*)$ such that $\mathrm{tr}_g(G(S)) = S$ for all $S \in \Sigma^2(V^*)$.

Such an operator must start with a symmetric 2-tensor and construct a 4-tensor, using only the given 2-tensor and the metric. It turns out that there is a natural way to construct an algebraic curvature tensor out of two symmetric 2-tensors, which we now describe. Given $h, k \in \Sigma^2(V^*)$, we define a covariant 4-tensor $h \owedge k$, called the **Kulkarni–Nomizu product of h and k**, by the following formula:

$$h \owedge k(w,x,y,z) = h(w,z)k(x,y) + h(x,y)k(w,z)$$
$$- h(w,y)k(x,z) - h(x,z)k(w,y). \quad (7.37)$$

In terms of any basis, the components of $h \owedge k$ are

$$(h \owedge k)_{ijlm} = h_{im}k_{jl} + h_{jl}k_{im} - h_{il}k_{jm} - h_{jm}k_{il}.$$

(It should be noted that the Kulkarni–Nomizu product must be defined as the negative of this expression when the Riemann curvature tensor is defined as the negative of ours.)

Lemma 7.22 (Properties of the Kulkarni–Nomizu Product). *Let V be an n-dimensional vector space endowed with a scalar product g, let h and k be symmetric 2-tensors on V, let T be an algebraic curvature tensor on V, and let tr_g denote the trace on the first and last indices.*

(a) $h \mathbin{\odot} k$ *is an algebraic curvature tensor.*
(b) $h \mathbin{\odot} k = k \mathbin{\odot} h$.
(c) $\mathrm{tr}_g (h \mathbin{\odot} g) = (n-2)h + (\mathrm{tr}_g h)g$.
(d) $\mathrm{tr}_g (g \mathbin{\odot} g) = 2(n-1)g$.
(e) $\langle T, h \mathbin{\odot} g \rangle_g = 4 \langle \mathrm{tr}_g T, h \rangle_g$.
(f) *In case g is positive definite,* $|g \mathbin{\odot} h|_g^2 = 4(n-2)|h|_g^2 + 4(\mathrm{tr}_g h)^2$.

Proof. It is evident from the definition that $h \mathbin{\odot} k$ has three of the four symmetries of an algebraic curvature tensor: it is antisymmetric in its first two arguments and also in its last two, and its value is unchanged when the first two and last two arguments are interchanged. Thus to prove (a), only the algebraic Bianchi identity needs to be checked. This is a straightforward computation: when $h \mathbin{\odot} k(w, x, y, z)$ is written three times with the arguments w, x, y cyclically permuted and the three expressions are added together, all the terms cancel due to the symmetry of h and k.

Part (b) is immediate from the definition. To prove (c), choose a basis and use the definition to compute

$$(\mathrm{tr}_g (h \mathbin{\odot} g))_{jk} = g^{im}\left(h_{im}g_{jl} + h_{jl}g_{im} - h_{il}g_{jm} - h_{jm}g_{il}\right)$$
$$= h_i{}^i g_{jl} + n h_{jl} - h_{jl} - h_{jl},$$

which is equivalent to (c). Then (d) follows from (c) and the fact that $\mathrm{tr}_g g = n$.

The proofs of (e) and (f) are left to Problem 7-9. $\qquad\square$

Here is the primary application of the Kulkarni–Nomizu product.

Proposition 7.23. *Let (V, g) be an n-dimensional scalar product space with $n \geq 3$, and define a linear map $G \colon \Sigma^2(V^*) \to \mathcal{R}(V^*)$ by*

$$G(h) = \frac{1}{n-2}\left(h - \frac{\mathrm{tr}_g h}{2(n-1)}g\right) \mathbin{\odot} g. \tag{7.38}$$

Then G is a right inverse for tr_g, and its image is the orthogonal complement of the kernel of tr_g in $\mathcal{R}(V^*)$.

Proof. The fact that G is a right inverse is a straightforward computation based on the definition and Lemma 7.22(c,d). This implies that G is injective and tr_g is surjective, so the dimension of $\mathrm{Im}\,G$ is equal to the codimension of $\mathrm{Ker}(\mathrm{tr}_g)$, which in turn is equal to the dimension of $\mathrm{Ker}(\mathrm{tr}_g)^\perp$. If $T \in \mathcal{R}(V^*)$ is an algebraic curvature tensor such that $\mathrm{tr}_g T = 0$, then Lemma 7.22(e) shows that $\langle T, G(h) \rangle = 0$, so it follows by dimensionality that $\mathrm{Im}\,G = \mathrm{Ker}(\mathrm{tr}_g)^\perp$. $\qquad\square$

Now suppose g is a Riemannian or pseudo-Riemannian metric. Define the **Schouten tensor of g**, denoted by P, to be the following symmetric 2-tensor field:

$$P = \frac{1}{n-2}\left(Rc - \frac{S}{2(n-1)}g\right);$$

and define the **Weyl tensor of g** to be the following algebraic curvature tensor field:

$$\begin{aligned}
W &= Rm - P \owedge g \\
&= Rm - \frac{1}{n-2}Rc \owedge g + \frac{S}{2(n-1)(n-2)}g \owedge g.
\end{aligned}$$

Proposition 7.24. *For every Riemannian or pseudo-Riemannian manifold (M,g) of dimension $n \geq 3$, the trace of the Weyl tensor is zero, and $Rm = W + P \owedge g$ is the orthogonal decomposition of Rm corresponding to $\mathcal{R}(V^*) = \mathrm{Ker}(\mathrm{tr}_g) \oplus \mathrm{Ker}(\mathrm{tr}_g)^\perp$.*

Proof. This follows immediately from Proposition 7.23 and the fact that $P \owedge g = G(Rc) = G(\mathrm{tr}_g Rm)$. $\qquad\square$

These results lead to some important simplifications in low dimensions.

Corollary 7.25 *Let V be an n-dimensional real vector space.*

(a) *If $n = 0$ or $n = 1$, then $\mathcal{R}(V^*) = \{0\}$.*
(b) *If $n = 2$, then $\mathcal{R}(V^*)$ is 1-dimensional, spanned by $g \owedge g$.*
(c) *If $n = 3$, then $\mathcal{R}(V^*)$ is 6-dimensional, and $G: \Sigma^2(V^*) \to \mathcal{R}(V^*)$ is an iso-morphism.*

Proof. The dimensional results follow immediately from Proposition 7.21. In the case $n = 2$, Lemma 7.22(d) shows that $\mathrm{tr}_g(g \owedge g) = 2g \neq 0$, which implies that $g \owedge g$ is nonzero and therefore spans the 1-dimensional space $\mathcal{R}(V^*)$.

Now consider $n = 3$. Proposition 7.23 shows that $\mathrm{tr}_g \circ G$ is the identity, which means that $G: \Sigma^2(V^*) \to \mathcal{R}(V^*)$ is injective. On the other hand, Proposition 7.21 shows that $\dim \mathcal{R}(V^*) = 6 = \dim \Sigma^2(V^*)$, so G is also surjective. $\qquad\square$

The next corollary shows that the entire curvature tensor is determined by the Ricci tensor in dimension 3.

Corollary 7.26 (The Curvature Tensor in Dimension 3). *On every Riemannian or pseudo-Riemannian manifold (M,g) of dimension 3, the Weyl tensor is zero, and the Riemann curvature tensor is determined by the Ricci tensor via the formula*

$$Rm = P \owedge g = Rc \owedge g - \tfrac{1}{4}Sg \owedge g. \tag{7.39}$$

Proof. Corollary 7.25 shows that $G: \Sigma^2(V^*) \to \mathcal{R}(V^*)$ is an isomorphism in dimension 3. Since $\mathrm{tr}_g \circ G$ is the identity, it follows that tr_g is also an isomorphism. Because $\mathrm{tr}_g W$ is always zero by Proposition 7.24, it follows that W is always zero. Formula (7.39) then follows from the definition of the Weyl tensor. $\qquad\square$

In dimension 2, the definitions of the Weyl and Schouten tensors do not make sense; but we have the following analogous result instead.

Corollary 7.27 (The Curvature Tensor in Dimension 2). *On every Riemannian or pseudo-Riemannian manifold* (M, g) *of dimension* 2, *the Riemann and Ricci tensors are determined by the scalar curvature as follows:*

$$Rm = \tfrac{1}{4} S g \otimes g, \qquad Rc = \tfrac{1}{2} S g.$$

Proof. In dimension 2, it follows from Corollary 7.25(b) that there is some scalar function f such that $Rm = fg \otimes g$. Taking traces, we find from Lemma 7.22(d) that $Rc = \mathrm{tr}_g(Rm) = 2fg$, and then $S = \mathrm{tr}_g(Rc) = 2f \, \mathrm{tr}_g(g) = 4f$. The results follow by substituting $f = \tfrac{1}{4} S$ back into these equations. □

Although the traceless Ricci tensor is always zero on a 2-manifold, this does not imply that S is constant, since the proof of Schur's lemma fails in dimension 2. Einstein metrics in dimension 2 are simply those with constant scalar curvature.

Returning now to dimensions greater than 2, we can use (7.27) to further decompose the Schouten tensor into a part determined by the traceless Ricci tensor and a purely scalar part. The next proposition is the analogue of Proposition 7.17 for the full curvature tensor.

Proposition 7.28 (The Ricci Decomposition of the Curvature Tensor). *Let* (M, g) *be a Riemannian or pseudo-Riemannian manifold of dimension* $n \geq 3$. *Then the* $(0, 4)$-*curvature tensor of* g *has the following orthogonal decomposition:*

$$Rm = W + \frac{1}{n-2} \overset{\circ}{Rc} \otimes g + \frac{1}{2n(n-1)} S \, g \otimes g. \tag{7.40}$$

Therefore, in the Riemannian case,

$$
\begin{aligned}
|Rm|_g^2 &= |W|_g^2 + \frac{1}{(n-2)^2} |\overset{\circ}{Rc} \otimes g|_g^2 + \frac{1}{4n^2(n-1)^2} |S \, g \otimes g|_g^2 \\
&= |W|_g^2 + \frac{4}{n-2} |\overset{\circ}{Rc}|_g^2 + \frac{2}{n(n-1)} S^2.
\end{aligned}
\tag{7.41}
$$

Proof. The decomposition (7.40) follows immediately by substituting (7.27) into the definition of the Weyl tensor and simplifying. The decomposition is orthogonal thanks to Lemma 7.22(e), and (7.41) follows from Lemma 7.22(f). □

Curvatures of Conformally Related Metrics

Recall that two Riemannian or pseudo-Riemannian metrics on the same manifold are said to be *conformal* to each other if one is a positive function times the other. For example, we have seen that the round metrics and the hyperbolic metrics are all conformal to Euclidean metrics, at least locally.

If g and $\tilde{g}$ are conformal metrics on a smooth manifold M, there is no reason to expect that the curvature tensors of g and $\tilde{g}$ should be closely related to each other.

But it is a remarkable fact that the Weyl tensor has a very simple transformation law under conformal changes of metric. In this section, we derive that law.

First we need to determine how the Levi-Civita connection changes when a metric is changed conformally. Given conformal metrics g and $\widetilde{g}$, we can always write $\widetilde{g} = e^{2f} g$ for some smooth real-valued function f.

Proposition 7.29 (Conformal Transformation of the Levi-Civita Connection).
Let (M,g) be a Riemannian or pseudo-Riemannian n-manifold (with or without boundary), and let $\widetilde{g} = e^{2f} g$ be any metric conformal to g. If ∇ and $\widetilde{\nabla}$ denote the Levi-Civita connections of g and $\widetilde{g}$, respectively, then

$$\widetilde{\nabla}_X Y = \nabla_X Y + (Xf)Y + (Yf)X - \langle X,Y \rangle_g \operatorname{grad} f, \qquad (7.42)$$

where the gradient on the right-hand side is that of g. In any local coordinates, the Christoffel symbols of the two connections are related by

$$\widetilde{\Gamma}_{ij}^k = \Gamma_{ij}^k + f_{;i}\delta_j^k + f_{;j}\delta_i^k - g^{kl} f_{;l} g_{ij}, \qquad (7.43)$$

where $f_{;i} = \partial_i f$ is the ith component of $\nabla f = df = f_{;i} dx^i$.

Proof. Formula (7.43) is a straightforward computation using formula (5.10) for the Christoffel symbols in coordinates, and then (7.42) follows by expanding everything in coordinates and using (7.43). $\qquad \square$

This result leads to transformation laws for the various curvature tensors.

Theorem 7.30 (Conformal Transformation of the Curvature). *Let g be a Riemannian or pseudo-Riemannian metric on an n-manifold M with or without boundary, $f \in C^\infty(M)$, and $\widetilde{g} = e^{2f} g$. In the Riemannian case, the curvature tensors of $\widetilde{g}$ (represented with tildes) are related to those of g by the following formulas:*

$$\widetilde{Rm} = e^{2f}\left(Rm - (\nabla^2 f)\owedge g + (df \otimes df)\owedge g - \tfrac{1}{2}|df|_g^2 (g \owedge g)\right), \qquad (7.44)$$

$$\widetilde{Rc} = Rc - (n-2)(\nabla^2 f) + (n-2)(df \otimes df) - \left(\Delta f + (n-2)|df|_g^2\right)g, \quad (7.45)$$

$$\widetilde{S} = e^{-2f}\left(S - 2(n-1)\Delta f - (n-1)(n-2)|df|_g^2\right), \qquad (7.46)$$

where the curvatures and covariant derivatives on the right are those of g, and $\Delta f = \operatorname{div}(\operatorname{grad} f)$ is the Laplacian of f defined in Chapter 2. If in addition $n \geq 3$, then

$$\widetilde{P} = P - \nabla^2 f + (df \otimes df) - \tfrac{1}{2}|df|_g^2 g, \qquad (7.47)$$

$$\widetilde{W} = e^{2f} W. \qquad (7.48)$$

In the pseudo-Riemannian case, the same formulas hold with each occurrence of $|df|_g^2$ replaced by $\langle df, df \rangle_g^2$.

Proof. We begin with (7.44). The plan is to choose local coordinates and insert formula (7.43) for the Christoffel symbols into the coordinate formula (7.8) for the coefficients of the curvature tensor. As in the proof of the differential Bianchi identity, we can make the computations much more tractable by computing the components of these tensors at a point $p \in M$ in normal coordinates for g centered at p, so that the equations $g_{ij} = \delta_{ij}$, $\partial_k g_{ij} = 0$, and $\Gamma_{ij}^k = 0$ hold at p. This has the following consequences at p:

$$f_{;ij} = \partial_j \partial_i f,$$
$$\widetilde{\Gamma}_{ij}^k = f_{;i}\delta_j^k + f_{;j}\delta_i^k - g^{kl} f_{;l}g_{ij},$$
$$\partial_m \widetilde{\Gamma}_{ij}^k = \partial_m \Gamma_{ij}^k + f_{;im}\delta_j^k + f_{;jm}\delta_i^k - g^{kl} f_{;lm}g_{ij},$$
$$R_{ijk}{}^l = \partial_i \Gamma_{jk}^l - \partial_j \Gamma_{ik}^l.$$

Now start with

$$\widetilde{R}_{ijkl} = e^{2f} g_{lm}\left(\partial_i \widetilde{\Gamma}_{jk}^m - \partial_j \widetilde{\Gamma}_{ik}^m + \widetilde{\Gamma}_{jk}^p \widetilde{\Gamma}_{ip}^m - \widetilde{\Gamma}_{ik}^p \widetilde{\Gamma}_{jp}^m\right).$$

Inserting the relations above and simplifying, we eventually obtain

$$\widetilde{R}_{ijkl} = e^{2f}\Big(R_{ijkl} - (f_{;il}g_{jk} + f_{;jk}g_{il} - f_{;il}g_{jk} - f_{;jk}g_{il})$$
$$+ (f_{;i} f_{;l}g_{jk} + f_{;j} f_{;k}g_{il} - f_{;i} f_{;k}g_{jl} - f_{;j} f_{;l}g_{ik})$$
$$- g^{mp} f_{;m} f_{;p}(g_{il}g_{jk} - g_{ik}g_{jl})\Big),$$

which is the coordinate version of (7.44). (See Problem 5-14 for the formula for the Laplacian in terms of covariant derivatives.) The rest of the formulas follow easily from this, using the identities of Lemma 7.22 and the fact that $g^{ij} = e^{-2f}\widetilde{g}^{ij}$.   □

The next corollary begins to explain the geometric significance of the Weyl tensor. Recall that a Riemannian manifold is said to be ***locally conformally flat*** if every point has a neighborhood that is conformally equivalent to an open subset of Euclidean space. Similarly, a pseudo-Riemannian manifold is locally conformally flat if every point has a neighborhood conformally equivalent to an open set in a pseudo-Euclidean space.

Corollary 7.31. *Suppose (M,g) is a Riemannian or pseudo-Riemannian manifold of dimension $n \geq 3$. If g is locally conformally flat, then its Weyl tensor vanishes identically.*

Proof. Suppose (M,g) is locally conformally flat. Then for each $p \in M$ there exist a neighborhood U and an embedding $\varphi: U \to \mathbb{R}^n$ such that φ pulls back a flat (Riemannian or pseudo-Riemannian) metric on $\mathbb{R}^n$ to a metric of the form $\widetilde{g} = e^{2f} g$. This implies that $\widetilde{g}$ has zero Weyl tensor, because its entire curvature tensor is zero. By virtue of (7.48), the Weyl tensor of g is also zero.   □

Thus a necessary condition for a smooth manifold of dimension at least 3 to be locally conformally flat is that its Weyl tensor vanish identically. As Theorem 7.37 below will show, in dimensions 4 and higher, this condition is also sufficient. But in 3 dimensions, Corollary 7.26 shows that the Weyl tensor vanishes identically on *every* manifold, so to understand that case, we must introduce one more tensor field. On a Riemannian or pseudo-Riemannian manifold, the **Cotton tensor** is defined as the negative of the exterior covariant derivative of the Schouten tensor: $C = -DP$. This is the 3-tensor field whose expression in terms of any local frame is

$$C_{ijk} = P_{ij;k} - P_{ik;j}. \tag{7.49}$$

Proposition 7.32. *Suppose (M,g) is a Riemannian or pseudo-Riemannian manifold of dimension $n \geq 3$, and let W and C denote its Weyl and Cotton tensors, respectively. Then*

$$\mathrm{tr}_g(\nabla W) = (n-3)C, \tag{7.50}$$

where the trace is on the first and last indices of the 5-tensor ∇W.

Proof. Writing $W = Rm - P \owedge g$ and using the component form of the first contracted Bianchi identity (7.32), we obtain

$$W_{ijkl;}{}^i = R_{jk;l} - R_{jl;k} - P_{il;}{}^i g_{jk} - P_{jk;}{}^i g_{il} + P_{ik;}{}^i g_{jl} + P_{jl;}{}^i g_{ik}. \tag{7.51}$$

Note that $P_{jk;}{}^i g_{il} = P_{jk;l}$ and $P_{jl;}{}^i g_{ik} = P_{jl;k}$. To simplify the other two terms, we use the definition of the Schouten tensor and the second contracted Bianchi identity (7.33) to obtain that

$$P_{il;}{}^i = \frac{1}{n-2}\left(R_{il;}{}^i - \frac{S^{;i}}{2(n-1)}g_{il}\right) = \frac{1}{2(n-1)}S_{;l}.$$

The analogous formula holds for $P_{ik;}{}^i$. When we insert these expressions into (7.51) and simplify, we find that the terms involving derivatives of the Ricci and scalar curvatures combine to yield

$$W_{ijkl;}{}^i = (n-2)(P_{jk;l} - P_{jl;k}) - P_{jk;l} + P_{jl;k} = (n-3)C_{jkl},$$

which is the component version of (7.50). $\square$

Corollary 7.33. *Suppose (M,g) is a Riemannian or pseudo-Riemannian manifold. If $\dim M \geq 4$ and the Weyl tensor vanishes identically, then so does the Cotton tensor.* $\square$

The next proposition expresses another important feature of the Cotton tensor.

Proposition 7.34 (Conformal Invariance of the Cotton Tensor in Dimension 3). *Suppose (M,g) is a Riemannian or pseudo-Riemannian 3-manifold, and $\tilde{g} = e^{2f}g$ for some $f \in C^\infty(M)$. If C and $\tilde{C}$ denote the Cotton tensors of g and $\tilde{g}$, respectively, then $\tilde{C} = C$.*

Proof. Problem 7-10.   □

Corollary 7.35. *If (M, g) is a locally conformally flat 3-manifold, then the Cotton tensor of g vanishes identically.*

▶ **Exercise 7.36.** Prove this corollary.

The real significance of the Weyl and Cotton tensors is explained by the following important theorem.

Theorem 7.37 (Weyl–Schouten). *Suppose (M, g) is a Riemannian or pseudo-Riemannian manifold of dimension $n \geq 3$.*

(a) *If $n \geq 4$, then (M, g) is locally conformally flat if and only if its Weyl tensor is identically zero.*

(b) *If $n = 3$, then (M, g) is locally conformally flat if and only if its Cotton tensor is identically zero.*

Proof. The necessity of each condition was proved in Corollaries 7.31 and 7.35. To prove sufficiency, suppose (M, g) satisfies the hypothesis appropriate to its dimension; then it follows from Corollaries 7.26 and 7.33 that the Weyl and Cotton tensors of g are both identically zero. Every metric $\widetilde{g} = e^{2f} g$ conformal to g also has zero Weyl tensor, and therefore its curvature tensor is $\widetilde{Rm} = \widetilde{P} \otimes \widetilde{g}$. We will show that in a neighborhood of each point, the function f can be chosen to make $\widetilde{P} \equiv 0$, which implies that $\widetilde{Rm} \equiv 0$ and therefore $\widetilde{g}$ is flat.

From (7.47), it follows that $\widetilde{P} \equiv 0$ if and only if

$$P - \nabla^2 f + (df \otimes df) - \tfrac{1}{2} \langle df, df \rangle_g^2 g = 0. \tag{7.52}$$

This equation can be written in the form $\nabla(df) = A(df)$, where A is the map from 1-tensors to symmetric 2-tensors given by

$$A(\xi) = (\xi \otimes \xi) - \tfrac{1}{2} \langle \xi, \xi \rangle_g^2 g + P,$$

or in components,

$$A(\xi)_{ij} = \xi_i \xi_j - \tfrac{1}{2} \xi_m \xi^m g_{ij} + P_{ij}. \tag{7.53}$$

We will solve this equation by first looking for a smooth 1-form ξ that satisfies $\nabla \xi = A(\xi)$. In any local coordinates, if we write $\xi = \xi_j \, dx^j$, this becomes an overdetermined system of first-order partial differential equations for the n unknown functions $\xi_1, \ldots, \xi_n$. (A system of differential equations is said to be **overdetermined** if there are more equations than unknown functions.) If ξ is a solution to $\nabla \xi = A(\xi)$ on an open subset of M, then the Ricci identity (7.21) shows that $A(\xi)$ satisfies

$$A(\xi)_{ij;k} - A(\xi)_{ik;j} = \xi_{i;jk} - \xi_{i;kj} = R_{jki}{}^l \xi_l. \tag{7.54}$$

Lemma 7.38 below shows that this condition is actually sufficient: more precisely, $\nabla \xi = A(\xi)$ has a smooth solution in a neighborhood of each point, provided that for every smooth covector field ξ, the covariant derivatives of $A(\xi)$ satisfy $A(\xi)_{ij;k} - A(\xi)_{ik;j} = R_{jki}{}^l \xi_l$ when $A(\xi)_{ij}$ is substituted for $\xi_{i;j}$ wherever it appears.

To see that this condition is satisfied, differentiate (7.53) to obtain

$$A_{ij;k} - A_{ik;j}$$
$$= \xi_{i;k}\xi_j + \xi_i\xi_{j;k} - \xi_{m;k}\xi^m g_{ij} - \xi_{i;j}\xi_k - \xi_i\xi_{k;j} + \xi_{m;j}\xi^m g_{ik} + C_{ijk}.$$

Now substitute the right-hand side of (7.53) (with appropriate index substitutions) for $\xi_{i;j}$ wherever it appears, subtract $R_{jki}{}^l\xi_l$, and use the relation $Rm = W + P \otimes g$. After extensive cancellations, we obtain

$$A(\xi)_{ij;k} - A(\xi)_{ik;j} - R_{jki}{}^l\xi_l = -W_{jki}{}^l\xi_l + C_{ijk}. \tag{7.55}$$

Our hypotheses guarantee that the right-hand side is identically zero, so there is a solution ξ to $\nabla\xi = A(\xi)$ in a neighborhood of each point.

Because $A(\xi)_{ij}$ is symmetric in i and j, it follows that the ordinary derivatives $\partial_j\xi_i = \xi_{i;j} + \Gamma_{ij}^k\xi_k$ are also symmetric, and thus ξ is a closed 1-form. By the Poincaré lemma, in some (possibly smaller) neighborhood of each point, there is a smooth function f such that $\xi = df = \nabla f$; this f is the function we seek. $\square$

Here is the lemma that was used in the proof of the preceding theorem.

Lemma 7.38. *Let (M,g) be a Riemannian or pseudo-Riemannian manifold, and consider the overdetermined system of equations*

$$\nabla\xi = A(\xi), \tag{7.56}$$

where $A\colon T^*M \to T^2T^*M$ is a smooth map satisfying the following compatibility condition: for any smooth covector field ξ, the 3-tensor field $\nabla(A(\xi))$ satisfies the following identity when $A(\xi)$ is substituted for $\nabla\xi$ wherever it appears:

$$A(\xi)_{ij;k} - A(\xi)_{ik;j} = R_{jki}{}^l\xi_l. \tag{7.57}$$

Then for every $p \in M$ and every covector $\eta_0 \in T_p^*M$, there is a smooth solution to (7.56) on a neighborhood of p satisfying $\xi_p = \eta_0$.

Proof. Let $p \in M$ be given. In smooth local coordinates (x^i) on a neighborhood of p, (7.56) is equivalent to the overdetermined system

$$\frac{\partial\xi_i(x)}{\partial x^j} = a_{ij}(x,\xi(x)),$$

where

$$a_{ij}(x,\xi) = \Gamma_{ij}^k(x)\xi_k + A(\xi)_{ij}.$$

An application of the Frobenius theorem [LeeSM, Prop. 19.29] shows that there is a smooth solution to this overdetermined system in a neighborhood of p with ξ_p arbitrary, provided that

$$\frac{\partial a_{ij}}{\partial x^k} + a_{kl}\frac{\partial a_{ij}}{\partial \xi_l} = \frac{\partial a_{ik}}{\partial x^j} + a_{jl}\frac{\partial a_{ik}}{\partial \xi_l}. \tag{7.58}$$

By virtue of the chain rule, this identity means exactly that the first derivatives $\partial\big(a_{ij}(x,\xi(x))\big)/\partial x^k$, after substituting $a_{kl}(x,\xi(x))$ in place of $\partial\xi_k(x)/\partial x^l$, are symmetric in the indices j,k. Because $\xi_{k;l} = \partial_l\xi_k - \Gamma^m_{kl}\xi_m$, this substitution is equivalent to substituting $A(\xi)_{kl}$ for $\xi_{k;l}$ wherever it occurs. After expanding the hypothesis (7.57) in terms of Christoffel symbols, we find after some manipulation that it reduces to (7.58). □

Because all Riemannian 1-manifolds are flat, the only nontrivial case that is not addressed by the Weyl–Schouten theorem is that of dimension 2. Smooth coordinates that provide a conformal equivalence between an open subset of a Riemannian or pseudo-Riemannian 2-manifold and an open subset of $\mathbb{R}^2$ are called **isothermal coordinates**, and it is a fact that such coordinates always exist locally. For the Riemannian case, there are various proofs available, all of which involve more machinery from partial differential equations and complex analysis than we have at our disposal; see [Che55] for a reasonably elementary proof. For the pseudo-Riemannian case, see Problem 7-15. Thus every Riemannian or pseudo-Riemannian 2-manifold is locally conformally flat.

Problems

7-1. Complete the proof of Proposition 7.3 by showing that $R(X,Y)(fZ) = fR(X,Y)Z$ for all smooth vector fields X,Y,Z and smooth real-valued functions f.

7-2. Prove Proposition 7.4 (the formula for the curvature tensor in coordinates).

7-3. Show that the curvature tensor of a Riemannian locally symmetric space is parallel: $\nabla Rm \equiv 0$. (*Used on pp. 297, 351.*)

7-4. Let M be a Riemannian or pseudo-Riemannian manifold, and let (x^i) be normal coordinates centered at $p \in M$. Show that the following holds at p:

$$R_{ijkl} = \frac{1}{2}\big(\partial_j\partial_l g_{ik} + \partial_i\partial_k g_{jl} - \partial_i\partial_l g_{jk} - \partial_j\partial_k g_{il}\big).$$

7-5. Let ∇ be the Levi-Civita connection on a Riemannian or pseudo-Riemannian manifold (M,g), and let $\omega_i{}^j$ be its connection 1-forms with respect to a local frame (E_i) (Problem 4-14). Define a matrix of 2-forms $\Omega_i{}^j$, called the **curvature 2-forms**, by

$$\Omega_i{}^j = \frac{1}{2}R_{kli}{}^j\,\varepsilon^k \wedge \varepsilon^l,$$

where (ε^i) is the coframe dual to (E_i). Show that the curvature 2-forms satisfy **Cartan's second structure equation**:

$$\Omega_i{}^j = d\omega_i{}^j - \omega_i{}^k \wedge \omega_k{}^j.$$

[Hint: Expand $R(E_k, E_l)E_i$ in terms of ∇ and $\omega_i{}^j$.]

7-6. Suppose (M_1, g_1) and (M_2, g_2) are Riemannian manifolds, and $M_1 \times M_2$ is endowed with the product metric $g = g_1 \oplus g_2$ as in (2.12). Show that the Riemann curvature, Ricci curvature, and scalar curvature of g are given by the following formulas:

$$Rm = \pi_1^* Rm_1 + \pi_2^* Rm_2,$$
$$Rc = \pi_1^* Rc_1 + \pi_2^* Rc_2,$$
$$S = \pi_1^* S_1 + \pi_1^* S_2,$$

where Rm_i, Rc_i, and S_i are the Riemann, Ricci, and scalar curvatures of (M_i, g_i), and $\pi_i : M_1 \times M_2 \to M_i$ is the projection. (*Used on pp. 257, 261.*)

7-7. Suppose (M, g) is a Riemannian manifold and $u \in C^\infty(M)$. Use (5.29) and (7.21) to prove **Bochner's formula**:

$$\Delta\left(\tfrac{1}{2}|\operatorname{grad} u|^2\right) = \left|\nabla^2 u\right|^2 + \left\langle \operatorname{grad}(\Delta u), \operatorname{grad} u\right\rangle + Rc\left(\operatorname{grad} u, \operatorname{grad} u\right).$$

7-8. LICHNEROWICZ'S THEOREM: Suppose (M, g) is a compact Riemannian n-manifold, and there is a positive constant κ such that the Ricci tensor of g satisfies $Rc(v, v) \geq \kappa |v|^2$ for all tangent vectors v. If λ is any positive eigenvalue of M, then $\lambda \geq n\kappa/(n-1)$. [Hint: Use Probs. 2-23(c), 5-15, and 7-7.]

7-9. Prove parts (e) and (f) of Lemma 7.22 (properties of the Kulkarni–Nomizu product).

7-10. Prove Proposition 7.34 (conformal invariance of the Cotton tensor in dimension 3).

7-11. Let (M, g) be a Riemannian manifold of dimension $n > 2$. Define the **conformal Laplacian** $L : C^\infty(M) \to C^\infty(M)$ by the formula

$$Lu = -\frac{4(n-1)}{(n-2)}\Delta u + Su,$$

where Δ is the Laplace–Beltrami operator of g and S is its scalar curvature. Prove that if $\widetilde{g} = e^{2f} g$ for some $f \in C^\infty(M)$, and $\widetilde{L}$ denotes the conformal Laplacian with respect to $\widetilde{g}$, then for every $u \in C^\infty(M)$,

$$e^{\frac{n+2}{2}f} \widetilde{L}u = L\left(e^{\frac{n-2}{2}f} u\right).$$

Conclude that a metric $\widetilde{g}$ conformal to g has constant scalar curvature λ if and only if it can be expressed in the form $\widetilde{g} = \varphi^{4/(n-2)} g$, where φ is a smooth positive solution to the **Yamabe equation**:

$$L\varphi = \lambda \varphi^{\frac{n+2}{n-2}}. \tag{7.59}$$

7-12. Let M be a smooth manifold and let ∇ be any connection on TM. We can define the **curvature endomorphism of** ∇ by the same formula as in the Riemannian case: $R(X,Y)Z = \nabla_X \nabla_Y Z - \nabla_Y \nabla_X Z - \nabla_{[X,Y]}Z$. Then ∇ is said to be a **flat connection** if $R(X,Y)Z \equiv 0$. Prove that the following are equivalent:

(a) ∇ is flat.
(b) For every point $p \in M$, there exists a parallel local frame defined on a neighborhood of p.
(c) For all $p,q \in M$, parallel transport along an admissible curve segment γ from p to q depends only on the path-homotopy class of γ.
(d) Parallel transport around any sufficiently small closed curve is the identity; that is, for every $p \in M$, there exists a neighborhood U of p such that if $\gamma : [a,b] \to U$ is an admissible curve in U starting and ending at p, then $P_{ab} : T_p M \to T_p M$ is the identity map.

7-13. Let G be a Lie group with a bi-invariant metric g. Show that the following formula holds whenever X, Y, Z are left-invariant vector fields on G:

$$R(X,Y)Z = \frac{1}{4}[Z,[X,Y]].$$

(See Problem 5-8.)

7-14. Suppose $\pi : (\widetilde{M}, \widetilde{g}) \to (M,g)$ is a Riemannian submersion. Using the notation and results of Problem 5-6, prove **O'Neill's formula**:

$$Rm(W,X,Y,Z) \circ \pi = \widetilde{Rm}(\widetilde{W}, \widetilde{X}, \widetilde{Y}, \widetilde{Z}) - \frac{1}{2}\left\langle [\widetilde{W}, \widetilde{X}]^V, [\widetilde{Y}, \widetilde{Z}]^V \right\rangle$$
$$- \frac{1}{4}\left\langle [\widetilde{W}, \widetilde{Y}]^V, [\widetilde{X}, \widetilde{Z}]^V \right\rangle + \frac{1}{4}\left\langle [\widetilde{W}, \widetilde{Z}]^V, [\widetilde{X}, \widetilde{Y}]^V \right\rangle.$$

(*Used on p. 258.*)

7-15. Suppose (M,g) is a 2-dimensional pseudo-Riemannian manifold of signature $(1,1)$, and $p \in M$.

(a) Show that there is a smooth local frame (E_1, E_2) in a neighborhood of p such that $g(E_1, E_1) = g(E_2, E_2) = 0$.
(b) Show that there are smooth coordinates (x,y) in a neighborhood of p such that $(dx)^2 - (dy)^2 = fg$ for some smooth, positive real-valued function f. [Hint: Use Prop. A.45 to show that there exist coordinates (t,u) in which $E_1 = \partial/\partial t$, and coordinates (v,w) in which $E_2 = \partial/\partial v$, and set $x = u + w$, $y = u - w$.]
(c) Show that (M,g) is locally conformally flat.

Chapter 8
Riemannian Submanifolds

This chapter has a dual purpose: first to apply the theory of curvature to Riemannian submanifolds, and then to use these concepts to derive a precise quantitative interpretation of the curvature tensor.

After introducing some basic definitions and terminology concerning Riemannian submanifolds, we define a vector-valued bilinear form called the *second fundamental form*, which measures the way a submanifold curves within the ambient manifold. We then prove the fundamental relationships between the intrinsic and extrinsic geometries of a submanifold, including the *Gauss formula*, which relates the Riemannian connection on the submanifold to that of the ambient manifold, and the *Gauss equation*, which relates their curvatures. We then show how the second fundamental form measures the extrinsic curvature of submanifold geodesics.

Using these tools, we focus on the special case of hypersurfaces, and use the second fundamental form to define some real-valued curvature quantities called the *principal curvatures*, *mean curvature*, and *Gaussian curvature*. Specializing even more to hypersurfaces in Euclidean space, we describe various concrete geometric interpretations of these quantities. Then we prove Gauss's *Theorema Egregium*, which shows that the Gaussian curvature of a surface in $\mathbb{R}^3$ can be computed intrinsically from the curvature tensor of the induced metric.

In the last section, we introduce the promised quantitative geometric interpretation of the curvature tensor. It allows us to compute *sectional curvatures*, which are just the Gaussian curvatures of 2-dimensional submanifolds swept out by geodesics tangent to 2-planes in the tangent space. Finally, we compute the sectional curvatures of our frame-homogeneous model Riemannian manifolds—Euclidean spaces, spheres, and hyperbolic spaces.

The Second Fundamental Form

Suppose (M, g) is a Riemannian submanifold of a Riemannian manifold $(\widetilde{M}, \widetilde{g})$. Recall that this means that M is a submanifold of $\widetilde{M}$ endowed with the induced metric

© Springer International Publishing AG 2018

J. M. Lee, *Introduction to Riemannian Manifolds*, Graduate Texts
in Mathematics 176, https://doi.org/10.1007/978-3-319-91755-9_8

$g = \iota_M^* \widetilde{g}$ (where $\iota_M : M \hookrightarrow \widetilde{M}$ is the inclusion map). Our goal in this chapter is to study the relationship between the geometry of M and that of $\widetilde{M}$.

Although we focus our attention in this chapter on embedded submanifolds for simplicity, the results we present are all local, so they apply in much greater generality. In particular, if $M \subseteq \widetilde{M}$ is an immersed submanifold, then every point of M has a neighborhood in M that is embedded in $\widetilde{M}$, so the results of this chapter can be applied by restricting to such a neighborhood. Even more generally, if (M, g) is any Riemannian manifold and $F : M \to \widetilde{M}$ is an isometric immersion (meaning that $F^* \widetilde{g} = g$), then again every point $p \in M$ has a neighborhood $U \subseteq M$ such that $F|_U$ is an embedding, so the results apply to $F(U) \subseteq M$. We leave it to the reader to sort out the minor modifications in notation and terminology needed to handle these more general situations.

The results in the first section of this chapter apply virtually without modification to *Riemannian* submanifolds of pseudo-Riemannian manifolds (ones on which the induced metric is positive definite), so we state most of our theorems in that case. Recall that when the ambient metric $\widetilde{g}$ is indefinite, this includes the assumption that the induced metric $g = \iota_M^* \widetilde{g}$ is positive definite. Some of the results can also be extended to pseudo-Riemannian submanifolds of mixed signature, but there are various pitfalls to watch out for in that case; so for simplicity we restrict to the case of Riemannian submanifolds. See [dC92] for a thorough treatment of pseudo-Riemannian submanifolds.

Also, most of these results can be adapted to manifolds and submanifolds with boundary, simply by embedding everything in slightly larger manifolds without boundary, but one might need to be careful about the statements of some of the results when the submanifold intersects the boundary. Since the interaction of submanifolds with boundaries is not our primary concern here, for simplicity we state all of these results in the case of empty boundary only.

Throughout this chapter, therefore, we assume that $(\widetilde{M}, \widetilde{g})$ is a Riemannian or pseudo-Riemannian manifold of dimension m, and (M, g) is an embedded n-dimensional Riemannian submanifold of $\widetilde{M}$. We call $\widetilde{M}$ the **ambient manifold**. We will denote covariant derivatives and curvatures associated with (M, g) in the usual way (∇, R, Rm, etc.), and write those associated with $(\widetilde{M}, \widetilde{g})$ with tildes ($\widetilde{\nabla}$, $\widetilde{R}$, $\widetilde{Rm}$, etc.). We can unambiguously use the inner product notation $\langle v, w \rangle$ to refer either to g or to $\widetilde{g}$, since g is just the restriction of $\widetilde{g}$ to pairs of vectors in TM.

Our first main task is to compare the Levi-Civita connection of M with that of $\widetilde{M}$. The starting point for doing so is the orthogonal decomposition of sections of the ambient tangent bundle $T\widetilde{M}|_M$ into tangential and orthogonal components. Just as we did for submanifolds of $\mathbb{R}^n$ in Chapter 4, we define orthogonal projection maps called **tangential** and **normal projections**:

$$\pi^\top : T\widetilde{M}|_M \to TM,$$
$$\pi^\perp : T\widetilde{M}|_M \to NM.$$

In terms of an adapted orthonormal frame $(E_1, \ldots, E_m)$ for M in $\widetilde{M}$, these are just the usual projections onto $\mathrm{span}(E_1, \ldots, E_n)$ and $\mathrm{span}(E_{n+1}, \ldots, E_m)$ respectively,

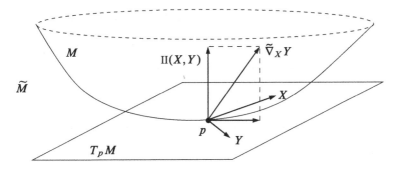

Fig. 8.1: The second fundamental form

so both projections are smooth bundle homomorphisms (i.e., they are linear on fibers and map smooth sections to smooth sections). If X is a section of $T\widetilde{M}|_M$, we often use the shorthand notations $X^\top = \pi^\top X$ and $X^\perp = \pi^\perp X$ for its tangential and normal projections.

If X, Y are vector fields in $\mathfrak{X}(M)$, we can extend them to vector fields on an open subset of $\widetilde{M}$ (still denoted by X and Y), apply the ambient covariant derivative operator $\widetilde{\nabla}$, and then decompose at points of M to get

$$\widetilde{\nabla}_X Y = \left(\widetilde{\nabla}_X Y\right)^\top + \left(\widetilde{\nabla}_X Y\right)^\perp. \tag{8.1}$$

We wish to interpret the two terms on the right-hand side of this decomposition.

Let us focus first on the normal component. We define the **second fundamental form of** M to be the map $\mathrm{II}\colon \mathfrak{X}(M) \times \mathfrak{X}(M) \to \Gamma(NM)$ (read "two") given by

$$\mathrm{II}(X,Y) = \left(\widetilde{\nabla}_X Y\right)^\perp,$$

where X and Y are extended arbitrarily to an open subset of $\widetilde{M}$ (Fig. 8.1). Since $\pi^\perp$ maps smooth sections to smooth sections, $\mathrm{II}(X, Y)$ is a smooth section of NM.

The term *first fundamental form*, by the way, was originally used to refer to the induced metric g on M. Although that usage has mostly been replaced by more descriptive terminology, we seem unfortunately to be stuck with the name "second fundamental form." The word "form" in both cases refers to bilinear form, not differential form.

Proposition 8.1 (Properties of the Second Fundamental Form). *Suppose (M, g) is an embedded Riemannian submanifold of a Riemannian or pseudo-Riemannian manifold $(\widetilde{M}, \widetilde{g})$, and let $X, Y \in \mathfrak{X}(M)$.*

(a) *$\mathrm{II}(X, Y)$ is independent of the extensions of X and Y to an open subset of $\widetilde{M}$.*
(b) *$\mathrm{II}(X, Y)$ is bilinear over $C^\infty(M)$ in X and Y.*
(c) *$\mathrm{II}(X, Y)$ is symmetric in X and Y.*
(d) *The value of $\mathrm{II}(X, Y)$ at a point $p \in M$ depends only on X_p and Y_p.*

Proof. Choose particular extensions of X and Y to a neighborhood of M in $\widetilde{M}$, and for simplicity denote the extended vector fields also by X and Y. We begin by proving that $\mathrm{II}(X, Y)$ is symmetric in X and Y when defined in terms of these extensions. The symmetry of the connection $\widetilde{\nabla}$ implies

$$\mathrm{II}(X, Y) - \mathrm{II}(Y, X) = \left(\widetilde{\nabla}_X Y - \widetilde{\nabla}_Y X\right)^{\perp} = [X, Y]^{\perp}.$$

Since X and Y are tangent to M at all points of M, so is their Lie bracket (Cor. A.40). Therefore $[X, Y]^{\perp} = 0$, so II is symmetric.

Because $\widetilde{\nabla}_X Y|_p$ depends only on X_p, it follows that the value of $\mathrm{II}(X, Y)$ at p depends only on X_p, and in particular is independent of the extension chosen for X. Because $\widetilde{\nabla}_X Y$ is linear over $C^{\infty}(\widetilde{M})$ in X, and every $f \in C^{\infty}(M)$ can be extended to a smooth function on a neighborhood of M in $\widetilde{M}$, it follows that $\mathrm{II}(X, Y)$ is linear over $C^{\infty}(M)$ in X. By symmetry, the same claims hold for Y. □

As a consequence of the preceding proposition, for every $p \in M$ and all vectors $v, w \in T_p M$, it makes sense to interpret $\mathrm{II}(v, w)$ as the value of $\mathrm{II}(V, W)$ at p, where V and W are any vector fields on M such that $V_p = v$ and $W_p = W$, and we will do so from now on without further comment.

We have not yet identified the tangential term in the decomposition of $\widetilde{\nabla}_X Y$. Proposition 5.12(b) showed that in the special case of a submanifold of a Euclidean or pseudo-Euclidean space, it is none other than the covariant derivative with respect to the Levi-Civita connection of the induced metric on M. The following theorem shows that the same is true in the general case. Therefore, we can interpret the second fundamental form as a measure of the difference between the intrinsic Levi-Civita connection on M and the ambient Levi-Civita connection on $\widetilde{M}$.

Theorem 8.2 (The Gauss Formula). *Suppose (M, g) is an embedded Riemannian submanifold of a Riemannian or pseudo-Riemannian manifold $(\widetilde{M}, \widetilde{g})$. If $X, Y \in \mathfrak{X}(M)$ are extended arbitrarily to smooth vector fields on a neighborhood of M in $\widetilde{M}$, the following formula holds along M:*

$$\widetilde{\nabla}_X Y = \nabla_X Y + \mathrm{II}(X, Y).$$

Proof. Because of the decomposition (8.1) and the definition of the second fundamental form, it suffices to show that $\left(\widetilde{\nabla}_X Y\right)^{\top} = \nabla_X Y$ at all points of M.

Define a map $\nabla^{\top} : \mathfrak{X}(M) \times \mathfrak{X}(M) \to \mathfrak{X}(M)$ by

$$\nabla^{\top}_X Y = \left(\widetilde{\nabla}_X Y\right)^{\top},$$

where X and Y are extended arbitrarily to an open subset of $\widetilde{M}$. We examined a special case of this construction, in which $\widetilde{g}$ is a Euclidean or pseudo-Euclidean metric, in Example 4.9. It follows exactly as in that example that $\nabla^{\top}$ is a connection on M, and exactly as in the proofs of Propositions 5.8 and 5.9 that it is symmetric and compatible with g. The uniqueness of the Riemannian connection on M then shows that $\nabla^{\top} = \nabla$. □

The Gauss formula can also be used to compare intrinsic and extrinsic covariant derivatives along curves. If $\gamma\colon I \to M$ is a smooth curve and X is a vector field along γ that is everywhere tangent to M, then we can regard X as either a vector field along γ in $\widetilde{M}$ or a vector field along γ in M. We let $\widetilde{D}_t X$ and $D_t X$ denote its covariant derivatives along γ as a curve in $\widetilde{M}$ and as a curve in M, respectively. The next corollary shows how the two covariant derivatives are related.

Corollary 8.3 (The Gauss Formula Along a Curve). *Suppose (M,g) is an embedded Riemannian submanifold of a Riemannian or pseudo-Riemannian manifold $(\widetilde{M},\widetilde{g})$, and $\gamma\colon I \to M$ is a smooth curve. If X is a smooth vector field along γ that is everywhere tangent to M, then*

$$\widetilde{D}_t X = D_t X + \mathrm{II}(\gamma', X). \tag{8.2}$$

Proof. For each $t_0 \in I$, we can find an adapted orthonormal frame $(E_1, \ldots, E_m)$ in a neighborhood of $\gamma(t_0)$. (Recall that our default assumption is that $\dim \widetilde{M} = m$ and $\dim M = n$.) In terms of this frame, X can be written $X(t) = \sum_{i=1}^{n} X^i(t) E_i|_{\gamma(t)}$. Applying the product rule and the Gauss formula, and using the fact that each vector field E_i is extendible, we get

$$\widetilde{D}_t X = \sum_{i=1}^{n} \left(\dot{X}^i E_i + X^i \widetilde{\nabla}_{\gamma'} E_i \right)$$

$$= \sum_{i=1}^{n} \left(\dot{X}^i E_i + X^i \nabla_{\gamma'} E_i + X^i \mathrm{II}(\gamma', E_i) \right)$$

$$= D_t X + \mathrm{II}(\gamma', X). \qquad \square$$

Although the second fundamental form is defined in terms of covariant derivatives of vector fields *tangent* to M, it can also be used to evaluate extrinsic covariant derivatives of *normal* vector fields, as the following proposition shows. To express it concisely, we introduce one more notation. For each normal vector field $N \in \Gamma(NM)$, we obtain a scalar-valued symmetric bilinear form $\mathrm{II}_N\colon \mathfrak{X}(M) \times \mathfrak{X}(M) \to C^\infty(M)$ by

$$\mathrm{II}_N(X, Y) = \langle N, \mathrm{II}(X, Y) \rangle. \tag{8.3}$$

Let $W_N\colon \mathfrak{X}(M) \to \mathfrak{X}(M)$ denote the self-adjoint linear map associated with this bilinear form, characterized by

$$\langle W_N(X), Y \rangle = \mathrm{II}_N(X, Y) = \langle N, \mathrm{II}(X, Y) \rangle. \tag{8.4}$$

The map W_N is called the **Weingarten map in the direction of** N. Because the second fundamental form is bilinear over $C^\infty(M)$, it follows that W_N is linear over $C^\infty(M)$ and thus defines a smooth bundle homomorphism from TM to itself.

Proposition 8.4 (The Weingarten Equation). *Suppose (M, g) is an embedded Riemannian submanifold of a Riemannian or pseudo-Riemannian manifold $(\widetilde{M},\widetilde{g})$. For every $X \in \mathfrak{X}(M)$ and $N \in \Gamma(NM)$, the following equation holds:*

$$\left(\tilde{\nabla}_X N\right)^\top = -W_N(X), \tag{8.5}$$

when N is extended arbitrarily to an open subset of $\widetilde{M}$.

Proof. Note that at points of M, the covariant derivative $\tilde{\nabla}_X N$ is independent of the choice of extensions of X and N by Proposition 4.26. Let $Y \in \mathfrak{X}(M)$ be arbitrary, extended to a vector field on an open subset of $\widetilde{M}$. Since $\langle N, Y \rangle$ vanishes identically along M and X is tangent to M, the following holds at points of M:

$$\begin{aligned}
0 &= X \langle N, Y \rangle \\
&= \langle \tilde{\nabla}_X N, Y \rangle + \langle N, \tilde{\nabla}_X Y \rangle \\
&= \langle \tilde{\nabla}_X N, Y \rangle + \langle N, \nabla_X Y + \mathrm{II}(X,Y) \rangle \\
&= \langle \tilde{\nabla}_X N, Y \rangle + \langle N, \mathrm{II}(X,Y) \rangle \\
&= \langle \tilde{\nabla}_X N, Y \rangle + \langle W_N(X), Y \rangle.
\end{aligned}$$

Since Y was an arbitrary vector field tangent to M, this implies

$$0 = \left(\tilde{\nabla}_X N + W_N(X)\right)^\top = \left(\tilde{\nabla}_X N\right)^\top + W_N(X),$$

which is equivalent to (8.5). □

In addition to describing the difference between the intrinsic and extrinsic connections, the second fundamental form plays an even more important role in describing the difference between the curvature tensors of $\widetilde{M}$ and M. The explicit formula, also due to Gauss, is given in the following theorem.

Theorem 8.5 (The Gauss Equation). *Suppose (M, g) is an embedded Riemannian submanifold of a Riemannian or pseudo-Riemannian manifold $\left(\widetilde{M}, \tilde{g}\right)$. For all $W, X, Y, Z \in \mathfrak{X}(M)$, the following equation holds:*

$$\widetilde{Rm}(W, X, Y, Z) = Rm(W, X, Y, Z) - \langle \mathrm{II}(W,Z), \mathrm{II}(X,Y) \rangle + \langle \mathrm{II}(W,Y), \mathrm{II}(X,Z) \rangle.$$

Proof. Let W, X, Y, Z be extended arbitrarily to an open subset of $\widetilde{M}$. At points of M, using the definition of the curvature and the Gauss formula, we get

$$\begin{aligned}
\widetilde{Rm}(W, X, Y, Z) &= \langle \tilde{\nabla}_W \tilde{\nabla}_X Y - \tilde{\nabla}_X \tilde{\nabla}_W Y - \tilde{\nabla}_{[W,X]} Y, \ Z \rangle \\
&= \langle \tilde{\nabla}_W (\nabla_X Y + \mathrm{II}(X,Y)) - \tilde{\nabla}_X (\nabla_W Y + \mathrm{II}(W,Y)) \\
&\quad - \tilde{\nabla}_{[W,X]} Y, \ Z \rangle.
\end{aligned}$$

Apply the Weingarten equation to each of the terms involving II (with $\mathrm{II}(X,Y)$ or $\mathrm{II}(W,Y)$ playing the role of N) to get

$$\begin{aligned}
\widetilde{Rm}(W, X, Y, Z) &= \langle \tilde{\nabla}_W \nabla_X Y, Z \rangle - \langle \mathrm{II}(X,Y), \mathrm{II}(W,Z) \rangle \\
&\quad - \langle \tilde{\nabla}_X \nabla_W Y, Z \rangle + \langle \mathrm{II}(W,Y), \mathrm{II}(X,Z) \rangle - \langle \tilde{\nabla}_{[W,X]} Y, Z \rangle.
\end{aligned}$$

Decomposing each term involving $\widetilde{\nabla}$ into its tangential and normal components, we see that only the tangential component survives, because Z is tangent to M. The Gauss formula allows each such term to be rewritten in terms of ∇, giving

$$
\begin{aligned}
\widetilde{Rm}(W, X, Y, Z) &= \langle \nabla_W \nabla_X Y, Z \rangle - \langle \nabla_X \nabla_W Y, Z \rangle - \langle \nabla_{[W,X]} Y, Z \rangle \\
&\quad - \langle \mathrm{II}(X, Y), \mathrm{II}(W, Z) \rangle + \langle \mathrm{II}(W, Y), \mathrm{II}(X, Z) \rangle \\
&= \langle R(W, X)Y, Z \rangle \\
&\quad - \langle \mathrm{II}(X, Y), \mathrm{II}(W, Z) \rangle + \langle \mathrm{II}(W, Y), \mathrm{II}(X, Z) \rangle. \qquad \square
\end{aligned}
$$

There is one other fundamental submanifold equation, which relates the normal part of the ambient curvature endomorphism to derivatives of the second fundamental form. We will not have need for it, but we include it here for completeness. To state it, we need to introduce a connection on the normal bundle of a Riemannian submanifold.

If (M, g) is a Riemannian submanifold of a Riemannian or pseudo-Riemannian manifold $(\widetilde{M}, \widetilde{g})$, the **normal connection** $\nabla^\perp \colon \mathfrak{X}(M) \times \Gamma(NM) \to \Gamma(NM)$ is defined by

$$
\nabla^\perp_X N = \left(\widetilde{\nabla}_X N \right)^\perp,
$$

where N is extended to a smooth vector field on a neighborhood of M in $\widetilde{M}$.

Proposition 8.6. *If (M, g) is an embedded Riemannian submanifold of a Riemannian or pseudo-Riemannian manifold $(\widetilde{M}, \widetilde{g})$, then $\nabla^\perp$ is a well-defined connection in NM, which is compatible with $\widetilde{g}$ in the sense that for any two sections N_1, N_2 of NM and every $X \in \mathfrak{X}(M)$, we have*

$$
X \langle N_1, N_2 \rangle = \left\langle \nabla^\perp_X N_1, N_2 \right\rangle + \left\langle N_1, \nabla^\perp_X N_2 \right\rangle.
$$

▶ **Exercise 8.7.** Prove the preceding proposition.

We need the normal connection primarily to make sense of tangential covariant derivatives of the second fundamental form. To do so, we make the following definitions. Let $F \to M$ denote the bundle whose fiber at each point $p \in M$ is the set of bilinear maps $T_p M \times T_p M \to N_p M$. It is easy to check that F is a smooth vector bundle over M, and that smooth sections of F correspond to smooth maps $\mathfrak{X}(M) \times \mathfrak{X}(M) \to \Gamma(NM)$ that are bilinear over $C^\infty(M)$, such as the second fundamental form. Define a connection ∇^F in F as follows: if B is any smooth section of F, let $\nabla^F_X B$ be the smooth section of F defined by

$$
\left(\nabla^F_X B \right)(Y, Z) = \nabla^\perp_X (B(Y, Z)) - B\left(\nabla_X Y, \ Z \right) - B\left(Y, \ \nabla_X Z \right).
$$

▶ **Exercise 8.8.** Prove that ∇^F is a connection in F.

Now we are ready to state the last of the fundamental equations for submanifolds. This equation was independently discovered (in the special case of surfaces in $\mathbb{R}^3$) by Karl M. Peterson (1853), Gaspare Mainardi (1856), and Delfino Codazzi (1868–1869), and is sometimes designated by various combinations of these three names.

For the sake of simplicity we use the traditional but historically inaccurate name
Codazzi equation.

Theorem 8.9 (The Codazzi Equation). *Suppose (M, g) is an embedded Riemannian submanifold of a Riemannian or pseudo-Riemannian manifold $(\widetilde{M}, \widetilde{g})$. For all
$W, X, Y \in \mathfrak{X}(M)$, the following equation holds:*

$$\left(\widetilde{R}(W, X)Y\right)^{\perp} = \left(\nabla_W^F \mathrm{II}\right)(X, Y) - \left(\nabla_X^F \mathrm{II}\right)(W, Y). \tag{8.6}$$

Proof. It suffices to show that both sides of (8.6) give the same result when we take
their inner products with an arbitrary smooth normal vector field N along M:

$$\left\langle \widetilde{R}(W, X)Y, \ N \right\rangle = \left\langle \left(\nabla_W^F \mathrm{II}\right)(X, Y), \ N \right\rangle - \left\langle \left(\nabla_X^F \mathrm{II}\right)(W, Y), \ N \right\rangle. \tag{8.7}$$

We begin as in the proof of the Gauss equation: after extending the vector fields
to a neighborhood of M and applying the Gauss formula, we obtain

$$\widetilde{Rm}(W, X, Y, N) = \left\langle \widetilde{\nabla}_W \left(\nabla_X Y + \mathrm{II}(X, Y)\right) - \widetilde{\nabla}_X \left(\nabla_W Y + \mathrm{II}(W, Y)\right) \right. \\ \left. - \widetilde{\nabla}_{[W, X]} Y, \ N \right\rangle.$$

Now when we expand the covariant derivatives, we need only pay attention to the
normal components. This yields

$$\widetilde{Rm}(W, X, Y, N)$$
$$= \left\langle \mathrm{II}(W, \nabla_X Y) + \left(\nabla_W^F \mathrm{II}\right)(X, Y) + \mathrm{II}(\nabla_W X, Y) + \mathrm{II}(X, \nabla_W Y), \ N \right\rangle$$
$$- \left\langle \mathrm{II}(X, \nabla_W Y) + \left(\nabla_X^F \mathrm{II}\right)(W, Y) + \mathrm{II}(\nabla_X W, Y) + \mathrm{II}(W, \nabla_X Y), \ N \right\rangle$$
$$- \left\langle \mathrm{II}([W, X], Y), \ N \right\rangle.$$

The terms involving $\nabla_X Y$ and $\nabla_W Y$ cancel each other in pairs, and three other
terms sum to zero because $\nabla_W X - \nabla_X W - [W, X] = 0$. What remains is (8.7). $\qquad \square$

Curvature of a Curve

By studying the curvatures of curves, we can give a more geometric interpretation of the second fundamental form. Suppose (M, g) is a Riemannian or pseudo-
Riemannian manifold, and $\gamma \colon I \to M$ is a smooth unit-speed curve in M. We define
the (*geodesic*) *curvature of γ* as the length of the acceleration vector field, which is
the function $\kappa \colon I \to \mathbb{R}$ given by

$$\kappa(t) = \left| D_t \gamma'(t) \right|.$$

If γ is an arbitrary regular curve in a Riemannian manifold (not necessarily of unit
speed), we first find a unit-speed reparametrization $\widetilde{\gamma} = \gamma \circ \varphi$, and then define the
curvature of γ at t to be the curvature of $\widetilde{\gamma}$ at $\varphi^{-1}(t)$. In a pseudo-Riemannian

manifold, the same approach works, but we have to restrict the definition to curves γ such that $|\gamma'(t)|$ is everywhere nonzero. Problem 8-6 gives a formula that can be used in the Riemannian case to compute the geodesic curvature directly without explicitly finding a unit-speed reparametrization.

From the definition, it follows that a smooth unit-speed curve has vanishing geodesic curvature if and only if it is a geodesic, so the geodesic curvature of a curve can be regarded as a quantitative measure of how far it deviates from being a geodesic. If $M = \mathbb{R}^n$ with the Euclidean metric, the geodesic curvature agrees with the notion of curvature introduced in advanced calculus courses.

Now suppose $(\widetilde{M}, \widetilde{g})$ is a Riemannian or pseudo-Riemannian manifold and (M, g) is a Riemannian submanifold. Every regular curve $\gamma: I \to M$ has two distinct geodesic curvatures: its **intrinsic curvature** κ as a curve in M, and its **extrinsic curvature** $\widetilde{\kappa}$ as a curve in $\widetilde{M}$. The second fundamental form can be used to compute the relationship between the two.

Proposition 8.10 (Geometric Interpretation of II). *Suppose (M, g) is an embedded Riemannian submanifold of a Riemannian or pseudo-Riemannian manifold $(\widetilde{M}, \widetilde{g})$, $p \in M$, and $v \in T_p M$.*

(a) $\mathrm{II}(v, v)$ is the $\widetilde{g}$-acceleration at p of the g-geodesic γ_v.

(b) If v is a unit vector, then $|\mathrm{II}(v, v)|$ is the $\widetilde{g}$-curvature of γ_v at p.

Proof. Suppose $\gamma: (-\varepsilon, \varepsilon) \to M$ is any regular curve with $\gamma(0) = p$ and $\gamma'(0) = v$. Applying the Gauss formula (Corollary 8.3) to the vector field γ' along γ, we obtain

$$\widetilde{D}_t \gamma' = D_t \gamma' + \mathrm{II}(\gamma', \gamma').$$

If γ is a g-geodesic in M, this formula simplifies to

$$\widetilde{D}_t \gamma' = \mathrm{II}(\gamma', \gamma').$$

Both conclusions of the proposition follow from this. $\square$

Note that the second fundamental form is completely determined by its values of the form $\mathrm{II}(v, v)$ for unit vectors v, by the following lemma.

Lemma 8.11. *Suppose V is an inner product space, W is a vector space, and $B, B': V \times V \to W$ are symmetric and bilinear. If $B(v, v) = B'(v, v)$ for every unit vector $v \in V$, then $B = B'$.*

Proof. Every vector $v \in V$ can be written $v = \lambda \widehat{v}$ for some unit vector $\widehat{v}$, so the bilinearity of B and B' implies $B(v, v) = B'(v, v)$ for *every* v, not just unit vectors. The result then follows from the following polarization identity, which is proved in exactly the same way as its counterpart (2.2) for inner products:

$$B(v, w) = \tfrac{1}{4}\big(B(v + w, v + w) - B(v - w, v - w)\big).$$ $\square$

Because the intrinsic and extrinsic accelerations of a curve are usually different, it is generally not the case that a $\widetilde{g}$-geodesic that starts tangent to M stays in M; just think of a sphere in Euclidean space, for example. A Riemannian submanifold (M, g) of $(\widetilde{M}, \widetilde{g})$ is said to be **totally geodesic** if every $\widetilde{g}$-geodesic that is tangent to M at some time t_0 stays in M for all t in some interval $(t_0 - \varepsilon, t_0 + \varepsilon)$.

Proposition 8.12. *Suppose* (M, g) *is an embedded Riemannian submanifold of a Riemannian or pseudo-Riemannian manifold* $(\widetilde{M}, \widetilde{g})$, *The following are equivalent:*

(a) *M is totally geodesic in $\widetilde{M}$.*

(b) *Every g-geodesic in M is also a $\widetilde{g}$-geodesic in $\widetilde{M}$.*

(c) *The second fundamental form of M vanishes identically.*

Proof. We will prove (a) $\Rightarrow$ (b) $\Rightarrow$ (c) $\Rightarrow$ (a). First assume that M is totally geodesic. Let $\gamma \colon I \to M$ be a g-geodesic. For each $t_0 \in I$, let $\widetilde{\gamma} \colon \widetilde{I} \to \widetilde{M}$ be the $\widetilde{g}$-geodesic with $\widetilde{\gamma}(t_0) = \gamma(t_0)$ and $\widetilde{\gamma}'(t_0) = \gamma'(t_0)$. The hypothesis implies that there is some open interval I_0 containing t_0 such that $\widetilde{\gamma}(t) \in M$ for $t \in I_0$. On I_0, the Gauss formula (8.2) for $\widetilde{\gamma}'$ reads

$$0 = \widetilde{D}_t \widetilde{\gamma}' = D_t \widetilde{\gamma}' + \mathrm{II}(\widetilde{\gamma}', \widetilde{\gamma}').$$

Because the first term on the right is tangent to M and the second is normal, the two terms must vanish individually. In particular, $D_t \widetilde{\gamma}' \equiv 0$ on I_0, which means that $\widetilde{\gamma}$ is also a g-geodesic there. By uniqueness of geodesics, therefore, $\gamma = \widetilde{\gamma}$ on I_0, so it follows in turn that γ is a $\widetilde{g}$-geodesic there. Since the same is true in a neighborhood of every $t_0 \in I$, it follows that γ is a $\widetilde{g}$-geodesic on its whole domain.

Next assume that every g-geodesic is a $\widetilde{g}$-geodesic. Let $p \in M$ and $v \in T_p M$ be arbitrary, and let $\gamma = \gamma_v \colon I \to M$ be the g-geodesic with $\gamma(0) = p$ and $\gamma'(0) = v$. The hypothesis implies that γ is also a $\widetilde{g}$-geodesic. Thus $\widetilde{D}_t \gamma' = D_t \gamma' = 0$, so the Gauss formula yields $\mathrm{II}(\gamma', \gamma') = 0$ along γ. In particular, $\mathrm{II}(v, v) = 0$. By Lemma 8.11, this implies that II is identically zero.

Finally, assume that $\mathrm{II} \equiv 0$, and let $\widetilde{\gamma} \colon I \to \widetilde{M}$ be a $\widetilde{g}$-geodesic such that $\widetilde{\gamma}(t_0) \in M$ and $\widetilde{\gamma}'(t_0) \in TM$ for some $t_0 \in I$. Let $\gamma \colon I \to M$ be the g-geodesic with the same initial conditions: $\gamma(t_0) = \widetilde{\gamma}(t_0)$ and $\gamma'(t_0) = \widetilde{\gamma}'(t_0)$. The Gauss formula together with the hypothesis $\mathrm{II} \equiv 0$ implies that $\widetilde{D}_t \gamma = D_t \gamma = 0$, so γ is also a $\widetilde{g}$-geodesic. By uniqueness of geodesics, therefore, $\widetilde{\gamma} = \gamma$ on the intersection of their domains, which implies that $\widetilde{\gamma}(t)$ lies in M for t in some open interval around t_0.   □

Hypersurfaces

Now we specialize the preceding considerations to the case in which M is a **hypersurface** (i.e., a submanifold of codimension 1) in $\widetilde{M}$. Throughout this section, our default assumption is that (M, g) is an embedded n-dimensional Riemannian submanifold of an $(n + 1)$-dimensional Riemannian manifold $(\widetilde{M}, \widetilde{g})$. (The analogous formulas in the pseudo-Riemannian case are a little different; see Problem 8-19.)

In this situation, at each point of M there are exactly two unit normal vectors. In terms of any local adapted orthonormal frame $(E_1, \ldots, E_{n+1})$, the two choices are $\pm E_{n+1}$. In a small enough neighborhood of each point of M, therefore, we can always choose a smooth unit normal vector field along M.

If both M and $\widetilde{M}$ are orientable, we can use an orientation to pick out a global smooth unit normal vector field along all of M. In general, though, this might or might not be possible. Since all of our computations in this chapter are local, we will always assume that we are working in a small enough neighborhood that a smooth unit normal field exists. We will address as we go along the question of how various quantities depend on the choice of normal vector field.

The Scalar Second Fundamental Form and the Shape Operator

Having chosen a distinguished smooth unit normal vector field N on the hypersurface $M \subseteq \widetilde{M}$, we can replace the vector-valued second fundamental form II by a simpler scalar-valued form. The *scalar second fundamental form of M* is the symmetric covariant 2-tensor field $h \in \Gamma(\Sigma^2 T^* M)$ defined by $h = \mathrm{II}_N$ (see (8.3)); in other words,

$$h(X, Y) = \langle N, \mathrm{II}(X, Y) \rangle. \tag{8.8}$$

Using the Gauss formula $\widetilde{\nabla}_X Y = \nabla_X Y + \mathrm{II}(X, Y)$ and noting that $\nabla_X Y$ is orthogonal to N, we can rewrite the definition as

$$h(X, Y) = \langle N, \widetilde{\nabla}_X Y \rangle. \tag{8.9}$$

Also, since N is a unit vector spanning NM at each point, the definition of h is equivalent to

$$\mathrm{II}(X, Y) = h(X, Y) N. \tag{8.10}$$

Note that replacing N by $-N$ multiplies h by -1, so the sign of h depends on which unit normal is chosen; but h is otherwise independent of the choices.

The choice of unit normal field also determines a Weingarten map $W_N : \mathfrak{X}(M) \to \mathfrak{X}(M)$ by (8.4); in the case of a hypersurface, we use the notation $s = W_N$ and call it the *shape operator of M*. Alternatively, we can think of s as the $(1,1)$-tensor field on M obtained from h by raising an index. It is characterized by

$$\langle sX, Y \rangle = h(X, Y) \qquad \text{for all } X, Y \in \mathfrak{X}(M).$$

Because h is symmetric, s is a self-adjoint endomorphism of TM, that is,

$$\langle sX, Y \rangle = \langle X, sY \rangle \qquad \text{for all } X, Y \in \mathfrak{X}(M).$$

As with h, the sign of s depends on the choice of N.

In terms of the tensor fields h and s, the formulas of the last section can be rewritten somewhat more simply. For this purpose, we will use two tensor operations defined in Chapter 7: the *Kulkarni–Nomizu product* of symmetric 2-tensors h, k is

$$h \otimes k(w,x,y,z) = h(w,z)k(x,y) + h(x,y)k(w,z)$$
$$- h(w,y)k(x,z) - h(x,z)k(w,y),$$

and the **exterior covariant derivative** of a smooth symmetric 2-tensor field T is

$$(DT)(x,y,z) = -(\nabla T)(x,y,z) + (\nabla T)(x,z,y).$$

Theorem 8.13 (Fundamental Equations for a Hypersurface). *Suppose (M,g) is a Riemannian hypersurface in a Riemannian manifold $(\widetilde{M}, \widetilde{g})$, and N is a smooth unit normal vector field along M.*

(a) THE GAUSS FORMULA FOR A HYPERSURFACE: *If $X, Y \in \mathfrak{X}(M)$ are extended to an open subset of $\widetilde{M}$, then*

$$\widetilde{\nabla}_X Y = \nabla_X Y + h(X,Y)N.$$

(b) THE GAUSS FORMULA FOR A CURVE IN A HYPERSURFACE: *If $\gamma : I \to M$ is a smooth curve and $X : I \to TM$ is a smooth vector field along γ, then*

$$\widetilde{D}_t X = D_t X + h(\gamma', X)N.$$

(c) THE WEINGARTEN EQUATION FOR A HYPERSURFACE: *For every $X \in \mathfrak{X}(M)$,*

$$\widetilde{\nabla}_X N = -sX. \tag{8.11}$$

(d) THE GAUSS EQUATION FOR A HYPERSURFACE: *For all $W, X, Y, Z \in \mathfrak{X}(M)$,*

$$\widetilde{Rm}(W,X,Y,Z) = Rm(W,X,Y,Z) - \tfrac{1}{2}(h \otimes h)(W,X,Y,Z).$$

(e) THE CODAZZI EQUATION FOR A HYPERSURFACE: *For all $W, X, Y \in \mathfrak{X}(M)$,*

$$\widetilde{Rm}(W,X,Y,N) = (Dh)(Y,W,X). \tag{8.12}$$

Proof. Parts (a), (b), and (d) follow immediately from substituting (8.10) into the general versions of the Gauss formula and Gauss equation. To prove (c), note first that the general version of the Weingarten equation can be written $\left(\widetilde{\nabla}_X N\right)^{\top} = -sX$. Since $\left\langle \widetilde{\nabla}_X N, N \right\rangle = \tfrac{1}{2} X(|N|^2) = 0$, it follows that $\widetilde{\nabla}_X N$ is tangent to M, so (c) follows.

To prove the hypersurface Codazzi equation, note that the fact that N is a unit vector field implies

$$0 = X|N|_{\widetilde{g}}^2 = 2\left\langle \nabla_X^{\perp} N, N \right\rangle_{\widetilde{g}}.$$

Since N spans the normal bundle, this implies that N is parallel with respect to the normal connection. Moreover,

$$(\nabla_W^F \text{II})(X,Y) = \nabla_W^\perp (\text{II}(X,Y)) - \text{II}(\nabla_W X, Y) - \text{II}(X, \nabla_W Y)$$
$$= \nabla_W^\perp (h(X,Y)N) - \text{II}(\nabla_W X, Y) - \text{II}(X, \nabla_W Y)$$
$$= \Big(W(h(X,Y)) - h(\nabla_W X, Y) - h(X, \nabla_W Y) \Big) N$$
$$= \nabla_W (h)(X,Y)N.$$

Inserting this into the general form (8.6) of the Codazzi equation and using the fact that ∇h is symmetric in its first two indices yields

$$\widetilde{Rm}(W,X,Y,N) = \nabla_W(h)(X,Y) - \nabla_X(h)(W,Y)$$
$$= (\nabla h)(X,Y,W) - (\nabla h)(W,Y,X)$$
$$= (\nabla h)(Y,X,W) - (\nabla h)(Y,W,X),$$

which is equivalent to (8.12). $\qquad\square$

Principal Curvatures

At every point $p \in M$, we have seen that the shape operator s is a self-adjoint linear endomorphism of the tangent space $T_p M$. To analyze such an operator, we recall some linear-algebraic facts about self-adjoint endomorphisms.

Lemma 8.14. *Suppose V is a finite-dimensional inner product space and $s: V \to V$ is a self-adjoint linear endomorphism. Let C denote the set of unit vectors in V. There is a vector $v_0 \in C$ where the function $v \mapsto \langle sv, v \rangle$ achieves its maximum among elements of C, and every such vector is an eigenvector of s with eigenvalue $\lambda_0 = \langle sv_0, v_0 \rangle$.*

▶ **Exercise 8.15.** Use the Lagrange multiplier rule (Prop. A.29) to prove this lemma.

Proposition 8.16 (Finite-Dimensional Spectral Theorem). *Suppose V is a finite-dimensional inner product space and $s: V \to V$ is a self-adjoint linear endomorphism. Then V has an orthonormal basis of s-eigenvectors, and all of the eigenvalues are real.*

Proof. The proof is by induction on $n = \dim V$. The $n = 1$ result is easy, so assume that the theorem holds for some $n \geq 1$ and suppose $\dim V = n + 1$. Lemma 8.14 shows that s has a unit eigenvector b_0 with a real eigenvalue λ_0. Let $B \subseteq V$ be the span of b_0. Since $s(B) \subseteq B$, self-adjointness of s implies $s(B^\perp) \subseteq B^\perp$. The inductive hypothesis applied to $s|_{B^\perp}$ implies that $B^\perp$ has an orthonormal basis $(b_1, \ldots, b_n)$ of s-eigenvectors with real eigenvalues, and then $(b_0, b_1, \ldots, b_n)$ is the desired basis of V. $\qquad\square$

Applying this proposition to the shape operator $s: T_p M \to T_p M$, we see that s has real eigenvalues $\kappa_1, \ldots, \kappa_n$, and there is an orthonormal basis $(b_1, \ldots, b_n)$ for $T_p M$ consisting of s-eigenvectors, with $sb_i = \kappa_i b_i$ for each i (no summation). In this basis, both h and s are represented by diagonal matrices, and h has the expression

$$h(v, w) = \kappa_1 v^1 w^1 + \cdots + \kappa_n v^n w^n.$$

The eigenvalues of s at a point $p \in M$ are called the **principal curvatures of M at p**, and the corresponding eigenspaces are called the **principal directions**. The principal curvatures all change sign if we reverse the normal vector, but the principal directions and principal curvatures are otherwise independent of the choice of coordinates or bases.

There are two combinations of the principal curvatures that play particularly important roles for hypersurfaces. The **Gaussian curvature** is defined as $K = \det(s)$, and the **mean curvature** as $H = (1/n) \operatorname{tr}(s) = (1/n) \operatorname{tr}_g(h)$. Since the determinant and trace of a linear endomorphism are basis-independent, these are well defined once a unit normal is chosen. In terms of the principal curvatures, they are

$$K = \kappa_1 \kappa_2 \cdots \kappa_n, \qquad H = \frac{1}{n}(\kappa_1 + \cdots + \kappa_n),$$

as can be seen by expressing s in terms of an orthonormal basis of eigenvectors. If N is replaced by $-N$, then H changes sign, while K is multiplied by $(-1)^n$.

Computations in Semigeodesic Coordinates

Semigeodesic coordinates (Prop. 6.41) provide an extremely convenient tool for computing the invariants of hypersurfaces.

Let $(\widetilde{M}, \widetilde{g})$ be an $(n+1)$-dimensional Riemannian manifold, and let $(x^1, \ldots, x^n, v)$ be semigeodesic coordinates on an open subset $U \subseteq \widetilde{M}$. (For example, they might be Fermi coordinates for the hypersurface $M_0 = v^{-1}(0)$; see Example 6.43.) For each real number a such that $v^{-1}(a) \neq \varnothing$, the level set $M_a = v^{-1}(a)$ is a properly embedded hypersurface in U. Let g_a denote the induced metric on M_a. Corollary 6.42 shows that $\widetilde{g}$ is given by

$$\widetilde{g} = dv^2 + \sum_{\alpha, \beta = 1}^{n} g_{\alpha\beta}(x^1, \ldots, x^n, v) dx^\alpha dx^\beta. \tag{8.13}$$

The restrictions of $(x^1, \ldots, x^n)$ give smooth coordinates for each hypersurface M_a, and in those coordinates the induced metric g_a is given by $g_a = g_{\alpha\beta} dx^\alpha dx^\beta|_{v=a}$. (Here we use the summation convention with Greek indices running from 1 to n.) The vector field $\partial_v = \partial_{n+1}$ restricts to a unit normal vector field along each hypersurface M_a.

As the next proposition shows, semigeodesic coordinates give us a simple formula for the second fundamental forms of all of the submanifolds M_a at once.

Proposition 8.17. *With notation as above, the components in $(x^1, \ldots, x^n)$-coordinates of the scalar second fundamental form, the shape operator, and the mean* *curvature of* (M_a, g_a) *(denoted by* $h_a, s_a,$ *and* $H_a,$ *respectively) with respect to the normal* $N = \partial_v$ *are given by*

$$(h_a)_{\alpha\beta} = -\frac{1}{2}\partial_v g_{\alpha\beta}\Big|_{v=a},$$

$$(s_a)^\alpha_\beta = -\frac{1}{2}g^{\alpha\gamma}\partial_v g_{\gamma\beta}\Big|_{v=a},$$

$$H_a = -\frac{1}{2n}g^{\alpha\beta}\partial_v g_{\alpha\beta}\Big|_{v=a}.$$

Proof. The normal component of $\widetilde{\nabla}_{\partial_\alpha}\partial_\beta$ is $\widetilde{\Gamma}^{n+1}_{\alpha\beta}\partial_v$, which Corollary 6.42 shows is equal to $-\frac{1}{2}\partial_v g_{\alpha\beta}\partial_r$ (noting that the roles of g and $\hat{g}$ in that corollary are being played here by $\widetilde{g}$ and g_a, respectively). Equation (8.9) evaluated at points of M_a gives

$$(h_a)_{\alpha\beta} = \langle \widetilde{\nabla}_{\partial_\alpha}\partial_\beta, N \rangle_{\widetilde{g}} = \langle -\tfrac{1}{2}\partial_v g_{\alpha\beta}\partial_v, \partial_v \rangle_{\widetilde{g}} = -\tfrac{1}{2}\partial_v g_{\alpha\beta}.$$

The formulas for s_a and H_a follow by using $(g^{\alpha\gamma})$ (the inverse matrix of $(g_{\alpha\gamma})$) to raise an index and then taking the trace. $\square$

Minimal Hypersurfaces

A natural question that has received a great deal of attention over the past century is this: Given a simple closed curve C in $\mathbb{R}^3$, is there an embedded or immersed surface M with $\partial M = C$ that has least area among all surfaces with the same boundary? If so, what is it? Such surfaces are models of the soap films that are produced when a closed loop of wire is dipped in soapy water.

More generally, we can consider the analogous question for hypersurfaces in Riemannian manifolds. Suppose M is a compact codimension-1 submanifold with nonempty boundary in an $(n + 1)$-dimensional Riemannian manifold $(\widetilde{M}, \widetilde{g})$. By analogy with the case of surfaces in $\mathbb{R}^3$, it is traditional to use the term **area** to refer to the n-dimensional volume of M with its induced Riemannian metric, and to say that M is **area-minimizing** if it has the smallest area among all compact embedded hypersurfaces in $\widetilde{M}$ with the same boundary. One key observation is the following theorem, which is an analogue for hypersurfaces of Theorem 6.4 about length-minimizing curves.

Theorem 8.18. *Let M be a compact codimension-1 submanifold with nonempty boundary in an $(n + 1)$-dimensional Riemannian manifold $(\widetilde{M}, \widetilde{g})$. If M is area-minimizing, then its mean curvature is identically zero.*

Proof. Let g denote the induced metric on M. The fact that M minimizes area among hypersurfaces with the same boundary means, in particular, that it minimizes area among small perturbations of M in a neighborhood of a single point. We will exploit this idea to prove that M must have zero mean curvature everywhere.

Let $p \in \text{Int } M$ be arbitrary, let $(x^1, \ldots, x^n, v)$ be Fermi coordinates for M on an open set $\widetilde{U} \subseteq \widetilde{M}$ containing p (see Example 6.43), and let $U = \widetilde{U} \cap M$. By taking

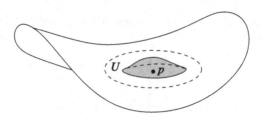

Fig. 8.2: The hypersurface M_t

$\widetilde{U}$ sufficiently small, we can arrange that U is a regular coordinate ball in M (see p. 374) and $\widetilde{U} \cap \partial M = \varnothing$. We use $(x^1, \dots, x^n)$ as coordinates on M, and observe the summation convention with Greek indices running from 1 to n.

Let φ be an arbitrary smooth real-valued function on M with compact support in U. For sufficiently small t, define a set $M_t \subseteq \widetilde{M}$ by

$$M_t = (M \smallsetminus U) \cup \{z \in \widetilde{U} : v(z) = t\varphi(x^1(z), \dots, x^n(z))\}.$$

(See Fig. 8.2.) Then M_t is an embedded smooth hypersurface in $\widetilde{M}$, which agrees with M outside of U and which coincides with the graph of $v = t\varphi$ in $\widetilde{U}$. Let $f_t : U \to \widetilde{U}$ be the graph parametrization of $M_t \cap \widetilde{U}$, given in Fermi coordinates by

$$f_t(x^1, \dots, x^n) = (x^1, \dots, x^n, t\varphi(x)). \tag{8.14}$$

Using this map, for each t we can define a diffeomorphism $F_t : M \to M_t$ by

$$F_t(z) = \begin{cases} z, & z \in M \smallsetminus \operatorname{supp} \varphi, \\ f_t(z), & z \in U. \end{cases}$$

For each t, let $\widehat{g}_t = \iota_{M_t}^* \widetilde{g}$ denote the induced Riemannian metric on M_t, and let $g_t = F_t^* \widehat{g}_t = F_t^* \widetilde{g}$ denote the pulled-back metric on M. When $t = 0$, we have $M_0 = M$, and both g_0 and $\widehat{g}_0$ are equal to the induced metric g on M. Since $\widetilde{g}$ is given by (8.13) in Fermi coordinates, a simple computation shows that in U, $g_t = F_t^* \widetilde{g}$ has the coordinate expression $g_t = (g_t)_{\alpha\beta} \, dx^\alpha \, dx^\beta$, where

$$(g_t)_{\alpha\beta} = t^2 \frac{\partial \varphi}{\partial x^\alpha}(x) \frac{\partial \varphi}{\partial x^\beta}(x) + g_{\alpha\beta}(x, t\varphi(x)), \tag{8.15}$$

while on $M \smallsetminus U$, g_t is equal to g and thus is independent of t.

Since each M_t is a smooth hypersurface with the same boundary as M, our hypothesis guarantees that $\operatorname{Area}(M_t, \widehat{g}_t)$ achieves a minimum at $t = 0$. Because F_t is an isometry from (M, g_t) to $(M_t, \widehat{g}_t)$, we can express this area as follows:

$$\operatorname{Area}(M_t, \widehat{g}_t) = \operatorname{Area}(M, g_t) = \operatorname{Area}(M \smallsetminus U, g) + \operatorname{Area}(U, g_t).$$

The first term on the right is independent of t, and we can compute the second term explicitly in coordinates $(x^1, \ldots, x^n)$ on U:

$$\text{Area}(U, g_t) = \int_U \sqrt{\det g_t}\, dx^1 \cdots dx^n,$$

where $\det g_t$ denotes the determinant of the matrix $((g_t)_{\alpha\beta})$ defined by (8.15). The integrand above is a smooth function of t and $(x^1, \ldots, x^n)$, so the area is a smooth function of t. We have

$$
\begin{aligned}
\frac{d}{dt}\bigg|_{t=0} \text{Area}(M_t, \hat g_t) &= \int_U \frac{\partial}{\partial t}\bigg|_{t=0} \sqrt{\det g_t}\, dx^1 \cdots dx^n \\
&= \int_U \frac{1}{2}(\det g)^{-1/2} \frac{\partial}{\partial t}\bigg|_{t=0} (\det g_t)\, dx^1 \cdots dx^n.
\end{aligned}
\tag{8.16}
$$

(The differentiation under the integral sign is justified because the integrand is smooth and has compact support in U.)

To compute the derivative of the determinant, note that the expansion by minors along, say, row α shows that the partial derivative of $\det$ with respect to the matrix entry in position (α, β) is equal to the cofactor $\text{cof}^{\alpha\beta}$, and thus by the chain rule,

$$\frac{\partial}{\partial t}\bigg|_{t=0} (\det g_t) = \text{cof}^{\alpha\beta} \frac{\partial}{\partial t}\bigg|_{t=0} (g_t)_{\alpha\beta}.
\tag{8.17}$$

On the other hand, Cramer's rule shows that the (α, β) component of the inverse matrix is given by $g^{\alpha\beta} = (\det g)^{-1} \text{cof}^{\alpha\beta}$. Thus from (8.17) and (8.15) we obtain

$$\frac{\partial}{\partial t}\bigg|_{t=0} (\det g_t) = (\det g)g^{\alpha\beta} \frac{\partial}{\partial t}\bigg|_{t=0} (g_t)_{\alpha\beta} = (\det g)g^{\alpha\beta} \frac{\partial g_{\alpha\beta}}{\partial v}\varphi.$$

Inserting this into (8.16) and using the result of Proposition 8.17, we conclude that

$$
\begin{aligned}
\frac{d}{dt}\bigg|_{t=0} \text{Area}(M_t, \hat g_t) &= \int_U \frac{1}{2}(\det g)^{1/2} g^{\alpha\beta} \frac{\partial g_{\alpha\beta}}{\partial v}\varphi\, dx^1 \cdots dx^n \\
&= -n \int_U H\varphi\, dV_g,
\end{aligned}
\tag{8.18}
$$

where H is the mean curvature of (M, g). Since $\text{Area}(M_t, \hat g_t)$ attains a minimum at $t = 0$, we conclude that $\int_U H\varphi\, dV_g = 0$ for every such φ.

Now suppose for the sake of contradiction that $H(p) \neq 0$. If $H(p) > 0$, we can let φ be a smooth nonnegative bump function that is positive at p and supported in a small neighborhood of p on which $H > 0$. The argument above shows that $\int_U H\varphi\, dV_g = 0$, which is impossible because the integrand is nonnegative on U and positive on an open set. A similar argument rules out $H(p) < 0$. Since p was an arbitrary point in $\text{Int}\, M$, we conclude that $H \equiv 0$ on $\text{Int}\, M$, and then by continuity on all of M. $\square$

Because of the result of Theorem 8.18, a hypersurface (immersed or embedded, with or without boundary) that has mean curvature identically equal to zero is called a *minimal hypersurface* (or a *minimal surface* when it has dimension 2). It is an unfortunate historical accident that the term "minimal hypersurface" is defined in this way, because in fact, a minimal hypersurface is just a critical point for the area, not necessarily area-minimizing. It can be shown that as in the case of geodesics, a small enough piece of every minimal hypersurface is area-minimizing.

As a complement to the above theorem about hypersurfaces that minimize area with fixed boundary, we have the following result about hypersurfaces that minimize area while enclosing a fixed volume.

Theorem 8.19. *Suppose* $(\widetilde{M}, g)$ *is a Riemannian* $(n+1)$-*manifold,* $D \subseteq \widetilde{M}$ *is a compact regular domain, and* $M = \partial D$. *If* M *has the smallest surface area among boundaries of compact regular domains with the same volume as* D, *then* M *has constant mean curvature (computed with respect to the outward unit normal).*

Proof. Let g denote the induced metric on M. Assume for the sake of contradiction that the mean curvature H of M is not constant, and let $p, q \in M$ be points such that $H(p) < H(q)$.

Since M is compact, it has an ε-tubular neighborhood for some $\varepsilon > 0$ by Theorem 5.25. As in the previous proof, let $(x^1, \ldots, x^n, v)$ be Fermi coordinates for M on an open set $\widetilde{U} \subseteq \widetilde{M}$ containing p, and let $U = \widetilde{U} \cap M$. We may assume that U is a regular coordinate ball in M and the image of the chart is a set of the form $\widehat{U} \times (-\varepsilon, \varepsilon)$ for some open subset $\widehat{U} \subseteq \mathbb{R}^n$. Similarly, let $(y^1, \ldots, y^n, w)$ be Fermi coordinates for M on an open set $\widetilde{W} \subseteq \widetilde{M}$ containing q and satisfying the analogous conditions, and let $W = \widetilde{W} \cap M$. By replacing v with its negative if necessary, we can arrange that $D \cap \widetilde{U}$ is the set where $v \le 0$, and similarly for w. Also, by shrinking both domains, we can assume that the mean curvature of M satisfies $H \le H_1$ on U and $H \ge H_2$ on W, where H_1, H_2 are constants such that $H(p) < H_1 < H_2 < H(q)$.

Let φ and ψ be smooth real-valued functions on M, with φ compactly supported in U and ψ compactly supported in W, and satisfying $\int_U \varphi \, dV_g = \int_W \psi \, dV_g = 1$. For sufficiently small $s, t \in \mathbb{R}$, define a subset $D_{s,t} \subseteq \widetilde{M}$ as follows:

$$
\begin{aligned}
D_{s,t} = &\left\{ z \in \widetilde{U} : v(z) \le s\varphi\big(x^1(z), \ldots, x^n(z)\big) \right\} \\
&\cup \left\{ z \in \widetilde{W} : w(z) \le t\psi\big(y^1(z), \ldots, y^n(z)\big) \right\} \\
&\cup \big(D \smallsetminus (\widetilde{U} \cup \widetilde{W})\big),
\end{aligned}
$$

and let $M_{s,t} = \partial D_{s,t}$, so $D_{0,0} = D$ and $M_{0,0} = M$ (see Fig. 8.3). For sufficiently small s and t, the set $D_{s,t}$ is a regular domain and $M_{s,t}$ is a compact smooth hypersurface, and $\mathrm{Vol}(D_{s,t})$ and $\mathrm{Area}(M_{s,t})$ are both smooth functions of (s,t). For convenience, write $V(s,t) = \mathrm{Vol}(D_{s,t})$ and $A(s,t) = \mathrm{Area}(M_{s,t})$.

The same argument that led to (8.18) shows that

$$
\frac{\partial A}{\partial s}(0,0) = -n \int_U H\varphi \, dV_g, \qquad \frac{\partial A}{\partial t}(0,0) = -n \int_W H\psi \, dV_g.
$$

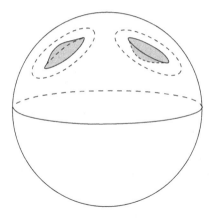

Fig. 8.3: The domain $D_{s,t}$

To compute the partial derivatives of the volume, we just note that if we hold $t = 0$ fixed and let s vary, the only change in volume occurs in the part of $D_{s,t}$ contained in $\widetilde{U}$, so the fundamental theorem of calculus gives

$$
\begin{aligned}
\frac{\partial V}{\partial s}(0,0) &= \frac{d}{ds}\bigg|_{s=0} \operatorname{Vol}\big(D_{s,0} \cap \widetilde{U}\big) \\
&= \frac{d}{ds}\bigg|_{s=0} \int_U \left(\int_{-\varepsilon}^{s\varphi(x)} \sqrt{\det \widetilde{g}(x,v)}\, dv \right) dx^1 \cdots dx^n \\
&= \int_U \left(\frac{d}{ds}\bigg|_{s=0} \int_{-\varepsilon}^{s\varphi(x)} \sqrt{\det \widetilde{g}(x,v)}\, dv \right) dx^1 \cdots dx^n \\
&= \int_U \varphi(x) \sqrt{\det \widetilde{g}(x,0)}\, dx^1 \cdots dx^n = \int_U \varphi\, dV_g = 1,
\end{aligned}
$$

where the differentiation under the integral sign in the third line is justified just like (8.16), and in the last line we used the fact that $g_{\alpha\beta}(x) = \widetilde{g}_{\alpha\beta}(x,0)$ in these coordinates. Similarly, $\partial V/\partial t (0,0) = 1$.

Because $V(0,0) = \operatorname{Vol}(D)$ and $\partial V/\partial t (0,0) \neq 0$, the implicit function theorem guarantees that there is a smooth function $\lambda \colon (-\delta,\delta) \to \mathbb{R}$ for some $\delta > 0$ such that $V(s, \lambda(s)) \equiv \operatorname{Vol}(D)$. The chain rule then implies

$$
0 = \frac{d}{ds}\bigg|_{s=0} V(s, \lambda(s)) = \frac{\partial V}{\partial s}(0,0) + \lambda'(0)\frac{\partial V}{\partial t}(0,0) = 1 + \lambda'(0).
$$

Thus $\lambda'(0) = -1$.

Our hypothesis that M minimizes area implies that

$$0 = \frac{d}{ds}\bigg|_{s=0} A(s, \lambda(s)) = \frac{\partial A}{\partial s}(0,0) + \lambda'(0)\frac{\partial A}{\partial t}(0,0)$$

$$= -n\int_U H\varphi \, dV_g + n\int_W H\psi \, dV_g,$$

and thus $\int_U H\varphi \, dV_g = \int_W H\psi \, dV_g$. But our choice of U and V together with the fact that $\int_U \varphi \, dV_g = \int_W \psi \, dV_g = 1$ guarantees that $\int_U H\varphi \, dV_g \leq H_1 < H_2 \leq \int_W H\psi \, dV_g$, which is a contradiction. $\square$

We do not pursue minimal or constant-mean-curvature hypersurfaces any further in this book, but you can find a good introductory treatment in [CM11].

Hypersurfaces in Euclidean Space

Now we specialize even further, to hypersurfaces in Euclidean space. In this section, we assume that $M \subseteq \mathbb{R}^{n+1}$ is an embedded n-dimensional submanifold with the induced Riemannian metric. The Euclidean metric will be denoted as usual by $\bar{g}$, and covariant derivatives and curvatures associated with $\bar{g}$ will be indicated by a bar. The induced metric on M will be denoted by g.

In this setting, because $\overline{Rm} \equiv 0$, the Gauss and Codazzi equations take even simpler forms:

$$\tfrac{1}{2} h \owedge h = Rm, \tag{8.19}$$

$$Dh = 0, \tag{8.20}$$

or in terms of a local frame for M,

$$h_{il}h_{jk} - h_{ik}h_{jl} = R_{ijkl}, \tag{8.21}$$

$$h_{ij;k} - h_{ik;j} = 0. \tag{8.22}$$

In particular, this means that the Riemann curvature tensor of a hypersurface in $\mathbb{R}^{n+1}$ is completely determined by the second fundamental form. A symmetric 2-tensor field that satisfies $Dh = 0$ is called a **Codazzi tensor**, so (8.20) can be expressed succinctly by saying that h is a Codazzi tensor.

▶ **Exercise 8.20.** Show that a smooth 2-tensor field h on a Riemannian manifold is a Codazzi tensor if and only if both h and ∇h are symmetric.

The equations (8.19)–(8.20) can be viewed as compatibility conditions for the existence of an embedding or immersion into Euclidean space with prescribed first and second fundamental forms. If (M, g) is a Riemannian n-manifold and h is a given smooth symmetric 2-tensor field on M, then Theorem 8.13 shows that these two equations are necessary conditions for the existence of an isometric immersion $M \to \mathbb{R}^{n+1}$ for which h is the scalar second fundamental form. (Note that an immersion is locally an embedding, so the theorem applies in a neighborhood of each

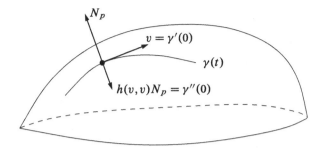

Fig. 8.4: Geometric interpretation of $h(v, v)$

point.) It is a remarkable fact that the Gauss and Codazzi equations are actually sufficient, at least locally. A sketch of a proof of this fact, called the *fundamental theorem of hypersurface theory*, can be found in [Pet16, pp. 108–109].

In the setting of a hypersurface $M \subseteq \mathbb{R}^{n+1}$, we can give some very concrete geometric interpretations of the quantities we have defined so far. We begin with curves. For every unit vector $v \in T_p M$, let $\gamma = \gamma_v : I \to M$ be the g-geodesic in M with initial velocity v. Then the Gauss formula shows that the ordinary Euclidean acceleration of γ at 0 is $\gamma''(0) = \bar{D}_t \gamma'(0) = h(v, v) N_p$ (Fig. 8.4). Thus $|h(v, v)|$ is the Euclidean curvature of γ at 0, and $h(v, v) = \langle \gamma''(0), N_p \rangle > 0$ if and only if $\gamma''(0)$ points in the same direction as N_p. In other words, $h(v, v)$ is positive if γ is curving in the direction of N_p, and negative if it is curving away from N_p.

The next proposition shows that this Euclidean curvature can be interpreted in terms of the radius of the "best circular approximation," as mentioned in Chapter 1.

Proposition 8.21. *Suppose $\gamma : I \to \mathbb{R}^m$ is a unit-speed curve, $t_0 \in I$, and $\kappa(t_0) \neq 0$.*

(a) *There is a unique unit-speed parametrized circle $c : \mathbb{R} \to \mathbb{R}^m$, called the **osculating circle at $\gamma(t_0)$**, with the property that c and γ have the same position, velocity, and acceleration at $t = t_0$.*

(b) *The Euclidean curvature of γ at t_0 is $\kappa(t_0) = 1/R$, where R is the radius of the osculating circle.*

Proof. An easy geometric argument shows that every circle in $\mathbb{R}^m$ with center q and radius R has a unit-speed parametrization of the form

$$c(t) = q + R \cos\left(\frac{t - t_0}{R}\right) v + R \sin\left(\frac{t - t_0}{R}\right) w,$$

where (v, w) is a pair of orthonormal vectors in $\mathbb{R}^m$. By direct computation, such a parametrization satisfies

$$c(t_0) = q + Rv, \qquad c'(t_0) = w, \qquad c''(t_0) = -\frac{1}{R} v.$$

Thus if we put

$$R = \frac{1}{|\gamma''(t_0)|} = \frac{1}{\kappa(t_0)}, \quad v = -R\gamma''(t_0), \quad w = \gamma'(t_0), \quad q = \gamma(t_0) - Rv,$$

we obtain a circle satisfying the required conditions, and its radius is equal to $1/\kappa(t_0)$ by construction. Uniqueness is left as an exercise.   □

▶ **Exercise 8.22.** Complete the proof of the preceding proposition by proving uniqueness of the osculating circle.

Computations in Euclidean Space

When we wish to *compute* the invariants of a Euclidean hypersurface $M \subseteq \mathbb{R}^{n+1}$, it is usually unnecessary to go to all the trouble of computing Christoffel symbols. Instead, it is usually more effective to use either a defining function or a parametrization to compute the scalar second fundamental form, and then use (8.21) to compute the curvature. Here we describe several contexts in which this computation is not too hard.

Usually the computations are simplest if the hypersurface is presented in terms of a local parametrization. Suppose $M \subseteq \mathbb{R}^{n+1}$ is a smooth embedded hypersurface, and let $X \colon U \to \mathbb{R}^{n+1}$ be a smooth local parametrization of M. The coordinates $(u^1, \ldots, u^n)$ on $U \subseteq \mathbb{R}^n$ thus give local coordinates for M. The coordinate vector fields $\partial_i = \partial/\partial u^i$ push forward to vector fields $dX(\partial_i)$ on M, which we can view as sections of the restricted tangent bundle $T\mathbb{R}^{n+1}|_M$, or equivalently as $\mathbb{R}^{n+1}$-valued functions. If we think of $X(u) = (X^1(u), \ldots, X^{n+1}(u))$ as a vector-valued function of u, these vectors can be written as

$$dX_u(\partial_i) = \partial_i X(u) = (\partial_i X^1(u), \ldots, \partial_i X^{n+1}(u)).$$

For simplicity, write $X_i = \partial_i X$.

Once these vector fields are computed, a unit normal field can be computed as follows: Choose any coordinate vector field $\partial/\partial x^{j_0}$ that is not contained in $\mathrm{span}(X_1, \ldots, X_n)$ (there will always be one, at least in a neighborhood of each point). Then apply the Gram–Schmidt algorithm to the local frame $(X_1, \ldots, X_n, \partial/\partial x^{j_0})$ along M to obtain an adapted orthonormal frame $(E_1, \ldots, E_{n+1})$. The two choices of unit normal are $N = \pm E_{n+1}$.

The next proposition gives a formula for the second fundamental form that is often easy to use for computation.

Proposition 8.23. *Suppose $M \subseteq \mathbb{R}^{n+1}$ is an embedded hypersurface, $X \colon U \to M$ is a smooth local parametrization of M, $(X_1, \ldots, X_n)$ is the local frame for TM determined by X, and N is a unit normal field on M. Then the scalar second fundamental form is given by*

$$h(X_i, X_j) = \left\langle \frac{\partial^2 X}{\partial u^i \partial u^j}, N \right\rangle. \tag{8.23}$$

Proof. Let $u_0 = (u_0^1, \ldots, u_0^n)$ be an arbitrary point of U and let $p = X(u_0) \in M$. For each $i \in \{1, \ldots, n\}$, the curve $\gamma(t) = X(u_0^1, \ldots, u_0^i + t, \ldots, u_0^n)$ is a smooth curve in M whose initial velocity is X_i. Regarding the normal field N as a smooth map from M to $\mathbb{R}^{n+1}$, we have

$$\frac{\partial}{\partial u^i} N(X(u_0)) = (N \circ \gamma)'(0) = \bar{\nabla}_{X_i} N(X(u_0)).$$

Because $X_j = \partial X/\partial u^j$ is tangent to M and N is normal, the following expression is zero for all $u \in U$:

$$\left\langle \frac{\partial X}{\partial u^j}(u), N(X(u)) \right\rangle.$$

Differentiating with respect to u^i and using the product rule for ordinary inner products in $\mathbb{R}^{n+1}$ yields

$$0 = \left\langle \frac{\partial^2 X}{\partial u^i \partial u^j}(u), N(X(u)) \right\rangle + \left\langle \frac{\partial X}{\partial u^j}(u), \bar{\nabla}_{X_i(u)} N(X(u)) \right\rangle.$$

By the Weingarten equation (8.11), the last term on the right becomes

$$\langle X_j(u), -s(X_i(u)) \rangle = -h(X_j(u), X_i(u)).$$

Inserting this above yields (8.23). $\qquad\qquad\qquad\qquad\qquad\qquad\qquad\qquad\qquad\qquad$ □

Here is an application of this formula: it shows how the principal curvatures give a concise description of the local shape of an embedded hypersurface by approximating the surface with the graph of a quadratic function.

Proposition 8.24. *Suppose $M \subseteq \mathbb{R}^{n+1}$ is a Riemannian hypersurface. Let $p \in M$, and let $\kappa_1, \ldots, \kappa_n$ denote the principal curvatures of M at p with respect to some choice of unit normal. Then there is an isometry $\varphi \colon \mathbb{R}^{n+1} \to \mathbb{R}^{n+1}$ that takes p to the origin and takes a neighborhood of p in M to a graph of the form $x^{n+1} = f(x^1, \ldots, x^n)$, where*

$$f(x) = \frac{1}{2}\left(\kappa_1(x^1)^2 + \cdots + \kappa_n(x^n)^2\right) + O(|x|^3). \tag{8.24}$$

Proof. Replacing M by its image under a translation and a rotation (which are Euclidean isometries), we may assume that p is the origin and $T_p M$ is equal to the span of $(\partial_1, \ldots, \partial_n)$. Then after reflecting in the $(x^1, \ldots, x^n)$-hyperplane if necessary, we may assume that the chosen unit normal is $(0, \ldots, 0, 1)$. By an orthogonal transformation in the first n variables, we can also arrange that the scalar second fundamental form at 0 is diagonal with respect to the basis $(\partial_1, \ldots, \partial_n)$, with diagonal entries $(\kappa_1, \ldots, \kappa_n)$.

It follows from the implicit function theorem that there is some neighborhood U of 0 such that $M \cap U$ is the graph of a smooth function of the form $x^{n+1} = f(x^1, \ldots, x^n)$ with $f(0) = 0$. A smooth local parametrization of M is then given by

$X(u) = (u^1, \ldots, u^n, f(u))$, and the fact that $T_0 M$ is spanned by $(\partial/\partial x^1, \ldots, \partial/\partial x^n)$ guarantees that $\partial_1 f(0) = \cdots = \partial_n f(0) = 0$. Because $X_i = \partial/\partial x^i$ at 0, Proposition 8.23 then yields

$$h\left(\frac{\partial}{\partial x^i}, \frac{\partial}{\partial x^j}\right) = \left\langle \left(0, \ldots, 0, \frac{\partial^2 f}{\partial x^i \partial x^j}(0)\right), (0, \ldots, 0, 1)\right\rangle = \frac{\partial^2 f}{\partial x^i \partial x^j}(0).$$

It follows from our normalization that the matrix of second derivatives of f at 0 is diagonal, and its diagonal entries are the principal curvatures of M at that point. Then (8.24) follows from Taylor's theorem.   □

Here is another approach. When it is practical to write down a smooth vector field $N = N^i \partial_i$ on an open subset of $\mathbb{R}^{n+1}$ that restricts to a unit normal vector field along M, then the shape operator can be computed straightforwardly using the Weingarten equation and observing that the Euclidean covariant derivatives of N are just ordinary directional derivatives in Euclidean space. Thus for every vector $X = X^j \partial_j$ tangent to M, we have

$$sX = -\bar{\nabla}_X N = -\sum_{i,j=1}^{n+1} X^j (\partial_j N^i) \partial_i. \tag{8.25}$$

One common way to produce such a smooth vector field is to work with a local defining function for M: Recall that this is a smooth real-valued function defined on some open subset $U \subseteq \mathbb{R}^{n+1}$ such that $U \cap M$ is a regular level set of F (see Prop. A.27). The definition ensures that $\mathrm{grad}\, F$ (the gradient of F with respect to $\bar{g}$) is nonzero on some neighborhood of $M \cap U$, so a convenient choice for a unit normal vector field along M is

$$N = \frac{\mathrm{grad}\, F}{|\mathrm{grad}\, F|}.$$

Here is an application.

Example 8.25 (Shape Operators of Spheres). The function $F: \mathbb{R}^{n+1} \to \mathbb{R}$ defined by $F(x) = |x|^2$ is a smooth defining function for each sphere $\mathbb{S}^n(R)$. The gradient of this function is $\mathrm{grad}\, F = 2\sum_i x^i \partial_i$, which has length $2R$ along $\mathbb{S}^n(R)$. The smooth vector field

$$N = \frac{1}{R} \sum_{i=1}^{n+1} x^i \partial_i$$

thus restricts to a unit normal along $\mathbb{S}^n(R)$. (It is the outward pointing normal.) The shape operator is now easy to compute:

$$sX = -\frac{1}{R} \sum_{i,j=1}^{n+1} X^j (\partial_j x^i) \partial_i = -\frac{1}{R} X.$$

Therefore $s = (-1/R)\mathrm{Id}$. The principal curvatures, therefore, are all equal to $-1/R$, and it follows that the mean curvature is $H = -1/R$ and the Gaussian curvature is $(-1/R)^n$.   //

For surfaces in $\mathbb{R}^3$, either of the above methods can be used. When a parametrization X is given, the normal vector field is particularly easy to compute: because X_1 and X_2 span the tangent space to M at each point, their cross product is a nonzero normal vector, so one choice of unit normal is

$$N = \frac{X_1 \times X_2}{|X_1 \times X_2|}. \tag{8.26}$$

Problems 8-1, 8-2, and 8-3 will give you practice in carrying out these computations for surfaces presented in various ways.

The Gaussian Curvature of a Surface Is Intrinsic

Because the Gaussian and mean curvatures are defined in terms of a particular embedding of M into $\mathbb{R}^{n+1}$, there is little reason to suspect that they have much to do with the intrinsic Riemannian geometry of (M, g). The next exercise illustrates the fact that the mean curvature has no intrinsic meaning.

▶ **Exercise 8.26.** Let $M_1 \subseteq \mathbb{R}^3$ be the plane $\{z = 0\}$, and let $M_2 \subseteq \mathbb{R}^3$ be the cylinder $\{x^2 + y^2 = 1\}$. Show that M_1 and M_2 are locally isometric, but the former has mean curvature zero, while the latter has mean curvature $\pm\frac{1}{2}$, depending on which normal is chosen.

The amazing discovery made by Gauss was that the Gaussian curvature of a surface in $\mathbb{R}^3$ is actually an intrinsic invariant of the Riemannian manifold (M, g). He was so impressed with this discovery that he called it *Theorema Egregium*, Latin for "excellent theorem."

Theorem 8.27 (Gauss's *Theorema Egregium*). *Suppose (M, g) is an embedded 2-dimensional Riemannian submanifold of $\mathbb{R}^3$. For every $p \in M$, the Gaussian curvature of M at p is equal to one-half the scalar curvature of g at p, and thus the Gaussian curvature is a local isometry invariant of (M, g).*

Proof. Let $p \in M$ be arbitrary, and choose an orthonormal basis (b_1, b_2) for T_pM. In this basis g is represented by the identity matrix, and the shape operator has the same matrix as the scalar second fundamental form. Thus $K(p) = \det(s^i_j) = \det(h_{ij})$, and the Gauss equation (8.21) reads

$$Rm_p(b_1, b_2, b_2, b_1) = h_{11}h_{22} - h_{12}h_{21} = \det(h_{ij}) = K(p).$$

On the other hand, Corollary 7.27 shows that $Rm = \frac{1}{4}Sg \otimes g$, and thus by the definition of the Kulkarni–Nomizu product we have

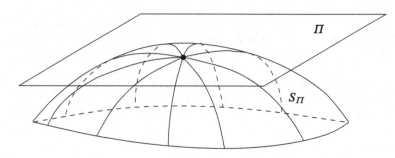

Fig. 8.5: A plane section

$$Rm_p(b_1, b_2, b_2, b_1) = \tfrac{1}{4} S(p)\big(2g_{11}g_{22} - 2g_{12}g_{21}\big) = \tfrac{1}{2} S(p). \qquad \square$$

Motivated by the *Theorema Egregium*, for an abstract Riemannian 2-manifold (M, g), not necessarily embedded in $\mathbb{R}^3$, we define the **Gaussian curvature** to be $K = \tfrac{1}{2} S$, where S is the scalar curvature. If M is a Riemannian submanifold of $\mathbb{R}^3$, then the *Theorema Egregium* shows that this new definition agrees with the original definition of K as the determinant of the shape operator. The following result is a restatement of Corollary 7.27 using this new definition.

Corollary 8.28. *If (M, g) is a Riemannian 2-manifold, the following relationships hold:*

$$Rm = \tfrac{1}{2} K g \otimes g, \qquad Rc = Kg, \qquad S = 2K. \qquad \square$$

Sectional Curvatures

Now, finally, we can give a quantitative geometric interpretation to the curvature tensor in dimensions higher than 2. Suppose M is a Riemannian n-manifold (with $n \geq 2$), p is a point of M, and $V \subseteq T_p M$ is a star-shaped neighborhood of zero on which $\exp_p$ is a diffeomorphism onto an open set $U \subseteq M$. Let Π be any 2-dimensional linear subspace of $T_p M$. Since $\Pi \cap V$ is an embedded 2-dimensional submanifold of V, it follows that $S_\Pi = \exp_p(\Pi \cap V)$ is an embedded 2-dimensional submanifold of $U \subseteq M$ containing p (Fig. 8.5), called the **plane section determined by Π**. Note that S_Π is just the set swept out by geodesics whose initial velocities lie in Π, and $T_p S_\Pi$ is exactly Π.

We define the **sectional curvature of Π**, denoted by $\sec(\Pi)$, to be the intrinsic Gaussian curvature at p of the surface S_Π with the metric induced from the embedding $S_\Pi \subseteq M$. If (v, w) is any basis for Π, we also use the notation $\sec(v, w)$ for $\sec(\Pi)$.

The next theorem shows how to compute the sectional curvatures in terms of the curvature of (M, g). To make the formula more concise, we introduce the following notation. Given vectors v, w in an inner product space V, we set

$$|v \wedge w| = \sqrt{|v|^2 |w|^2 - \langle v, w \rangle^2}. \tag{8.27}$$

It follows from the Cauchy–Schwarz inequality that $|v \wedge w| \geq 0$, with equality if and only if v and w are linearly dependent, and $|v \wedge w| = 1$ when v and w are orthonormal. (One can define an inner product on the space $\Lambda^2(V)$ of contravariant alternating 2-tensors, analogous to the inner product on forms defined in Problem 2-16, and this is the associated norm; see Problem 8-33(a).)

Proposition 8.29 (Formula for the Sectional Curvature). *Let (M, g) be a Riemannian manifold and $p \in M$. If v, w are linearly independent vectors in $T_p M$, then the sectional curvature of the plane spanned by v and w is given by*

$$\sec(v, w) = \frac{Rm_p(v, w, w, v)}{|v \wedge w|^2}. \tag{8.28}$$

Proof. Let $\Pi \subseteq T_p M$ be the subspace spanned by (v, w). For this proof, we denote the induced metric on S_Π by $\hat{g}$, and its associated curvature tensor by $\widehat{Rm}$. By definition, $\sec(v, w)$ is equal to $\hat{K}(p)$, the Gaussian curvature of $\hat{g}$ at p.

We show first that the second fundamental form of S_Π in M vanishes at p. To see why, let $z \in \Pi$ be arbitrary, and let $\gamma = \gamma_z$ be the g-geodesic with initial velocity z, whose image lies in S_Π for t sufficiently near 0. By the Gauss formula for vector fields along curves,

$$0 = D_t \gamma' = \hat{D}_t \gamma' + \mathrm{II}(\gamma', \gamma').$$

Since the two terms on the right-hand side are orthogonal, each must vanish identically. Evaluating at $t = 0$ gives $\mathrm{II}(z, z) = 0$. Since z was an arbitrary element of $\Pi = T_p(S_\Pi)$ and II is symmetric, polarization shows that $\mathrm{II} = 0$ at p. (We cannot in general expect II to vanish at other points of S_Π—it is only at p that all g-geodesics starting tangent to S_Π remain in S_Π.) The Gauss equation then tells us that the curvature tensors of M and S_Π are related at p by

$$Rm_p(u, v, w, x) = \widehat{Rm}_p(u, v, w, x)$$

whenever $u, v, w, x \in \Pi$.

Now choose an orthonormal basis (b_1, b_2) for Π. Based on the observations above, we see that the sectional curvature of Π is

$$\hat{K}(p) = \tfrac{1}{2} \hat{S}(p)$$
$$= \tfrac{1}{2} \sum_{i,j=1}^{2} \widehat{Rm}_p (b_i, b_j, b_j, b_i)$$
$$= \tfrac{1}{2} \widehat{Rm}_p (b_1, b_2, b_2, b_1) + \tfrac{1}{2} \widehat{Rm}_p (b_2, b_1, b_1, b_2)$$
$$= \widehat{Rm}_p (b_1, b_2, b_2, b_1) = Rm_p (b_1, b_2, b_2, b_1).$$

To see how to compute this in terms of an arbitrary basis, let (v, w) be any basis for Π. The Gram–Schmidt algorithm yields an orthonormal basis as follows:

$$b_1 = \frac{v}{|v|};$$

$$b_2 = \frac{w - \langle w, b_1 \rangle b_1}{|w - \langle w, b_1 \rangle b_1|}.$$

Then by the preceding computation,

$$\hat{K}(p) = Rm_p(b_1, b_2, b_2, b_1)$$

$$= Rm_p\left(\frac{v}{|v|}, \frac{w - \langle w, b_1 \rangle b_1}{|w - \langle w, b_1 \rangle b_1|}, \frac{w - \langle w, b_1 \rangle b_1}{|w - \langle w, b_1 \rangle b_1|}, \frac{v}{|v|}\right) \qquad (8.29)$$

$$= \frac{Rm_p(v, w, w, v)}{|v|^2 |w - \langle w, b_1 \rangle b_1|^2},$$

where we have used the fact that $Rm(v, b_1, \cdot, \cdot) = Rm(\cdot, \cdot, b_1, v) = 0$ because b_1 is a multiple of v. To simplify the denominator of this last expression, we substitute $b_1 = v/|v|$ to obtain

$$|v|^2 |w - \langle w, b_1 \rangle b_1|^2 = |v|^2 \left(|w|^2 - 2\frac{\langle w, v \rangle^2}{|v|^2} + \frac{\langle w, v \rangle^2}{|v|^2}\right)$$

$$= |v|^2 |w|^2 - \langle v, w \rangle^2 = |v \wedge w|^2.$$

Inserting this into (8.29) proves the theorem.   □

▶ **Exercise 8.30.** Suppose (M, g) is a Riemannian manifold and $\tilde{g} = \lambda g$ for some positive constant λ. Use Theorem 7.30 to prove that for every $p \in M$ and plane $\Pi \subseteq T_p M$, the sectional curvatures of Π with respect to $\tilde{g}$ and g are related by $\widetilde{\sec}(\Pi) = \lambda^{-1} \sec(\Pi)$.

Proposition 8.29 shows that one important piece of quantitative information provided by the curvature tensor is that it encodes the sectional curvatures of all plane sections. It turns out, in fact, that this is *all* of the information contained in the curvature tensor: as the following proposition shows, the sectional curvatures completely determine the curvature tensor.

Proposition 8.31. *Suppose R_1 and R_2 are algebraic curvature tensors on a finite-dimensional inner product space V. If for every pair of linearly independent vectors $v, w \in V$,*

$$\frac{R_1(v, w, w, v)}{|v \wedge w|^2} = \frac{R_2(v, w, w, v)}{|v \wedge w|^2},$$

then $R_1 = R_2$.

Proof. Let R_1 and R_2 be tensors satisfying the hypotheses, and set $D = R_1 - R_2$. Then D is an algebraic curvature tensor, and $D(v, w, w, v) = 0$ for all $v, w \in V$. (This is true by hypothesis when v and w are linearly independent, and it is true by the symmetries of D when they are not.) We need to show that $D = 0$.

For all vectors v, w, x, the symmetries of D give

$$0 = D(v+w,x,x,v+w)$$
$$= D(v,x,x,v) + D(v,x,x,w) + D(w,x,x,v) + D(w,x,x,w)$$
$$= 2D(v,x,x,w).$$

From this it follows that

$$0 = D(v,x+u,x+u,w)$$
$$= D(v,x,x,w) + D(v,x,u,w) + D(v,u,x,w) + D(v,u,u,w)$$
$$= D(v,x,u,w) + D(v,u,x,w).$$

Therefore D is antisymmetric in every adjacent pair of arguments. Now the algebraic Bianchi identity yields

$$0 = D(v,w,x,u) + D(w,x,v,u) + D(x,v,w,u)$$
$$= D(v,w,x,u) - D(w,v,x,u) - D(v,x,w,u)$$
$$= 3D(v,w,x,u). \qquad \square$$

We can also give a geometric interpretation of the Ricci and scalar curvatures on a Riemannian manifold. Since the Ricci tensor is symmetric and bilinear, Lemma 8.11 shows that it is completely determined by its values of the form $Rc(v,v)$ for unit vectors v.

Proposition 8.32 (Geometric Interpretation of Ricci and Scalar Curvatures).
Let (M,g) be a Riemannian n-manifold and $p \in M$.

(a) *For every unit vector $v \in T_pM$, $Rc_p(v,v)$ is the sum of the sectional curvatures of the 2-planes spanned by $(v,b_2),\dots,(v,b_n)$, where $(b_1,\dots,b_n)$ is any orthonormal basis for T_pM with $b_1 = v$.*

(b) *The scalar curvature at p is the sum of all sectional curvatures of the 2-planes spanned by ordered pairs of distinct basis vectors in any orthonormal basis.*

Proof. Given any unit vector $v \in T_pM$, let $(b_1,\dots,b_n)$ be as in the hypothesis. Then $Rc_p(v,v)$ is given by

$$Rc_p(v,v) = R_{11}(p) = R_{k11}{}^k(p) = \sum_{k=1}^{n} Rm_p(b_k,b_1,b_1,b_k) = \sum_{k=2}^{n} \sec(b_1,b_k).$$

For the scalar curvature, we let $(b_1,\dots,b_n)$ be any orthonormal basis for T_pM, and compute

$$S(p) = R_j{}^j(p) = \sum_{j=1}^{n} Rc_p(b_j,b_j) = \sum_{j,k=1}^{n} Rm_p(b_k,b_j,b_j,b_k)$$
$$= \sum_{j \neq k} \sec(b_j,b_k). \qquad \square$$

One consequence of this proposition is that if (M, g) is a Riemannian manifold in which all sectional curvatures are positive, then the Ricci and scalar curvatures are both positive as well. The analogous statement holds if "positive" is replaced by "negative," "nonpositive," or "nonnegative."

If the opposite sign convention is chosen for the curvature tensor, then the right-hand side of formula (8.28) has to be adjusted accordingly, with $Rm_p(v, w, v, w)$ taking the place of $Rm_p(v, w, w, v)$. This is so that whatever sign convention is chosen for the curvature tensor, the notion of positive or negative sectional, Ricci, or scalar curvature has the same meaning for everyone.

Sectional Curvatures of the Model Spaces

We can now compute the sectional curvatures of our three families of frame-homogeneous model spaces. A Riemannian metric or Riemannian manifold is said to have **constant sectional curvature** if the sectional curvatures are the same for all planes at all points.

Lemma 8.33. *If a Riemannian manifold (M, g) is frame-homogeneous, then it has constant sectional curvature.*

Proof. Frame homogeneity implies, in particular, that given two 2-planes at the same or different points, there is an isometry taking one to the other. The result follows from the isometry invariance of the curvature tensor.   □

Thus to compute the sectional curvature of one of our model spaces, it suffices to compute the sectional curvature for one plane at one point in each space.

Theorem 8.34 (Sectional Curvatures of the Model Spaces). *The following Riemannian manifolds have the indicated constant sectional curvatures:*

(a) $\left(\mathbb{R}^n, \bar{g}\right)$ *has constant sectional curvature* 0.

(b) $\left(\mathbb{S}^n(R), \mathring{g}_R\right)$ *has constant sectional curvature* $1/R^2$.

(c) $\left(\mathbb{H}^n(R), \breve{g}_R\right)$ *has constant sectional curvature* $-1/R^2$.

Proof. First we consider the simplest case: Euclidean space. Since the curvature tensor of $\mathbb{R}^n$ is identically zero, clearly all sectional curvatures are zero. This is also easy to see geometrically, since each plane section is actually a plane, which has zero Gaussian curvature.

Next consider the sphere $\mathbb{S}^n(R)$. We need only compute the sectional curvature of the plane Π spanned by (∂_1, ∂_2) at the point $(0, \ldots, 0, 1)$. The geodesics with initial velocities in Π are great circles in the (x^1, x^2, x^{n+1}) subspace. Therefore S_Π is isometric to the round 2-sphere of radius R embedded in $\mathbb{R}^3$. As Example 8.25 showed, $\mathbb{S}^2(R)$ has Gaussian curvature $1/R^2$. Therefore $\mathbb{S}^n(R)$ has constant sectional curvature equal to $1/R^2$.

Finally, the proof for hyperbolic spaces is left to Problem 8-28.   □

Note that for every real number c, exactly one of the model spaces listed above has constant sectional curvature c. These spaces will play vital roles in our comparison and local-to-global theorems in the last two chapters of the book.

▶ **Exercise 8.35.** Show that the metric on real projective space $\mathbb{RP}^n$ defined in Example 2.34 has constant positive sectional curvature.

Since the sectional curvatures determine the curvature tensor, one would expect to have an explicit formula for the full curvature tensor when the sectional curvature is constant. Such a formula is provided in the following proposition.

Proposition 8.36. *A Riemannian metric g has constant sectional curvature c if and only if its curvature tensor satisfies*

$$Rm = \tfrac{1}{2}cg \odot g.$$

In this case, the Ricci tensor and scalar curvature of g are given by the formulas

$$Rc = (n-1)cg; \qquad S = n(n-1)c,$$

and the Riemann curvature endomorphism is

$$R(v,w)x = c(\langle w,x\rangle v - \langle v,x\rangle w).$$

In terms of any basis,

$$R_{ijkl} = c(g_{il}g_{jk} - g_{ik}g_{jl}); \qquad R_{ij} = (n-1)cg_{ij}.$$

Proof. Problem 8-29. □

The basic concepts of Riemannian metrics and sectional curvatures in arbitrary dimensions were introduced by Bernhard Riemann in a famous 1854 lecture at Göttingen University [Spi97, Vol. II, Chap. 4]. These were just a few of the seminal accomplishments in his tragically short life.

Problems

8-1. Suppose U is an open set in $\mathbb{R}^n$ and $f : U \to \mathbb{R}$ is a smooth function. Let $M = \{(x, f(x)) : x \in U\} \subseteq \mathbb{R}^{n+1}$ be the graph of f, endowed with the induced Riemannian metric and upward unit normal (see Example 2.19).

  (a) Compute the components of the shape operator in graph coordinates, in terms of f and its partial derivatives.
  (b) Let $M \subseteq \mathbb{R}^{n+1}$ be the n-dimensional paraboloid defined as the graph of $f(x) = |x|^2$. Compute the principal curvatures of M.

  (*Used on p. 361.*)

8-2. Let (M, g) be an embedded Riemannian hypersurface in a Riemannian manifold $(\widetilde{M}, \widetilde{g})$, let F be a local defining function for M, and let $N = \operatorname{grad} F / |\operatorname{grad} F|$.

(a) Show that the scalar second fundamental form of M with respect to the unit normal N is given by

$$h(X,Y) = -\frac{\widetilde{\nabla}^2 F(X,Y)}{|\operatorname{grad} F|}$$

for all $X,Y \in \mathfrak{X}(M)$.

(b) Show that the mean curvature of M is given by

$$H = -\frac{1}{n} \operatorname{div}_{\widetilde{g}}\left(\frac{\operatorname{grad} F}{|\operatorname{grad} F|}\right),$$

where $n = \dim M$ and $\operatorname{div}_{\widetilde{g}}$ is the divergence operator of $\widetilde{g}$ (see Problem 5-14). [Hint: First prove the following linear-algebra lemma: *If V is a finite-dimensional inner product space, $w \in V$ is a unit vector, and $A: V \to V$ is a linear map that takes $w^\perp$ to $w^\perp$, then $\operatorname{tr}(A|_{w^\perp}) = \operatorname{tr} B$, where $B: V \to V$ is defined by $Bx = Ax - \langle x, w \rangle Aw$.*]

8-3. Let C be an embedded smooth curve in the half-plane $H = \{(r,z) : r > 0\}$, and let $S_C \subseteq \mathbb{R}^3$ be the surface of revolution determined by C as in Example 2.20. Let $\gamma(t) = (a(t), b(t))$ be a unit-speed local parametrization of C, and let X be the local parametrization of S_C given by (2.11).

(a) Compute the shape operator and principal curvatures of S_C in terms of a and b, and show that the principal directions at each point are tangent to the meridians and latitude circles.

(b) Show that the Gaussian curvature of S_C at a point $X(t,\theta)$ is equal to $-a''(t)/a(t)$.

8-4. Show that there is a surface of revolution in $\mathbb{R}^3$ that has constant Gaussian curvature equal to 1 but does not have constant principal curvatures. [Remark: We will see in Chapter 10 that this surface is locally isometric to $\mathbb{S}^2$ (see Cor. 10.15), so this gives an example of two nonflat hypersurfaces in $\mathbb{R}^3$ that are locally isometric but have different principal curvatures.]

8-5. Let $S \subseteq \mathbb{R}^3$ be the paraboloid given by $z = x^2 + y^2$, with the induced metric. Prove that S is isotropic at only one point. (*Used on p. 56.*)

8-6. Suppose (M,g) is a Riemannian manifold, and $\gamma: I \to M$ is a regular (but not necessarily unit-speed) curve in M. Show that the geodesic curvature of γ at $t \in I$ is

$$\kappa(t) = \frac{|\gamma'(t) \wedge D_t \gamma'(t)|}{|\gamma'(t)|^3},$$

where the norm in the numerator is the one defined by (8.27). Show also that in $\mathbb{R}^3$ with the Euclidean metric, the formula can be written

$$\kappa(t) = \frac{|\gamma'(t) \times \gamma''(t)|}{|\gamma'(t)|^3}.$$

8-7. For $w > 0$, let $M_w \subseteq \mathbb{R}^3$ be the surface of revolution obtained by revolving the curve $\gamma(t) = (w \cosh(t/w), t)$ around the z-axis, called a **catenoid**. Show that M_w is a minimal surface for each w.

8-8. For $h > 0$, let $C_h \subseteq \mathbb{R}^3$ be the one-dimensional submanifold $\{(x, y, z) : x^2 + y^2 = 1, z = \pm h\}$; it is the union of two unit circles lying in the $z = \pm h$ planes. Let $h_0 = 1/\sinh c$, where c is the unique positive solution to the equation $c \tanh c = 1$. Prove that if $h < h_0$, then there are two distinct positive values $w_1 < w_2$ such that the portions of the catenoids M_{w_1} and M_{w_2} lying in the region $|z| \le h$ are minimal surfaces with boundary C_h (using the notation of Problem 8-7), while if $h = h_0$, there is exactly one such value, and if $h > h_0$, there are none. [Remark: This phenomenon can be observed experimentally. If you dip two parallel wire circles into soapy water, as long as they are close together they will form an area-minimizing soap film in the shape of a portion of a catenoid (the one with the larger "waist," M_{w_2} in the notation above). As you pull the circles apart, the "waist" of the catenoid gets smaller, until the ratio of the distance between the circles to their diameter reaches $h_0 \approx 0.6627$, at which point the film will burst and the only soap film that can be formed is two disjoint flat disks.]

8-9. Let $M \subseteq \mathbb{R}^{n+1}$ be a Riemannian hypersurface, and let N be a smooth unit normal vector field along M. At each point $p \in M$, $N_p \in T_p\mathbb{R}^{n+1}$ can be thought of as a unit vector in $\mathbb{R}^{n+1}$ and therefore as a point in $\mathbb{S}^n$. Thus each choice of unit normal vector field defines a smooth map $\nu \colon M \to \mathbb{S}^n$, called the **Gauss map of M**. Show that $\nu^* dV_{\mathring{g}} = (-1)^n K \, dV_g$, where K is the Gaussian curvature of M.

8-10. Suppose $g = g_1 \oplus g_2$ is a product metric on $M_1 \times M_2$ as in (2.12). (See also Problem 7-6.)

(a) Show that for each point $p_i \in M_i$, the submanifolds $M_1 \times \{p_2\}$ and $\{p_1\} \times M_2$ are totally geodesic.

(b) Show that $\sec(\Pi) = 0$ for every 2-plane $\Pi \subseteq T_{(p_1, p_2)}(M_1 \times M_2)$ spanned by $v_1 \in T_{p_1} M_1$ and $v_2 \in T_{p_2} M_2$.

(c) Show that if M_1 and M_2 both have nonnegative sectional curvature, then $M_1 \times M_2$ does too; and if M_1 and M_2 both have nonpositive sectional curvature, then $M_1 \times M_2$ does too.

(d) Now suppose (M_i, g_i) has constant sectional curvature c_i for $i = 1, 2$; let $n_i = \dim M_i$. Show that (M, g) is Einstein if and only if $c_1(n_1 - 1) = c_2(n_2 - 1)$; and (M, g) has constant sectional curvature if and only if $n_1 = n_2 = 1$ or $c_1 = c_2 = 0$.

(*Used on p. 366.*)

8-11. Suppose $M \subseteq \mathbb{R}^3$ is a smooth surface, $p \in M$, and N_p is a normal vector to M at p. Show that if $\Pi \subseteq \mathbb{R}^3$ is any affine plane containing p and parallel to N_p, then there is a neighborhood U of p in $\mathbb{R}^3$ such that $M \cap \Pi \cap U$ is a smooth embedded curve in Π, and the principal curvatures of M at p are equal to the minimum and maximum signed Euclidean curvatures of these curves as Π varies. [Remark: This justifies the informal recipe for computing principal curvatures given in Chapter 1.]

8-12. Suppose $\pi\colon \left(\widetilde{M},\widetilde{g}\right) \to (M,g)$ is a Riemannian submersion, and $\left(\widetilde{M},\widetilde{g}\right)$ has all sectional curvatures bounded below by a constant c. Use O'Neill's formula (Problem 7-14) to show that the sectional curvatures of (M,g) are bounded below by the same constant.

8-13. Let $p\colon \mathbb{S}^{2n+1} \to \mathbb{CP}^n$ be the Riemannian submersion described in Example 2.30. In this problem, we identify $\mathbb{C}^{n+1}$ with $\mathbb{R}^{2n+2}$ by means of coordinates $(x^1,y^1,\dots,x^{n+1},y^{n+1})$ defined by $z^j = x^j + iy^j$.

(a) Show that the vector field

$$S = x^j \frac{\partial}{\partial y^j} - y^j \frac{\partial}{\partial x^j}$$

on $\mathbb{C}^{n+1}$ is tangent to $\mathbb{S}^{2n+1}$ and spans the vertical space V_z at each point $z \in \mathbb{S}^{2n+1}$. (The implicit summation here is from 1 to $n+1$.)

(b) Show that for all horizontal vector fields W, Z on $\mathbb{S}^{2n+1}$,

$$[W,Z]^V = -d\omega(W,Z)S = 2\langle W, JZ\rangle S,$$

where ω is the 1-form on $\mathbb{C}^{n+1}$ given by

$$\omega = S^\flat = \sum_j x^j\,dy^j - y^j\,dx^j,$$

and $J\colon T\mathbb{C}^{n+1} \to T\mathbb{C}^{n+1}$ is the real-linear orthogonal map given by

$$J\left(a^j \frac{\partial}{\partial x^j} + b^j \frac{\partial}{\partial y^j}\right) = a^j \frac{\partial}{\partial y^j} - b^j \frac{\partial}{\partial x^j}.$$

(This is just multiplication by $i = \sqrt{-1}$ in complex coordinates. Notice that $J \circ J = -\mathrm{Id}$.)

(c) Using O'Neill's formula (Problem 7-14), show that the curvature tensor of $\mathbb{CP}^n$ satisfies

$$\begin{aligned} Rm(w,x,y,z) = {}& \langle\widetilde{w},\widetilde{z}\rangle\langle\widetilde{x},\widetilde{y}\rangle - \langle\widetilde{w},\widetilde{y}\rangle\langle\widetilde{x},\widetilde{z}\rangle \\ & - 2\langle\widetilde{w},J\widetilde{x}\rangle\langle\widetilde{y},J\widetilde{z}\rangle - \langle\widetilde{w},J\widetilde{y}\rangle\langle\widetilde{x},J\widetilde{z}\rangle \\ & + \langle\widetilde{w},J\widetilde{z}\rangle\langle\widetilde{x},J\widetilde{y}\rangle, \end{aligned}$$

for every $q \in \mathbb{CP}^n$ and $w,x,y,z \in T_q\mathbb{CP}^n$, where $\widetilde{w},\widetilde{x},\widetilde{y},\widetilde{z}$ are horizontal lifts of w,x,y,z to an arbitrary point $\widetilde{q} \in p^{-1}(q) \subseteq \mathbb{S}^{2n+1}$.

(d) Using the notation of part (c), show that for orthonormal vectors $w,x \in T_q\mathbb{CP}^n$, the sectional curvature of the plane spanned by $\{w,x\}$ is

$$\sec(w,x) = 1 + 3\langle\widetilde{w},J\widetilde{x}\rangle^2.$$

(e) Show that for $n \geq 2$, the sectional curvatures at each point of $\mathbb{C}\mathbb{P}^n$ take on all values between 1 and 4, inclusive, and conclude that $\mathbb{C}\mathbb{P}^n$ is not frame-homogeneous.

(f) Compute the Gaussian curvature of $\mathbb{C}\mathbb{P}^1$.

(*Used on pp. 56, 262, 367.*)

8-14. Show that a Riemannian 3-manifold is Einstein if and only if it has constant sectional curvature.

8-15. Suppose (M,g) is a 3-dimensional Riemannian manifold that is homogeneous and isotropic. Show that g has constant sectional curvature. Show that the analogous result in dimension 4 is not true. [Hint: See Problem 8-13.]

8-16. For each $a > 0$, let g_a be the Berger metric on SU(2) defined in Problem 3-10. Compute the sectional curvatures with respect to g_a of the planes spanned by (X,Y), (Y,Z), and (Z,X). Prove that if $a \neq 1$, then $(SU(2), g_a)$ is homogeneous but not isotropic anywhere. (*Used on p. 56.*)

8-17. Let G be a Lie group with a bi-invariant metric g (see Problem 7-13).

(a) Suppose X and Y are orthonormal elements of Lie(G). Show that $\sec(X_p, Y_p) = \frac{1}{4}\big|[X,Y]\big|^2$ for each $p \in G$, and conclude that the sectional curvatures of (G,g) are all nonnegative.

(b) Show that every Lie subgroup of G is totally geodesic in G.

(c) Now suppose G is connected. Show that G is flat if and only if it is abelian.

(*Used on p. 282.*)

8-18. Suppose (M,g) is a Riemannian hypersurface in a Riemannian manifold $(\widetilde{M}, \widetilde{g})$, and N is a unit normal vector field along M. We say that M is *convex* (with respect to N) if its scalar second fundamental form satisfies $h(v,v) \leq 0$ for all $v \in TM$. Show that if M is convex and $\widetilde{M}$ has sectional curvatures bounded below by a constant c, then all sectional curvatures of M are bounded below by c.

8-19. Suppose (M,g) is a Riemannian hypersurface in an $(n+1)$-dimensional Lorentz manifold $(\widetilde{M}, \widetilde{g})$, and N is a smooth unit normal vector field along M. Define the *scalar second fundamental form* h and the *shape operator* s by requiring that $\mathrm{II}(X,Y) = h(X,Y)N$ and $\langle sX,Y \rangle = h(X,Y)$ for all $X,Y \in \mathfrak{X}(M)$. Prove the following Lorentz analogues of the formulas of Theorem 8.13 (with notation as in that theorem):

(a) GAUSS FORMULA: $\widetilde{\nabla}_X Y = \nabla_X Y + h(X,Y)N$.

(b) GAUSS FORMULA FOR A CURVE: $\widetilde{D}_t X = D_t X + h(\gamma', X)N$.

(c) WEINGARTEN EQUATION: $\widetilde{\nabla}_X N = sX$.

(d) GAUSS EQUATION: $\widetilde{Rm}(W,X,Y,Z) = \big(Rm + \frac{1}{2}h \otimes h\big)(W,X,Y,Z)$.

(e) CODAZZI EQUATION: $\widetilde{Rm}(W,X,Y,N) = -(Dh)(Y,W,X)$.

8-20. Suppose $(\widetilde{M}, \widetilde{g})$ is an $(n+1)$-dimensional Lorentz manifold, and assume that $\widetilde{g}$ satisfies the Einstein field equation with a cosmological constant:

$\widetilde{Rc} - \frac{1}{2}\widetilde{S}\widetilde{g} + \Lambda\widetilde{g} = T$, where Λ is a constant and T is a smooth symmetric 2-tensor field (see p. 211). Let (M, g) be a Riemannian hypersurface in $\widetilde{M}$, and let h be its scalar second fundamental form as defined in Problem 8-19. Use the results of Problem 8-19 to show that g and h satisfy the following **Einstein constraint equations** on M:

$$S - 2\Lambda - |h|_g^2 + \left(\operatorname{tr}_g h\right)^2 = 2\rho,$$
$$\operatorname{div} h - d\left(\operatorname{tr}_g h\right) = J,$$

where S is the scalar curvature of g, $\operatorname{div} h$ the divergence of h with respect to g as defined in Problem 5-16, ρ is the function $\rho = T(N, N)$, and J is the 1-form $J(X) = T(N, X)$. [Remark: When J and ρ are both zero, these equations, known as the **vacuum Einstein constraint equations**, are necessary conditions for g and h to be the metric and second fundamental form of a Riemannian hypersurface in an $(n + 1)$-dimensional Lorentz manifold satisfying the vacuum Einstein field equation with cosmological constant Λ. It was proved in 1950 by Yvonne Choquet-Bruhat that these conditions are also sufficient; for a proof, see [CB09, pp. 166–168].]

8-21. For every linear endomorphism $A \colon \mathbb{R}^n \to \mathbb{R}^n$, the **associated quadratic form** is the function $Q \colon \mathbb{R}^n \to \mathbb{R}$ defined by $Q(x) = \langle Ax, x \rangle$. Prove that

$$\int_{\mathbb{S}^{n-1}} Q \, dV_{\mathring{g}} = \frac{1}{n}(\operatorname{tr} A)\operatorname{Vol}(\mathbb{S}^{n-1}).$$

[Hint: Show that $\int_{\mathbb{S}^{n-1}} x^i x^j \, dV_{\mathring{g}} = 0$ when $i \neq j$ by examining the effect of the isometry

$$(x^1, \dots, x^i, \dots, x^j, \dots, x^n) \mapsto (x^1, \dots, x^j, \dots, -x^i, \dots, x^n),$$

and compute $\int_{\mathbb{S}^{n-1}} \left(x^i\right)^2 dV_{\mathring{g}}$ using the fact that $\sum_i \left(x^i\right)^2 = 1$ on the sphere.] [Remark: It is a standard fact of linear algebra that the trace of A is independent of the choice of basis. This gives a geometric interpretation to the trace as n times the average value of the associated quadratic form Q.] *(Used on p. 313.)*

8-22. Let (M, g) be a Riemannian n-manifold and $p \in M$. Proposition 8.32 gave a geometric interpretation of the Ricci curvature at p based on a choice of orthonormal basis. This problem describes an interpretation that does not refer to a basis. For each unit vector $v \in T_p M$, prove that

$$Rc_p(v, v) = \frac{n-1}{\operatorname{Vol}(\mathbb{S}^{n-2})} \int_{w \in S_v^\perp} \operatorname{sec}(v, w) \, dV_{\widehat{g}},$$

where $S_v^\perp$ denotes the set of unit vectors in $T_p M$ that are orthogonal to v and $\widehat{g}$ denotes the Riemannian metric on $S_v^\perp$ induced from the flat metric g_p on $T_p M$. [Hint: Use Problem 8-21.]

8-23. Suppose (M,g) is a connected n-dimensional Riemannian manifold, and a Lie group G acts isometrically and effectively on M. Problem 5-12 showed that $\dim G \le n(n+1)/2$. Prove that if equality holds, then g has constant sectional curvature.

8-24. Let (M,g) be a connected Riemannian manifold. Recall the definition of *k-point homogeneous* from Problem 6-18. Prove the following:

   (a) If (M,g) is homogeneous, then it has constant scalar curvature.
   (b) If (M,g) is 2-point homogeneous, then it is Einstein.
   (c) If (M,g) is 3-point homogeneous, then it has constant sectional curvature.

8-25. For $i = 1,2$, suppose (M_i, g_i) is a Riemannian manifold of dimension $n_i \ge 2$ with constant sectional curvature c_i; and let $g = g_1 \oplus g_2$ be the product metric on $M = M_1 \times M_2$. Show that the Weyl tensor of g is given by

$$W = \frac{c_1 + c_2}{2(n-1)(n-2)}\Big(n_2(n_2 - 1)h_1 \oslash h_1$$
$$- 2(n_1 - 1)(n_2 - 1)h_1 \oslash h_2 + n_1(n_1 - 1)h_2 \oslash h_2 \Big),$$

where $n = n_1 + n_2$ and $h_i = \pi_i^* g_i$ for $i = 1,2$. Conclude that (M,g) is locally conformally flat if and only if $c_2 = -c_1$. (See also Problem 7-6.) (*Used on p. 67.*)

8-26. Let (M,g) be a 4-dimensional Riemannian manifold. Given $p \in M$ and an orthonormal basis (b_i) for $T_p M$, prove that the Weyl tensor at p satisfies

$$W_{1221} = \frac{1}{3}(k_{12} + k_{34}) - \frac{1}{6}(k_{13} + k_{14} + k_{23} + k_{24}),$$

where k_{ij} is the sectional curvature of the plane spanned by (b_i, b_j).

8-27. Prove that the Fubini–Study metric on $\mathbb{CP}^2$ is not locally conformally flat. [Hint: Use Problems 8-13 and 8-26.]

8-28. Complete the proof of Theorem 8.34 by showing in two ways that the hyperbolic space of radius R has constant sectional curvature equal to $-1/R^2$.

   (a) In the hyperboloid model, compute the second fundamental form of $\mathbb{H}^n(R) \subseteq \mathbb{R}^{n,1}$ at the point $(0,\dots,0,R)$, and use either the general form of the Gauss equation (Thm. 8.5) or the formulas of Problem 8-19.
   (b) In the Poincaré ball model, use formula (7.44) for the conformal transformation of the curvature to compute the Riemann curvature tensor at the origin.

8-29. Prove Proposition 8.36 (curvature tensors on constant-curvature spaces).

8-30. Suppose $M \subseteq \mathbb{R}^{n+1}$ is a hypersurface with the induced Riemannian metric. Show that the Ricci tensor of M satisfies

$$Rc(v,w) = \langle nHsv - s^2v,\ w\rangle,$$

where H and s are the mean curvature and shape operator of M, respectively, and $s^2v = s(sv)$.

8-31. For $1 < k \le n$, show that any k points in the hyperbolic space $\mathbb{H}^n(R)$ lie in a totally geodesic $(k-1)$-dimensional submanifold, which is isometric to $\mathbb{H}^{k-1}(R)$.

8-32. Suppose (M,g) is a Riemannian manifold and G is a Lie group acting smoothly and isometrically on M. Let $S \subseteq M$ be the fixed point set of G, that is, the set of points $p \in M$ such that $\varphi(p) = p$ for all $\varphi \in G$. Show that each connected component of S is a smoothly embedded totally geodesic submanifold of M. (The reason for restricting to a single connected component is that different components may have different dimensions.)

8-33. Suppose (M,g) is a Riemannian manifold. Let $\Lambda^2(TM)$ denote the bundle of 2-vectors (alternating contravariant 2-tensors) on M.

(a) Show that there is a unique fiber metric on $\Lambda^2(TM)$ whose associated norm satisfies
$$|w \wedge x|^2 = |w|^2|x|^2 - \langle w,x\rangle^2$$
for all tangent vectors w, x at every point $q \in M$.

(b) Show that there is a unique bundle endomorphism $\mathscr{R}\colon \Lambda^2(TM) \to \Lambda^2(TM)$, called the **curvature operator** of g, that satisfies
$$\langle \mathscr{R}(w \wedge x), y \wedge z\rangle = -Rm(w,x,y,z) \tag{8.30}$$
for all tangent vectors w, x, y, z at a point of M, where the inner product on the left-hand side is the one described in part (a).

(c) A Riemannian metric is said to have **positive curvature operator** if $\mathscr{R}$ is positive definite, and **negative curvature operator** if $\mathscr{R}$ is negative definite. Show that positive curvature operator implies positive sectional curvature, and negative curvature operator implies negative sectional curvature. [Remark: This is the reason for the negative sign in (8.30). If the Riemann curvature tensor is defined as the negative of ours, the negative sign should be omitted.]

(d) Show that the converse need not be true, by using the results of Problem 8-13 to compute the following expression on $\mathbb{CP}^2$ with the Fubini–Study metric:
$$\langle \mathscr{R}(w \wedge x + y \wedge z),\ w \wedge x + y \wedge z\rangle,$$
where w, x, y, z are orthonormal vectors at an arbitrary point of $\mathbb{CP}^2$, chosen so that their horizontal lifts satisfy $\tilde{y} = J\tilde{w}$ and $\tilde{z} = J\tilde{x}$.

Chapter 9
The Gauss–Bonnet Theorem

All the work we have done so far has been focused on purely local properties of Riemannian manifolds and their submanifolds. We are finally in a position to prove our first major local-to-global theorem in Riemannian geometry: the Gauss–Bonnet theorem. The grandfather of all such theorems in Riemannian geometry, it is a local-to-global theorem par excellence, because it asserts the equality of two very differently defined quantities on a compact Riemannian 2-manifold: the integral of the Gaussian curvature, which is determined by the local geometry, and 2π times the Euler characteristic, which is a global topological invariant. Although in this form it applies only in two dimensions, it has provided a model and an inspiration for innumerable local-to-global results in higher-dimensional geometry, some of which we will prove in Chapter 12.

This chapter begins with some not-so-elementary notions from plane geometry, leading up to a proof of Hopf's rotation index theorem, which expresses the intuitive idea that the velocity vector of a simple closed plane curve, or more generally of a "curved polygon," makes a net rotation through an angle of exactly 2π as one traverses the curve counterclockwise. Then we investigate curved polygons on Riemannian 2-manifolds, leading to a far-reaching generalization of the rotation index theorem called the Gauss–Bonnet formula, which relates the exterior angles, geodesic curvature of the boundary, and Gaussian curvature in the interior of a curved polygon. Finally, we use the Gauss–Bonnet formula to prove the global statement of the Gauss–Bonnet theorem.

Some Plane Geometry

Look back for a moment at the three local-to-global theorems about plane geometry stated in Chapter 1: the angle-sum theorem, the circumference theorem, and the total curvature theorem. When looked at correctly, these three theorems all turn out to be manifestations of the same phenomenon: as one traverses a simple closed plane

© Springer International Publishing AG 2018
J. M. Lee, *Introduction to Riemannian Manifolds*, Graduate Texts
in Mathematics 176, https://doi.org/10.1007/978-3-319-91755-9_9

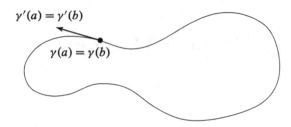

Fig. 9.1: A closed curve with $\gamma'(a) = \gamma'(b)$

curve, the velocity vector makes a net rotation through an angle of exactly 2π. Our task in the first part of this chapter is to make these notions precise.

Throughout this section, $\gamma: [a,b] \to \mathbb{R}^2$ is an admissible curve in the plane. We say that γ is a **simple closed curve** if $\gamma(a) = \gamma(b)$ but γ is injective on $[a,b)$. We do not assume that γ has unit speed; instead, we define the **unit tangent vector field of γ** as the vector field T along each smooth segment of γ given by

$$T(t) = \frac{\gamma'(t)}{|\gamma'(t)|}.$$

Since each tangent space to $\mathbb{R}^2$ is naturally identified with $\mathbb{R}^2$ itself, we can think of T as a map into $\mathbb{R}^2$, and since T has unit length, it takes its values in $\mathbb{S}^1$.

If γ is smooth (or at least continuously differentiable), we define a **tangent angle function for γ** to be a continuous function $\theta: [a,b] \to \mathbb{R}$ such that $T(t) = (\cos\theta(t), \sin\theta(t))$ for all $t \in [a,b]$. It follows from the theory of covering spaces that such a function exists: the map $q: \mathbb{R} \to \mathbb{S}^1$ given by $q(s) = (\cos s, \sin s)$ is a smooth covering map, and the path-lifting property of covering maps (Prop. A.54(b)) ensures that there is a continuous function $\theta: [a,b] \to \mathbb{R}$ that satisfies $q(\theta(t)) = T(t)$. By the unique lifting property (Prop. A.54(a)), a lift is uniquely determined once its value at any single point is determined, and thus any two lifts differ by a constant integral multiple of 2π.

If γ is a continuously differentiable simple closed curve such that $\gamma'(a) = \gamma'(b)$ (Fig. 9.1), then $(\cos\theta(a), \sin\theta(a)) = (\cos\theta(b), \sin\theta(b))$, so $\theta(b) - \theta(a)$ must be an integral multiple of 2π. For such a curve, we define the **rotation index of γ** to be the following integer:

$$\rho(\gamma) = \frac{1}{2\pi}(\theta(b) - \theta(a)),$$

where θ is any tangent angle function for γ. For any other choice of tangent angle function, $\theta(a)$ and $\theta(b)$ would change by addition of the same constant, so the rotation index is independent of the choice of θ.

We would also like to extend the definition of the rotation index to certain piecewise regular closed curves. For this purpose, we have to take into account the "jumps" in the tangent angle function at corners. To do so, suppose $\gamma: [a,b] \to \mathbb{R}^2$

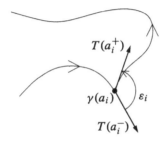
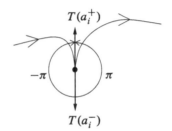

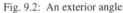

Fig. 9.2: An exterior angle   Fig. 9.3: A cusp vertex

is an admissible simple closed curve, and let $(a_0,\ldots,a_k)$ be an admissible partition of $[a,b]$. The points $\gamma(a_i)$ are called the **vertices of γ**, and the curve segments
$\gamma|_{[a_{i-1},a_i]}$ are called its **edges** or **sides**.

At each vertex $\gamma(a_i)$, recall that γ has left-hand and right-hand velocity vectors denoted by $\gamma'(a_i^-)$ and $\gamma'(a_i^+)$, respectively; let $T(a_i^-)$ and $T(a_i^+)$ denote the
corresponding unit vectors. We classify each vertex into one of three categories:

- If $T(a_i^-) \neq \pm T(a_i^+)$, then $\gamma(a_i)$ is an **ordinary vertex**.
- If $T(a_i^-) = T(a_i^+)$, then $\gamma(a_i)$ is a **flat vertex**.
- If $T(a_i^-) = -T(a_i^+)$, then $\gamma(a_i)$ is a **cusp vertex**.

At each ordinary vertex, define the **exterior angle at $\gamma(a_i)$** to be the oriented measure ε_i of the angle from $T(a_i^-)$ to $T(a_i^+)$, chosen to be in the interval $(-\pi,\pi)$, with
a positive sign if $\left(T(a_i^-),T(a_i^+)\right)$ is an oriented basis for $\mathbb{R}^2$, and a negative sign
otherwise (Fig. 9.2). At a flat vertex, the exterior angle is defined to be zero. At a
cusp vertex, there is no simple way to choose unambiguously between π and $-\pi$
(Fig. 9.3), so we leave the exterior angle undefined. The vertex $\gamma(a) = \gamma(b)$ is han-
dled in the same way, with $T(b)$ and $T(a)$ playing the roles of $T(a_i^-)$ and $T(a_i^+)$,
respectively. If $\gamma(a_i)$ is an ordinary or a flat vertex, the **interior angle at $\gamma(a_i)$** is
defined to be $\theta_i = \pi - \varepsilon_i$; our conventions ensure that $0 < \theta_i < 2\pi$.

The curves we wish to consider are of the following type: a **curved polygon** in
the plane is an admissible simple closed curve without cusp vertices, whose image
is the boundary of a precompact open set $\Omega \subseteq \mathbb{R}^2$. The set Ω is called the **interior
of γ** (not to be confused with the topological interior of its image as a subset of $\mathbb{R}^2$,
which is the empty set).

Suppose $\gamma \colon [a,b] \to \mathbb{R}^2$ is a curved polygon. If γ is parametrized so that at
smooth points, γ' is positively oriented with respect to the induced orientation on
$\partial\Omega$ in the sense of Stokes's theorem, we say that γ is **positively oriented** (Fig. 9.4).
Intuitively, this means that γ is parametrized in the counterclockwise direction, or
that Ω is always to the left of γ.

We define a **tangent angle function for a curved polygon γ** to be a piecewise
continuous function $\theta \colon [a,b] \to \mathbb{R}$ that satisfies $T(t) = (\cos\theta(t),\sin\theta(t))$ at each
point t where γ is smooth, that is continuous from the right at each a_i with

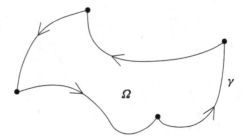

Fig. 9.4: A positively oriented curved polygon

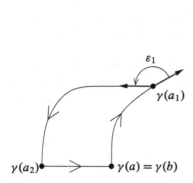

Fig. 9.5: Tangent angle at a vertex

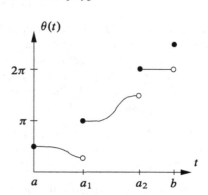

Fig. 9.6: Tangent angle function

$$\theta(a_i) = \lim_{t \nearrow a_i} \theta(t) + \varepsilon_i, \tag{9.1}$$

and that satisfies

$$\theta(b) = \lim_{t \nearrow b} \theta(t) + \varepsilon_k, \tag{9.2}$$

where ε_k is the exterior angle at $\gamma(b)$. (See Figs. 9.5 and 9.6.) Such a function always exists: start by defining $\theta(t)$ for $t \in [a, a_1)$ to be any lift of T on that interval; then on $[a_1, a_2]$ define $\theta(t)$ to be the unique lift that satisfies (9.1), and continue by induction, ending with $\theta(b)$ defined by (9.2). Once again, the difference between any two such functions is a constant integral multiple of 2π. We define the **rotation index of** γ to be $\rho(\gamma) = \frac{1}{2\pi}\big(\theta(b) - \theta(a)\big)$ just as in the smooth case. As before, $\rho(\gamma)$ is an integer, because the definition ensures that $(\cos \theta(b), \sin \theta(b)) = (\cos \theta(a), \sin \theta(a))$.

The following theorem was first proved by Heinz Hopf in 1935. (For a readable version of Hopf's proof, see [Hop89, p. 42].) It is frequently referred to by the German name given to it by Hopf, the *Umlaufsatz*.

Theorem 9.1 (Rotation Index Theorem). *The rotation index of a positively oriented curved polygon in the plane is* $+1$.

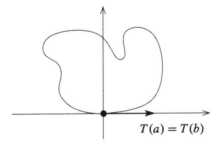

Fig. 9.7: The curve γ after changing the parameter interval and translating $\gamma(a)$ to the origin

Proof. Let $\gamma: [a,b] \to \mathbb{R}^2$ be such a curved polygon. Assume first that all the vertices of γ are flat. This means, in particular, that γ' is continuous and $\gamma'(a) = \gamma'(b)$. Since $\gamma(a) = \gamma(b)$, we can extend γ to a continuous map from $\mathbb{R}$ to $\mathbb{R}^2$ by requiring it to be periodic of period $b - a$, and our hypothesis $\gamma'(a) = \gamma'(b)$ guarantees that the extended map still has continuous first derivatives. Define $T(t) = \gamma'(t)/|\gamma'(t)|$ as before.

Let $\theta: \mathbb{R} \to \mathbb{R}$ be any lift of $T: \mathbb{R} \to \mathbb{S}^1$. Then $\theta|_{[a,b]}$ is a tangent angle function for γ, and thus $\theta(b) = \theta(a) + 2\pi\rho(\gamma)$. If we set $\widetilde{\theta}(t) = \theta(t+b-a) - 2\pi\rho(\gamma)$, then

$$\left(\cos\widetilde{\theta}(t), \sin\widetilde{\theta}(t)\right) = \left(\cos\theta(t+b-a), \sin\theta(t+b-a)\right) = T(t+b-a) = T(t),$$

so $\widetilde{\theta}$ is also a lift of T. Because $\widetilde{\theta}(a) = \theta(a)$, it follows that $\widetilde{\theta} \equiv \theta$, or in other words the following equation holds for all $t \in \mathbb{R}$:

$$\theta(t+b-a) = \theta(t) + 2\pi\rho(\gamma). \tag{9.3}$$

If a_1 is an arbitrary point in $[a,b]$ and $b_1 = a_1 + b - a$, then $\gamma|_{[a_1,b_1]}$ is also a positively oriented curved polygon with only flat vertices, and $\theta|_{[a_1,b_1]}$ is a tangent angle function for it. Note that (9.3) implies

$$\theta(b_1) - \theta(a_1) = \theta(a_1+b-a) - \theta(a_1) = (\theta(a_1) + 2\pi\rho(\gamma)) - \theta(a_1) = 2\pi\rho(\gamma),$$

so $\gamma|_{[a_1,b_1]}$ has the same rotation index as $\gamma|_{[a,b]}$. Thus we obtain the same result by restricting γ to any closed interval of length $b - a$.

Using this freedom, we can assume that the parameter interval $[a,b]$ has been chosen so that the y-coordinate of γ achieves its minimum at $t = a$. Moreover, by a translation in the xy-plane (which does not change γ' or θ), we may as well assume that $\gamma(a)$ is the origin. With these adjustments, the image of γ remains in the closed upper half-plane, and $T(a) = T(b) = (1,0)$ (Fig. 9.7). By adding a constant integral multiple of 2π to θ if necessary, we can also assume that $\theta(a) = 0$.

Next, we define a continuous *secant angle function*, denoted by $\varphi(t_1,t_2)$, representing the angle between the positive x-direction and the vector from $\gamma(t_1)$ to $\gamma(t_2)$. To be precise, let $\Delta \subseteq \mathbb{R}^2$ be the triangular region $\Delta = \{(t_1,t_2) : a \le t_1 \le t_2 \le b\}$

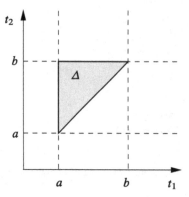

Fig. 9.8: The domain of φ Fig. 9.9: The secant angle function

(Fig. 9.8), and define a map $V : \Delta \to \mathbb{S}^1$ by

$$
V(t_1, t_2) = \begin{cases}
\dfrac{\gamma(t_2) - \gamma(t_1)}{|\gamma(t_2) - \gamma(t_1)|}, & t_1 < t_2 \text{ and } (t_1, t_2) \neq (a, b); \\[2mm]
T(t_1), & t_1 = t_2; \\[1mm]
-T(b), & (t_1, t_2) = (a, b).
\end{cases}
$$

The function V is continuous at points where $t_1 < t_2$ and $(t_1, t_2) \neq (a, b)$, because γ is continuous and injective there. To see that it is continuous elsewhere, note that for $t_1 < t_2$, the fundamental theorem of calculus gives

$$
\gamma(t_2) - \gamma(t_1) = \int_0^1 \frac{d}{ds} \gamma\big(t_1 + s(t_2 - t_1)\big)\, ds = \int_0^1 \gamma'\big(t_1 + s(t_2 - t_1)\big)(t_2 - t_1)\, ds,
$$

and thus

$$
\left| \frac{\gamma(t_2) - \gamma(t_1)}{t_2 - t_1} - \gamma'(t) \right| \leq \int_0^1 \left| \gamma'\big(t_1 + s(t_2 - t_1)\big) - \gamma'(t) \right| ds.
$$

Because γ' is uniformly continuous on the compact set $[a, b]$, this last expression can be made as small as desired by taking (t_1, t_2) close to (t, t). It follows that

$$
\lim_{\substack{(t_1, t_2) \to (t, t) \\ t_1 < t_2}} \frac{\gamma(t_2) - \gamma(t_1)}{t_2 - t_1} = \gamma'(t),
$$

and therefore

$$\lim_{\substack{(t_1,t_2)\to(t,t)\\t_1<t_2}} V(t_1,t_2) = \lim_{\substack{(t_1,t_2)\to(t,t)\\t_1<t_2}} \frac{\gamma(t_2)-\gamma(t_1)}{|\gamma(t_2)-\gamma(t_1)|}$$

$$= \lim_{\substack{(t_1,t_2)\to(t,t)\\t_1<t_2}} \frac{\gamma(t_2)-\gamma(t_1)}{t_2-t_1} \Bigg/ \left|\frac{\gamma(t_2)-\gamma(t_1)}{t_2-t_1}\right|$$

$$= \frac{\gamma'(t)}{|\gamma'(t)|} = T(t) = V(t,t).$$

Similarly, because T is continuous,

$$\lim_{\substack{(t_1,t_2)\to(t,t)\\t_1=t_2}} V(t_1,t_2) = \lim_{t_1\to t} T(t_1) = T(t) = V(t,t).$$

It follows that V is continuous at (t,t).

To prove that V is continuous at (a,b), recall that we have extended γ to be periodic of period $b-a$. The argument above gives

$$\lim_{\substack{(t_1,t_2)\to(a,b)\\t_1<t_2}} V(t_1,t_2) = \lim_{\substack{(t_1,t_2)\to(a,b)\\t_1<t_2}} \frac{\gamma(t_2)-\gamma(t_1+b-a)}{|\gamma(t_2)-\gamma(t_1+b-a)|}$$

$$= \lim_{\substack{(s_1,s_2)\to(b,b)\\s_1>s_2}} \frac{\gamma(s_2)-\gamma(s_1)}{|\gamma(s_2)-\gamma(s_1)|} = -T(b) = V(a,b).$$

Thus V is continuous.

Since Δ is simply connected, Corollary A.57 guarantees that $V: \Delta \to \mathbb{S}^1$ has a continuous lift $\varphi: \Delta \to \mathbb{R}$, which is unique if we require $\varphi(a,a) = 0$ (Fig. 9.9). This is our secant angle function.

We can express the rotation index in terms of the secant angle function as follows:

$$\rho(\gamma) = \frac{1}{2\pi}\big(\theta(b)-\theta(a)\big) = \frac{1}{2\pi}\big(\varphi(b,b)-\varphi(a,a)\big) = \frac{1}{2\pi}\varphi(b,b).$$

Observe that along the side of Δ where $t_1 = a$ and $t_2 \in [a,b]$, the vector $V(a,t_2)$ has its tail at the origin and its head in the upper half-plane. Since we stipulate that $\varphi(a,a) = 0$, we must have $\varphi(a,t_2) \in [0,\pi]$ on this segment. By continuity, therefore, $\varphi(a,b) = \pi$ (since $\varphi(a,b)$ represents the tangent angle of $-T(b) = (-1,0)$). Similarly, on the side where $t_2 = b$, the vector $V(t_1,b)$ has its head at the origin and its tail in the upper half-plane, so $\varphi(t_1,b) \in [\pi,2\pi]$. Therefore, since $\varphi(b,b)$ represents the tangent angle of $T(b) = (1,0)$, we must have $\varphi(b,b) = 2\pi$ and therefore $\rho(\gamma) = 1$. This completes the proof for the case in which γ' is continuous.

Now suppose γ has one or more ordinary vertices. It suffices to show there is a curve with a continuous velocity vector field that has the same rotation index as γ. We will construct such a curve by "rounding the corners" of γ. It will simplify the proof somewhat if we choose the parameter interval $[a,b]$ so that $\gamma(a) = \gamma(b)$ is not one of the ordinary vertices.

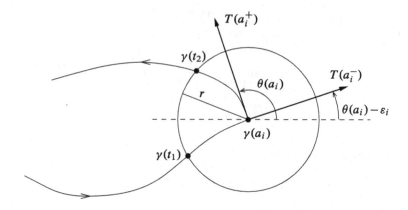

Fig. 9.10: Isolating the change in the tangent angle at a vertex

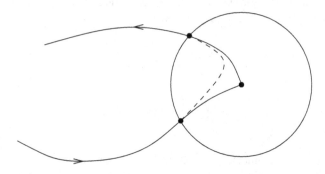

Fig. 9.11: Rounding a corner

Let $\gamma(a_i)$ be any ordinary vertex, let ε_i be its exterior angle, and let α be a positive number less than $\frac{1}{2}(\pi - |\varepsilon_i|)$. Recall that θ is continuous from the right at a_i and $\lim_{t \nearrow a_i} \theta(t) = \theta(a_i) - \varepsilon_i$. Therefore, we can choose δ small enough that $|\theta(t) - (\theta(a_i) - \varepsilon_i)| < \alpha$ when $t \in (a_i - \delta, a_i)$, and $|\theta(t) - \theta(a_i)| < \alpha$ when $t \in (a_i, a_i + \delta)$.

The image under γ of $[a,b] \smallsetminus (a_i - \delta, a_i + \delta)$ is a compact set that does not contain $\gamma(a_i)$, so we can choose r small enough that γ does not enter $\bar{B}_r(\gamma(a_i))$ except when $t \in (a_i - \delta, a_i + \delta)$. Let $t_1 \in (a_i - \delta, a_i)$ denote a time when γ enters $\bar{B}_r(\gamma(a_i))$, and $t_2 \in (a_i, a_i + \delta)$ a time when it leaves (Fig. 9.10). By our choice of δ, the total change in $\theta(t)$ is not more than α when $t \in [t_1, a_i)$, and again not more than α when $t \in (a_i, t_2]$. Therefore, the total change $\Delta\theta$ in $\theta(t)$ during the time interval $[t_1, t_2]$ is between $\varepsilon_i - 2\alpha$ and $\varepsilon_i + 2\alpha$, which implies $-\pi < \Delta\theta < \pi$.

Now we simply replace $\gamma|_{[t_1, t_2]}$ with a smooth curve segment σ that has the same velocity as γ at $\gamma(t_1)$ and $\gamma(t_2)$, and whose tangent angle increases or decreases monotonically from $\theta(t_1)$ to $\theta(t_2)$; an arc of a hyperbola will do (Fig. 9.11). Since

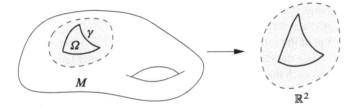

Fig. 9.12: A curved polygon on a surface

the change in tangent angle of σ is between $-\pi$ and π and represents the angle between $T(t_1)$ and $T(t_2)$, it must be exactly $\Delta\theta$. Repeating this process for each vertex, we obtain a new curved polygon with a continuous velocity vector field whose rotation index is the same as that of γ, thus proving the theorem. $\quad\square$

From the rotation index theorem, it is not hard to deduce the three local-to-global theorems mentioned at the beginning of the chapter as corollaries. (The angle-sum theorem is immediate; for the total curvature theorem, the trick is to show that $\theta'(t)$ is equal to the signed curvature of γ; the circumference theorem follows from the total curvature theorem as mentioned in Chapter 1.) However, instead of proving them directly, we will prove a general formula, called the *Gauss–Bonnet formula*, from which these results and more follow easily. You will easily see how the statement and proof of Theorem 9.3 below can be simplified in case the metric is Euclidean.

The Gauss–Bonnet Formula

We now direct our attention to the case of an oriented Riemannian 2-manifold (M, g). In this setting, an admissible simple closed curve $\gamma \colon [a,b] \to M$ is called a *curved polygon in M* if the image of γ is the boundary of a precompact open set $\Omega \subseteq M$, and there is an oriented smooth coordinate disk containing $\overline{\Omega}$ under whose image γ is a curved polygon in the plane (Fig. 9.12). As in the planar case, we call Ω the *interior of γ*. A curved polygon whose edges are all geodesic segments is called a *geodesic polygon.*

For a curved polygon γ in M, our previous definitions go through almost unchanged. We say that γ is *positively oriented* if it is parametrized in the direction of its Stokes orientation as the boundary of Ω. On each smooth segment of γ, we define the *unit tangent vector field* $T(t) = \gamma'(t)/|\gamma'(t)|_g$. If $\gamma(a_i)$ is an ordinary or flat vertex, we define the *exterior angle of γ at $\gamma(a_i)$* as the oriented measure ε_i of the angle from $T(a_i^-)$ to $T(a_i^+)$ with respect to the g-inner product and the given orientation of M; explicitly, this is

$$\varepsilon_i = \frac{dV_g\big(T(a_i^-), T(a_i^+)\big)}{\big|dV_g\big(T(a_i^-), T(a_i^+)\big)\big|} \arccos\big\langle T(a_i^-), T(a_i^+)\big\rangle_g. \tag{9.4}$$

The corresponding *interior angle of γ at $\gamma(a_i)$* is $\theta_i = \pi - \varepsilon_i$. Exterior and interior angles at $\gamma(a) = \gamma(b)$ are defined similarly.

We need a version of the rotation index theorem for curved polygons in M. Suppose $\gamma : [a,b] \to M$ is a curved polygon and Ω is its interior, and let (U,φ) be an oriented smooth chart such that U contains $\overline{\Omega}$. Using the coordinate map φ to transfer γ, Ω, and g to the plane, we may as well assume that g is a metric on some open subset $\hat{U} \subseteq \mathbb{R}^2$, and γ is a curved polygon in $\hat{U}$. Let (E_1, E_2) be the oriented orthonormal frame for g obtained by applying the Gram–Schmidt algorithm to (∂_x, ∂_y), so that E_1 is a positive scalar multiple of ∂_x everywhere in $\hat{U}$.

We define a *tangent angle function for γ* to be a piecewise continuous function $\theta : [a,b] \to \mathbb{R}$ that satisfies

$$T(t) = \cos\theta(t) E_1|_{\gamma(t)} + \sin\theta(t) E_2|_{\gamma(t)}$$

at each t where γ' is continuous, and that is continuous from the right and satisfies (9.1) and (9.2) at vertices. The existence of such a function follows as in the planar case, using the fact that

$$T(t) = u_1(t) E_1|_{\gamma(t)} + u_2(t) E_2|_{\gamma(t)} \tag{9.5}$$

for a pair of piecewise continuous functions $u_1, u_2 : [a,b] \to \mathbb{R}$ that can be regarded as the coordinate functions of a map $(u_1, u_2) : [a,b] \to \mathbb{S}^1$ because T has unit length.

The *rotation index of γ* is $\rho(\gamma) = \frac{1}{2\pi}\big(\theta(b) - \theta(a)\big)$. Because of the role played by the specific frame (E_1, E_2) in the definition, it is not obvious that the rotation index has any coordinate-independent meaning; however, the following easy consequence of the rotation index theorem shows that it does not depend on the choice of coordinates.

Lemma 9.2. *If M is an oriented Riemannian 2-manifold, the rotation index of every positively oriented curved polygon in M is $+1$.*

Proof. If we use the given oriented coordinate chart to regard γ as a curved polygon in the plane, we can compute its tangent angle function either with respect to g or with respect to the Euclidean metric $\bar{g}$. In either case, $\rho(\gamma)$ is an integer because $\theta(a)$ and $\theta(b)$ both represent the angle between the same two vectors, calculated with respect to some inner product. Now for $0 \le s \le 1$, let $g_s = sg + (1-s)\bar{g}$. By the same reasoning, the rotation index $\rho_{g_s}(\gamma)$ with respect to g_s is also an integer for each s, so the function $f(s) = \rho_{g_s}(\gamma)$ is integer-valued.

In fact, the function f is continuous in s, as can be deduced easily from the following observations: (1) Our preferred g_s-orthonormal frame $\big(E_1^{(s)}, E_2^{(s)}\big)$ depends continuously on s, as can be seen from the formulas (2.5)–(2.6) used to implement the Gram–Schmidt algorithm. (2) On every interval $[a_{i-1}, a_i]$ where γ is smooth, the functions u_1 and u_2 satisfying the g_s-analogue of (9.5) can be expressed as $u_j(t,s) = \big\langle T_s(t), E_j^{(s)}|_{\gamma(t)}\big\rangle_{g_s}$, where $T_s(t) = \gamma'(t)/|\gamma'(t)|_{g_s}$. Thus u_1 and u_2 depend continuously on $(t,s) \in [a_{i-1}, a_i] \times [0,1]$, so the function $(u_1, u_2) : [a_{i-1}, a_i] \times$

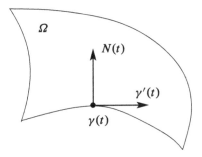

Fig. 9.13: $N(t)$ is the inward-pointing normal

$[0, 1] \to \mathbb{S}^1$ has a continuous lift $\theta : [a_{i-1}, a_i] \times [0, 1] \to \mathbb{R}$, uniquely determined by its value at one point. (3) At each vertex, it follows from formula (9.4) that the exterior angle depends continuously on g_s.

Because f is continuous and integer-valued, it follows that $\rho_g(\gamma) = f(1) = f(0) = \rho_{\bar{g}}(\gamma) = 1$, which was to be proved. $\qquad \square$

From this point onward, we assume for convenience that our curved polygon γ is given a unit-speed parametrization, so the unit tangent vector field $T(t)$ is equal to $\gamma'(t)$. There is a unique unit normal vector field N along the smooth portions of γ such that $(\gamma'(t), N(t))$ is an oriented orthonormal basis for $T_{\gamma(t)}M$ for each t. If γ is positively oriented as the boundary of Ω, this is equivalent to N being the inward-pointing normal to $\partial\Omega$ (Fig. 9.13). We define the **signed curvature of γ** at smooth points of γ by

$$\kappa_N(t) = \langle D_t \gamma'(t), N(t) \rangle_g.$$

By differentiating $|\gamma'(t)|_g^2 \equiv 1$, we see that $D_t \gamma'(t)$ is orthogonal to $\gamma'(t)$, and therefore we can write $D_t \gamma'(t) = \kappa_N(t) N(t)$, and the (unsigned) geodesic curvature of γ is $\kappa(t) = |\kappa_N(t)|$. The sign of $\kappa_N(t)$ is positive if γ is curving toward Ω, and negative if it is curving away.

Theorem 9.3 (The Gauss–Bonnet Formula). *Let (M, g) be an oriented Riemannian 2-manifold. Suppose γ is a positively oriented curved polygon in M, and Ω is its interior. Then*

$$\int_\Omega K \, dA + \int_\gamma \kappa_N \, ds + \sum_{i=1}^{k} \varepsilon_i = 2\pi, \tag{9.6}$$

where K is the Gaussian curvature of g, dA is its Riemannian volume form, $\varepsilon_1, \ldots, \varepsilon_k$ are the exterior angles of γ, and the second integral is taken with respect to arc length (Problem 2-32).

Proof. Let $(a_0, \ldots, a_k)$ be an admissible partition of $[a, b]$, and let (x, y) be oriented smooth coordinates on an open set U containing $\overline{\Omega}$. Let $\theta : [a, b] \to \mathbb{R}$ be a tangent angle function for γ. Using the rotation index theorem and the fundamental theorem of calculus, we can write

$$2\pi = \theta(b) - \theta(a) = \sum_{i=1}^{k} \varepsilon_i + \sum_{i=1}^{k} \int_{a_{i-1}}^{a_i} \theta'(t)\,dt. \tag{9.7}$$

To prove (9.6), we need to derive a relationship among θ', κ_N, and K.

Let (E_1, E_2) be the oriented g-orthonormal frame obtained by applying the Gram–Schmidt algorithm to $(\partial/\partial x, \partial/\partial y)$ as before. Then by definition of θ and N, the following formulas hold at smooth points of γ:

$$\gamma'(t) = \cos\theta(t)E_1|_{\gamma(t)} + \sin\theta(t)E_2|_{\gamma(t)};$$
$$N(t) = -\sin\theta(t)E_1|_{\gamma(t)} + \cos\theta(t)E_2|_{\gamma(t)}.$$

Differentiating γ' (and omitting the t dependence from the notation for simplicity), we get

$$\begin{aligned} D_t\gamma' &= -(\sin\theta)\theta'E_1 + (\cos\theta)\nabla_{\gamma'}E_1 + (\cos\theta)\theta'E_2 + (\sin\theta)\nabla_{\gamma'}E_2 \\ &= \theta'N + (\cos\theta)\nabla_{\gamma'}E_1 + (\sin\theta)\nabla_{\gamma'}E_2. \end{aligned} \tag{9.8}$$

Next we analyze the covariant derivatives of E_1 and E_2. Because (E_1, E_2) is an orthonormal frame, for every vector v we have

$$0 = \nabla_v|E_1|^2 = 2\langle \nabla_v E_1, E_1\rangle,$$
$$0 = \nabla_v|E_2|^2 = 2\langle \nabla_v E_2, E_2\rangle,$$
$$0 = \nabla_v\langle E_1, E_2\rangle = \langle \nabla_v E_1, E_2\rangle + \langle E_1, \nabla_v E_2\rangle.$$

The first two equations show that $\nabla_v E_1$ is a multiple of E_2 and $\nabla_v E_2$ is a multiple of E_1. Define a 1-form ω by

$$\omega(v) = \langle E_1, \nabla_v E_2\rangle = -\langle \nabla_v E_1, E_2\rangle.$$

It follows that the covariant derivatives of the basis vectors are given by

$$\begin{aligned} \nabla_v E_1 &= -\omega(v)E_2; \\ \nabla_v E_2 &= \omega(v)E_1. \end{aligned} \tag{9.9}$$

Thus the 1-form ω completely determines the connection in U. (In fact, when the connection is expressed in terms of the local frame (E_1, E_2) as in Problem 4-14, this computation shows that the connection 1-forms are just $\omega_2{}^1 = -\omega_1{}^2 = \omega$, $\omega_1{}^1 = \omega_2{}^2 = 0$; but it is simpler in this case just to derive the result directly as we have done.)

Using (9.8) and (9.9), we can compute

$$\kappa_N = \langle D_t \gamma', N \rangle$$
$$= \langle \theta' N, N \rangle + \cos\theta \langle \nabla_{\gamma'} E_1, N \rangle + \sin\theta \langle \nabla_{\gamma'} E_2, N \rangle$$
$$= \theta' - \cos\theta \langle \omega(\gamma') E_2, N \rangle + \sin\theta \langle \omega(\gamma') E_1, N \rangle$$
$$= \theta' - \cos^2\theta \, \omega(\gamma') - \sin^2\theta \, \omega(\gamma')$$
$$= \theta' - \omega(\gamma').$$

Therefore, (9.7) becomes

$$2\pi = \sum_{i=1}^{k} \varepsilon_i + \sum_{i=1}^{k} \int_{a_{i-1}}^{a_i} \kappa_N(t) \, dt + \sum_{i=1}^{k} \int_{a_{i-1}}^{a_i} \omega(\gamma'(t)) \, dt$$
$$= \sum_{i=1}^{k} \varepsilon_i + \int_{\gamma} \kappa_N \, ds + \int_{\gamma} \omega.$$

The theorem will therefore be proved if we can show that

$$\int_{\gamma} \omega = \int_{\Omega} K \, dA. \tag{9.10}$$

Because Ω is a smooth manifold with corners (see [LeeSM, pp. 415–419]), we can apply Stokes's theorem and conclude that the left-hand side of (9.10) is equal to $\int_{\Omega} d\omega$. The last step of the proof is to show that $d\omega = K \, dA$. This follows from the general formula relating the curvature tensor and the connection 1-forms given in Problem 7-5; but in the case of two dimensions we can give an easy direct proof. Since (E_1, E_2) is an oriented orthonormal frame, it follows from Proposition 2.41 that $dA(E_1, E_2) = 1$. Using (9.9), we compute

$$K \, dA(E_1, E_2) = K = Rm(E_1, E_2, E_2, E_1)$$
$$= \langle \nabla_{E_1} \nabla_{E_2} E_2 - \nabla_{E_2} \nabla_{E_1} E_2 - \nabla_{[E_1, E_2]} E_2, \ E_1 \rangle$$
$$= \langle \nabla_{E_1} (\omega(E_2) E_1) - \nabla_{E_2}(\omega(E_1) E_1) - \omega([E_1, E_2]) E_1, \ E_1 \rangle$$
$$= \langle E_1(\omega(E_2)) E_1 + \omega(E_2) \nabla_{E_1} E_1 - E_2(\omega(E_1)) E_1$$
$$\qquad - \omega(E_1) \nabla_{E_2} E_1 - \omega([E_1, E_2]) E_1, \ E_1 \rangle$$
$$= E_1(\omega(E_2)) - E_2(\omega(E_1)) - \omega([E_1, E_2])$$
$$= d\omega(E_1, E_2).$$

This completes the proof.   □

The three local-to-global theorems of plane geometry stated in Chapter 1 follow from the Gauss–Bonnet formula as easy corollaries.

Corollary 9.4 (Angle-Sum Theorem). *The sum of the interior angles of a Euclidean triangle is π.*   □

Corollary 9.5 (Circumference Theorem). *The circumference of a Euclidean circle of radius R is $2\pi R$.*   □

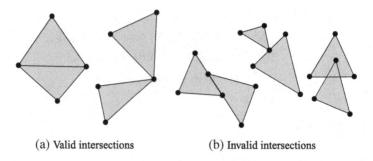

(a) Valid intersections   (b) Invalid intersections

Fig. 9.14: Valid and invalid intersections of triangles in a triangulation

Corollary 9.6 (Total Curvature Theorem). *If $\gamma \colon [a,b] \to \mathbb{R}^2$ is a smooth, unit-speed, simple closed curve such that $\gamma'(a) = \gamma'(b)$, and N is the inward-pointing normal, then*

$$\int_a^b \kappa_N(t)\,dt = 2\pi.$$

$\square$

The Gauss–Bonnet Theorem

It is now a relatively easy matter to "globalize" the Gauss–Bonnet formula to obtain the Gauss–Bonnet theorem. The link between the local and global results is provided by *triangulations*, so we begin by discussing this construction borrowed from algebraic topology.

Let M be a smooth, compact 2-manifold. A *curved triangle in M* is a curved polygon with exactly three edges and three vertices. A *smooth triangulation of M* is a finite collection of curved triangles with disjoint interiors such that the union of the triangles with their interiors is M, and the intersection of any pair of triangles (if not empty) is either a single vertex of each or a single edge of each (Fig. 9.14). Every smooth, compact surface possesses a smooth triangulation. In fact, it was proved by Tibor Radó [Rad25] in 1925 that every compact *topological* 2-manifold possesses a triangulation (without the assumption of smoothness of the edges, of course), in which every edge belongs to exactly two triangles. There is a proof for the smooth case that is not terribly hard, based on choosing geodesic triangles contained in convex geodesic balls (see Problem 9-5).

If M is a triangulated 2-manifold, the *Euler characteristic of M* (with respect to the given triangulation) is defined to be

$$\chi(M) = V - E + F,$$

where V is the number of vertices in the triangulation, E is the number of edges, and F is the number of faces (the interiors of the triangles). It is an important result of algebraic topology that the Euler characteristic is in fact a topological invariant, and is independent of the choice of triangulation (see [LeeTM, Cor. 10.25]), but we do not need that result here.

Theorem 9.7 (The Gauss–Bonnet Theorem). *If (M, g) is a smoothly triangulated compact Riemannian 2-manifold, then*

$$\int_M K \, dA = 2\pi \chi(M),$$

where K is the Gaussian curvature of g and dA is its Riemannian density.

Proof. We may as well assume that M is connected, because if not we can prove the theorem for each connected component and add up the results.

First consider the case in which M is orientable. In this case, we can choose an orientation for M, and then $\int_M K \, dA$ gives the same result whether we interpret dA as the Riemannian density or as the Riemannian volume form, so we will use the latter interpretation for the proof. Let $\{\Omega_i : i = 1, \ldots, F\}$ denote the faces of the triangulation, and for each i, let $\{\gamma_{ij} : j = 1, 2, 3\}$ be the edges of Ω_i and $\{\theta_{ij} : j = 1, 2, 3\}$ its interior angles. Since each exterior angle is π minus the corresponding interior angle, applying the Gauss–Bonnet formula to each triangle and summing over i gives

$$\sum_{i=1}^{F} \int_{\Omega_i} K \, dA + \sum_{i=1}^{F} \sum_{j=1}^{3} \int_{\gamma_{ij}} \kappa_N \, ds + \sum_{i=1}^{F} \sum_{j=1}^{3} (\pi - \theta_{ij}) = \sum_{i=1}^{F} 2\pi. \qquad (9.11)$$

Note that each edge integral appears exactly twice in the above sum, with opposite orientations, so the integrals of κ_N all cancel out. Thus (9.11) becomes

$$\int_M K \, dA + 3\pi F - \sum_{i=1}^{F} \sum_{j=1}^{3} \theta_{ij} = 2\pi F. \qquad (9.12)$$

Note also that each interior angle θ_{ij} appears exactly once. At each vertex, the angles that touch that vertex must have interior measures that add up to 2π (Fig. 9.15); thus the angle sum can be rearranged to give exactly $2\pi V$. Equation (9.12) thus can be written

$$\int_M K \, dA = 2\pi V - \pi F. \qquad (9.13)$$

Finally, since each edge appears in exactly two triangles, and each triangle has exactly three edges, the total number of edges *counted with multiplicity* is $2E = 3F$, where we count each edge once for each triangle in which it appears. This means that $F = 2E - 2F$, so (9.13) finally becomes

$$\int_M K \, dA = 2\pi V - 2\pi E + 2\pi F = 2\pi \chi(M).$$

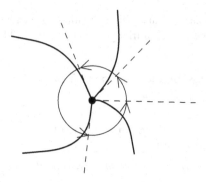

Fig. 9.15: Interior angles at a vertex add up to 2π

Now suppose M is nonorientable. Then Proposition B.18 shows that there is an orientable connected smooth manifold $\widehat{M}$ that admits a 2-sheeted smooth covering $\widehat{\pi} \colon \widehat{M} \to M$, and Exercise A.62 shows that $\widehat{M}$ is compact. If we endow $\widehat{M}$ with the metric $\widehat{g} = \widehat{\pi}^* g$, then the Riemannian density of $\widehat{g}$ is given by $\widehat{dA} = \widehat{\pi}^* dA$, and its Gaussian curvature is $\widehat{K} = \widehat{\pi}^* K$, so $\widehat{\pi}^*(K\,dA) = \widehat{K}\,\widehat{dA}$. The result of Problem 2-14 shows that $\int_{\widehat{M}} \widehat{K}\,\widehat{dA} = 2\int_M K\,dA$.

To compare Euler characteristics, we will show that the given triangulation of M "lifts" to a smooth triangulation of $\widehat{M}$. To see this, let γ be any curved triangle in M and let Ω be its interior. By definition, this means that there exists a smooth chart (U, φ) whose domain contains $\overline{\Omega}$ and whose image is a disk $D \subseteq \mathbb{R}^2$, and such that $\varphi(\overline{\Omega}) = \overline{\Omega}_0$, where Ω_0 is the interior of a curved triangle γ_0 in $\mathbb{R}^2$. Then φ^{-1} is an embedding of D into M, which restricts to a diffeomorphism $F \colon \overline{\Omega}_0 \to \overline{\Omega}$. Because D is simply connected, it follows from Corollary A.57 that φ^{-1} (and therefore also F) has a lift to $\widehat{M}$, which is smooth because $\widehat{\pi}$ is a local diffeomorphism; and because the covering is two-sheeted, there are exactly two such lifts F_1, F_2. Each lift is injective because $\widehat{\pi} \circ F_i = F$, which is injective, and their images are disjoint because if two lifts agree at a point, they must be identical. From this it is straightforward to verify that the lifted curved triangles form a triangulation of $\widehat{M}$ with twice as many vertices, edges, and faces as that of M, and thus $\chi(\widehat{M}) = 2\chi(M)$. Substituting these relations into the Gauss–Bonnet theorem for $\widehat{M}$ and dividing through by 2, we obtain the analogous relation for M. $\square$

The significance of this theorem cannot be overstated. Together with the classification theorem for compact surfaces, it gives us very detailed information about the possible Gaussian curvatures for metrics on compact surfaces. The classification theorem [LeeTM, Thms. 6.15 and 10.22] says that every compact, connected, orientable 2-manifold M is homeomorphic to a sphere or a connected sum of n tori, and every nonorientable one is homeomorphic to a connected sum of n copies of the real projective plane $\mathbb{RP}^2$; the number n is called the ***genus of M***. (The sphere is said to have genus zero.) By constructing simple triangulations, one can show that the Euler characteristic of an orientable surface of genus n is $2 - 2n$, and that of a nonorientable one is $2 - n$. The following corollary follows immediately from the Gauss–Bonnet theorem.

Corollary 9.8 *Let* (M, g) *be a compact Riemannian 2-manifold and let* K *be its Gaussian curvature.*

(a) *If* M *is homeomorphic to the sphere or the projective plane, then* $K > 0$ *somewhere.*

(b) *If* M *is homeomorphic to the torus or the Klein bottle, then either* $K \equiv 0$ *or* K *takes on both positive and negative values.*

(c) *If* M *is any other compact surface, then* $K < 0$ *somewhere.* ☐

This corollary has a remarkable converse, proved in the mid-1970s by Jerry Kazdan and Frank Warner: *If* K *is any smooth function on a compact 2-manifold* M *satisfying the necessary sign condition of Corollary 9.8, then there exists a Riemannian metric on* M *for which* K *is the Gaussian curvature.* The proof is a deep application of the theory of nonlinear partial differential equations. (See [Kaz85] for a nice expository account.)

In Corollary 9.8 we assumed we knew the topology of M and drew conclusions about the possible curvatures it could support. In the following corollary we reverse our point of view, and use assumptions about the curvature to draw conclusions about the topology of the manifold.

Corollary 9.9 *Let* (M, g) *be a compact Riemannian 2-manifold and* K *its Gaussian curvature.*

(a) *If* $K > 0$ *everywhere on* M, *then the universal covering manifold of* M *is homeomorphic to* $\mathbb{S}^2$, *and* $\pi_1(M)$ *is either trivial or isomorphic to the two-element group* $\mathbb{Z}/2$.

(b) *If* $K \leq 0$ *everywhere on* M, *then the universal covering manifold of* M *is homeomorphic to* $\mathbb{R}^2$, *and* $\pi_1(M)$ *is infinite.*

Proof. Suppose first that M has positive Gaussian curvature. From the Gauss–Bonnet theorem, M has positive Euler characteristic. The classification theorem for compact surfaces shows that the only such surfaces are the sphere (with trivial fundamental group) and the projective plane (with fundamental group isomorphic to $\mathbb{Z}/2$), both of which are covered by the sphere.

On the other hand, suppose M has nonpositive Gaussian curvature. Then its Euler characteristic is nonpositive, so it is either an orientable surface of genus $n \geq 1$ or a nonorientable one of genus $n \geq 2$. Thus the universal covering space of M is $\mathbb{R}^2$ if M is the torus or the Klein bottle, and $\mathbb{B}^2$ in all other cases (see [LeeTM, Thm. 12.29]), both of which are homeomorphic to $\mathbb{R}^2$. The fact that the universal covering space is noncompact implies that the universal covering map has infinitely many sheets by the result of Exercise A.62, and then Proposition A.61 shows that $\pi_1(M)$ is infinite. ☐

Much of the effort in contemporary Riemannian geometry is aimed at generalizing the Gauss–Bonnet theorem and its topological consequences to higher dimensions. As we will see in the next few chapters, most of the interesting results have required the development of different methods.

However, there is one rather direct generalization of the Gauss–Bonnet theorem that deserves mention. For a compact manifold M of any dimension, the **Euler characteristic of** M, denoted by $\chi(M)$, can be defined analogously to that of a surface and is a topological invariant (see [LeeTM, Thm. 13.36]). It turns out that it is always zero for an odd-dimensional compact manifold, but it is a nontrivial invariant in each even-dimensional case.

The **Chern–Gauss–Bonnet theorem** equates the Euler characteristic of an even-dimensional compact oriented Riemannian manifold to a certain curvature integral. Versions of this theorem were proved by Heinz Hopf in 1925 for an embedded Riemannian hypersurface in Euclidean space, and independently by Carl Allendoerfer and Werner Fenchel in 1940 for an embedded Riemannian submanifold of any Euclidean space (well before Nash's 1956 proof that every Riemannian manifold has such an embedding). Finally, an intrinsic proof for the general case was discovered by Shiing-Shen Chern in 1944 (see [Spi79, Vol. 5] for a complete discussion with references). The theorem asserts that for each $2n$-dimensional oriented inner product space V, there exists a basis-independent function

$$\mathrm{Pf} \colon \mathscr{R}(V^*) \to \Lambda^{2n}(V^*),$$

called the **Pfaffian**, with the property that for every oriented compact Riemannian $2n$-manifold M,

$$\int_M \mathrm{Pf}(Rm) = (2\pi)^n \chi(M).$$

(Depending on how the Pfaffian is defined, you will see different choices of normalization constants on the right-hand side of this equation.) For example, in four dimensions, the theorem can be written in terms of familiar curvature quantities as follows:

$$\int_M \left(|Rm|^2 - 4|Rc|^2 + S^2 \right) dV_g = 32\pi^2 \chi(M). \tag{9.14}$$

In a certain sense, this might be considered a very satisfactory generalization of Gauss–Bonnet. The only problem with this result is that the relationship between the Pfaffian and sectional curvatures is obscure in higher dimensions, so it is very hard to interpret the theorem geometrically. For example, after he proved the version of the theorem for Euclidean hypersurfaces, Hopf conjectured in the 1930s that every compact even-dimensional manifold that admits a metric with strictly positive sectional curvature must have positive Euler characteristic; to date, the conjecture is known to be true in dimensions 2 and 4, but it is still open in higher dimensions (see [Pet16, p. 320]).

Problems

9-1. Let (M, g) be an oriented Riemannian 2-manifold with nonpositive Gaussian curvature everywhere. Prove that there are no geodesic polygons with exactly 0, 1, or 2 ordinary vertices. Give examples of all three if the curvature hypothesis is not satisfied.

9-2. Let (M, g) be a Riemannian 2-manifold. If γ is a geodesic polygon in M with n vertices, the **angle excess of γ** is defined as

$$E(\gamma) = \left(\sum_{i=1}^{n} \theta_i \right) - (n-2)\pi,$$

where $\theta_1, \ldots, \theta_n$ are the interior angles of γ. Show that if M has constant Gaussian curvature K, then every geodesic polygon has angle excess equal to K times the area of the region bounded by the polygon.

9-3. Given $h \in (-R, R)$, let C_h be the circle in $\mathbb{S}^2(R) \subseteq \mathbb{R}^3$ where $z = h$ (where we label the standard coordinates in $\mathbb{R}^3$ as (x, y, z)), and let Ω be the subset of $\mathbb{S}^2(R)$ where $z > h$. Compute the signed curvature of C_h and verify the Gauss–Bonnet formula in this case.

9-4. Let $T \subseteq \mathbb{R}^3$ be the torus of revolution obtained by revolving the circle $(r-2)^2 + z^2 = 1$ around the z-axis (see p. 19). Compute the Gaussian curvature of T and verify the Gauss–Bonnet theorem in this case.

9-5. This problem outlines a proof that every compact smooth 2-manifold has a smooth triangulation.

    (a) Show that it suffices to prove that there exist finitely many convex geodesic polygons whose interiors cover M, and each of which lies in a uniformly normal convex geodesic ball. (A curved polygon is called convex if the union of the polygon and its interior is a geodesically convex subset of M.)

    (b) Using Theorem 6.17, show that there exist finitely many points $v_1, \ldots, v_k$ and a positive number ε such that the geodesic balls $B_{3\varepsilon}(v_i)$ are geodesically convex and uniformly normal, and the balls $B_\varepsilon(v_i)$ cover M.

    (c) For each i, show that there is a convex geodesic polygon in $B_{3\varepsilon}(v_i)$ whose interior contains $B_\varepsilon(v_i)$. [Hint: Let the vertices be sufficiently nearby points on the circle of radius 2ε around v_i.]

    (d) Prove the result.

(Used on p. 276.)

9-6. Let $M \subseteq \mathbb{R}^3$ be a compact, embedded, 2-dimensional Riemannian submanifold. Show that M cannot have $K \leq 0$ everywhere. [Hint: Look at a point where the distance from the origin takes a maximum.]

9-7. Suppose M is either the 2-sphere of radius R or the hyperbolic plane of radius R for some $R > 0$. Show that similar triangles in M are congruent. More precisely, if γ_1 and γ_2 are geodesic triangles in M such that corresponding side lengths are proportional and corresponding interior angles are equal, then there exists an isometry of M taking γ_1 to γ_2.

9-8. Use the Gauss–Bonnet theorem to prove that every compact connected Lie group of dimension 2 is isomorphic to the direct product group $\mathbb{S}^1 \times \mathbb{S}^1$. [Hint: See Problem 8-17.]

9-9. (a) Show that there is an upper bound for the areas of geodesic triangles in the hyperbolic plane $\mathbb{H}^2(R)$, and compute the least upper bound.

 (b) Two distinct maximal geodesics in the hyperbolic plane $\mathbb{H}^2$ are said to be **asymptotically parallel** if they have unit-speed parametrizations $\gamma_1, \gamma_2 : \mathbb{R} \to \mathbb{H}^2$ such that $d_{\breve{g}}(\gamma_1(t), \gamma_2(2))$ remains bounded as $t \to +\infty$ or as $t \to -\infty$. An **ideal triangle** in $\mathbb{H}^2$ is a region whose boundary consists of three distinct maximal geodesics, any two of which are asymptotically parallel to each other. Show that all ideal triangles have the same finite area, and compute it. Be careful to justify any limits.

9-10. THE GAUSS–BONNET THEOREM FOR SURFACES WITH BOUNDARY: Suppose (M, g) is a compact Riemannian 2-manifold with boundary, endowed with a smooth triangulation such that the intersection of each curved triangle with ∂M, if not empty, is either a single vertex or a single edge. Then

$$\int_M K \, dA + \int_{\partial M} \kappa_N \, ds = 2\pi \chi(M),$$

where κ_N is the signed geodesic curvature of ∂M with respect to the inward-pointing normal N.

9-11. Suppose g is a Riemannian metric on the cylinder $\mathbb{S}^1 \times [0, 1]$ such that both boundary curves are totally geodesic. Prove that the Gaussian curvature of g either is identically zero or attains both positive and negative values. Give examples of both possibilities.

9-12. Prove the plane curve classification theorem (Theorem 1.5). [Hint: Show that every smooth unit-speed plane curve $\gamma(t) = (x(t), y(t))$ satisfies the second-order ODE $\gamma''(t) = \kappa_N(t) N(t)$, where N is the unit normal vector field given by $N(t) = (-y'(t), x'(t))$.] (*Used on p. 4.*)

9-13. Use the four-dimensional Chern–Gauss–Bonnet formula (9.14) to prove that a compact 4-dimensional Einstein manifold must have positive Euler characteristic unless it is flat.

Chapter 10
Jacobi Fields

Our goal for the remainder of this book is to generalize to higher dimensions some of the geometric and topological consequences of the Gauss–Bonnet theorem. We need to develop a new approach: instead of using Stokes's theorem and differential forms to relate the curvature to global topology as in the proof of the Gauss–Bonnet theorem, we study the way that curvature affects the behavior of nearby geodesics. Roughly speaking, positive curvature causes nearby geodesics to converge, while negative curvature causes them to spread out (Fig. 10.1). In order to draw topological consequences from this fact, we need a quantitative way to measure the effect of curvature on a one-parameter family of geodesics.

We begin by deriving the *Jacobi equation*, which is an ordinary differential equation satisfied by the variation field of any one-parameter family of geodesics. A vector field satisfying this equation along a geodesic is called a *Jacobi field*. We then introduce *conjugate points*, which are pairs of points along a geodesic where some nontrivial Jacobi field vanishes. Intuitively, if p and q are conjugate along a geodesic, one expects to find a one-parameter family of geodesic segments that start at p and end (almost) at q.

After defining conjugate points, we prove a simple but essential fact: the points conjugate to p are exactly the points where $\exp_p$ fails to be a local diffeomorphism. We then derive an expression for the second derivative of the length functional with respect to proper variations of a geodesic, called the *second variation formula*. Using this formula, we prove another essential fact about conjugate points: once a geodesic passes its first conjugate point, it is no longer minimizing. The converse of this statement, however, is untrue: a geodesic can cease to be minimizing before it reaches its first conjugate point. In the last section of the chapter, we study the set of points where geodesics starting at a given point p cease to minimize, called the *cut locus of p*.

In the next two chapters, we will derive geometric and topological consequences of these facts.

© Springer International Publishing AG 2018
J. M. Lee, *Introduction to Riemannian Manifolds*, Graduate Texts
in Mathematics 176, https://doi.org/10.1007/978-3-319-91755-9_10

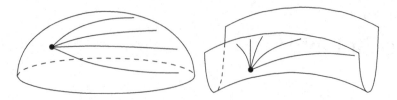

Fig. 10.1: Geodesics converge in positive curvature, and spread out in negative curvature

The Jacobi Equation

Let (M, g) be an n-dimensional Riemannian or pseudo-Riemannian manifold. In order to study the effect of curvature on nearby geodesics, we focus on variations through geodesics. Suppose, therefore, that $I, K \subseteq \mathbb{R}$ are intervals, $\gamma \colon I \to M$ is a geodesic, and $\Gamma \colon K \times I \to M$ is a variation of γ (as defined in Chapter 6). We say that Γ is a **variation through geodesics** if each of the main curves $\Gamma_s(t) = \Gamma(s, t)$ is also a geodesic. (In particular, this requires that Γ be smooth.) Our first goal is to derive an equation that must be satisfied by the variation field of a variation through geodesics.

Theorem 10.1 (The Jacobi Equation). *Let (M, g) be a Riemannian or pseudo-Riemannian manifold, let γ be a geodesic in M, and let J be a vector field along γ. If J is the variation field of a variation through geodesics, then J satisfies the following equation, called the **Jacobi equation**:*

$$D_t^2 J + R(J, \gamma')\gamma' = 0. \tag{10.1}$$

Proof. Write $T(s, t) = \partial_t \Gamma(s, t)$ and $S(s, t) = \partial_s \Gamma(s, t)$ as in Chapter 6. The geodesic equation tells us that

$$D_t T \equiv 0$$

for all (s, t). We can take the covariant derivative of this equation with respect to s, yielding

$$D_s D_t T \equiv 0.$$

Using Proposition 7.5 to commute the covariant derivatives along Γ, we compute

$$\begin{aligned}
0 &= D_s D_t T \\
&= D_t D_s T + R(S, T)T \\
&= D_t D_t S + R(S, T)T,
\end{aligned}$$

where the last step follows from the symmetry lemma. Evaluating at $s = 0$, where $S(0, t) = J(t)$ and $T(0, t) = \gamma'(t)$, we get (10.1). $\square$

A smooth vector field along a geodesic that satisfies the Jacobi equation is called a *Jacobi field*. As the following proposition shows, the Jacobi equation can be written as a system of second-order linear ordinary differential equations, so it has a unique solution given initial values for J and $D_t J$ at one point.

Proposition 10.2 (Existence and Uniqueness of Jacobi Fields). *Let (M,g) be a Riemannian or pseudo-Riemannian manifold. Suppose $I \subseteq \mathbb{R}$ is an interval, $\gamma: I \to M$ is a geodesic, $a \in I$, and $p = \gamma(a)$. For every pair of vectors $v, w \in T_p M$, there is a unique Jacobi field J along γ satisfying the initial conditions*

$$J(a) = v, \qquad D_t J(a) = w.$$

Proof. Choose a parallel orthonormal frame (E_i) along γ, and write $v = v^i E_i(a)$, $w = w^i E_i(a)$, and $\gamma'(t) = y^i(t) E_i(t)$ in terms of this frame. Writing an unknown vector field $J \in \mathfrak{X}(\gamma)$ as $J(t) = J^i(t) E_i(t)$, we can express the Jacobi equation as

$$\ddot{J}^i(t) + R_{jkl}{}^i(\gamma(t)) J^j(t) y^k(t) y^l(t) = 0.$$

This is a system of n linear second-order ODEs for the n functions $J^i: I \to \mathbb{R}$. Making the substitution $W^i = \dot{J}^i$ converts it to the following equivalent first-order linear system for the $2n$ unknown functions (J^i, W^i):

$$\dot{J}^i(t) = W^i(t),$$
$$\dot{W}^i(t) = -R_{jkl}{}^i(\gamma(t)) J^j(t) y^k(t) y^l(t).$$

Then Theorem 4.31 guarantees the existence and uniqueness of a smooth solution on the whole interval I with arbitrary initial conditions $J^i(a) = v^i$, $W^i(a) = w^i$. Since $D_t J(a) = \dot{J}^i(a) E_i(a) = W^i(a) E_i(a) = w$, it follows that $J(t) = J^i(t) E_i(t)$ is the desired Jacobi field. $\qquad\square$

Given a geodesic γ, let $\mathcal{J}(\gamma) \subseteq \mathfrak{X}(\gamma)$ denote the set of Jacobi fields along γ.

Corollary 10.3. *Suppose (M,g) is a Riemannian or pseudo-Riemannian manifold of dimension n, and γ is any geodesic in M. Then $\mathcal{J}(\gamma)$ is a $2n$-dimensional linear subspace of $\mathfrak{X}(\gamma)$.*

Proof. Because the Jacobi equation is linear, $\mathcal{J}(\gamma)$ is a linear subspace of $\mathfrak{X}(\gamma)$. Let $p = \gamma(a)$ be any point on γ, and consider the linear map from $\mathcal{J}(\gamma)$ to $T_p M \oplus T_p M$ by sending J to $(J(a), D_t J(a))$. The preceding proposition says precisely that this map is bijective. $\qquad\square$

The following proposition is a converse to Theorem 10.1; it shows that each Jacobi field along a geodesic segment tells us how some family of geodesics behaves, at least to first order along γ.

Proposition 10.4. *Let (M,g) be a Riemannian or pseudo-Riemannian manifold, and let $\gamma: I \to M$ be a geodesic. If M is complete or I is a compact interval, then every Jacobi field along γ is the variation field of a variation of γ through geodesics.*

Proof. Let J be a Jacobi field along γ. After applying a translation in t (which does not affect either the fact that γ is a geodesic or the fact that J is a Jacobi field), we can assume that the interval I contains 0, and write $p = \gamma(0)$ and $v = \gamma'(0)$. Note that this implies $\gamma(t) = \exp_p(tv)$ for all $t \in I$.

Choose a smooth curve $\sigma: (-\varepsilon, \varepsilon) \to M$ and a smooth vector field V along σ satisfying

$$\sigma(0) = p, \qquad\qquad\qquad V(0) = v,$$
$$\sigma'(0) = J(0), \qquad\qquad D_s V(0) = D_t J(0),$$

where D_s and D_t denote covariant differentiation along σ and γ, respectively. (They are easily constructed in local coordinates around p.) We wish to define a variation of γ by setting

$$\Gamma(s,t) = \exp_{\sigma(s)}\big(t V(s)\big). \tag{10.2}$$

If M is geodesically complete, this is defined for all $(s,t) \in (-\varepsilon, \varepsilon) \times I$. On the other hand, if I is compact, the fact that the domain of the exponential map is an open subset of TM that contains the compact set $\{(p, tv) : t \in I\}$ guarantees that there is some $\delta > 0$ such that $\Gamma(s,t)$ is defined for all $(s,t) \in (-\delta, \delta) \times I$.

Note that

$$\Gamma(0,t) = \exp_{\sigma(0)}\big(t V(0)\big) = \exp_p(tv) = \gamma(t), \tag{10.3}$$
$$\Gamma(s,0) = \exp_{\sigma(s)}(0) = \sigma(s). \tag{10.4}$$

In particular, (10.3) shows that Γ is a variation of γ. The properties of the exponential map guarantee that Γ is a variation through geodesics, and therefore its variation field $W(t) = \partial_s \Gamma(0,t)$ is a Jacobi field along γ.

Now, (10.4) implies

$$W(0) = \frac{\partial}{\partial s}\bigg|_{s=0} \Gamma(s,0) = \sigma'(0) = J(0).$$

If we can show that $D_t W(0) = D_t J(0)$ as well, it then follows from the uniqueness of Jacobi fields that $W \equiv J$, and the proposition is proved.

Formula (10.2) shows that each main curve $\Gamma_s(t)$ is a geodesic whose initial velocity is $V(s)$, so

$$\partial_t \Gamma(s,0) = \frac{\partial}{\partial t}\bigg|_{t=0} \Gamma_s(t) = V(s).$$

It follows from the symmetry lemma that $D_t \partial_s \Gamma = D_s \partial_t \Gamma$, and our choice of V gives

$$D_t W(0) = D_t \partial_s \Gamma(0,0) = D_s \partial_t \Gamma(0,0) = D_s V(0) = D_t J(0).$$

It follows that $W \equiv J$, as claimed. $\square$

Fig. 10.2: Tangential Jacobi fields

Proposition 10.5 (Local Isometry Invariance of Jacobi Fields). *Suppose* (M, g) *and* $\big(\widetilde{M}, \widetilde{g}\big)$ *are Riemannian or pseudo-Riemannian manifolds and* $\varphi \colon M \to \widetilde{M}$ *is a local isometry. Let* $\gamma \colon I \to M$ *and* $\widetilde{\gamma} \colon I \to \widetilde{M}$ *be geodesics related by* $\widetilde{\gamma} = \varphi \circ \gamma$, *and let* $J \in \mathfrak{X}(\gamma)$, $\widetilde{J} \in \mathfrak{X}(\widetilde{\gamma})$ *be related by* $d\varphi_{\gamma(t)}(J(t)) = \widetilde{J}(t)$ *for all* $t \in I$. *Then* J *is a Jacobi field if and only if* $\widetilde{J}$ *is.*

▶ **Exercise 10.6.** Prove the preceding proposition.

Basic Computations with Jacobi Fields

There are various situations in which Jacobi fields can be computed explicitly. We begin by describing the most important of these.

Tangential and Normal Jacobi Fields

Along every geodesic $\gamma \colon I \to M$, there are always two Jacobi fields that we can write down immediately (see Fig. 10.2). Because $D_t \gamma' \equiv 0$ and $R(\gamma', \gamma')\gamma' \equiv 0$ by antisymmetry of R, the vector fields $J_0(t) = \gamma'(t)$ and $J_1(t) = t\gamma'(t)$ both satisfy the Jacobi equation by direct computation. If I is compact or M is complete, the vector field J_0 is the variation field of the variation $\Gamma(s,t) = \gamma(s+t)$, while J_1 is the variation field of $\Gamma(s,t) = \gamma((1+s)t)$. Therefore, these two Jacobi fields just reflect the possible reparametrizations of γ, and do not tell us anything about the behavior of geodesics other than γ itself.

To distinguish these trivial cases from more informative ones, we make the following definitions. Given a regular curve $\gamma \colon I \to M$, for each $t \in I$ we let $T^{\top}_{\gamma(t)} M \subseteq T_{\gamma(t)} M$ denote the one-dimensional subspace spanned by $\gamma'(t)$, and $T^{\perp}_{\gamma(t)} M$ its orthogonal complement. A **tangential vector field along** γ is a vector field $V \in \mathfrak{X}(\gamma)$ such that $V(t) \in T^{\top}_{\gamma(t)} M$ for all t, and a **normal vector field along** γ is one such that $V(t) \in T^{\perp}_{\gamma(t)} M$ for all t. Thus a **normal Jacobi field along** γ is a Jacobi field J satisfying $J(t) \perp \gamma'(t)$ for all t. Let $\mathfrak{X}^{\perp}(\gamma)$ and $\mathfrak{X}^{\top}(\gamma)$ denote the spaces of smooth normal and tangential vector fields along γ, respectively. When γ is a geodesic, $\mathcal{J}^{\perp}(\gamma)$ and $\mathcal{J}^{\top}(\gamma)$ denote the spaces of normal and tangential Jacobi fields along γ, respectively.

Proposition 10.7. *Let (M, g) be a Riemannian or pseudo-Riemannian manifold. Suppose $\gamma: I \to M$ is a geodesic and J is a Jacobi field along γ. Then the following are equivalent:*

(a) *J is a normal Jacobi field.*
(b) *J is orthogonal to γ' at two distinct points.*
(c) *Both J and $D_t J$ are orthogonal to γ' at one point.*
(d) *Both J and $D_t J$ are orthogonal to γ' everywhere along γ.*

Proof. Define a function $f: I \to \mathbb{R}$ by $f(t) = \langle J(t), \gamma'(t) \rangle$, so that $f(t) = 0$ if and only if $J(t) \perp \gamma'(t)$. Using compatibility with the metric and the fact that $D_t \gamma' \equiv 0$, we compute

$$
\begin{aligned}
f'' &= \langle D_t^2 J, \gamma' \rangle \\
&= -\langle R(J, \gamma') \gamma', \gamma' \rangle \\
&= -Rm(J, \gamma', \gamma', \gamma') = 0
\end{aligned}
$$

by the symmetries of the curvature tensor. Thus, by elementary calculus, f is an affine function of t.

Note that $f'(t) = \langle D_t J(t), \gamma'(t) \rangle$, which vanishes at t if and only if $D_t J(t) \perp \gamma'(t)$. It follows that $J(a)$ and $D_t J(a)$ are orthogonal to $\gamma'(a)$ for some $a \in I$ if and only if f and its first derivative vanish at a, which happens if and only if $f \equiv 0$. Similarly, J is orthogonal to γ' at two points if and only if f vanishes at two points, which happens if and only if f is identically zero. If this is the case, then $f' \equiv 0$ as well, which implies that both J and $D_t J$ are orthogonal to γ' for all t. $\square$

Corollary 10.8. *Suppose (M, g) is a Riemannian or pseudo-Riemannian n-manifold and $\gamma: I \to M$ is any nonconstant geodesic. Then $\mathcal{J}^\perp(\gamma)$ is a $(2n-2)$-dimensional subspace of $\mathcal{J}(\gamma)$, and $\mathcal{J}^\top(\gamma)$ is a 2-dimensional subspace. Every Jacobi field can be uniquely decomposed as a sum of a tangential Jacobi field plus a normal Jacobi field.*

Proof. As we noted in the proof of Corollary 10.3, for every $a \in I$, the map from $\mathcal{J}(\gamma)$ to $T_{\gamma(a)}M \oplus T_{\gamma(a)}M$ given by $J \mapsto (J(a), D_t J(a))$ is an isomorphism, and Proposition 10.7 shows that $\mathcal{J}^\perp(\gamma)$ is exactly the preimage of the subspace consisting of all pairs $(v, w) \in T_{\gamma(a)}M \oplus T_{\gamma(a)}M$ such that $\langle v, \gamma'(a) \rangle = \langle w, \gamma'(a) \rangle = 0$. Because this subspace has dimension $2n - 2$, it follows that $\mathcal{J}^\perp(\gamma)$ has the same dimension.

On the other hand, $\mathcal{J}^\top(\gamma)$ contains $J_0(t) = \gamma'(t)$ and $J_1(t) = t\gamma'(t)$, which are linearly independent over $\mathbb{R}$ because $\gamma'(t)$ never vanishes, so it is a subspace of dimension at least 2. Because $\mathcal{J}^\perp(\gamma) \cap \mathcal{J}^\top(\gamma) = \{0\}$, the dimension of $\mathcal{J}^\top(\gamma)$ must be exactly 2, and we have a direct sum decomposition $\mathcal{J}(\gamma) = \mathcal{J}^\perp(\gamma) \oplus \mathcal{J}^\top(\gamma)$. This implies that every $J \in \mathcal{J}(\gamma)$ has a unique decomposition $J = J^\perp + J^\top$, with $J^\perp \in \mathcal{J}^\perp(\gamma)$ and $J^\top \in \mathcal{J}^\top(\gamma)$. $\square$

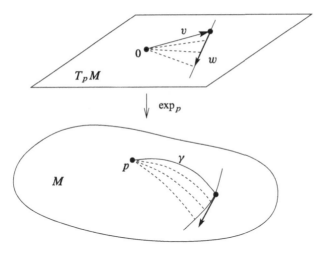

Fig. 10.3: The variation of Lemma 10.9

Jacobi Fields Vanishing at a Point

For many purposes, we will be primarily interested in Jacobi fields that vanish at a particular point. For these, there are some useful explicit formulas.

Lemma 10.9. *Let (M, g) be a Riemannian or pseudo-Riemannian manifold, $I \subseteq \mathbb{R}$ an interval containing 0, and $\gamma : I \to M$ a geodesic. Suppose $J : I \to M$ is a Jacobi field such that $J(0) = 0$. If M is geodesically complete or I is compact, then J is the variation field of the following variation of γ through geodesics (Fig. 10.3):*

$$\Gamma(s,t) = \exp_p\big(t(v + sw)\big), \tag{10.5}$$

where $p = \gamma(0)$, $v = \gamma'(0)$, and $w = D_t J(0)$.

Proof. The proof of Proposition 10.4 showed that J is the variation field of a variation Γ of the form (10.2), with σ any smooth curve satisfying $\sigma(0) = p$ and $\sigma'(0) = 0$, and V a smooth vector field along σ with $V(0) = v$ and $D_s V(0) = w$. In this case, we can take $\sigma(s) \equiv p$ and $V(s) = v + sw \in T_p M$, yielding (10.5). $\square$

This result leads to some explicit formulas for all of the Jacobi fields vanishing at a point.

Proposition 10.10 (Jacobi Fields Vanishing at a Point). *Let (M, g) be a Riemannian or pseudo-Riemannian n-manifold and $p \in M$. Suppose $\gamma : I \to M$ is a geodesic such that $0 \in I$ and $\gamma(0) = p$. For every $w \in T_p M$, the Jacobi field J along γ such that $J(0) = 0$ and $D_t J(0) = w$ is given by*

$$J(t) = d\big(\exp_p\big)_{tv}(tw), \tag{10.6}$$

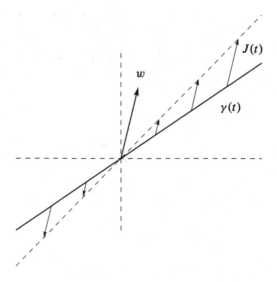

Fig. 10.4: A Jacobi field in normal coordinates

where $v = \gamma'(0)$, and we regard tw as an element of $T_{tv}(T_pM)$ by means of the canonical identification $T_{tv}(T_pM) \cong T_pM$. If (x^i) are normal coordinates on a normal neighborhood of p containing the image of γ, then J is given by the formula

$$J(t) = tw^i \partial_i\big|_{\gamma(t)}, \tag{10.7}$$

where $w^i \partial_i\big|_0$ is the coordinate representation of w.

Proof. Under the given hypotheses, Lemma 10.9 showed that the restriction of J to any compact interval containing 0 is the variation field of a variation Γ through geodesics of the form (10.5). Using the chain rule to compute $J(t) = \partial_s \Gamma(0,t)$, we arrive at (10.6). Because every t in the domain of γ is contained in some such compact interval, the formula holds for all such t.

In normal coordinates, the coordinate representation of the exponential map is the identity, so Γ can be written explicitly in coordinates as

$$\Gamma(s,t) = \big(t(v^1 + sw^1), \dots, t(v^n + sw^n)\big).$$

(See Fig. 10.4.) Differentiating $\Gamma(s,t)$ with respect to s and setting $s = 0$ shows that its variation field J is given by (10.7). □

Corollary 10.11. *Suppose (M,g) is a Riemannian or pseudo-Riemannian manifold and U is a normal neighborhood of $p \in M$. For each $q \in U \smallsetminus \{p\}$, every vector in T_qM is the value of a Jacobi field J along a radial geodesic such that J vanishes at p.*

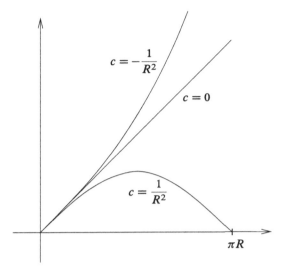

Fig. 10.5: The graph of s_c

Proof. Let (x^i) be normal coordinates on U. Given $q = (q^1, \ldots, q^n) \in U \smallsetminus \{p\}$ and $w = w^i \partial_i|_q \in T_q M$, the curve $\gamma(t) = (tq^1, \ldots, tq^n)$ is a radial geodesic satisfying $\gamma(0) = p$ and $\gamma(1) = q$. The previous proposition showed that $J(t) = t w^i \partial_i|_{\gamma(t)}$ is a Jacobi field along γ. Because $J(0) = 0$ and $J(1) = w$, the result follows. $\square$

Jacobi Fields in Constant-Curvature Spaces

For metrics with constant sectional curvature, we have a different kind of explicit formula for Jacobi fields—this one expresses a Jacobi field as a scalar multiple of a parallel vector field. To handle the various cases concisely, for each $c \in \mathbb{R}$, let us define a function $s_c : \mathbb{R} \to \mathbb{R}$ (Fig. 10.5) by

$$
s_c(t) = \begin{cases}
t, & \text{if } c = 0; \\[2mm]
R \sin \dfrac{t}{R}, & \text{if } c = \dfrac{1}{R^2} > 0; \\[2mm]
R \sinh \dfrac{t}{R}, & \text{if } c = -\dfrac{1}{R^2} < 0.
\end{cases}
\tag{10.8}
$$

Proposition 10.12 (Jacobi Fields in Constant Curvature). *Suppose (M, g) is a Riemannian manifold with constant sectional curvature c, and γ is a unit-speed geodesic in M. The normal Jacobi fields along γ vanishing at $t = 0$ are the vector fields of the form*

$$
J(t) = k s_c(t) E(t),
\tag{10.9}
$$

where E is any parallel unit normal vector field along γ, k is an arbitrary constant, and s_c is defined by (10.8). The initial derivative of such a Jacobi field is

$$D_t J(0) = k E(0), \tag{10.10}$$

and its norm is

$$|J(t)| = |s_c(t)||D_t J(0)|. \tag{10.11}$$

Proof. Since g has constant curvature, its curvature endomorphism is given by the formula of Proposition 8.36:

$$R(v, w)x = c(\langle w, x\rangle v - \langle v, x\rangle w).$$

Substituting this into the Jacobi equation, we find that a normal Jacobi field J satisfies

$$\begin{aligned}
0 &= D_t^2 J + c(\langle\gamma', \gamma'\rangle J - \langle J, \gamma'\rangle\gamma') \\
&= D_t^2 J + cJ,
\end{aligned} \tag{10.12}$$

where we have used the facts that $|\gamma'|^2 = 1$ and $\langle J, \gamma'\rangle = 0$.

Since (10.12) says that the second covariant derivative of J is a multiple of J itself, it is reasonable to try to construct a solution by choosing an arbitrary parallel unit normal vector field E along γ and setting $J(t) = u(t)E(t)$ for some function u to be determined. Plugging this into (10.12), we find that J is a Jacobi field if and only if u is a solution to the differential equation

$$u''(t) + cu(t) = 0.$$

It is an easy matter to solve this ODE explicitly. In particular, the solutions satisfying $u(0) = 0$ are constant multiples of s_c. This construction yields all the normal Jacobi fields vanishing at 0, since there is an $(n-1)$-dimensional space of them, and the space of parallel normal vector fields has the same dimension.

To prove the last two statements, suppose J is given by (10.9), and compute

$$D_t J(0) = k s_c'(0) E(0) = k E(0),$$

since $s_c'(0) = 1$ in every case. Because E is a unit vector field, $|D_t J(0)| = |k|$, and (10.11) follows. $\qquad\square$

Here is our first significant application of Jacobi fields. Because every tangent vector in a normal neighborhood is the value of a Jacobi field vanishing at the origin by Corollary 10.11, Proposition 10.12 yields explicit formulas for constant-curvature metrics in normal coordinates. To set the stage, we will rewrite the Euclidean metric on $\mathbb{R}^n$ in a form that is somewhat more convenient for these computations.

Let $\pi : \mathbb{R}^n \smallsetminus \{0\} \to \mathbb{S}^{n-1}$ be the radial projection

$$\pi(x) = \frac{x}{|x|}, \tag{10.13}$$

and define a symmetric 2-tensor field on $\mathbb{R}^n \smallsetminus \{0\}$ by

$$\hat{g} = \pi^* \mathring{g}, \tag{10.14}$$

where $\mathring{g}$ is the round metric of radius 1 on $\mathbb{S}^{n-1}$.

Lemma 10.13. *On* $\mathbb{R}^n \smallsetminus \{0\}$, *the metric* $\hat{g}$ *defined by* (10.14) *and the Euclidean metric* $\bar{g}$ *are related by*

$$\bar{g} = dr^2 + r^2 \hat{g}, \tag{10.15}$$

where $r(x) = |x|$ *is the Euclidean distance from the origin.*

Proof. Example 2.24 observed that the map $\Phi \colon \mathbb{R}^+ \times_\rho \mathbb{S}^{n-1} \to \mathbb{R}^n \smallsetminus \{0\}$ given by

$$\Phi(\rho, \omega) = \rho \omega \tag{10.16}$$

is an isometry when $\mathbb{R}^+ \times_\rho \mathbb{S}^{n-1}$ has the warped product metric $d\rho^2 \oplus \rho^2 \mathring{g}$ and $\mathbb{R}^n \smallsetminus \{0\}$ has the Euclidean metric (see also Problem 2-4). Because $\Phi^{-1}(x) = (r(x), \pi(x))$, this means that $\bar{g} = (\Phi^{-1})^* (d\rho^2 \oplus \rho^2 \mathring{g}) = dr^2 + r^2 \hat{g}$. $\qquad\square$

Theorem 10.14 (Constant-Curvature Metrics in Normal Coordinates). *Suppose* (M, g) *is a Riemannian manifold with constant sectional curvature* c. *Given* $p \in M$, *let* (x^i) *be normal coordinates on a normal neighborhood* U *of* p; *let* r *be the radial distance function on* U *defined by* (6.4); *and let* $\hat{g}$ *be the symmetric 2-tensor defined in* x-*coordinates by* (10.14). *On* $U \smallsetminus \{p\}$, *the metric* g *can be written*

$$g = dr^2 + s_c(r)^2 \hat{g}, \tag{10.17}$$

where s_c *is defined by* (10.8).

Proof. Let $\bar{g}$ denote the Euclidean metric in x-coordinates, and let g_c denote the metric defined by the formula on the right-hand side of (10.17). By the properties of normal coordinates, at points of $U \smallsetminus \{p\}$, all three metrics g, $\bar{g}$, and g_c make the radial vector field ∂_r a unit vector orthogonal to the level sets of r. Thus we need only show that $g(w_1, w_2) = g_c(w_1, w_2)$ when w_1, w_2 are tangent to a level set of r, and by polarization it suffices to show that $g(w, w) = g_c(w, w)$ for every such vector w. Note that if w is tangent to a level set $r = b$, then formulas (10.17) and (10.15) imply

$$g_c(w, w) = s_c(b)^2 \hat{g}(w, w) = \frac{s_c(b)^2}{b^2} \bar{g}(w, w).$$

Let $q \in U \smallsetminus \{p\}$ and $w \in T_q M$, and assume that w is tangent to the r-level set containing q. Let $b = d_g(p, q)$, and let $\gamma \colon [0, b] \to U$ be the unit-speed radial geodesic from p to q, so the coordinate representation of γ is

$$\gamma(t) = \left(\frac{t}{b} q^1, \ldots, \frac{t}{b} q^n \right),$$

where $(q^1, \ldots, q^n)$ is the coordinate representation of q. Let $J \in \mathfrak{X}(\gamma)$ be the vector field along γ given by

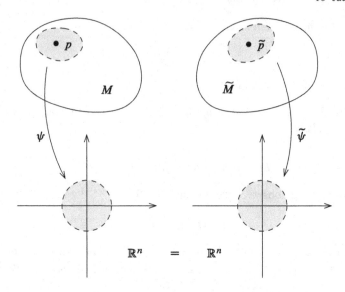

Fig. 10.6: Local isometry constructed from normal coordinate charts

$$J(t) = \frac{t}{b} w^i \partial_i \big|_{\gamma(t)}, \tag{10.18}$$

where $w^i \partial_i \big|_q$ is the coordinate representation for w. By Proposition 10.10, J is a Jacobi field satisfying $D_t J(0) = (1/b) w^i \partial_i \big|_p$, and it follows from the definition that $J(b) = w$. Because J is orthogonal to γ' at p and q, it is normal by Proposition 10.7. Thus by Proposition 10.12,

$$|w|_g^2 = |J(b)|_g^2 = s_c(b)^2 |D_t J(0)|_g^2$$
$$= s_c(b)^2 \frac{1}{b^2} \Big| w^i \partial_i \big|_p \Big|_g^2 = s_c(b)^2 \frac{1}{b^2} |w|_{\tilde{g}}^2 = |w|_{g_c}^2. \qquad \square$$

Corollary 10.15 (Local Uniqueness of Constant-Curvature Metrics). *Let (M, g) and $(\widetilde{M}, \tilde{g})$ be Riemannian manifolds of the same dimension with constant sectional curvature c. For all points $p \in M$, $\tilde{p} \in \widetilde{M}$, there exist neighborhoods U of p and $\widetilde{U}$ of $\tilde{p}$ and an isometry $\varphi: U \to \widetilde{U}$.*

Proof. Choose $p \in M$ and $\tilde{p} \in \widetilde{M}$, and let U and $\widetilde{U}$ be geodesic balls of small radius ε around p and $\tilde{p}$, respectively. Riemannian normal coordinates give maps $\psi: U \to B_\varepsilon(0) \subseteq \mathbb{R}^n$ and $\tilde{\psi}: \widetilde{U} \to B_\varepsilon(0) \subseteq \mathbb{R}^n$, under which both metrics are given by formula (10.17) on the complement of the origin (Fig. 10.6). At the origin, $g_{ij} = \tilde{g}_{ij} = \delta_{ij}$. Therefore $\tilde{\psi}^{-1} \circ \psi$ is the required local isometry. $\square$

Corollary 10.16 (Constant-Curvature Metrics as Warped Products). *Suppose (M, g) is a Riemannian manifold with constant sectional curvature c, and U is a geodesic ball of radius b centered at $p \in M$. Then $U \smallsetminus \{p\}$ is isometric to a warped* *product of the form* $(0,b) \times_{s_c} \mathbb{S}^{n-1}$, *where* $(0,b) \subseteq \mathbb{R}$ *has the Euclidean metric, and* $\mathbb{S}^{n-1}$ *is the unit sphere with the round metric* $\mathring{g}$.

Proof. By virtue of Theorem 10.14, we may consider g to be a metric on the ball of radius b in $\mathbb{R}^n$ given by formula (10.17). Let $\Phi \colon (0,b) \times \mathbb{S}^{n-1} \to U \smallsetminus \{p\}$ and $\pi \colon \mathbb{R}^n \smallsetminus \{0\} \to \mathbb{S}^{n-1}$ be the maps defined by (10.16) and (10.13). Because $\pi \circ \Phi$ restricts to the identity on $\{\rho\} \times \mathbb{S}^{n-1}$ for each fixed ρ, it follows that $\Phi^* \mathring{g} = \Phi^* \pi^* \mathring{g} = \mathring{g}$, and thus

$$\Phi^* g = d\rho^2 \oplus s_c(\rho)^2 \mathring{g}. \qquad \square$$

The next corollary describes a formula for integration in polar coordinates with respect to a constant-curvature metric. It will be useful in our proofs of volume comparison theorems in the next chapter.

Corollary 10.17 (Polar Decomposition of Integrals). *Suppose* (M,g) *is a Riemannian manifold with constant sectional curvature c, and U is an open or closed geodesic ball of radius b around a point $p \in M$. If $f \colon U \to \mathbb{R}$ is any bounded integrable function, then the integral of f over U can be expressed as*

$$\int_U f \, dV_g = \int_{\mathbb{S}^{n-1}} \int_0^b f \circ \Phi(\rho, \omega) s_c(\rho)^{n-1} \, d\rho \, dV_{\mathring{g}},$$

where dV_g is the Riemannian density of g, and $\Phi \colon (0,b) \times \mathbb{S}^{n-1} \to U \smallsetminus \{p\}$ is defined in normal coordinates by (10.16).

Proof. Because every geodesic ball is orientable, we might as well choose an orientation on U and interpret dV_g as a differential form. Since the boundary of a geodesic ball has measure zero, it does not matter whether U is open or closed. Similarly, integrating over $U \smallsetminus \{p\}$ instead of U does not change the value of the integral. The claim therefore follows from Corollary 10.16 together with the result of Problem 2-15, which shows that the volume form of the warped product metric $d\rho^2 \oplus s_c(\rho)^2 \mathring{g}$ can be written $s_c(\rho)^{n-1} d\rho \wedge dV_{\mathring{g}}$. $\qquad \square$

Locally Symmetric Spaces

Here is another application of the theory of Jacobi fields. Recall that a Riemannian manifold is a *locally symmetric space* if every $p \in M$ has a neighborhood that admits a point reflection at p. Problem 7-3 showed that every locally symmetric space has parallel curvature tensor. Now we can prove the converse. The key is the following lemma due to Élie Cartan. We will use the lemma again in Chapter 12 to prove a more global result (see Thm. 12.6).

Lemma 10.18. *Suppose (M,g) is a Riemannian manifold with parallel curvature tensor, and for some points $p, \hat{p} \in M$ we are given a linear map $A \colon T_p M \to T_{\hat{p}} M$ satisfying $A^*(g_{\hat{p}}) = g_p$ and $A^*(Rm_{\hat{p}}) = Rm_p$. Then there exist a neighborhood U of p and a local isometry $\varphi \colon U \to M$ such that $\varphi(p) = \hat{p}$ and $d\varphi_p = A$. If M is complete, then U can be taken to be any normal neighborhood of p.*

Proof. The hypothesis means that $\nabla Rm \equiv 0$, and because covariant differentiation commutes with raising and lowering indices, the curvature endomorphism is also parallel. If M is complete, let U be any normal neighborhood of p; otherwise choose $U = B_\varepsilon(p)$, where $\varepsilon > 0$ is chosen small enough that both $B_\varepsilon(p)$ and $B_\varepsilon(\hat{p})$ are geodesic balls. Our choice guarantees that $\varphi = \exp_{\hat{p}} \circ A \circ \exp_p^{-1}$ is a smooth map from U into M, and it satisfies $\varphi(p) = \hat{p}$ and $d\varphi_p = A$. We will show that $|d\varphi_q(x)|_g = |x|_g$ for every tangent vector x at every point $q \in U$. It then follows by polarization that $\langle d\varphi_q(x), d\varphi_q(y) \rangle_g = \langle x, y \rangle_g$ for all $x, y \in T_q M$, and thus $\varphi^* g = g$.

Because $d\varphi_p = A$ is a linear isometry, we have $|d\varphi_p(x)|_g = |x|_g$ for $x \in T_p M$, so we need only consider points $q \neq p$. Let $q \in U \smallsetminus \{p\}$ and $x \in T_q M$ be arbitrary. Let $\gamma \colon [0,1] \to U$ be the radial geodesic from p to q, given explicitly by $\gamma(t) = \exp_p(tv)$ for some $v \in T_p M$. It follows from Corollary 10.11 that there is a Jacobi field J along γ such that $J(0) = 0$ and $J(1) = x$; and Proposition 10.10 shows that it is of the form $J(t) = d(\exp_p)_{tv}(tw)$ for some $w \in T_p M$. Let $\hat{v} = A(v)$ and $\hat{w} = A(w) \in T_{\hat{p}} M$, and define $\hat{\gamma}(t) = \exp_{\hat{p}}(t\hat{v})$ and $\hat{J}(t) = d(\exp_{\hat{p}})_{t\hat{v}}(t\hat{w})$ for $t \in [0,1]$. Then $\hat{\gamma}$ is a geodesic from $\hat{p} = \varphi(p)$ to $\varphi(q)$, and $\hat{J}$ is a Jacobi field along $\hat{\gamma}$. It follows from the definition of φ and the chain rule (using the fact that $dA_v = A$ because A is linear) that $d\varphi_q \circ d(\exp_p)_v = d(\exp_{\hat{p}})_{\hat{v}} \circ A$, and thus

$$\hat{J}(1) = d(\exp_{\hat{p}})_{\hat{v}}(\hat{w}) = d(\exp_{\hat{p}})_{\hat{v}} \circ A(w)$$
$$= d\varphi_q \circ d(\exp_p)_v(w) = d\varphi_q(J(1)) = d\varphi_q(x),$$

so to prove the theorem it suffices to show that $\left|\hat{J}(1)\right|_g = |J(1)|_g$.

Let $(E_1, \dots, E_n)$ be a parallel orthonormal frame along γ, and let $(\hat{E}_1, \dots, \hat{E}_n)$ be the parallel orthonormal frame along $\hat{\gamma}$ such that $\hat{E}_i(0) = A(E_i(0))$. At points of γ, we can express the curvature endomorphism in terms of the frame (E_i) as

$$R\big(E_i(t), E_j(t)\big)E_k(t) = R_{ijk}{}^l(t) E_l(t),$$

for some smooth functions $R_{ijk}{}^l \colon [0,1] \to \mathbb{R}$. The parallel curvature hypothesis and the fact that each E_i is parallel imply

$$0 = (D_t R)\big(E_i(t), E_j(t)\big)E_k(t) = D_t\big(R(E_i(t), E_j(t))E_k(t)\big) = \big(R_{ijk}{}^l\big)'(t) E_l(t),$$

so in fact the coefficients $R_{ijk}{}^l(t)$ are constant in t. The same argument shows that the curvature endomorphism has constant coefficients along $\hat{\gamma}$ in terms of the frame $(\hat{E}_i)$. Because our hypotheses guarantee that the coefficients of the two curvature endomorphisms agree at $t = 0$, they are in fact the same constants along both geodesics; we write those constants henceforth as $R_{ijk}{}^l$.

Also, we can write $J(t) = J^i(t)E_i(t)$ and $\hat{J}(t) = \hat{J}^i(t)\hat{E}_i(t)$ for some smooth functions $J^i, \hat{J}^i \colon [0,1] \to \mathbb{R}$. The Jacobi equations for J and $\hat{J}$, written in terms of our parallel frames, read

$$\left(J^l\right)''(t) + R_{ijk}{}^l J^i(t) v^j v^k = 0,$$
$$\left(\hat{J}^l\right)''(t) + R_{ijk}{}^l \hat{J}^i(t) v^j v^k = 0.$$

Proposition 10.10 shows that $D_t J(0) = w$, which we can write as $w = w^l E_l(0)$. It also follows that $\hat{D}_t \hat{J}(0) = \hat{w} = A(w) = w^l \hat{E}_l(0)$, so the functions J^l and $\hat{J}^l$ satisfy the same system of differential equations with the same initial conditions $J^l(0) = \hat{J}^l(0) = 0$ and $\left(J^l\right)'(0) = \left(\hat{J}^l\right)'(0) = w^l$. Uniqueness of ODE solutions implies $J^l(t) = \hat{J}^l(t)$ for all t, and in particular $J^l(1) = \hat{J}^l(1)$. Because the frames (E_i) and $\left(\hat{E}_i\right)$ are orthonormal, we have

$$|J(1)|_g^2 = \sum_{i=1}^n \left(J^i(1)\right)^2 = \sum_{i=1}^n \left(\hat{J}^i(1)\right)^2 = \left|\hat{J}(1)\right|_g^2,$$

thus completing the proof.   □

Theorem 10.19 (Characterization of Locally Symmetric Spaces). *A Riemannian manifold is a locally symmetric space if and only if its curvature tensor is parallel.*

Proof. One direction is taken care of by Problem 7-3. To prove the converse, suppose (M, g) is a Riemannian manifold with $\nabla Rm \equiv 0$. Let $p \in M$ be arbitrary, and let U be a geodesic ball centered at p. The linear map $A = -\mathrm{Id} \colon T_p M \to T_p M$ satisfies $A^* g_p = g_p$, and for all $w, x, y, z \in T_p M$, we have

$$(A^* Rm_p)(w, x, y, z) = Rm_p(-w, -x, -y, -z) = Rm_p(w, x, y, z).$$

It follows that $A = -\mathrm{Id}$ satisfies the hypotheses of Lemma 10.18 with $p = \hat{p}$, and thus there is a local isometry $\varphi \colon U \to M$ such that $\varphi(p) = p$ and $d\varphi_p = -\mathrm{Id}$. Since a local isometry takes geodesics to geodesics, $\varphi(U)$ is also a geodesic ball centered at p of the same radius as U, so φ actually maps U to U. If we take the radius of U small enough, then φ is an isometry from U to itself.   □

Conjugate Points

Our next application of Jacobi fields is to study the question of when the exponential map is a local diffeomorphism.

Suppose (M, g) is a Riemannian or pseudo-Riemannian manifold and $p \in M$. The restricted exponential map $\exp_p$ is defined on an open subset $\mathcal{E}_p \subseteq T_p M$, and because it is a smooth map between manifolds of the same dimension, the inverse function theorem guarantees that it is a local diffeomorphism in a neighborhood of each of its regular points (points $v \in T_p M$ where $d\left(\exp_p\right)_v$ is surjective and thus invertible). To see where this fails, we need to identify the critical points of $\exp_p$ (the points where its differential is singular). Proposition 5.19(d) guarantees that 0 is a regular point, but it may well happen that it has critical points elsewhere in $\mathcal{E}_p$.

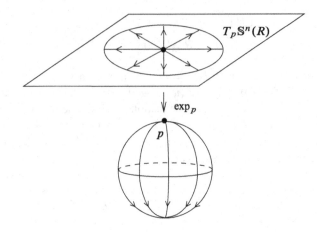

Fig. 10.7: The exponential map of the sphere

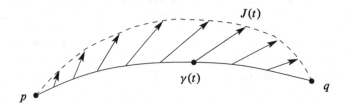

Fig. 10.8: Conjugate points

An enlightening example is provided by the sphere $\mathbb{S}^n(R)$ (Fig. 10.7). All geodesics starting at a given point p meet at the antipodal point, which is at a distance of πR along each geodesic. The exponential map is a diffeomorphism on the ball $B_{\pi R}(0) \subseteq T_p\mathbb{S}^n(R)$, but every point on the boundary of that ball is a critical point. Moreover, Proposition 10.12 shows that each Jacobi field on $\mathbb{S}^n(R)$ vanishing at p has its first zero precisely at distance πR.

On the other hand, formula (10.7) shows that if U is a normal neighborhood of p (the image of a star-shaped open set on which $\exp_p$ is a diffeomorphism), then no Jacobi field that vanishes at p can vanish at any other point in U. We might thus be led to expect a relationship between zeros of Jacobi fields and critical points of the exponential map.

Let (M, g) be a Riemannian or pseudo-Riemannian manifold, $\gamma \colon I \to M$ a geodesic, and $p = \gamma(a)$, $q = \gamma(b)$ for some $a, b \in I$. We say that p and q are *conjugate along γ* if there is a Jacobi field along γ vanishing at $t = a$ and $t = b$ but not identically zero (Fig. 10.8). The *order* (or *multiplicity*) *of conjugacy* is the dimension of the space of Jacobi fields vanishing at a and b. From the existence and uniqueness theorem for Jacobi fields, there is an n-dimensional space of Jacobi fields that vanish at a; since tangential Jacobi fields vanish at most at one point, the order of conjugacy of two points along γ can be at most $n - 1$. This bound is sharp: Proposition 10.12 shows that if γ is a geodesic joining antipodal points p and q on $\mathbb{S}^n(R)$, then there is a Jacobi field vanishing at p and q for each parallel normal vector field along γ; thus in that case p and q are conjugate to order exactly $n - 1$.

The most important fact about conjugate points is that they are the images of critical points of the exponential map, as the following proposition shows.

Proposition 10.20. *Suppose (M, g) is a Riemannian or pseudo-Riemannian manifold, $p \in M$, and $v \in \mathcal{E}_p \subseteq T_p M$. Let $\gamma = \gamma_v \colon [0, 1] \to M$ be the geodesic segment $\gamma(t) = \exp_p(tv)$, and let $q = \gamma(1) = \exp_p(v)$. Then v is a critical point of $\exp_p$ if and only if q is conjugate to p along γ.*

Proof. Suppose first that v is a critical point of $\exp_p$. Then there is a nonzero vector $w \in T_v(T_p M)$ such that $d\left(\exp_p\right)_v(w) = 0$. Because $T_p M$ is a vector space, we can identify $T_v(T_p M)$ with $T_p M$ as usual and regard w as a vector in $T_p M$. Let Γ be the variation of γ defined by (10.5), and let $J(t) = \partial_s \Gamma(0, t)$ be its variation field. We can compute $J(1)$ as follows:

$$J(1) = \partial_s \Gamma(0, 1) = \left. \frac{\partial}{\partial s} \right|_{s=0} \exp_p(v + sw) = d\left(\exp_p\right)_v(w) = 0.$$

Thus J is a nontrivial Jacobi field vanishing at $t = 0$ and $t = 1$, so q is conjugate to p along γ.

Conversely, if q is conjugate to p along γ, then there is some nontrivial Jacobi field J along γ such that $J(0) = 0$ and $J(1) = 0$. Lemma 10.9 shows that J is the variation field of a variation of γ of the form (10.5) with $w = D_t J(0) \in T_p M$, and the computation in the preceding paragraph shows that $d\left(\exp_p\right)_v(w) = J(1) = 0$. Thus v is a critical point for $\exp_p$. □

As Proposition 10.2 shows, the "natural" way to specify a unique Jacobi field is by giving its initial value and initial derivative. However, in Corollary 10.11 and Proposition 10.20, we had to construct Jacobi fields along a geodesic satisfying $J(0) = 0$ and $J(1) = w$ for some specific vector w. More generally, one can pose the **two-point boundary problem** for Jacobi fields: given $v \in T_{\gamma(a)} M$ and $w \in T_{\gamma(b)} M$, find a Jacobi field J along γ such that $J(a) = v$ and $J(b) = w$. Another interesting property of conjugate points is that they are the obstructions to solving the two-point boundary problem, as the next proposition shows.

Proposition 10.21 (The Two-Point Boundary Problem for Jacobi Fields). *Suppose (M, g) is a Riemannian or pseudo-Riemannian manifold, and $\gamma \colon [a, b] \to M$ is a geodesic segment. The two-point boundary problem for Jacobi fields along γ is uniquely solvable for every pair of vectors $v \in T_{\gamma(a)} M$ and $w \in T_{\gamma(b)} M$ if and only if $\gamma(a)$ and $\gamma(b)$ are not conjugate along γ.*

Proof. Problem 10-8. □

The Second Variation Formula

Our next task is to study the question of which geodesic segments are minimizing. In the remainder of the chapter, because of the complications involved in studying lengths on pseudo-Riemannian manifolds, we restrict our attention to the Riemannian case.

In our proof that every minimizing curve is a geodesic, we imitated the first-derivative test of elementary calculus: if a geodesic γ is minimizing, then the first derivative of the length functional must vanish for every proper variation of γ. Now we imitate the second-derivative test: if γ is minimizing, the second derivative must be nonnegative. First, we must compute this second derivative. In keeping with classical terminology, we call it the **second variation** of the length functional.

Theorem 10.22 (Second Variation Formula). *Suppose (M,g) is a Riemannian manifold. Let $\gamma : [a,b] \to M$ be a unit-speed geodesic segment, $\Gamma : J \times [a,b] \to M$ a proper variation of γ, and V its variation field. The second variation of $L_g(\Gamma_s)$ is given by the following formula:*

$$\frac{d^2}{ds^2}\bigg|_{s=0} L_g(\Gamma_s) = \int_a^b \left(\left| D_t V^\perp \right|^2 - Rm\left(V^\perp, \gamma', \gamma', V^\perp \right) \right) dt, \qquad (10.19)$$

where $V^\perp$ is the normal component of V.

Proof. As usual, write $T = \partial_t \Gamma$ and $S = \partial_s \Gamma$, and let $(a_0, \dots, a_k)$ be an admissible partition for Γ. We begin, as we did when computing the first variation formula, by restricting to a rectangle $J \times [a_{i-1}, a_i]$ where Γ is smooth. From (6.3) we have, for every s,

$$\frac{d}{ds} L_g\left(\Gamma_s|_{[a_{i-1}, a_i]} \right) = \int_{a_{i-1}}^{a_i} \frac{\langle D_t S, T \rangle}{\langle T, T \rangle^{1/2}} dt.$$

Differentiating again with respect to s, and using the symmetry lemma and Proposition 7.5, we obtain

$$\frac{d^2}{ds^2} L_g\left(\Gamma_s|_{[a_{i-1}, a_i]} \right)$$
$$= \int_{a_{i-1}}^{a_i} \left(\frac{\langle D_s D_t S, T \rangle}{\langle T, T \rangle^{1/2}} + \frac{\langle D_t S, D_s T \rangle}{\langle T, T \rangle^{1/2}} - \frac{1}{2} \frac{\langle D_t S, T \rangle 2 \langle D_s T, T \rangle}{\langle T, T \rangle^{3/2}} \right) dt$$
$$= \int_{a_{i-1}}^{a_i} \left(\frac{\langle D_t D_s S + R(S,T)S, T \rangle}{|T|} + \frac{\langle D_t S, D_t S \rangle}{|T|} - \frac{\langle D_t S, T \rangle^2}{|T|^3} \right) dt.$$

Now restrict to $s = 0$, where $|T| = 1$:

$$\frac{d^2}{ds^2}\bigg|_{s=0} L_g\left(\Gamma_s|_{[a_{i-1}, a_i]} \right) = \int_{a_{i-1}}^{a_i} \left(\langle D_t D_s S, T \rangle - Rm(S, T, T, S) \right.$$
$$\left. + |D_t S|^2 - \langle D_t S, T \rangle^2 \right) dt \bigg|_{s=0}. \qquad (10.20)$$

Because $D_t T = D_t \gamma' = 0$ when $s = 0$, the first term in (10.20) can be integrated as follows:

$$\int_{a_{i-1}}^{a_i} \langle D_t D_s S, T \rangle \, dt = \int_{a_{i-1}}^{a_i} \frac{d}{dt} \langle D_s S, T \rangle \, dt = \langle D_s S, T \rangle \Big|_{t=a_{i-1}}^{t=a_i}. \qquad (10.21)$$

Notice that $S(s,t) = 0$ for all s at the endpoints $t = a_0 = a$ and $t = a_k = b$ because Γ is a proper variation, so $D_s S = 0$ there. Moreover, along the boundaries $\{t = a_i\}$ of the smooth regions, $D_s S = D_s(\partial_s \Gamma)$ depends only on the values of Γ when $t = a_i$, and it is smooth up to the line $\{t = a_i\}$ from both sides; therefore $D_s S$ is continuous for all (s,t). Thus when we insert (10.21) into (10.20) and sum over i, the boundary contributions from the first term all cancel, and we get

$$\frac{d^2}{ds^2}\Big|_{s=0} L_g(\Gamma_s) = \int_a^b \left(|D_t S|^2 - \langle D_t S, T \rangle^2 - Rm(S, T, T, S) \right) dt \Big|_{s=0} \quad (10.22)$$

$$= \int_a^b \left(|D_t V|^2 - \langle D_t V, \gamma' \rangle^2 - Rm(V, \gamma', \gamma', V) \right) dt.$$

Every vector field V along γ can be written uniquely as $V = V^\top + V^\perp$, where $V^\top$ is tangential and $V^\perp$ is normal. Explicitly,

$$V^\top = \langle V, \gamma' \rangle \gamma'; \qquad V^\perp = V - V^\top.$$

Because $D_t \gamma' = 0$, it follows that

$$D_t(V^\top) = \langle D_t V, \gamma' \rangle \gamma' = (D_t V)^\top; \qquad D_t(V^\perp) = (D_t V)^\perp.$$

Therefore,

$$|D_t V|^2 = |(D_t V)^\top|^2 + |(D_t V)^\perp|^2 = \langle D_t V, \gamma' \rangle^2 + |D_t V^\perp|^2.$$

Also, the fact that $Rm(\gamma', \gamma', \cdot, \cdot) = Rm(\cdot, \cdot, \gamma', \gamma') = 0$ implies

$$Rm(V, \gamma', \gamma', V) = Rm(V^\perp, \gamma', \gamma', V^\perp).$$

Substituting these relations into (10.22) gives (10.19). $\qquad\qquad\qquad\qquad\qquad\square$

It should come as no surprise that the second variation depends only on the normal component of V, because the tangential component of V contributes only to a reparametrization of γ, and length is independent of parametrization. For this reason, we will generally restrict our attention to variations of the following type: if γ is an admissible curve, a variation of γ is called a **normal variation** if its variation field is a normal vector field along γ.

Given a geodesic segment $\gamma : [a,b] \to M$, we define a symmetric bilinear form I, called the **index form of γ**, on the space of normal vector fields along γ by

$$I(V,W) = \int_a^b \left(\langle D_t V, D_t W \rangle - Rm(V, \gamma', \gamma', W) \right) dt. \tag{10.23}$$

You should think of $I(V,W)$ as a sort of "Hessian" or second derivative of the length functional. Because every proper normal vector field along γ is the variation field of some proper normal variation, the preceding theorem can be rephrased in terms of the index form in the following way.

Corollary 10.23. *Suppose (M,g) is a Riemannian manifold. Let $\gamma : [a,b] \to M$ be a unit-speed geodesic, Γ a proper normal variation of γ, and V its variation field. The second variation of $L_g(\Gamma_s)$ is $I(V,V)$. If γ is minimizing, then $I(V,V) \geq 0$ for every proper normal vector field along γ.* $\qquad\square$

The next proposition gives another expression for I, which makes the role of the Jacobi equation more evident.

Proposition 10.24. *Let (M,g) be a Riemannian manifold and let $\gamma : [a,b] \to M$ be a geodesic segment. For every pair of piecewise smooth normal vector fields V, W along γ,*

$$I(V,W) = -\int_a^b \langle D_t^2 V + R(V, \gamma')\gamma', W \rangle dt$$

$$+ \langle D_t V, W \rangle \Big|_{t=a}^{t=b} - \sum_{i=1}^{k-1} \langle \Delta_i D_t V, W(a_i) \rangle, \tag{10.24}$$

where $(a_0, \ldots, a_k)$ is an admissible partition for V and W, and $\Delta_i D_t V$ is the jump in $D_t V$ at $t = a_i$.

Proof. On every subinterval $[a_{i-1}, a_i]$ where V and W are smooth,

$$\frac{d}{dt} \langle D_t V, W \rangle = \langle D_t^2 V, W \rangle + \langle D_t V, D_t W \rangle.$$

Thus, by the fundamental theorem of calculus,

$$\int_{a_{i-1}}^{a_i} \langle D_t V, D_t W \rangle dt = -\int_{a_{i-1}}^{a_i} \langle D_t^2 V, W \rangle dt + \langle D_t V, W \rangle \Big|_{a_{i-1}}^{a_i}.$$

Summing over i, and noting that W is continuous at $t = a_i$ for $i = 1, \ldots, k-1$, we get (10.24). $\qquad\square$

Corollary 10.25. *If γ is a geodesic segment and V is a proper normal piecewise smooth vector field along γ, then $I(V,W) = 0$ for every proper normal piecewise smooth vector field W along γ if and only if V is a Jacobi field.*

Proof. Problem 10-11. $\qquad\square$

Geodesics Do Not Minimize Past Conjugate Points

We can use the second variation formula to prove another extremely important fact about conjugate points: no geodesic is minimizing past its first conjugate point. The geometric intuition is as follows. Suppose $\gamma \colon [a,c] \to M$ is a minimizing geodesic segment, and $\gamma(b)$ is conjugate to $\gamma(a)$ along γ for some $a < b < c$. If J is a Jacobi field along γ that vanishes at $t = a$ and $t = b$, then there is a variation of γ through geodesics, all of which start at $\gamma(a)$. Since $J(b) = 0$, we can expect them to end "almost" at $\gamma(b)$. If they really did all end at $\gamma(b)$, we could construct a broken geodesic by following some Γ_s from $\gamma(a)$ to $\gamma(b)$ and then following γ from $\gamma(b)$ to $\gamma(c)$, which would have the same length and thus would also be a minimizing curve. But this is impossible: as the proof of Theorem 6.4 shows, a broken geodesic can always be shortened by rounding the corner.

The problem with this heuristic argument is that there is no guarantee that we can construct a variation through geodesics that actually end at $\gamma(b)$. The proof of the following theorem is based on an "infinitesimal" version of rounding the corner to obtain a shorter curve.

Given a geodesic segment $\gamma \colon [a,c] \to M$, we say that *$\gamma$ has a conjugate point* if there is some $b \in (a,c]$ such that $\gamma(b)$ is conjugate to $\gamma(a)$ along γ, and *γ has an interior conjugate point* if there is such a $b \in (a,c)$.

Theorem 10.26. *Let (M,g) be a Riemannian manifold and $p,q \in M$. If γ is a unit-speed geodesic segment from p to q that has an interior conjugate point, then there exists a proper normal vector field X along γ such that $I(X,X) < 0$. Therefore, γ is not minimizing.*

Proof. Suppose $\gamma \colon [a,c] \to M$ is a unit-speed geodesic segment, and $\gamma(b)$ is conjugate to $\gamma(a)$ along γ for some $a < b < c$. This means that there is a nontrivial normal Jacobi field J along γ that vanishes at $t = a$ and $t = b$. Define a vector field V along all of γ by

$$V(t) = \begin{cases} J(t), & t \in [a,b]; \\ 0, & t \in [b,c]. \end{cases}$$

This is a proper, normal, piecewise smooth vector field along γ.

Let W be a smooth proper normal vector field along γ such that $W(b)$ is equal to the jump $\Delta D_t V$ at $t = b$ (Fig. 10.9). Such a vector field is easily constructed with the help of an orthonormal frame along γ and a bump function. Note that $\Delta D_t V = -D_t J(b)$ is not zero, because otherwise J would be a Jacobi field satisfying $J(b) = D_t J(b) = 0$, and thus would be identically zero.

For small positive ε, let $X_\varepsilon = V + \varepsilon W$. Then

$$I(X_\varepsilon, X_\varepsilon) = I(V + \varepsilon W, V + \varepsilon W)$$
$$= I(V,V) + 2\varepsilon I(V,W) + \varepsilon^2 I(W,W).$$

Since V satisfies the Jacobi equation on each subinterval $[a,b]$ and $[b,c]$, and $V(b) = 0$, (10.24) gives

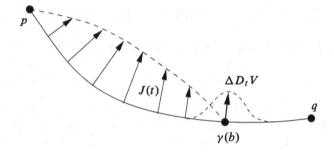

Fig. 10.9: Constructing a vector field X with $I(X,X) < 0$

$$I(V,V) = -\langle \Delta D_t V, V(b) \rangle = 0.$$

Similarly,

$$I(V,W) = -\langle \Delta D_t V, W(b) \rangle = -|W(b)|^2.$$

Thus

$$I(X_\varepsilon, X_\varepsilon) = -2\varepsilon |W(b)|^2 + \varepsilon^2 I(W,W).$$

If we choose ε small enough, this is strictly negative.   □

There is a partial converse to the preceding theorem, which says that a geodesic without conjugate points has the shortest length among all nearby curves in any proper variation. Before we prove it, we need the following technical lemma.

Lemma 10.27. *Let* $\gamma \colon [a,b] \to M$ *be a geodesic segment, and suppose* J_1 *and* J_2 *are Jacobi fields along* γ. *Then* $\langle D_t J_1(t), J_2(t) \rangle - \langle J_1(t), D_t J_2(t) \rangle$ *is constant along* γ.

Proof. Let $f(t) = \langle D_t J_1(t), J_2(t) \rangle - \langle J_1(t), D_t J_2(t) \rangle$. Using the Jacobi equation, we compute

$$
\begin{aligned}
f'(t) &= \langle D_t^2 J_1(t), J_2(t) \rangle + \langle D_t J_1(t), D_t J_2(t) \rangle \\
&\quad - \langle D_t J_1(t), D_t J_2(t) \rangle - \langle J_1(t), D_t^2 J_2(t) \rangle \\
&= -Rm\big(J_1(t), \gamma'(t), \gamma'(t), J_2(t)\big) + Rm\big(J_2(t), \gamma'(t), \gamma'(t), J_1(t)\big) = 0,
\end{aligned}
$$

where the last equality follows from the symmetries of the curvature tensor.   □

Theorem 10.28. *Let* (M,g) *be a Riemannian manifold. Suppose* $\gamma \colon [a,b] \to M$ *is a unit-speed geodesic segment without interior conjugate points. If* V *is any proper normal piecewise smooth vector field along* γ, *then* $I(V,V) \geq 0$, *with equality if and only if* V *is a Jacobi field. In particular, if* $\gamma(b)$ *is not conjugate to* $\gamma(a)$ *along* γ, *then* $I(V,V) > 0$.

Proof. To simplify the notation, we can assume (after replacing t by $t-a$) that $a = 0$. Let $p = \gamma(0)$, and let $(w_1, \dots, w_n)$ be an orthonormal basis for $T_p M$, chosen so that $w_1 = \gamma'(0)$. For each $\alpha = 2,\ldots,n$, let J_α be the unique normal Jacobi field along γ satisfying $J_\alpha(0) = 0$ and $D_t J_\alpha(0) = w_\alpha$.

Our assumption that γ has no interior conjugate points guarantees that no linear combination of the $J_\alpha(t)$'s can vanish for any $t \in (0,b)$, and thus $(J_\alpha(t))$ forms a basis for the orthogonal complement of $\gamma'(t)$ in $T_{\gamma(t)}M$ for each such t. Thus, given V as in the statement of the theorem, for $t \in (0,b)$ we can write

$$V(t) = v^\alpha(t) J_\alpha(t) \tag{10.25}$$

for some piecewise smooth functions $v^\alpha : (0,b) \to M$. (Here and in the remainder of this proof, the summation convention is in effect, with Greek indices running from 2 to n.)

In fact, each function v^α has a piecewise smooth extension to $[0,b]$. To see why, let (x^i) be the normal coordinates centered at p determined by the basis (w_i). For sufficiently small $t > 0$, we can express $J_\alpha(t)$ and $V(t)$ in normal coordinates as

$$J_\alpha(t) = t \left.\frac{\partial}{\partial x_\alpha}\right|_{\gamma(t)}, \quad \alpha = 2,\ldots,n,$$

$$V(t) = v^\alpha(t) J_\alpha(t) = t v^\alpha(t) \left.\frac{\partial}{\partial x_\alpha}\right|_{\gamma(t)}.$$

(The formula for $J_\alpha(t)$ follows from Prop. 10.10.) Because V is smooth on $[0,\delta)$ for some $\delta > 0$ and $V(0) = 0$, it follows from Taylor's theorem that the components of $V(t)/t$ extend smoothly to $[0,\delta)$, which shows that v^α is smooth there. Because $V(b) = 0$, it follows similarly that v^α extends smoothly to $t = b$ as well. (If $J_\alpha(b) = 0$, the argument is the same as for $t = 0$, while if not, it is even easier.)

Let $(a_0,\ldots,a_k)$ be an admissible partition for V. On each subinterval (a_{i-1},a_i) where V is smooth, define vector fields X and Y along γ by

$$X = v^\alpha D_t J_\alpha, \quad Y = \dot{v}^\alpha J_\alpha.$$

Then $D_t V = X + Y$ on each such interval, and the fact that V is piecewise smooth implies that $D_t V$, X, and Y extend smoothly to $[a_{i-1},a_i]$ for each i, with one-sided derivatives at the endpoints.

To compute $I(V,V)$, we will use the following identity, which holds on each subinterval $[a_{i-1},a_i]$:

$$|D_t V|^2 - Rm(V,\gamma',\gamma',V) = \frac{d}{dt}\langle V, X\rangle + |Y|^2. \tag{10.26}$$

Granting this for now, we use the fundamental theorem of calculus to compute

$$I(V,V) = \sum_{i=1}^{k} \int_{a_{i-1}}^{a_i} \left(|D_t V|^2 - Rm(V, \gamma', \gamma', V) \right) dt$$

$$= \sum_{i=1}^{k} \langle V, X \rangle \Big|_{t=a_{i-1}}^{t=a_i} + \int_0^b |Y|^2 \, dt,$$

where the boundary terms are to be interpreted as limits from above and below. Because X and V are both continuous on $[0,b]$, the boundary terms at $t = a_1, \ldots, a_{k-1}$ all cancel, and because $V(0) = V(b) = 0$, the boundary terms at $t = 0$ and $t = b$ are both zero. It follows that $I(V,V) = \int_0^b |Y|^2 \, dt \geq 0$. If $I(V,V) = 0$, then Y is identically zero on $(0,b)$. Since the J_α's are linearly independent there, this implies that $\dot{v}^\alpha \equiv 0$ for each α, so each v^α is constant. Thus V is a linear combination of Jacobi fields with constant coefficients, so it is a Jacobi field.

It remains only to prove (10.26). Note that

$$\frac{d}{dt} \langle V, X \rangle = \langle D_t V, X \rangle + \langle V, D_t X \rangle = \langle X + Y, X \rangle + \langle V, D_t X \rangle. \tag{10.27}$$

The Jacobi equation gives

$$D_t X = \dot{v}^\alpha D_t J_\alpha + v^\alpha D_t^2 J_\alpha = \dot{v}^\alpha D_t J_\alpha - v^\alpha R(J_\alpha, \gamma')\gamma' = \dot{v}^\alpha D_t J_\alpha - R(V, \gamma')\gamma'.$$

Therefore,

$$\langle D_t X, V \rangle = \langle \dot{v}^\alpha D_t J_\alpha, v^\beta J_\beta \rangle - Rm(V, \gamma', \gamma', V). \tag{10.28}$$

Because $\langle D_t J_\alpha, J_\beta \rangle - \langle J_\alpha, D_t J_\beta \rangle = 0$ at $t = 0$, it follows from Lemma 10.27 that $\langle D_t J_\alpha, J_\beta \rangle = \langle J_\alpha, D_t J_\beta \rangle$ all along γ. Thus we can simplify the first term in (10.28) as follows:

$$\langle \dot{v}^\alpha D_t J_\alpha, v^\beta J_\beta \rangle = \dot{v}^\alpha v^\beta \langle D_t J_\alpha, J_\beta \rangle = \dot{v}^\alpha v^\beta \langle J_\alpha, D_t J_\beta \rangle$$

$$= \langle \dot{v}^\alpha J_\alpha, v^\beta D_t J_\beta \rangle = \langle Y, X \rangle.$$

Inserting this into (10.28), and then inserting the latter into (10.27) yields

$$\frac{d}{dt} \langle V, X \rangle = \langle X + Y, X \rangle + \langle Y, X \rangle - Rm(V, \gamma', \gamma', V)$$

$$= |X + Y|^2 - |Y|^2 - Rm(V, \gamma', \gamma', V),$$

which is equivalent to (10.26).   □

The next corollary summarizes the results of Theorems 10.26 and 10.28.

Corollary 10.29. *Let (M,g) be a Riemannian manifold, and let $\gamma \colon [a,b] \to M$ be a unit-speed geodesic segment.*

(a) *If γ has an interior conjugate point, then it is not minimizing.*

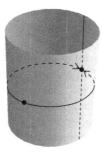

Fig. 10.10: Geodesics on the cylinder

(b) *If $\gamma(a)$ and $\gamma(b)$ are conjugate but γ has no interior conjugate points, then for every proper normal variation Γ of γ, the curve Γ_s is strictly longer than γ for all sufficiently small nonzero s unless the variation field of Γ is a Jacobi field.*
(c) *If γ has no conjugate points, then for every proper normal variation Γ of γ, the curve Γ_s is strictly longer than γ for all sufficiently small nonzero s.* □

There is a far-reaching quantitative generalization of Theorems 10.26 and 10.28, called the *Morse index theorem,* which we do not treat here. The **index of a geodesic segment** is defined to be the maximum dimension of a linear space of proper normal vector fields along the segment on which I is negative definite. Roughly speaking, the index is the number of independent directions in which γ can be deformed to decrease its length. (Analogously, the index of a critical point of a function on $\mathbb{R}^n$ is defined as the number of negative eigenvalues of its Hessian.) The Morse index theorem says that the index of every geodesic segment is finite, and is equal to the number of its interior conjugate points counted with multiplicity. (Proofs can be found in [Mil63, CE08, dC92].)

Cut Points

Theorem 10.26 showed that once a geodesic passes its first conjugate point, it ceases to be minimizing. The converse, however, is not true: a geodesic can cease to be minimizing without reaching a conjugate point. For example, on the cylinder $\mathbb{S}^1 \times \mathbb{R}$ with the product metric, there are no conjugate points along any geodesic; but no geodesic segment that wraps more than halfway around the cylinder is minimizing (Fig. 10.10).

Therefore it is useful to make the following definitions. Suppose (M, g) is a complete, connected Riemannian manifold, p is a point of M, and $v \in T_p M$. Define the **cut time of** (p, v) by

$$t_{\mathrm{cut}}(p, v) = \sup \{ b > 0 : \text{the restriction of } \gamma_v \text{ to } [0, b] \text{ is minimizing} \},$$

where γ_v is the maximal geodesic starting at p with initial velocity v. Because γ_v is minimizing as long as its image stays inside a geodesic ball (Prop. 6.11), $t_{\mathrm{cut}}(p, v)$ is always positive; but it might be $+\infty$.

If $t_{\mathrm{cut}}(p, v) < \infty$, the **cut point of p along γ_v** is the point $\gamma_v\big(t_{\mathrm{cut}}(p, v)\big) \in M$. The **cut locus of p**, denoted by $\mathrm{Cut}(p)$, is the set of all $q \in M$ such that q is the cut point of p along some geodesic. Because the question whether a geodesic is minimizing is independent of parametrization, the cut point of p along γ_v is the same as the cut point along $\gamma_{\lambda v}$ for every positive constant λ, so we may as well restrict attention to unit vectors v. Theorem 10.26 says that the cut point (if it exists) occurs at or before the first conjugate point along every geodesic.

The determination of the cut locus of a point is typically very difficult; but the next example gives some special cases in which it is relatively simple.

Example 10.30 (Cut Loci).

(a) If $\big(\mathbb{S}^n(R), \overset{\circ}{g}_R\big)$ is a sphere with a round metric, the cut locus of every point $p \in \mathbb{S}^n(R)$ is the singleton set containing only the antipodal point $-p$.

(b) On a product space $\mathbb{S}^n(R) \times \mathbb{R}^m$ with the product metric, the cut locus of every point (p, x) is the set $\{-p\} \times \mathbb{R}^m$. The case $n = m = 1$ is illustrated in Figure 10.10. //

▶ **Exercise 10.31.** Verify the claims in the preceding example.

Proposition 10.32 (Properties of Cut Times). *Suppose (M, g) is a complete, connected Riemannian manifold, $p \in M$, and v is a unit vector in $T_p M$. Let $c = t_{\mathrm{cut}}(p, v) \in (0, \infty]$.*

(a) If $0 < b < c$, then $\gamma_v|_{[0,b]}$ has no conjugate points and is the unique unit-speed minimizing curve between its endpoints.

(b) If $c < \infty$, then $\gamma_v|_{[0,c]}$ is minimizing, and one or both of the following conditions are true:

- *$\gamma_v(c)$ is conjugate to p along γ_v.*
- *There are two or more unit-speed minimizing geodesics from p to $\gamma_v(c)$.*

Proof. Suppose first that $0 < b < c$. By definition of $t_{\mathrm{cut}}(p, v)$, there is a time b' such that $b < b' < c$ and $\gamma_v|_{[0,b']}$ is minimizing. Then $\gamma_v(t)$ cannot be conjugate to p along γ_v for any $0 < t \le b$ (Thm. 10.26), and $\gamma_v|_{[0,b]}$ is minimizing because a shorter admissible curve from p to $\gamma_v(b)$ could be combined with $\gamma_v|_{[b,b']}$ to produce a shorter admissible curve from p to $\gamma_v(b')$, contradicting the fact that $\gamma_v|_{[0,b']}$ is minimizing.

To see that $\gamma_v|_{[0,b]}$ is the unique unit-speed minimizing curve between its endpoints, suppose for the sake of contradiction that $\sigma: [0, b] \to M$ is another. Note that $\sigma'(b) \ne \gamma_v'(b)$, since otherwise σ and γ_v would agree on $[0, b]$ by uniqueness of geodesics. Define a new unit-speed admissible curve $\widetilde{\gamma}: [0, b'] \to M$ that is equal to $\sigma(t)$ for $t \in [0, b]$ and equal to $\gamma_v(t)$ for $t \in [b, b']$. Then $\widetilde{\gamma}$ has length b', so it is also a minimizing curve from p to $\gamma_v(b')$; but it is not smooth at $t = b$, contradicting the fact that minimizing curves are smooth geodesics. This completes the proof of (a).

Now suppose $c < \infty$. By definition of $t_{\text{cut}}(p,v)$, there is a sequence of times $b_i \nearrow c$ such that the restriction of γ_v to $[0,b_i]$ is minimizing. By continuity of the distance function, therefore,

$$d_g(p,\gamma_v(c)) = \lim_{i\to\infty} d_g(p,\gamma_v(b_i)) = \lim_{i\to\infty} b_i = c,$$

which shows that γ_v is minimizing on $[0,c]$. To prove that one of the options in (b) must hold, assume that $\gamma_v(c)$ is not conjugate to p along γ_v. We will prove the existence of a second unit-speed minimizing geodesic from p to $\gamma_v(c)$.

Let (b_i) be a sequence of real numbers such that $b_i \searrow c$. By definition of cut time, $\gamma_v|_{[0,b_i]}$ is not minimizing, so for each i there is a unit-speed minimizing geodesic $\sigma_i : [0,a_i] \to M$ such that $\sigma_i(0) = p$, $\sigma_i(a_i) = \gamma_v(b_i)$, and $a_i < b_i$. Set $w_i = \sigma_i'(0) \in T_pM$, so each w_i is a unit vector. By compactness of the unit sphere, after passing to a subsequence we may assume that w_i converges to some unit vector w. Since the a_i's are all positive and bounded above by b_1, by passing to a further subsequence, we may also assume that a_i converges to some number a. Then by continuity of the exponential map, $\sigma_i(a_i) = \exp_p(a_i w_i)$ converges to $\exp_p(aw)$. But we also know that $\sigma_i(a_i) = \gamma_v(b_i)$, which converges to $\gamma_v(c)$, so $\exp_p(aw) = \gamma_v(c)$. Moreover, by continuity of the distance function,

$$c = d_g(p,\gamma_v(c)) = \lim_{i\to\infty} d_g(p,\sigma_i(a_i)) = \lim_{i\to\infty} a_i = a.$$

Thus $\sigma : [0,c] \to M$ given by $\sigma(t) = \exp_p(tw)$ is also a unit-speed minimizing geodesic from p to $\gamma_v(c)$. We need to show that it is not equal to γ_v.

The assumption that $\gamma_v(c)$ is not conjugate to p along γ_v implies that cv is a regular point of $\exp_p$ (Prop. 10.20), so $\exp_p$ is injective in some neighborhood V of cv. Note that $\exp_p(a_i w_i) = \exp_p(b_i v)$ for each i, while $a_i w_i \neq b_i v$, since w_i and v are unit vectors and $a_i < b_i$. Since $b_i v$ converges to cv, we conclude that $b_i v \in V$ for sufficiently large i, and thus by injectivity $a_i w_i \notin V$ for these values of i. Therefore $cw = \lim_{i\to\infty} a_i w_i \neq cv$, which implies $w \neq v$ and thus $\sigma \neq \gamma_v$, as claimed. $\square$

Next we examine how the cut time varies as the initial point and initial velocity of the geodesic vary. Recall that the *unit tangent bundle* of a Riemannian manifold (M,g) is the subset $UTM = \{(p,v) \in TM : |v|_g = 1\} \subseteq TM$. In the next theorem, we interpret continuity of a function into $(0,\infty]$ using the usual definition of infinite limits as in ordinary calculus.

Theorem 10.33. *Suppose (M,g) is a complete, connected Riemannian manifold. The function $t_{cut}: UTM \to (0,\infty]$ is continuous.*

Proof. Let $(p,v) \in UTM$ be arbitrary, and let (p_i,v_i) be any sequence in UTM converging to (p,v). Put $c_i = t_{\text{cut}}(p_i,v_i)$, and

$$b = \liminf_{i\to\infty} c_i, \qquad c = \limsup_{i\to\infty} c_i.$$

We will show that $c \leq t_{\text{cut}}(p, v) \leq b$, which implies $c_i \to t_{\text{cut}}(p, v)$.

To show that $c \leq t_{\text{cut}}(p, v)$, suppose first that $c < \infty$. By passing to a subsequence, we may assume that c_i is finite for each i and $c_i \to c$. Proposition 10.32 shows that γ_{v_i} is minimizing on $[0, c_i]$. By continuity of the exponential map, $\exp(p_i, c_i v_i) \to \exp(p, cv)$ as $i \to \infty$, and therefore by continuity of the distance function we have

$$d_g(p, \exp(p, cv)) = \lim_{i \to \infty} d_g(p_i, \exp(p_i, c_i v_i)) = \lim_{i \to \infty} c_i = c.$$

This shows that γ_v is minimizing on $[0, c]$, and therefore $t_{\text{cut}}(p, v) \geq c$, as claimed.

Now suppose $c = \infty$. Again, by passing to a subsequence, we may assume $c_i \to \infty$. It follows that for every positive number c_0, the geodesic γ_{v_i} is minimizing on $[0, c_0]$ for i sufficiently large, and it follows by continuity as above that γ_v is minimizing on $[0, c_0]$. Since c_0 was arbitrary, this means that $t_{\text{cut}}(p, v) = \infty$.

Next we show that $t_{\text{cut}}(p, v) \leq b$. If $b = \infty$, there is nothing to prove, so assume $b < \infty$. Again by passing to a subsequence, we may assume that c_i is finite for each i and $c_i \to b$. By virtue of Proposition 10.32, either there are infinitely many indices i for which $\gamma_{v_i}(c_i)$ is conjugate to p_i along γ_{v_i}, or there are infinitely many i for which there exists a second minimizing unit-speed geodesic σ_i from p_i to $\gamma_{v_i}(c_i)$.

In the first case, because conjugate points are critical values of the restricted exponential map, which can be detected in coordinates by the vanishing of a determinant of a matrix of first derivatives, it follows by continuity that $\gamma_v(b)$ is also a critical value, and thus $\gamma_v(b)$ is conjugate to p along γ_v. Then Theorem 10.26 shows that $t_{\text{cut}}(p, v) \leq b$.

In the second case, let w_i be the unit vector in $T_{p_i} M$ such that $\sigma_i = \gamma_{w_i}$. Because the components of w_i with respect to a local orthonormal frame lie in $\mathbb{S}^{n-1}$, by passing to a subsequence we may assume $(p_i . w_i) \to (p, w)$. If $\gamma_v(b)$ is conjugate to p along γ_v, then $t_{\text{cut}}(p, v) \leq b$ as above, so we may assume that $\gamma_v(b)$ is not a conjugate point. This means that bv is a regular point of the restricted exponential map $\exp_p$. Since the set of such regular points is an open subset of TM, there is some $\varepsilon > 0$ such that $\exp_{p_i}$ is injective on the ε-neighborhood of $c_i w_i$ for all i sufficiently large. This implies that $|c_i w_i - c_i v_i|_g \geq \varepsilon$ for all such i, and therefore the limits bw and bv are distinct. Thus $\gamma_w |_{[0,b]}$ is another minimizing geodesic from p to $\gamma_v(b)$, which by Proposition 10.32 implies that $t_{\text{cut}}(p, v) \leq b$. $\square$

Given $p \in M$, we define two subsets of $T_p M$ as follows: the **tangent cut locus of p** is the set

$$\text{TCL}(p) = \left\{ v \in T_p M : |v| = t_{\text{cut}}(p, v/|v|) \right\},$$

and the **injectivity domain of p** is

$$\text{ID}(p) = \left\{ v \in T_p M : |v| < t_{\text{cut}}(p, v/|v|) \right\}.$$

It is immediate that $\text{TCL}(p) = \partial \text{ID}(p)$, and $\text{Cut}(p) = \exp_p(\text{TCL}(p))$. Further properties of $\text{Cut}(p)$ and $\text{ID}(p)$ are described in the following theorem.

Theorem 10.34. *Let (M, g) be a complete, connected Riemannian manifold and $p \in M$.*

(a) *The cut locus of p is a closed subset of M of measure zero.*

(b) *The restriction of $\exp_p$ to $\overline{\mathrm{ID}(p)}$ is surjective.*

(c) *The restriction of $\exp_p$ to $\mathrm{ID}(p)$ is a diffeomorphism onto $M \smallsetminus \mathrm{Cut}(p)$.*

Proof. To prove that the cut locus is closed, suppose (q_i) is a sequence of points in $\mathrm{Cut}(p)$ converging to some $q \in M$. Write $q_i = \exp_p(t_{\mathrm{cut}}(p, v_i)v_i)$ for unit vectors v_i. By compactness of the unit sphere, we may assume after passing to a subsequence that v_i converges to some unit vector v, and by Theorem 10.33, $t_{\mathrm{cut}}(p, v) = \lim_{i \to \infty} t_{\mathrm{cut}}(p, v_i)$. Because convergent sequences in a metric space are bounded, the sequence $(t_{\mathrm{cut}}(p, v_i))$ is bounded, and therefore $t_{\mathrm{cut}}(p, v) < \infty$. By continuity of the exponential map, therefore, q must be equal to $\exp_p(t_{\mathrm{cut}}(p, v))$, which shows that $q \in \mathrm{Cut}(p)$, and thus $\mathrm{Cut}(p)$ is closed.

To see that $\mathrm{Cut}(p)$ has measure zero, note first that in any polar coordinates $(\theta^1, \dots, \theta^{n-1}, r)$ on $T_p M$, the set $\mathrm{TCL}(p)$ can be expressed locally as the graph of the continuous function $r = t_{\mathrm{cut}}(p, (\theta^1, \dots, \theta^{n-1}))$, using the fact that $(\theta^1, \dots, \theta^{n-1})$ form smooth local coordinates for the unit sphere in $T_p M$. Since graphs of continuous functions have measure zero (see, for example, [LeeSM, Prop. 6.3]), it follows that $\mathrm{TCL}(p)$ is a union of finitely many sets of measure zero and thus has measure zero in $T_p M$; and because smooth maps take sets of measure zero to sets of measure zero (see [LeeSM, Prop. 6.4]), $\mathrm{Cut}(p) = \exp_p(\mathrm{TCL}(p))$ has measure zero in M. This proves (a).

Part (b) follows from the fact that every point of M can be connected to p by a minimizing geodesic. To prove (c), note that it follows easily from the definitions that $\exp_p(\mathrm{ID}(p)) = M \smallsetminus \mathrm{Cut}(p)$. Also, the definition of $\mathrm{ID}(p)$ guarantees that no point in $\exp_p(\mathrm{ID}(p))$ can be a cut point of p, and thus no such point can be a conjugate point either. The absence of cut points implies that $\exp_p$ is injective on $\mathrm{ID}(p)$, and the absence of conjugate points implies that it is a local diffeomorphism there. Together these two facts imply that it is a diffeomorphism onto its image. $\square$

The preceding theorem leads to the following intriguing topological result about compact manifolds.

Corollary 10.35. *Every compact, connected, smooth n-manifold is homeomorphic to a quotient space of $\overline{\mathbb{B}}^n$ by an equivalence relation that identifies only points on the boundary.*

Proof. Let M be a compact, connected, smooth n-manifold, let p be any point of M, and let g be any Riemannian metric on M. Because a compact metric space has finite diameter, every unit vector in $T_p M$ has a finite cut time, no greater than the diameter of M. Let $\overline{B}_1(0) \subseteq T_p M$ denote the closed unit ball in $T_p M$, and define a map $f : \overline{B}_1(0) \to \overline{\mathrm{ID}(p)}$ by

$$f(v) = \begin{cases} t_{\mathrm{cut}}\left(p, \dfrac{v}{|v|_g}\right) v, & v \neq 0, \\ 0 & v = 0. \end{cases}$$

It follows from Theorem 10.33 that f is continuous, and it is easily seen to be bijective, so it is a homeomorphism by the closed map lemma (Lemma A.4). Since every orthonormal basis for $T_p M$ yields a homeomorphism of $\bar{B}_1(0)$ with $\bar{\mathbb{B}}^n$, it follows that $\overline{\mathrm{ID}(p)}$ is homeomorphic to $\bar{\mathbb{B}}^n$ and the homeomorphism takes $\mathrm{TCL}(p) = \partial \mathrm{ID}(p)$ to $\mathbb{S}^{n-1}$.

By Theorem 10.34, $\exp_p$ restricts to a surjective map from $\overline{\mathrm{ID}(p)}$ to M, and it is a quotient map by the closed map lemma. It follows that M is homeomorphic to the quotient of $\overline{\mathrm{ID}(p)}$ by the equivalence relation $v \sim w$ if and only if $\exp_p(v) = \exp_p(w)$. Since $\exp_p$ is injective on $\mathrm{ID}(p)$ and the images of $\mathrm{ID}(p)$ and $\partial \mathrm{ID}(p) = \mathrm{TCL}(p)$ are disjoint, the equivalence relation identifies only points on the boundary of $\mathrm{ID}(p)$. □

Recall from Chapter 6 that the *injectivity radius of M at p*, denoted by $\mathrm{inj}(p)$, is the supremum of all positive numbers a such that $\exp_p$ is a diffeomorphism from $B_a(0) \subseteq T_p M$ to its image. The injectivity radius is closely related to the cut locus, as the next proposition shows.

Proposition 10.36. *Let (M, g) be a complete, connected Riemannian manifold. For each $p \in M$, the injectivity radius at p is the distance from p to its cut locus if the cut locus is nonempty, and infinite otherwise.*

Proof. Given $p \in M$, let d denote the distance from p to its cut locus, with the convention that $d = \infty$ if the cut locus is empty. Let $a \in (0, \infty]$ be arbitrary, and let $B_a \subseteq T_p M$ denote the set of vectors $v \in T_p M$ with $|v|_g < a$ (so $B_a = T_p M$ if $a = \infty$). We will show that the restriction of $\exp_p$ to B_a is a diffeomorphism onto its image if and only if $a \leq d$, from which the result follows.

First suppose $a \leq d$. By definition of d, no point of the form $\exp_p(v)$ with $v \in B_a$ can be a cut point of p, so $B_a \subseteq \mathrm{ID}(p)$. It follows from Theorem 10.34(c) that $\exp_p$ is a diffeomorphism from B_a onto its image.

On the other hand, if $a > d$, then p has a cut point q whose distance from p is less than a. It follows from the definition of cut points that the radial geodesic from p to q is not minimizing past q, so Proposition 6.11 shows that there is no geodesic ball of radius greater than $d_g(p, q)$. In particular, the restriction of $\exp_p$ to B_a cannot be a diffeomorphism onto its image. □

Proposition 10.37. *Let (M, g) be a complete, connected Riemannian manifold. The function* $\mathrm{inj} \colon M \to (0, \infty]$ *is continuous.*

Proof. Let $p \in M$ be arbitrary. Proposition 10.32(b) shows that for each point $q \in \mathrm{Cut}(p)$, there is a minimizing unit-speed geodesic γ_v from p to q whose length is $t_{\mathrm{cut}}(p, v)$, and therefore the distance from p to $\mathrm{Cut}(p)$ is the infimum of the cut times of unit-speed geodesics starting at p. By the previous proposition, therefore,

$$\mathrm{inj}(p) = \inf \{ t_{\mathrm{cut}}(p, v) : v \in T_p M \text{ with } |v|_g = 1 \}.$$

Suppose (p_i) is a sequence in M converging to a point $p \in M$. As in the proof of Theorem 10.33, we will prove that $\mathrm{inj}(p_i) \to \mathrm{inj}(p)$ by showing that $c \leq \mathrm{inj}(p) \leq b$, where

$$b = \liminf_{i \to \infty} \mathrm{inj}(p_i), \qquad c = \limsup_{i \to \infty} \mathrm{inj}(p_i).$$

First we show that $\mathrm{inj}(p) \le b$. By passing to a subsequence, we may assume $\mathrm{inj}(p_i) \to b$. By compactness of the unit sphere, for each i there is a unit vector $v_i \in T_{p_i} M$ such that $\mathrm{inj}(p_i) = t_{\mathrm{cut}}(p_i, v_i)$, and after passing to a further subsequence, we may assume $(p_i, v_i) \to (p, v)$ for some $v \in T_p M$. By continuity of t_{cut}, we have $t_{\mathrm{cut}}(p, v) = \lim_i t_{\mathrm{cut}}(p_i, v_i) = b$, so $\mathrm{inj}(p) \le b$.

Next we show that $\mathrm{inj}(p) \ge c$. Once again, by passing to a subsequence of the original sequence (p_i), we may assume $\mathrm{inj}(p_i) \to c$. Suppose for the sake of contradiction that $\mathrm{inj}(p) < c$, and choose c_0 such that $\mathrm{inj}(p) < c_0 < c$. Let w be a unit vector in $T_p M$ such that $t_{\mathrm{cut}}(p, w) = \mathrm{inj}(p)$. We can choose some sequence of unit vectors $w_i \in T_{p_i} M$ such that $(p_i, w_i) \to (p, w)$, so $t_{\mathrm{cut}}(p_i, w_i) \to t_{\mathrm{cut}}(p, w) = \mathrm{inj}(p)$. For i sufficiently large, this implies $t_{\mathrm{cut}}(p_i, w_i) < c_0 < c$, contradicting the facts that $t_{\mathrm{cut}}(p_i, w_i) \ge \mathrm{inj}(p_i)$ and $\mathrm{inj}(p_i) \to c$. $\qquad\square$

Problems

10-1. Suppose (M, g) is a Riemannian manifold and $p \in M$. Show that the second-order Taylor series of g in normal coordinates centered at p is

$$g_{ij}(x) = \delta_{ij} - \frac{1}{3} \sum_{k,l} R_{iklj}(p) x^k x^l + O\big(|x|^3\big).$$

[Hint: Let $\gamma(t) = (tv^1, \dots, tv^n)$ be a radial geodesic starting at p, let $J(t) = tw^i \partial_i |_{\gamma(t)}$ be a Jacobi field along γ, and compute the first four t-derivatives of $|J(t)|^2$ at $t = 0$ in two ways.]

10-2. Suppose (M, g) is a Riemannian manifold and $p \in M$. Let $S_\Pi \subseteq M$ be the 2-dimensional submanifold obtained by applying the exponential map to a plane $\Pi \subseteq T_p M$, as on page 250. For sufficiently small r, let $A(r)$ be the area of the geodesic disk of radius r about p in S_Π with its induced metric. Using the results of Problems 10-1 and 8-21, find the Taylor series of $A(r)$ to fourth order in r. Then use this result to express $\sec(\Pi)$ in terms of a limit involving the difference $\pi r^2 - A(r)$.

10-3. Extend the result of Proposition 10.12 by finding a basis for the space of *all* Jacobi fields along a geodesic in the constant-curvature case, not just the normal ones that vanish at 0.

10-4. Prove that the volume of the round sphere $\mathbb{S}^n(R)$ is given for $n \ge 0$ by $\mathrm{Vol}\big(\mathbb{S}^n(R)\big) = \alpha_n R^n$, where

$$\alpha_n = \begin{cases} \dfrac{2^{2k+1}\pi^k k!}{(2k)!}, & n = 2k,\ k \in \mathbb{Z}, \\[2ex] \dfrac{2\pi^{k+1}}{k!}, & n = 2k+1,\ k \in \mathbb{Z}. \end{cases}$$

(The volume of a compact 0-manifold is just the number of points.)

(a) First show that it suffices to prove the volume formula for $R = 1$.

(b) Use Corollary 10.17 to prove the recurrence relation

$$\mathrm{Vol}\big(\mathbb{S}^{n+1}\big) = \sigma_n \,\mathrm{Vol}\big(\mathbb{S}^n\big),$$

where

$$\sigma_n = \int_0^\pi (\sin r)^n \, dr.$$

(c) By differentiating the function $(\sin r)^n \cos r$, prove that

$$\sigma_{n+1} = \frac{n}{n+1}\sigma_{n-1},$$

and use this to prove that

$$\sigma_n \sigma_{n-1} = \frac{2\pi}{n}.$$

(d) Prove the result by induction on k.

(*Used on p. 342.*)

10-5. For $r > 0$, let $B_r(0)$ denote the ball of radius r in Euclidean space $\big(\mathbb{R}^n, \bar{g}\big)$, $n \geq 1$. Prove that $\mathrm{Vol}\big(B_r(0)\big) = \frac{1}{n}\alpha_{n-1}r^n$, with α_{n-1} as in Problem 10-4.

10-6. Let p be a point in a Riemannian n-manifold (M, g). Use the results of Problems 10-1 and 8-21 to show that as $r \searrow 0$,

$$\mathrm{Vol}(B_r(p)) = \frac{1}{n}\alpha_{n-1}r^n\left(1 - \frac{S(p)r^2}{6(n+2)} + O\big(r^3\big)\right),$$

where $S(p)$ is the scalar curvature of g at p and α_{n-1} is as in Problem 10-4.

10-7. Suppose (M, g) is a Riemannian manifold with nonpositive sectional curvature. Prove that no point of M has conjugate points along any geodesic. [Hint: Consider derivatives of $|J(t)|^2$ when J is a Jacobi field.] (*Used on p. 333.*)

10-8. Prove Proposition 10.21 (solvability of the two-point boundary problem).

10-9. Suppose (M, g) is a Riemannian manifold and $M_1, M_2 \subseteq M$ are embedded submanifolds. Let $\gamma \colon [a, b] \to M$ be a unit-speed geodesic segment that meets M_1 orthogonally at $t = a$ and meets M_2 orthogonally at $t = b$, and let $\Gamma \colon K \times [a, b] \to M$ be a normal variation of γ such that $\Gamma(s, a) \in M_1$

and $\Gamma(s,b) \in M_2$ for all s. Prove the following generalization of the second variation formula:

$$\frac{d^2}{ds^2}\bigg|_{s=0} L_g(\Gamma_s) = \int_a^b \left(|D_t V|^2 - Rm\left(V, \gamma', \gamma', V\right)\right) dt$$
$$+ \langle \mathrm{II}_2(V(b), V(b)), \gamma'(b)\rangle - \langle \mathrm{II}_1(V(a), V(a)), \gamma'(a)\rangle,$$

where V is the variation field of Γ, and II_i is the second fundamental form of M_i for $i = 1, 2$. (*Used on p. 365*.)

10-10. Prove the following theorem of Theodore Frankel [Fra61], generalizing the well-known fact that any two great circles on $\mathbb{S}^2$ must intersect: Suppose (M,g) is a complete, connected Riemannian manifold with positive sectional curvature. If $M_1, M_2 \subseteq M$ are compact, totally geodesic submanifolds such that $\dim M_1 + \dim M_2 \geq \dim M$, then $M_1 \cap M_2 \neq \varnothing$. [Hint: Assuming that the intersection is empty, show that there exist a shortest geodesic segment γ connecting M_1 and M_2 and a parallel vector field along γ that is tangent to M_1 and M_2 at the endpoints; then apply the second variation formula of Problem 10-9 to derive a contradiction.]

10-11. Prove Corollary 10.25 ($I(V,W) = 0$ for all W if and only if V is a Jacobi field). [Hint: Adapt the proof of Theorem 6.4.]

10-12. Let (M,g) be a Riemannian manifold. Suppose $\gamma \colon [a,b] \to M$ is a unit-speed geodesic segment with no interior conjugate points, J is a normal Jacobi field along γ, and V is any other piecewise smooth normal vector field along γ such that $V(a) = J(a)$ and $V(b) = J(b)$.

   (a) Show that $I(V, V) \geq I(J, J)$.
   (b) Now assume in addition that $\gamma(b)$ is not conjugate to $\gamma(a)$ along γ. Show that $I(V, V) = I(J, J)$ if and only if $V = J$.

10-13. Suppose (M,g) is a Riemannian manifold and $X \in \mathfrak{X}(M)$ is a Killing vector field (see Problems 5-22 and 6-24). Show that if $\gamma \colon [a,b] \to M$ is any geodesic segment, then X restricts to a Jacobi field along γ.

10-14. Suppose P is an embedded submanifold of a Riemannian manifold M, and $\gamma \colon I \to M$ is a geodesic that meets P orthogonally at $t = a$ for some $a \in I$. A Jacobi field J along γ is said to be **transverse to P** if its restriction to each compact subinterval of I containing a is the variation field of a variation of γ through geodesics that all meet P orthogonally at $t = a$.

   (a) Prove that the tangential Jacobi field $J(t) = (t - a)\gamma'(t)$ along γ is transverse to P.
   (b) Prove that a normal Jacobi field J along γ is transverse to P if and only if
   $$D_t J(a) = -W_{\gamma'(a)}(J(a)),$$
   where $W_{\gamma'(a)}$ is the Weingarten map of P in the direction of $\gamma'(a)$.

(c) When M has dimension n, prove that the set of transverse Jacobi fields along γ is an n-dimensional linear subspace of $\mathcal{J}(\gamma)$, and the set of transverse normal Jacobi fields is an $(n-1)$-dimensional subspace of that.

(*Used on p. 342.*)

10-15. Let $(x^1, \dots, x^n)$ be any semigeodesic coordinates on an open subset U in a Riemannian n-manifold (M, g) (see Prop. 6.41 and Examples 6.43–6.46), and let $\gamma(t) = (x^1, \dots, x^{n-1}, t)$ be an x^n-coordinate curve defined on some interval I. Prove that for all constants $(a^1, \dots, a^{n-1})$, the following vector field along γ is a normal Jacobi field along γ that is transverse to each of the level sets of x^n (in the sense defined in Problem 10-14):

$$J(t) = \sum_{i=1}^{n-1} a^i \left. \frac{\partial}{\partial x^i} \right|_{\gamma(t)}.$$

10-16. Suppose P is an embedded k-dimensional submanifold of a Riemannian n-manifold (M, g), and $(x^1, \dots, x^k, v^1, \dots, v^{n-k})$ are Fermi coordinates for P on some open subset $U_0 \subseteq M$. For fixed (x^i, v^j), let $\gamma : I \to M$ be the curve with coordinate representation $\gamma(t) = (x^1, \dots, x^k, tv^1, \dots, tv^{n-k})$; Proposition 5.26 shows that γ is a geodesic that meets P orthogonally at $t = 0$. Prove that the Jacobi fields along γ that are transverse to P are exactly the vector fields of the form

$$J(t) = \sum_{i=1}^{k} a^i \left. \frac{\partial}{\partial x^i} \right|_{\gamma(t)} + \sum_{j=1}^{n-k} tb^j \left. \frac{\partial}{\partial v^j} \right|_{\gamma(t)},$$

for arbitrary constants $a^1, \dots, a^k, b^1, \dots, b^{n-k}$.

10-17. Suppose P is an embedded submanifold of a Riemannian manifold (M, g), and U is a normal neighborhood of P in M. Prove that every tangent vector to $U \smallsetminus P$ is the value of a transverse Jacobi field: more precisely, for each $q \in U \smallsetminus P$ and each $w \in T_q M$, there is a g-geodesic segment $\gamma : [0, b] \to U$ such that $\gamma(0) \in P$, $\gamma'(0) \perp T_{\gamma(0)} P$, and $\gamma(b) = q$, and a Jacobi field J along γ that is transverse to P and satisfies $J(b) = w$. [Hint: Use Problem 10-16.]

10-18. Suppose (M_1, g_1) and (M_2, g_2) are Riemannian manifolds, $f : M_1 \to \mathbb{R}^+$ is a smooth positive function, and $M_1 \times_f M_2$ is the resulting warped product manifold. Let $q_0 \in M_2$ be arbitrary, let $\gamma : I \to M_1$ be a g_1-geodesic, and let $\tilde{\gamma} : I \to M_1 \times_f M_2$ be the curve $\tilde{\gamma}(t) = (\gamma(t), q_0)$. It follows from the result of Problem 5-7 that $\tilde{\gamma}$ is a geodesic meeting each submanifold $\{\gamma(t)\} \times M_2$ orthogonally. Given any fixed vector $w \in T_{q_0} M_2$, define a vector field J along $\tilde{\gamma}$ by $J(t) = (0, w) \in T_{\gamma(t)} M_1 \oplus T_{q_0} M_2$. Prove that J is a Jacobi field that is transverse to each submanifold $\{\gamma(t)\} \times M_2$ (in the sense defined in Problem 10-14).

10-19. Suppose P is an embedded submanifold in a Riemannian manifold (M, g). A point $q \in M$ is said to be a *focal point of* P if it is a critical value of the normal exponential map $E \colon \mathcal{E}_P \to M$ (see p. 133). Show that q is a focal point of P if and only if there exist a geodesic segment $\gamma \colon [0, b] \to M$ that starts normal to P and ends at q and a nontrivial Jacobi field $J \in \mathcal{J}(\gamma)$ that is transverse to P in the sense defined in Problem 10-14 and satisfies $J(b) = 0$. (If this is the case, we say that q is a *focal point of* P *along* γ.)

10-20. Suppose (M, g) is a Riemannian manifold with nonpositive sectional curvature, and $P \subseteq M$ is a totally geodesic embedded submanifold. Prove that P has no focal points. [Hint: See Problem 10-7.] (*Used on p. 370.*)

10-21. Determine the cut locus of an arbitrary point in the n-torus $\mathbb{T}^n$ with the flat metric of Examples 2.21 and 7.1.

10-22. Suppose (M, g) is a connected, compact Riemannian manifold, $p \in M$, and $C \subseteq M$ is the cut locus of p. Prove that C is homotopy equivalent to $M \smallsetminus \{p\}$.

10-23. Let (M, g) be a complete Riemannian manifold, and suppose $p, q \in M$ are points such that $d_g(p, q)$ is equal to the distance from p to its cut locus.

    (a) Prove that either q is conjugate to p along some minimizing geodesic segment, or there are exactly two minimizing geodesic segments from p to q, say $\gamma_1, \gamma_2 \colon [0, b] \to M$, such that $\gamma_1'(b) = -\gamma_2'(b)$.

    (b) Now suppose in addition that $\operatorname{inj}(p) = \operatorname{inj}(M)$. Prove that if q is not conjugate to p along any minimizing geodesic, then there is a closed geodesic $\gamma \colon [0, 2b] \to M$ such that $\gamma(0) = \gamma(2b) = p$ and $\gamma(b) = q$, where $b = d_g(p, q)$.

(*Used on p. 343.*)

10-24. Let (M, g) be a complete Riemannian manifold and $p \in M$. Show that $\operatorname{inj}(p)$ is equal to the radius of the largest open ball in $T_p M$ on which $\exp_p$ is injective.

(*Used on p. 166.*)

Chapter 11
Comparison Theory

The purpose of this chapter is to show how upper or lower bounds on curvature can be used to derive bounds on other geometric quantities such as lengths of tangent vectors, distances, and volumes. The intuition behind all the comparison theorems is that negative curvature forces geodesics to spread apart faster as you move away from a point, and positive curvature forces them to spread slower and eventually to begin converging.

One of the most useful comparison theorems is the *Jacobi field comparison theorem* (see Thm. 11.9 below), which gives bounds on the sizes of Jacobi fields based on curvature bounds. Its importance is based on four observations: first, in a normal neighborhood of a point p, every tangent vector can be represented as the value of a Jacobi field that vanishes at p (by Cor. 10.11); second, zeros of Jacobi fields correspond to conjugate points, beyond which geodesics cannot be minimizing; third, Jacobi fields represent the first-order behavior of families of geodesics; and fourth, each Jacobi field satisfies a differential equation that directly involves the curvature.

In the first section of the chapter, we set the stage for the comparison theorems by showing how the growth of Jacobi fields in a normal neighborhood is controlled by the Hessian of the radial distance function, which satisfies a first-order differential equation called a *Riccati equation*. We then state and prove a fundamental comparison theorem for Riccati equations.

Next we proceed to derive some of the most important geometric comparison theorems that follow from the Riccati comparison theorem. The first few comparison theorems are all based on upper or lower bounds on sectional curvatures. Then we explain how some comparison theorems can also be proved based on lower bounds for the Ricci curvature. In the next chapter, we will see how these comparison theorems can be used to prove significant local-to-global theorems in Riemannian geometry.

Since all of the results in this chapter are deeply intertwined with lengths and distances, we restrict attention throughout the chapter to the Riemannian case.

© Springer International Publishing AG 2018
J. M. Lee, *Introduction to Riemannian Manifolds*, Graduate Texts
in Mathematics 176, https://doi.org/10.1007/978-3-319-91755-9_11

Jacobi Fields, Hessians, and Riccati Equations

Our main aim in this chapter is to use curvature inequalities to derive consequences about how fast the metric grows or shrinks, based primarily on size estimates for Jacobi fields. But first, we need to make one last stop along the way.

The Jacobi equation is a second-order differential equation, but comparison theory for differential equations generally works much more smoothly for first-order equations. In order to get the sharpest results about Jacobi fields and other geometric quantities, we will derive a first-order equation, called a *Riccati equation*, that is closely related to the Jacobi equation.

Let (M, g) be an n-dimensional Riemannian manifold, let U be a normal neighborhood of a point $p \in M$, and let $r \colon U \to \mathbb{R}$ be the radial distance function as defined by (6.4). The Gauss lemma shows that the gradient of r on $U \smallsetminus \{p\}$ is the radial vector field ∂_r.

On $U \smallsetminus \{p\}$, we can form the symmetric covariant 2-tensor field $\nabla^2 r$ (the covariant Hessian of r) and the $(1, 1)$-tensor field $\mathcal{H}_r = \nabla(\partial_r)$. Because $\partial_r = \operatorname{grad} r = (\nabla r)^\sharp$ and ∇ commutes with the musical isomorphisms (Prop. 5.17), we have

$$\mathcal{H}_r = \nabla(\partial_r) = \nabla\big((\nabla r)^\sharp\big) = \big(\nabla^2 r\big)^\sharp.$$

In other words, $\mathcal{H}_r$ is obtained from $\nabla^2 r$ by raising one of its indices.

Using Proposition B.1, we can also interpret the $(1, 1)$-tensor field $\mathcal{H}_r$ as an element of $\Gamma\big(\operatorname{End}(TM|_{U \smallsetminus \{p\}})\big)$ (that is, a field of endomorphisms of TM over $U \smallsetminus \{p\}$), defined by

$$\mathcal{H}_r(w) = \nabla_w \partial_r \tag{11.1}$$

for all $w \in TM|_{U \smallsetminus \{p\}}$. The endomorphism field $\mathcal{H}_r$ is called the **Hessian operator of r**. It is related to the $(0, 2)$-Hessian by

$$\langle \mathcal{H}_r(v), w \rangle = \big(\nabla^2 r\big)(v, w), \quad \text{for all } q \in U \smallsetminus \{p\} \text{ and } v, w \in T_q M. \tag{11.2}$$

The next lemma summarizes some of its basic algebraic properties.

Lemma 11.1. *Let r, ∂_r, and $\mathcal{H}_r$ be defined as above.*

(a) *$\mathcal{H}_r$ is self-adjoint.*
(b) *$\mathcal{H}_r(\partial_r) \equiv 0$.*
(c) *The restriction of $\mathcal{H}_r$ to vectors tangent to a level set of r is equal to the shape operator of the level set associated with the normal vector field $N = -\partial_r$.*

Proof. Since the covariant Hessian $\nabla^2 r$ is symmetric, equation (11.2) shows that the Hessian operator is self-adjoint. Part (b) follows immediately from the fact that $\mathcal{H}_r(\partial_r) = \nabla_{\partial_r} \partial_r = 0$ because the integral curves of ∂_r are geodesics.

Next, note that ∂_r is a unit vector field normal to each level set of r by the Gauss lemma, so (c) follows from the Weingarten equation 8.11. $\qquad\square$

Problem 11-1 gives another geometric interpretation of $\nabla^2 r$, as the radial derivative of the nonconstant components of the metric in polar normal coordinates.

The Hessian operator also has a close relationship with Jacobi fields.

Proposition 11.2. *Suppose (M, g) is a Riemannian manifold, $U \subseteq M$ is a normal neighborhood of $p \in M$, and r is the radial distance function on U. If $\gamma : [0, b] \to U$ is a unit-speed radial geodesic segment starting at p, and $J \in \mathcal{J}^\perp(\gamma)$ is a normal Jacobi field along γ that vanishes at $t = 0$, then the following equation holds for all $t \in (0, b]$:*

$$D_t J(t) = \mathcal{H}_r(J(t)). \tag{11.3}$$

Proof. Let $v = \gamma'(0)$, so $|v|_g = 1$ and $\gamma(t) = \exp_p(tv)$. Proposition 10.10 shows that

$$J(t) = d(\exp_p)_{tv}(tw),$$

where $D_t J(0) = w = w^i \partial_i|_p$. Because we are assuming that J is normal, it follows from Proposition 10.7 that $w \perp v$.

Because $w \perp v$ ensures that w is tangent to the unit sphere in $T_p M$ at v, we can choose a smooth curve $\sigma : (-\varepsilon, \varepsilon) \to T_p M$ that satisfies $|\sigma(s)|_g = 1$ for all $s \in (-\varepsilon, \varepsilon)$, with initial conditions $\sigma(0) = v$ and $\sigma'(0) = w$. (As always, we are using the canonical identification between $T_v(T_p M)$ and $T_p M$.) Define a smooth family of curves $\Gamma : (-\varepsilon, \varepsilon) \times [0, b] \to M$ by $\Gamma(s, t) = \exp_p(t\sigma(s))$. Then $\Gamma(0, t) = \exp_p(tv) = \gamma(t)$, so Γ is a variation of γ. The stipulation that $|\sigma(s)|_g \equiv 1$ ensures that each main curve $\Gamma_s(t) = \Gamma(s, t)$ is a unit-speed radial geodesic, so its velocity satisfies the following identity for all (s, t):

$$\partial_t \Gamma(s, t) = (\Gamma_s)'(t) = \partial_r\big|_{\Gamma(s,t)}. \tag{11.4}$$

The chain rule yields

$$\partial_s \Gamma(0, t) = d(\exp_p)_{t\sigma(0)}(t\sigma'(0)) = d(\exp_p)_{tv}(tw) = J(t),$$

so J is the variation field of Γ. By the symmetry lemma,

$$D_t J(t) = D_t \partial_s \Gamma(0, t) = D_s \partial_t \Gamma(0, t). \tag{11.5}$$

This last expression is the covariant derivative of $\partial_t \Gamma(s, t)$ along the curve $\Gamma^{(t)}(s) = \Gamma(s, t)$ evaluated at $s = 0$. Since the velocity of this curve at $s = 0$ is $\partial_s \Gamma(0, t) = J(t)$ and $\partial_t \Gamma(s, t) = \partial_r$ is an extendible vector field by (11.4), we obtain

$$D_s \partial_t \Gamma(0, t) = \nabla_{\Gamma^{(t)\prime}(0)}(\partial_r) = \nabla_{J(t)}(\partial_r) = \mathcal{H}_r(J(t)).$$

Combining this with (11.5) yields the result. □

In order to compare the Hessian operator of an arbitrary metric with those of the constant-curvature models, we need the following explicit formula for the constant-curvature case.

Proposition 11.3. *Suppose (M, g) is a Riemannian manifold, $U \subseteq M$ is a normal neighborhood of $p \in M$, and r is the radial distance function on U. Then g has* *constant sectional curvature c on U if and only if the following formula holds at all points of $U \smallsetminus \{p\}$:*

$$\mathcal{H}_r = \frac{s_c'(r)}{s_c(r)} \pi_r, \qquad (11.6)$$

where s_c is defined by (10.8), and for each $q \in U \smallsetminus \{p\}$, $\pi_r \colon T_q M \to T_q M$ is the orthogonal projection onto the tangent space of the level set of r (equivalently, onto the orthogonal complement of $\partial_r|_q$).

Proof. First suppose g has constant sectional curvature c on U. Let $q \in U \smallsetminus \{p\}$, and let $\gamma \colon [0, b] \to U$ be the unit-speed radial geodesic from p to q, so $b = r(q)$. Let $(E_1, \ldots, E_n)$ be a parallel orthonormal frame along γ, chosen so that $E_n(t) = \gamma'(t) = \partial_r|_{\gamma(t)}$. It follows from Proposition 10.12 that for $i = 1, \ldots, n-1$, the vector fields $J_i(t) = s_c(t)E_i(t)$ are normal Jacobi fields along γ that vanish at $t = 0$. The assumption that U is a normal neighborhood of p means that $U = \exp_p(V)$ for some star-shaped neighborhood V of $0 \in T_p M$, and every point of V is a regular point for $\exp_p$. Thus Proposition 10.20 shows that p has no conjugate points along γ, which implies that $s_c(t) \neq 0$ for $t \in (0, b]$. (For $c \leq 0$, this is automatic, because s_c vanishes only at 0; but in the case $c = 1/R^2 > 0$, it means that $b < \pi R$.)

For $1 \leq i \leq n-1$, we use Proposition 11.2 to compute

$$D_t J_i(t) = \mathcal{H}_r(J_i(t)) = s_c(t)\mathcal{H}_r(E_i(t)).$$

On the other hand, because each E_i is parallel along γ,

$$D_t J_i(t) = s_c'(t)E_i(t).$$

Comparing these two equations at $t = b$ and dividing by $s_c(b)$, we obtain

$$\mathcal{H}_r(E_i(b)) = \frac{s_c'(b)}{s_c(b)} E_i(b) = \frac{s_c'(b)}{s_c(b)} \pi_r(E_i(b)).$$

On the other hand, Lemma 11.1(b) shows that

$$\mathcal{H}_r(E_n(b)) = \mathcal{H}_r(\partial_r|_q) = 0 = \frac{s_c'(b)}{s_c(b)} \pi_r(E_n(b)),$$

because $\pi_r(E_n(b)) = \pi_r(\partial_r|_q) = 0$. Since $(E_i(b))$ is a basis for $T_q M$, this proves (11.6).

Conversely, suppose $\mathcal{H}_r$ is given by (11.6). Let γ be a radial geodesic starting at p, and let J be a normal Jacobi field along γ that vanishes at $t = 0$. By Proposition 11.2, $D_t J(t) = \mathcal{H}_r J(t) = s_c'(t)J(t)/s_c(t)$. A straightforward computation then shows that $s_c(t)^{-1}J(t)$ is parallel along γ. Thus we can write every such Jacobi field in the form $J(t) = ks_c(t)E(t)$ for some constant k and some parallel unit normal vector field E along γ. Proceeding exactly as in the proof of Theorem 10.14, we conclude that g is given by formula (10.17) in these coordinates, and therefore has constant sectional curvature c. $\square$

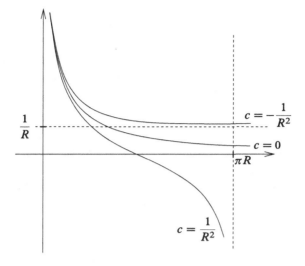

Fig. 11.1: The graph of s'_c/s_c

For convenience, we record the exact formulas for the quotient s'_c/s_c that appeared in the previous proposition (see Fig. 11.1):

$$\frac{s'_c(t)}{s_c(t)} = \begin{cases} \dfrac{1}{t}, & \text{if } c = 0; \\[2mm] \dfrac{1}{R}\cot\dfrac{t}{R}, & \text{if } c = \dfrac{1}{R^2} > 0; \\[2mm] \dfrac{1}{R}\coth\dfrac{t}{R}, & \text{if } c = -\dfrac{1}{R^2} < 0. \end{cases}$$

Now we are in a position to derive the first-order equation mentioned at the beginning of this section. (Problem 11-3 asks you to show, with a different argument, that the conclusion of the next theorem holds for the Hessian operator of *every* smooth local distance function, not just the radial distance function in a normal neighborhood.) This theorem concerns the covariant derivative of the endomorphism field $\mathcal{H}_r$ along a curve γ. We can compute the action of $D_t\mathcal{H}_r$ on every $V \in \mathfrak{X}(\gamma)$ by noting that $\mathcal{H}_r(V(t))$ is a contraction of $\mathcal{H}_r \otimes V(t)$, so the product rule implies $D_t(\mathcal{H}_r(V)) = (D_t\mathcal{H}_r)V + \mathcal{H}_r(D_tV)$.

Theorem 11.4 (The Riccati Equation). *Let (M, g) be a Riemannian manifold; let U be a normal neighborhood of a point $p \in M$; let $r: U \to \mathbb{R}$ be the radial distance function; and let $\gamma: [0, b] \to U$ be a unit-speed radial geodesic. The Hessian operator $\mathcal{H}_r$ satisfies the following equation along $\gamma|_{(0,b]}$, called a **Riccati equation**:*

$$D_t\mathcal{H}_r + \mathcal{H}_r^2 + R_{\gamma'} = 0, \tag{11.7}$$

where $\mathcal{H}_r^2$ and $R_{\gamma'}$ are the endomorphism fields along γ defined by $\mathcal{H}_r^2(w) = \mathcal{H}_r(\mathcal{H}_r(w))$ and $R_{\gamma'}(w) = R(w, \gamma')\gamma'$, with R the curvature endomorphism of g.

Proof. Let $t_0 \in (0, b]$ and $w \in T_{\gamma(t_0)}M$ be arbitrary. We can decompose w as $w = y + z$, where y is a multiple of ∂_r and z is tangent to a level set of r. Since (11.7) is an equation between linear operators, we can prove the equation by evaluating it separately on y and z.

Because γ is a unit-speed radial geodesic, its velocity is equal to ∂_r, and thus $D_t \partial_r = 0$ along γ. It follows that $(D_t \mathcal{H}_r)(\partial_r) = D_t(\mathcal{H}_r(\partial_r)) - \mathcal{H}_r(D_t \partial_r) = 0$. Since all three terms on the left-hand side of (11.7) annihilate ∂_r, the equation holds when applied to any multiple of ∂_r.

Next we consider a vector $z \in T_{\gamma(t_0)}M$ that is tangent to a level set of r, and thus by the Gauss lemma orthogonal to $\gamma'(t_0)$. By Corollary 10.11, z can be expressed as the value at $t = t_0$ of a Jacobi field J along γ vanishing at $t = 0$. Because $J(0)$ and $J(t_0)$ are orthogonal to γ', it follows that J is a normal Jacobi field, so Proposition 11.2 shows that $D_t J(t) = \mathcal{H}_r(J(t))$ for all $t \in [0, b]$. Differentiation yields

$$D_t^2 J = D_t(\mathcal{H}_r(J)) = (D_t \mathcal{H}_r)J + \mathcal{H}_r(D_t J) = (D_t \mathcal{H}_r)J + \mathcal{H}_r(\mathcal{H}_r(J)).$$

On the other hand, the Jacobi equation gives $D_t^2 J = -R_{\gamma'}(J)$, so

$$\left(D_t \mathcal{H}_r + \mathcal{H}_r^2 + R_{\gamma'}\right)(J) \equiv 0.$$

Evaluating this at $t = t_0$ proves the result.   □

The Riccati equation is named after Jacopo Riccati, an eighteenth-century Italian mathematician who studied scalar differential equations of the form $v' + pv^2 + qv + r = 0$, where p, q, r are known functions and v is an unknown function of one real variable. As is shown in some ODE texts, a linear second-order equation in one variable of the form $au'' + bu' + cu = 0$ can be transformed to a Riccati equation wherever $u \neq 0$ by making the substitution $v = u'/u$. The relation (11.3) generalizes this, and allows us to replace the analysis of the second-order linear Jacobi equation by an analysis of the first-order nonlinear Riccati equation.

The primary tool underlying all of our geometric comparison theorems is a fundamental comparison theorem for solutions to Riccati equations. It says, roughly, that a larger curvature term results in a smaller solution, and vice versa. When we apply this to (11.3), it will yield an analogous comparison for Jacobi fields.

In the statement and proof of this theorem, we will compare self-adjoint endomorphisms by comparing the quadratic forms they determine. Given a finite-dimensional inner product space V and self-adjoint endomorphisms $A, B \colon V \to V$, the notation $A \leq B$ means that $\langle Av, v \rangle \leq \langle Bv, v \rangle$ for all $v \in V$, or equivalently that $B - A$ is positive semidefinite. In particular, $B \geq 0$ means that B is positive semidefinite. Note that the square of every self-adjoint endomorphism is positive semidefinite, because $\langle B^2 v, v \rangle = \langle Bv, Bv \rangle \geq 0$ for all $v \in V$.

Theorem 11.5 (Riccati Comparison Theorem). *Suppose (M, g) is a Riemannian manifold and $\gamma \colon [a, b] \to M$ is a unit-speed geodesic segment. Suppose $\eta, \tilde{\eta}$*

are self-adjoint endomorphism fields along $\gamma|_{(a,b]}$ *that satisfy the following Riccati equations:*

$$D_t\eta + \eta^2 + \sigma = 0, \qquad D_t\tilde{\eta} + \tilde{\eta}^2 + \tilde{\sigma} = 0, \tag{11.8}$$

where σ *and* $\tilde{\sigma}$ *are continuous self-adjoint endomorphism fields along* γ *satisfying*

$$\tilde{\sigma}(t) \geq \sigma(t) \quad \text{for all } t \in [a,b]. \tag{11.9}$$

Suppose further that $\lim_{t\searrow a}\left(\tilde{\eta}(t) - \eta(t)\right)$ *exists and satisfies*

$$\lim_{t\searrow a}\left(\tilde{\eta}(t) - \eta(t)\right) \leq 0.$$

Then

$$\tilde{\eta}(t) \leq \eta(t) \quad \text{for all } t \in (a,b].$$

To prove this theorem, we will express the endomorphism fields η, $\tilde{\eta}$, σ, and $\tilde{\sigma}$ in terms of a parallel orthonormal frame along γ. In this frame, they become symmetric matrix-valued functions, and then the Riccati equations for η and $\tilde{\eta}$ become ordinary differential equations for these matrix-valued functions. The crux of the matter is the following comparison theorem for solutions to such matrix-valued equations.

Let $M(n,\mathbb{R})$ be the space of all $n \times n$ real matrices, viewed as linear endomorphisms of $\mathbb{R}^n$, and let $S(n,\mathbb{R}) \subseteq M(n,\mathbb{R})$ be the subspace of symmetric matrices, corresponding to self-adjoint endomorphisms of $\mathbb{R}^n$ with respect to the standard inner product.

Theorem 11.6 (Matrix Riccati Comparison Theorem). *Suppose* $H, \tilde{H}: (a,b] \to S(n,\mathbb{R})$ *satisfy the following matrix Riccati equations:*

$$H' + H^2 + S = 0, \qquad \tilde{H}' + \tilde{H}^2 + \tilde{S} = 0, \tag{11.10}$$

where $S, \tilde{S}: [a,b] \to S(n,\mathbb{R})$ *are continuous and satisfy*

$$\tilde{S}(t) \geq S(t) \quad \text{for all } t \in [a,b]. \tag{11.11}$$

Suppose further that $\lim_{t\searrow a}\left(\tilde{H}(t) - H(t)\right)$ *exists and satisfies*

$$\lim_{t\searrow a}\left(\tilde{H}(t) - H(t)\right) \leq 0. \tag{11.12}$$

Then

$$\tilde{H}(t) \leq H(t) \quad \text{for all } t \in (a,b]. \tag{11.13}$$

Proof. Define functions $A, B: (a,b] \to S(n,\mathbb{R})$ by

$$A(t) = \tilde{H}(t) - H(t), \qquad B(t) = -\frac{1}{2}\left(\tilde{H}(t) + H(t)\right).$$

The hypothesis implies that A extends to a continuous matrix-valued function (still denoted by A) on $[a,b]$ satisfying $A(a) \leq 0$. We need to show that $A(t) \leq 0$ for all $t \in (a,b]$.

Simple computations show that the following equalities hold on (a,b):

$$A' = BA + AB - \tilde{S} + S,$$

$$B' = \frac{1}{2}(\tilde{H}^2 + \tilde{S} + H^2 + S).$$

Our hypotheses applied to these formulas imply

$$A' \leq BA + AB, \tag{11.14}$$

$$B' \geq k \, \mathrm{Id}, \tag{11.15}$$

where the last inequality holds for some (possibly negative) real number k because H^2 and $\tilde{H}^2$ are positive semidefinite and S and $\tilde{S}$ are continuous on all of $[a,b]$ and thus bounded below. Therefore, for every $t \in (a,b]$,

$$B(t) = B(b) - \int_t^b B'(u)\,du \tag{11.16}$$
$$\leq B(b) - (b-t)k \, \mathrm{Id} \leq B(b) + |b-a|\,|k|\,\mathrm{Id},$$

which shows that $B(t)$ is bounded above on $(a,b]$. Let K be a constant large enough that $B(t) - K\,\mathrm{Id}$ is negative definite for all such t.

Define a continuous function $f : [a,b] \times \mathbb{S}^{n-1} \to \mathbb{R}$ by

$$f(t,x) = e^{-2Kt}\,\langle A(t)x, x \rangle.$$

To prove the theorem, we need to show that $f(t,x) \leq 0$ for all $(t,x) \in [a,b] \times \mathbb{S}^{n-1}$. Suppose this is not the case; then by compactness of $[a,b] \times \mathbb{S}^{n-1}$, f takes on a positive maximum at some $(t_0,x_0) \in [a,b] \times \mathbb{S}^{n-1}$. Since $f(a,x) = 0$ for all x, we must have $a < t_0 \leq b$. Because $\langle A(t_0)x, x \rangle \leq \langle A(t_0)x_0, x_0 \rangle$ for all $x \in \mathbb{S}^{n-1}$, it follows from Lemma 8.14 that x_0 is an eigenvector of $A(t_0)$ with eigenvalue $\lambda_0 = \langle A(t_0)x_0, x_0 \rangle > 0$.

Since f is differentiable at (t_0,x_0) and $f(t,x_0) \leq f(t_0,x_0)$ for $a < t < t_0$, we have

$$\frac{\partial f}{\partial t}(t_0,x_0) = \lim_{t \nearrow t_0} \frac{f(t,x_0) - f(t_0,x_0)}{t - t_0} \geq 0.$$

(We have to take a one-sided limit here to accommodate the fact that t_0 might equal b.) On the other hand, from (11.14) and the fact that $A(t_0)$ and $B(t_0)$ are self-adjoint, we have

$$\frac{\partial f}{\partial t}(t_0,x_0) = e^{-2Kt_0}\left(\langle A'(t_0)x_0, x_0 \rangle - 2K\,\langle A(t_0)x_0, x_0 \rangle\right)$$

$$\leq e^{-2Kt_0}\left(\langle B(t_0)A(t_0)x_0, x_0 \rangle + \langle A(t_0)B(t_0)x_0, x_0 \rangle - 2K\,\langle A(t_0)x_0, x_0 \rangle\right)$$

$$= e^{-2Kt_0}\left(2\,\langle A(t_0)x_0, B(t_0)x_0 \rangle - 2K\,\langle A(t_0)x_0, x_0 \rangle\right)$$

$$= e^{-2Kt_0}\left(2\lambda_0\,\langle x_0, (B(t_0) - K\,\mathrm{Id})x_0 \rangle\right) < 0.$$

These two inequalities contradict each other, thus completing the proof. □

Proof of Theorem 11.5. Suppose η, $\tilde{\eta}$, σ, and $\tilde{\sigma}$ are self-adjoint endomorphism fields along γ satisfying the hypotheses of the theorem. Let $(E_1, \ldots, E_n)$ be a parallel orthonormal frame along γ, and let $H, \tilde{H} \colon (a,b] \to S(n,\mathbb{R})$ and $S, \tilde{S} \colon [a,b] \to S(n,\mathbb{R})$ be the symmetric matrix-valued functions defined by

$$\eta(t)(E_i(t)) = H_i^j(t) E_j(t), \qquad \tilde{\eta}(t)(E_i(t)) = \tilde{H}_i^j(t) E_j(t),$$
$$\sigma(t)(E_i(t)) = S_i^j(t) E_j(t), \qquad \tilde{\sigma}(t)(E_i(t)) = \tilde{S}_i^j(t) E_j(t).$$

Because $D_t E_j \equiv 0$, the Riccati equations (11.8) reduce to the ordinary differential equations (11.10) for these matrix-valued functions. Theorem 11.6 shows that $\tilde{H}(t) \le H(t)$ for all $t \in (a,b]$, which in turn implies that $\tilde{\eta}(t) \le \eta(t)$. $\qquad\square$

Comparisons Based on Sectional Curvature

Now we are ready to establish some comparison theorems for metric quantities based on comparing sizes of Hessian operators and Jacobi fields for an arbitrary metric with those of the constant-curvature models.

The most fundamental comparison theorem is the following result, which compares the Hessian of the radial distance function with its counterpart for a constant-curvature metric.

Theorem 11.7 (Hessian Comparison). *Suppose (M,g) is a Riemannian n-manifold, $p \in M$, U is a normal neighborhood of p, and r is the radial distance function on U.*

(a) *If all sectional curvatures of M are bounded above by a constant c, then the following inequality holds in $U_0 \smallsetminus \{p\}$:*

$$\mathcal{H}_r \ge \frac{s_c'(r)}{s_c(r)} \pi_r, \tag{11.17}$$

where s_c and π_r are defined as in Proposition 11.3, and $U_0 = U$ if $c \le 0$, while $U_0 = \{q \in U : r(q) < \pi R\}$ if $c = 1/R^2 > 0$.

(b) *If all sectional curvatures of M are bounded below by a constant c, then the following inequality holds in all of $U \smallsetminus \{p\}$:*

$$\mathcal{H}_r \le \frac{s_c'(r)}{s_c(r)} \pi_r. \tag{11.18}$$

Proof. Let $(x^1, \ldots, x^n)$ be Riemannian normal coordinates on U centered at p, let r be the radial distance function on U, and let s_c be the function defined by (10.8). Let $U_0 \subseteq U$ be the subset on which $s_c(r) > 0$; when $c \le 0$, this is all of U, but when $c = 1/R^2 > 0$, it is the subset where $r < \pi R$. Let $\mathcal{H}_r^c$ be the endomorphism field on $U_0 \smallsetminus \{p\}$ given by

$$\mathcal{H}_r^c = \frac{s_c'(r)}{s_c(r)}\pi_r.$$

Let $q \in U_0 \smallsetminus \{p\}$ be arbitrary, and let $\gamma \colon [0,b] \to U_0$ be the unit-speed radial geodesic from p to q. Note that at every point $\gamma(t)$ for $0 < t \le b$, the endomorphism field π_r can be expressed as $\pi_r(w) = w - \langle w, \partial_r \rangle \partial_r = w - \langle w, \gamma' \rangle \gamma'$, and in this form it extends smoothly to an endomorphism field along all of γ. Moreover, since $D_t \gamma' = 0$ along γ, it follows that $D_t \pi_r = 0$ along γ as well. Therefore, direct computation using the facts that $s_c'' = -c s_c$ and $\pi_r^2 = \pi_r$ shows that $\mathcal{H}_r^c$ satisfies the following Riccati equation along $\gamma|_{(0,b]}$:

$$D_t \mathcal{H}_r^c + (\mathcal{H}_r^c)^2 + c\pi_r = 0.$$

On the other hand, Theorem 11.4 shows that $\mathcal{H}_r$ satisfies

$$D_t \mathcal{H}_r + \mathcal{H}_r^2 + R_{\gamma'} = 0.$$

The sectional curvature hypothesis implies that $R_{\gamma'} \le c\pi_r$ in case (a) and $R_{\gamma'} \ge c\pi_r$ in case (b), using the facts that $R_{\gamma'}(\gamma') = 0 = \pi_r(\gamma')$, and $\langle R_{\gamma'}(w), w \rangle = \sec(\gamma', w)$ if w is a unit vector orthogonal to γ'.

In order to apply the Riccati comparison theorem to $\mathcal{H}_r$ and $\mathcal{H}_r^c$, we need to show that $\mathcal{H}_r - \mathcal{H}_r^c$ has a finite limit along γ as $t \searrow 0$. A straightforward series expansion shows that no matter what c is,

$$s_c'(r)/s_c(r) = 1/r + O(r) \tag{11.19}$$

as $r \searrow 0$. We will show that $\mathcal{H}_r$ satisfies the analogous estimate:

$$\mathcal{H}_r = \frac{1}{r}\pi_r + O(r). \tag{11.20}$$

The easiest way to verify (11.20) is to note that on $U \smallsetminus \{p\}$, $\mathcal{H}_r$ has the following coordinate expression in normal coordinates:

$$\mathcal{H}_r = g^{ij}\left(\partial_j \partial_k r - \Gamma_{jk}^m \partial_m r\right)\partial_i \otimes dx^k.$$

Now $\partial_m r = x^m/r$, which is bounded on $U \smallsetminus \{p\}$, and $\partial_j \partial_k r = O(r^{-1})$. Moreover, $g^{ij} = \delta^{ij} + O(r^2)$ and $\Gamma_{jk}^m = O(r)$, so $\mathcal{H}_r$ is equal to $\delta^{ij}(\partial_j \partial_k r)\partial_i \otimes dx^k$ plus terms that are $O(r)$ in these coordinates. But this last expression is exactly the coordinate expression for $\mathcal{H}_r$ in the case of the Euclidean metric in normal coordinates, which Proposition 11.3 shows is equal to $(1/r)\pi_r$. This proves (11.20), from which we conclude that $\mathcal{H}_r - \mathcal{H}_r^c$ approaches zero along γ as $t \searrow 0$.

If the sectional curvatures of g are bounded above by c, then the arguments above show that the hypotheses of the Riccati comparison theorem are satisfied along $\gamma|_{[0,b]}$ with $\eta(t) = \mathcal{H}_r|_{\gamma(t)}$, $\sigma(t) = R_{\gamma'}|_{\gamma(t)}$, $\tilde{\eta}(t) = \mathcal{H}_r^c|_{\gamma(t)}$, and $\tilde{\sigma}(t) = c\pi_r|_{\gamma(t)}$. It follows that $\mathcal{H}_r \ge \mathcal{H}_r^c$ at $q = \gamma(b)$, thus proving (a).

On the other hand, if the sectional curvatures are bounded below by c, the same argument with the roles of $\mathcal{H}_r$ and $\mathcal{H}_r^c$ reversed shows that $\mathcal{H}_r \leq \mathcal{H}_r^c$ on U_0. It remains only to show that $U_0 = U$ in this case. If $c \leq 0$, this is automatic. If $c = 1/R^2 > 0$, then $s_c'(r)/s_c(r) \to -\infty$ as $r \nearrow \pi R$; since $\mathcal{H}_r$ is defined and smooth in all of $U \smallsetminus \{p\}$ and bounded above by $\mathcal{H}_r^c$, it must be the case that $r < \pi R$ in U, which implies that $U_0 = U$. $\qquad\square$

Corollary 11.8 (Principal Curvature Comparison). *Suppose (M, g) is a Riemannian n-manifold, $p \in M$, U is a normal neighborhood of p, r is the radial distance function on U, and s_c and π_r are defined as in Proposition 11.3.*

(a) *If all sectional curvatures of M are bounded above by a constant c, then the principal curvatures of the r-level sets in $U_0 \smallsetminus \{p\}$ (with respect to the inward unit normal) satisfy*

$$\kappa \geq \frac{s_c'(r)}{s_c(r)},$$

where $U_0 = U$ if $c \leq 0$, while $U_0 = \{q \in U : r(q) < \pi R\}$ if $c = 1/R^2 > 0$.

(b) *If all sectional curvatures of M are bounded below by a constant c, then the principal curvatures of the r-level sets in $U \smallsetminus \{p\}$ (with respect to the inward unit normal) satisfy*

$$\kappa \leq \frac{s_c'(r)}{s_c(r)}.$$

Proof. This follows immediately from the fact that the shape operator of each r-level set is the restriction of $\mathcal{H}_r$ by Lemma 11.1(c). $\qquad\square$

Because Jacobi fields describe the behavior of families of geodesics, the next theorem gives some substance to the intuitive notion that negative curvature tends to make nearby geodesics spread out, while positive curvature tends to make them converge. More precisely, an upper bound on curvature forces Jacobi fields to be at least as large as their constant-curvature counterparts, and a lower curvature bound constrains them to be no larger.

Theorem 11.9 (Jacobi Field Comparison). *Suppose (M, g) is a Riemannian manifold, $\gamma : [0, b] \to M$ is a unit-speed geodesic segment, and J is any normal Jacobi field along γ such that $J(0) = 0$. For each $c \in \mathbb{R}$, let s_c be the function defined by (10.8).*

(a) *If all sectional curvatures of M are bounded above by a constant c, then*

$$|J(t)| \geq s_c(t)|D_t J(0)| \tag{11.21}$$

for all $t \in [0, b_1]$, where $b_1 = b$ if $c \leq 0$, and $b_1 = \min(b, \pi R)$ if $c = 1/R^2 > 0$.

(b) *If all sectional curvatures of M are bounded below by a constant c, then*

$$|J(t)| \leq s_c(t)|D_t J(0)| \tag{11.22}$$

for all $t \in [0, b_2]$, where b_2 is chosen so that $\gamma(b_2)$ is the first conjugate point to $\gamma(0)$ along γ if there is one, and otherwise $b_2 = b$.

Proof. If $D_t J(0) = 0$, then J vanishes identically, so we may as well assume that $D_t J(0) \neq 0$. Let b_0 be the largest time in $(0, b]$ such that γ has no conjugate points in $(0, b_0)$ and $s_c(t) > 0$ for $t \in (0, b_0)$. Let $p = \gamma(0)$, and assume temporarily that $\gamma([0, b_0))$ is contained in a normal neighborhood U of p. Define a function $f : (0, b_0) \to \mathbb{R}$ by

$$f(t) = \log\left(s_c(t)^{-1}|J(t)|\right).$$

Differentiating with respect to t and using $D_t J = \mathcal{H}_r(J)$, we get

$$f' = \frac{d}{dt}\log|J| - \frac{d}{dt}\log s_c = \frac{\langle D_t J, J\rangle}{|J|^2} - \frac{s_c'}{s_c} = \frac{\langle \mathcal{H}_r(J), J\rangle}{|J|^2} - \frac{s_c'}{s_c}.$$

Under hypothesis (a), it follows from the Hessian comparison theorem that $f'(t) \geq 0$ for all $t \in (0, b_0)$, so $f(t)$ is nondecreasing, and thus so is $s_c(t)^{-1}|J(t)|$. Similarly, under hypothesis (b), we get $f'(t) \leq 0$, which implies that $s_c(t)^{-1}|J(t)|$ is nonincreasing.

Next we consider the limit of $s_c(t)^{-1}|J(t)|$ as $t \searrow 0$. Two applications of l'Hôpital's rule yield

$$\lim_{t \searrow 0} \frac{|J|^2}{s_c^2} = \lim_{t \searrow 0} \frac{2\langle D_t J, J\rangle}{2s_c' s_c} = \lim_{t \searrow 0} \frac{2\langle D_t^2 J, J\rangle + 2|D_t J|^2}{2s_c'' s_c + 2s_c'^2}$$

$$= \lim_{t \searrow 0} \frac{-2\langle R(J, \gamma')\gamma', J\rangle + 2|D_t J|^2}{2s_c'' s_c + 2s_c'^2},$$

provided the last limit exists. Since $J \to 0$, $s_c \to 0$, and $s_c' \to 1$ as $t \searrow 0$, this last limit does exist and is equal to $|D_t J(0)|^2$. Combined with the derivative estimates above, this shows that the appropriate conclusion (11.21) or (11.22) holds on $(0, b_0)$, and thus by continuity on $[0, b_0]$, when $\gamma([0, b_0))$ is contained in a normal neighborhood of p.

Now suppose γ is an arbitrary geodesic segment, not assumed to be contained in a normal neighborhood of p. Let $v = \gamma'(0)$, so that $\gamma(t) = \exp_p(tv)$ for $t \in [0, b]$, and define b_0 as above. The definition ensures that $\gamma(t)$ is not conjugate to $\gamma(0)$ for $t \in (0, b_0)$, and therefore $\exp_p$ is a local diffeomorphism on some neighborhood of the set $L = \{tv : 0 \leq t < b_0\}$. Let $W \subseteq T_p M$ be a convex open set containing L on which $\exp_p$ is a local diffeomorphism, and let $\tilde{g} = \exp_p^* g$, which is a Riemannian metric on W that satisfies the same curvature estimates as g (Fig. 11.2). By construction, W is a normal neighborhood of 0, and $\exp_p$ is a local isometry from $(W, \tilde{g})$ to (M, g). The curve $\tilde{\gamma}(t) = tv$ for $t \in [0, b_0)$ is a radial geodesic in W, and Proposition 10.5 shows that the vector field $\tilde{J}(t) = d\left(\exp_p\right)_{tv}^{-1}(J(t))$ is a Jacobi field along $\tilde{\gamma}$ that vanishes at $t = 0$. Therefore, for $t \in (0, b_0)$, the preceding argument implies that $|\tilde{J}(t)|_{\tilde{g}} \geq s_c(t)|\tilde{D}_t \tilde{J}(0)|_{\tilde{g}}$ in case (a) and $|\tilde{J}(t)|_{\tilde{g}} \leq s_c(t)|\tilde{D}_t \tilde{J}(0)|_{\tilde{g}}$ in case (b). This implies that the conclusions of the theorem hold for J on the interval $(0, b_0)$ and thus by continuity on $[0, b_0]$.

To complete the proof, we need to show that $b_0 \geq b_1$ in case (a) and $b_0 \geq b_2$ in case (b). Assuming the hypothesis of (a), suppose for contradiction that $b_0 < b_1$. The

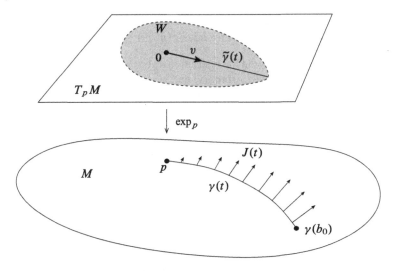

Fig. 11.2: Pulling the metric back to $T_p M$

only way this can occur is if $\gamma(b_0)$ is conjugate to $\gamma(0)$ along γ, while $s_c(b_0) > 0$. This means that there is a nontrivial normal Jacobi field $J \in \mathcal{J}(\gamma)$ satisfying $J(0) = 0 = J(b_0)$. But the argument above showed that every such Jacobi field satisfies $|J(t)| \ge s_c(t)|D_t J(0)|$ for $t \in [0, b_0]$ and thus $|J(b_0)| \ge s_c(b_0)|D_t J(0)| > 0$, which is a contradiction. Similarly, in case (b), suppose $b_0 < b_2$. Then $s_c(b_0) = 0$, but $\gamma(b_0)$ is not conjugate to $\gamma(0)$ along γ. If J is any nontrivial normal Jacobi field along γ that vanishes at $t = 0$, the argument above shows that $|J(t)| \le s_c(t)|D_t J(0)|$ for $t \in [0, b_0]$, so $|J(b_0)| \le s_c(b_0)|D_t J(0)| = 0$; but this is impossible because $\gamma(b_0)$ is not conjugate to $\gamma(0)$. $\qquad\qquad \square$

There is a generalization of the preceding theorem, called the *Rauch comparison theorem*, that allows for comparison of Jacobi fields in two different Riemannian manifolds when neither is assumed to have constant curvature. The statement and proof can be found in [CE08, Kli95].

Because all tangent vectors in a normal neighborhood are values of Jacobi fields along radial geodesics, the Jacobi field comparison theorem leads directly to the following comparison theorem for metrics.

Theorem 11.10 (Metric Comparison). *Let (M, g) be a Riemannian manifold, and let $(U, (x^i))$ be any normal coordinate chart for g centered at $p \in M$. For each $c \in \mathbb{R}$, let g_c denote the constant-curvature metric on $U \smallsetminus \{p\}$ given in the same coordinates by formula (10.17).*

(a) *Suppose all sectional curvatures of g are bounded above by a constant c. If $c \le 0$, then for all $q \in U \smallsetminus \{p\}$ and all $w \in T_q M$, we have $g(w, w) \ge g_c(w, w)$. If $c = 1/R^2 > 0$, then the same holds, provided that $d_g(p, q) < \pi R$.*

(b) *If all sectional curvatures of g are bounded below by a constant c, then for all*
$q \in U \smallsetminus \{p\}$ *and all* $w \in T_q M$, *we have* $g(w, w) \leq g_c(w, w)$.

Proof. Let $q \in U \smallsetminus \{p\}$, satisfying the restriction $d_g(p, q) < \pi R$ if we are in case (a)
and $c = 1/R^2 > 0$, but otherwise arbitrary; and let $b = d_g(p, q)$. Given $w \in T_q M$,
we can decompose w as a sum $w = y + z$, where y is a multiple of the radial
vector field ∂_r and z is tangent to the level set $r = b$. Then $g_c(y, z) = 0$ by direct
computation, and the Gauss lemma shows that $g(y, z) = 0$, so

$$g(w, w) = g(y, y) + g(z, z),$$
$$g_c(w, w) = g_c(y, y) + g_c(z, z).$$

Because ∂_r is a unit vector with respect to both g and g_c, it follows that $g(y, y) = g_c(y, y)$. So it suffices to prove the comparison for z.

There is a radial geodesic $\gamma : [0, b] \to U$ satisfying $\gamma(0) = p$ and $\gamma(b) = q$, and
Corollary 10.11 shows that $z = J(b)$ for some Jacobi field J along γ vanishing at
$t = 0$, which is normal because it is orthogonal to γ' at $t = 0$ and $t = b$. Proposition
10.10 shows that J has the coordinate formula $J(t) = t a^i \partial_i|_{\gamma(t)}$ for some constants
$(a^1, \ldots, a^n)$. Since the coordinates (x^i) are normal coordinates for both g and g_c,
it follows that γ is also a radial geodesic for g_c, and the same vector field J is also
a normal Jacobi field for g_c along γ. Therefore, $g(z, z) = |J(b)|_g^2$ and $g_c(z, z) = |J(b)|_{g_c}^2$. In case (a), our hypothesis guarantees that $s_c(b) > 0$, so

$$g(z, z) = |J(b)|_g^2 \geq |s_c(b)|^2 |D_t J(0)|_g^2 \qquad \text{(Jacobi field comparison theorem)}$$
$$= |s_c(b)|^2 |D_t J(0)|_{g_c}^2 \qquad \text{(since } g \text{ and } g_c \text{ agree at } p\text{)}$$
$$= |J(b)|_{g_c}^2 = g_c(z, z) \qquad \text{(by Prop. 10.12).}$$

In case (b), the same argument with the inequalities reversed shows that $g(z, z) \leq g_c(z, z)$.   □

The next three comparison theorems (Laplacian, conjugate point, and volume
comparisons) can be proved equally easily under the assumption of either an upper
bound or a lower bound for the sectional curvature, just like the preceding theorems.
However, we state these only for the case of an upper bound, because we will prove
stronger theorems later in the chapter based on lower bounds for the Ricci curvature
(see Thms. 11.15, 11.16, and 11.19).

The first of the three is a comparison of the Laplacian of the radial distance function with its constant-curvature counterpart. Our primary interest in the Laplacian of
the distance function stems from its role in volume and conjugate point comparisons
(see Thms. 11.14, 11.16, and 11.19 below); but it also plays an important role in the
study of various partial differential equations on Riemannian manifolds.

Theorem 11.11 (Laplacian Comparison I). *Suppose* (M, g) *is a Riemannian n-
manifold whose sectional curvatures are all bounded above by a constant c. Suppose*
$p \in M$, U *is a normal neighborhood of* p, r *is the radial distance function on* U,
and s_c *is defined as in Proposition 11.3. Then on* $U_0 \smallsetminus \{p\}$, *we have*

$$\Delta r \geq (n-1) \frac{s_c'(r)}{s_c(r)},$$

where $U_0 = U$ if $c \leq 0$, while $U_0 = \{q \in U : r(q) < \pi R\}$ if $c = 1/R^2 > 0$.

Proof. By the result of Problem 5-14, $\Delta r = \operatorname{tr}_g \left(\nabla^2 r \right) = \operatorname{tr} \left(\mathcal{H}_r \right)$. The result then follows from the Hessian comparison theorem, using the fact that $\operatorname{tr}(\pi_r) = n - 1$, which can be verified easily by expressing π_r locally in an adapted orthonormal frame for the r-level sets.   ☐

The next theorem shows how an upper curvature bound prevents the formation of conjugate points. It will play a decisive role in the proof of the Cartan–Hadamard theorem in the next chapter.

Theorem 11.12 (Conjugate Point Comparison I). *Suppose (M,g) is a Riemannian n-manifold whose sectional curvatures are all bounded above by a constant c. If $c \leq 0$, then no point of M has conjugate points along any geodesic. If $c = 1/R^2 > 0$, then there is no conjugate point along any geodesic segment shorter than πR.*

Proof. The case $c \leq 0$ is covered by Problem 10-7, so assume $c = 1/R^2 > 0$. Let $\gamma : [0,b] \to M$ be a unit-speed geodesic segment, and suppose J is a nontrivial normal Jacobi field along γ that vanishes at $t = 0$. The Jacobi field comparison theorem implies that $|J(t)| \geq (\text{constant}) \sin(t/R) > 0$ as long as $0 < t < \pi R$.   ☐

The last of our sectional curvature comparison theorems is a comparison of volume growth of geodesic balls. Before proving it, we need the following lemma, which shows how the Riemannian volume form is related to the Laplacian of the radial distance function.

Lemma 11.13. *Suppose (M,g) is a Riemannian manifold and (x^i) are Riemannian normal coordinates on a normal neighborhood U of $p \in M$. Let $\det g$ denote the determinant of the matrix (g_{ij}) in these coordinates, let r be the radial distance function, and let ∂_r be the unit radial vector field. The following identity holds on $U \smallsetminus \{p\}$:*

$$\Delta r = \partial_r \log \left(r^{n-1} \sqrt{\det g} \right). \tag{11.23}$$

Proof. Corollary 6.10 to the Gauss lemma shows that the vector fields $\operatorname{grad} r$ and ∂_r are equal on $U \smallsetminus \{p\}$. Comparing the components of these two vector fields in normal coordinates, we conclude (using the summation convention as usual) that

$$g^{ij} \partial_j r = \frac{x^i}{r}.$$

Based on the formula for Δr from Proposition 2.46, we compute

$$\Delta r = \frac{1}{\sqrt{\det g}} \partial_i \left(g^{ij} \sqrt{\det g} \, \partial_j r \right)$$

$$= \frac{1}{\sqrt{\det g}} \partial_i \left(\frac{x^i}{r} \sqrt{\det g} \right)$$

$$= \partial_i \left(\frac{x^i}{r} \right) + \frac{1}{\sqrt{\det g}} \frac{x^i}{r} \partial_i \sqrt{\det g}$$

$$= \frac{n-1}{r} + \frac{1}{\sqrt{\det g}} \partial_r \sqrt{\det g},$$

where the first term in the last line follows by direct computation using $r = \left(\sum_i (x^i)^2 \right)^{1/2}$. This is equivalent to (11.23). □

The following result was proved by Paul Günther in 1960 [Gün60]. (Günther also proved an analogous result in the case of a lower sectional curvature bound, but that result has been superseded by the Bishop–Gromov theorem, Thm. 11.19 below.)

Theorem 11.14 (Günther's Volume Comparison). *Suppose (M, g) is a connected Riemannian n-manifold whose sectional curvatures are all bounded above by a constant c. Given $p \in M$, let $\delta_0 = \mathrm{inj}(p)$ if $c \le 0$, and $\delta_0 = \min(\pi R, \mathrm{inj}(p))$ if $c = 1/R^2 > 0$. For every positive number $\delta \le \delta_0$, let $V_g(\delta)$ denote the volume of the geodesic ball $B_\delta(p)$ in (M, g), and let $V_c(\delta)$ denote the volume of a geodesic ball of radius δ in the n-dimensional Euclidean space, hyperbolic space, or sphere with constant sectional curvature c. Then for every $0 < \delta \le \delta_0$, we have*

$$V_g(\delta) \ge V_c(\delta), \tag{11.24}$$

and the quotient $V_g(\delta)/V_c(\delta)$ is a nondecreasing function of δ that approaches 1 as $\delta \searrow 0$. If equality holds in (11.24) for some $\delta \in (0, \delta_0]$, then g has constant sectional curvature c on the entire geodesic ball $B_\delta(p)$.

Proof. The volume estimate (11.24) follows easily from the metric comparison theorem, which implies that the determinants of the metrics g and g_c in normal coordinates satisfy $\sqrt{\det g} \ge \sqrt{\det g_c}$. If that were all we needed, we could stop here; but to prove the other statements, we need a more involved argument, which incidentally provides another proof of (11.24) that does not rely directly on the metric comparison theorem, and therefore can be adapted more easily to the case in which we have only an estimate of the Ricci curvature (see Thm. 11.19 below).

Let (x^i) be normal coordinates on $B_{\delta_0}(p)$ (interpreted as all of M if $\delta_0 = \infty$). Using these coordinates, we might as well consider g to be a Riemannian metric on an open subset of $\mathbb{R}^n$ and p to be the origin. Let $\bar g$ denote the Euclidean metric in these coordinates, and let g_c denote the constant-curvature metric in the same coordinates, given on the complement of the origin by (10.17).

The Laplacian comparison theorem together with Lemma 11.13 shows that

$$\partial_r \log \left(r^{n-1} \sqrt{\det g} \right) = \Delta r \ge (n-1) \frac{s'_c(r)}{s_c(r)} = \partial_r \log \left(s_c(r)^{n-1} \right). \tag{11.25}$$

Thus $\log\left(r^{n-1}\sqrt{\det g}/s_c(r)^{n-1}\right)$ is a nondecreasing function of r along each radial geodesic, and so is the ratio $r^{n-1}\sqrt{\det g}/s_c(r)^{n-1}$. To compute the limit as $r \searrow 0$, note that $g_{ij} = \delta_{ij}$ at the origin, so $\sqrt{\det g}$ converges uniformly to 1 as $r \searrow 0$. Also, for every c, we have $s_c(r)/r \to 1$ as $r \searrow 0$, so $r^{n-1}\sqrt{\det g}/s_c(r)^{n-1} \to 1$.

We can write $dV_g = \sqrt{\det g}\, dV_{\tilde{g}}$. Corollary 10.17 in the case $c = 0$ shows that

$$V_g(\delta) = \int_{\mathbb{S}^{n-1}} \int_0^\delta \left(\sqrt{\det g}\right) \circ \Phi(\rho,\omega)\rho^{n-1}\, d\rho\, dV_{\mathring{g}},$$

where $\Phi(\rho,\omega) = \rho\omega$ for $\rho \in (0,\delta_0)$ and $\omega \in \mathbb{S}^{n-1}$. The same corollary shows that

$$V_c(\delta) = \int_{\mathbb{S}^{n-1}} \int_0^\delta s_c(\rho)^{n-1}\, d\rho\, dV_{\mathring{g}} = \left(\int_0^\delta s_c(\rho)^{n-1}\, d\rho\right)\left(\int_{\mathbb{S}^{n-1}} dV_{\mathring{g}}\right).$$

Therefore,

$$
\begin{aligned}
\frac{V_g(\delta)}{V_c(\delta)} &= \frac{\displaystyle\int_{\mathbb{S}^{n-1}} \int_0^\delta \left(\sqrt{\det g}\right) \circ \Phi(\rho,\omega)\rho^{n-1}\, d\rho\, dV_{\mathring{g}}}{\displaystyle\left(\int_0^\delta s_c(\rho)^{n-1}\, d\rho\right)\left(\int_{\mathbb{S}^{n-1}} dV_{\mathring{g}}\right)} \\[2mm]
&= \frac{1}{\mathrm{Vol}(\mathbb{S}^{n-1})} \int_{\mathbb{S}^{n-1}} \left(\frac{\displaystyle\int_0^\delta \lambda(\rho,\omega)s_c(\rho)^{n-1}\, d\rho}{\displaystyle\int_0^\delta s_c(\rho)^{n-1}\, d\rho}\right) dV_{\mathring{g}}, \qquad (11.26)
\end{aligned}
$$

where we have written

$$\lambda(\rho,\omega) = \frac{\left(\sqrt{\det g}\right) \circ \Phi(\rho,\omega)\rho^{n-1}}{s_c(\rho)^{n-1}}.$$

The argument above (together with the fact that $r \circ \Phi = \rho$) shows that $\lambda(\rho,\omega)$ is a nondecreasing function of ρ for each ω, which approaches 1 uniformly as $\rho \searrow 0$.

We need to show that the quotient in parentheses in the last line of (11.26) is also nondecreasing as δ increases. Suppose $0 < \delta_1 < \delta_2$. Because λ is nondecreasing, we have

$$
\begin{aligned}
\int_0^{\delta_1} \lambda(\rho,\omega)s_c(\rho)^{n-1}\, d\rho &\int_{\delta_1}^{\delta_2} s_c(\rho)^{n-1}\, d\rho \\[2mm]
&\leq \left(\int_0^{\delta_1} s_c(\rho)^{n-1}\, d\rho\right)\lambda(\delta_1,\omega)\left(\int_{\delta_1}^{\delta_2} s_c(\rho)^{n-1}\, d\rho\right) \\[2mm]
&\leq \int_0^{\delta_1} s_c(\rho)^{n-1}\, d\rho \int_{\delta_1}^{\delta_2} \lambda(\rho,\omega)s_c(\rho)^{n-1}\, d\rho,
\end{aligned}
$$

and therefore (suppressing the dependence of the integrands on ρ and ω for brevity),

$$\int_0^{\delta_1} \lambda s_c^{n-1}\, d\rho \int_0^{\delta_2} s_c^{n-1}\, d\rho$$

$$= \int_0^{\delta_1} \lambda s_c^{n-1}\, d\rho \left(\int_0^{\delta_1} s_c^{n-1}\, d\rho + \int_{\delta_1}^{\delta_2} s_c^{n-1}\, d\rho \right)$$

$$\leq \int_0^{\delta_1} \lambda s_c^{n-1}\, d\rho \int_0^{\delta_1} s_c^{n-1}\, d\rho + \int_0^{\delta_1} s_c^{n-1}\, d\rho \int_{\delta_1}^{\delta_2} \lambda s_c^{n-1}\, d\rho$$

$$= \int_0^{\delta_1} s_c^{n-1}\, d\rho \int_0^{\delta_2} \lambda s_c^{n-1}\, d\rho.$$

Evaluating (11.26) at δ_1 and δ_2 and inserting the inequality above shows that the ratio $\mathrm{Vol}_g(\delta)/\mathrm{Vol}_c(\delta)$ is nondecreasing as a function of δ, and it approaches 1 as $\delta \searrow 0$ because $\lambda(\rho,\omega) \to 1$ as $\rho \searrow 0$. It follows that $V_g(\delta) \geq V_c(\delta)$ for all $\delta \in (0,\delta_0]$.

It remains only to consider the case in which the volume ratio is equal to 1 for some δ. If $\lambda(\rho,\omega)$ is not identically 1 on the set where $0 < \rho < \delta$, then it is strictly greater than 1 on a nonempty open subset, which implies that the volume ratio in (11.26) is strictly greater than 1; so $\mathrm{Vol}_g(\delta) = \mathrm{Vol}_c(\delta)$ implies $\lambda(\rho,\omega) \equiv 1$ on $(0,\delta) \times \mathbb{S}^{n-1}$, and pulling back to U via Φ^{-1} shows that $r^{n-1}\sqrt{\det g} \equiv s_c(r)^{n-1}$ on $B_\delta(p)$. By virtue of (11.25), we have $\Delta r \equiv (n-1)s_c'(r)/s_c(r)$, or in other words, $\mathrm{tr}(\mathcal{H}_r) = \mathrm{tr}((s_c'(r)/s_c(r))\pi_r)$. It follows from the Hessian comparison theorem that the endomorphism field $\mathcal{H}_r - \big(s_c'(r)/s_c(r)\big)\pi_r$ is positive semidefinite, so its eigenvalues are all nonnegative. Since its trace is zero, the eigenvalues must all be zero. In other words, $\mathcal{H}_r \equiv (s_c'(r)/s_c(r))\pi_r$ on the geodesic ball $B_\delta(p)$. It then follows from Proposition 11.3 that g has constant sectional curvature c on that ball. $\square$

Comparisons Based on Ricci Curvature

All of our comparison theorems so far have been based on assuming an upper or lower bound for the sectional curvature. It is natural to wonder whether anything can be said if we weaken the hypotheses and assume only bounds on other curvature quantities such as Ricci or scalar curvature.

It should be noted that except in very low dimensions, assuming a bound on Ricci or scalar curvature is a strictly weaker hypothesis than assuming one on sectional curvature. Recall Proposition 8.32, which says that on an n-dimensional Riemannian manifold, the Ricci curvature evaluated on a unit vector is a sum of $n-1$ sectional curvatures, and the scalar curvature is a sum of $n(n-1)$ sectional curvatures. Thus if (M,g) has sectional curvatures bounded below by c, then its Ricci curvature satisfies $Rc(v,v) \geq (n-1)c$ for all unit vectors v, and its scalar curvature satisfies $S \geq n(n-1)c$, with analogous inequalities if the sectional curvature is bounded above. However, the converse is not true: an upper or lower bound on the Ricci curvature implies nothing about individual sectional curvatures, except in dimensions 2 and 3, where the entire curvature tensor is determined by the Ricci curvature (see

Cors. 7.26 and 7.27). For example, in every even dimension greater than or equal to 4, there are compact Riemannian manifolds called **Calabi–Yau manifolds** that have zero Ricci curvature but nonzero sectional curvatures (see, for example, [Bes87, Chap. 11]).

In this section we investigate the extent to which bounds on the Ricci curvature lead to useful comparison theorems. The strongest theorems of the preceding section, such as the Hessian, Jacobi field, and metric comparison theorems, do not generalize to the case in which we merely have bounds on Ricci curvature. However, it is a remarkable fact that Laplacian, conjugate point, and volume comparison theorems can still be proved assuming only a lower (but not upper) bound on the Ricci curvature. (The problem of drawing global conclusions from scalar curvature bounds is far more subtle, and we do not pursue it here. A good starting point for learning about that problem is [Bes87].)

The next theorem is the analogue of Theorem 11.11.

Theorem 11.15 (Laplacian Comparison II). *Let (M, g) be a Riemannian n-manifold, and suppose there is a constant c such that the Ricci curvature of M satisfies $Rc(v, v) \geq (n-1)c$ for all unit vectors v. Given any point $p \in M$, let U be a normal neighborhood of p and let r be the radial distance function on U. Then the following inequality holds on $U_0 \smallsetminus \{p\}$:*

$$\Delta r \leq (n-1) \frac{s_c'(r)}{s_c(r)}, \tag{11.27}$$

where s_c is defined by (10.8), and $U_0 = U$ if $c \leq 0$, while $U_0 = \{q \in U : r(q) < \pi R\}$ if $c = 1/R^2 > 0$.

Proof. Let $q \in U_0 \smallsetminus \{p\}$ be arbitrary, and let $\gamma : [0, b] \to U_0$ be the unit-speed radial geodesic from p to q. We will show that (11.27) holds at $\gamma(t)$ for $0 < t \leq b$.

Because covariant differentiation commutes with the trace operator (since it is just a particular kind of contraction), the Riccati equation (11.7) for $\mathcal{H}_r$ implies the following scalar equation along γ for $\Delta r = \mathrm{tr}\left(\mathcal{H}_r\right)$:

$$\frac{d}{dt} \Delta r + \mathrm{tr}\left(\mathcal{H}_r^2\right) + \mathrm{tr}\, R_{\gamma'} = 0. \tag{11.28}$$

We need to analyze the last two terms on the left-hand side. We begin with the last term. In terms of any local orthonormal frame (E_i), we have

$$\mathrm{tr}\, R_{\gamma'} = \sum_{i=1}^{n} \langle R_{\gamma'}(E_i), E_i \rangle = \sum_{i=1}^{n} \langle R(E_i, \gamma')\gamma', E_i \rangle = \sum_{i=1}^{n} Rm(E_i, \gamma', \gamma', E_i)$$
$$= Rc(\gamma', \gamma').$$

To analyze the $\mathcal{H}_r^2$ term, let us set

$$\overset{\circ}{\mathcal{H}}_r = \mathcal{H}_r - \frac{\Delta r}{n-1} \pi_r. \tag{11.29}$$

We compute

$$\operatorname{tr}\left(\mathring{\mathcal{H}}_r^2\right) = \operatorname{tr}\left(\mathcal{H}_r^2\right) - \frac{\Delta r}{n-1}\operatorname{tr}(\mathcal{H}_r \circ \pi_r) - \frac{\Delta r}{n-1}\operatorname{tr}(\pi_r \circ \mathcal{H}_r) + \frac{(\Delta r)^2}{(n-1)^2}\operatorname{tr}\left(\pi_r^2\right).$$

To simplify this, note that $\pi_r^2 = \pi_r$ because π_r is a projection, and thus $\operatorname{tr}\left(\pi_r^2\right) = \operatorname{tr}(\pi_r) = n-1$. Also, $\mathcal{H}_r(\partial_r) \equiv 0$ implies that $\mathcal{H}_r \circ \pi_r = \mathcal{H}_r$; and since $\mathcal{H}_r$ is self-adjoint, $\langle \mathcal{H}_r v, \partial_r \rangle_g = \langle v, \mathcal{H}_r \partial_r \rangle_g = 0$ for all v, so the image of $\mathcal{H}_r$ is contained in the orthogonal complement of ∂_r, and it follows that $\pi_r \circ \mathcal{H}_r = \mathcal{H}_r$ as well. Thus the last three terms in the formula for $\operatorname{tr}\left(\mathring{\mathcal{H}}_r^2\right)$ combine to yield

$$\operatorname{tr}\left(\mathring{\mathcal{H}}_r^2\right) = \operatorname{tr}\left(\mathcal{H}_r^2\right) - \frac{(\Delta r)^2}{n-1}.$$

Solving this for $\operatorname{tr}\left(\mathcal{H}_r^2\right)$, substituting into (11.28), and dividing by $n-1$, we obtain

$$\frac{d}{dt}\left(\frac{\Delta r}{n-1}\right) + \left(\frac{\Delta r}{n-1}\right)^2 + \frac{\operatorname{tr}\left(\mathring{\mathcal{H}}_r^2\right) + Rc(\gamma',\gamma')}{n-1} = 0. \tag{11.30}$$

Let $H(t) = s_c'(t)/s_c(t)$, so that

$$H'(t) + H(t)^2 + c \equiv 0.$$

We wish to apply the 1×1 case of the matrix Riccati comparison theorem (Thm. 11.6) with $H(t)$ as above, $S(t) \equiv c$,

$$\widetilde{H}(t) = \frac{(\Delta r)|_{\gamma(t)}}{n-1}, \quad \text{and} \quad \widetilde{S}(t) = \frac{\operatorname{tr}\left(\mathring{\mathcal{H}}_r^2\right)|_{\gamma(t)} + Rc(\gamma'(t),\gamma'(t))}{n-1}.$$

Note that $\mathring{\mathcal{H}}_r^2$ is positive semidefinite, which means that all of its eigenvalues are nonnegative, so its trace (which is the sum of the eigenvalues) is also nonnegative. (This is the step that does not work in the case of an upper bound on Ricci curvature.) Thus our hypothesis on the Ricci curvature guarantees that $\widetilde{S}(t) \geq c$ for all $t \in (0,b]$.

To apply Theorem 11.6, we need to verify that $\widetilde{S}$ has a continuous extension to $[0,b]$ and that $\widetilde{H}(t) - H(t)$ has a nonnegative limit as $t \searrow 0$. Recall that we showed in (11.19) and (11.20) that $s_c'(r)/s_c(r) = 1/r + O(r)$ and $\mathcal{H}_r = (1/r)\pi_r + O(r)$ as $r \searrow 0$. This implies that $\Delta r = \operatorname{tr}(\mathcal{H}_r) = (n-1)/r + O(r)$, and therefore both $\mathring{\mathcal{H}}_r|_{\gamma(t)}$ and $\widetilde{H}(t) - H(t)$ approach 0 and $\widetilde{S}(t)$ approaches $Rc(\gamma'(0),\gamma'(0))/(n-1) \geq c$ as $t \searrow 0$. Therefore, we can apply Theorem 11.6 to conclude that $\widetilde{H}(t) \leq H(t)$ for $t \in (0,b]$. Since $\gamma(b) = q$ was arbitrary, this completes the proof. $\qquad \square$

The next theorem and its two corollaries will be crucial ingredients in the proofs of our theorems in the next chapter about manifolds with positive Ricci curvature (see Thms. 12.28 and 12.24).

Theorem 11.16 (Conjugate Point Comparison II). *Let (M, g) be a Riemannian n-manifold, and suppose there is a positive constant $c = 1/R^2$ such that the Ricci curvature of M satisfies $Rc(v, v) \geq (n-1)c$ for all unit vectors v. Then every geodesic segment of length at least πR has a conjugate point.*

Proof. Let U be a normal neighborhood of an arbitrary point $p \in M$. The second Laplacian comparison theorem (Thm. 11.15) combined with Lemma 11.13 shows that

$$\partial_r \log \left(r^{n-1} \sqrt{\det g} \right) = \Delta r \leq (n-1) \frac{s_c'(r)}{s_c(r)} = \partial_r \log \left(s_c(r)^{n-1} \right)$$

on the subset $U_0 \subseteq U$ where $r < \pi R$. Since $r^{n-1} \sqrt{\det g}/s_c(r)^{n-1} \to 1$ as $r \searrow 0$, this implies that $r^{n-1} \sqrt{\det g}/s_c(r)^{n-1} \leq 1$ everywhere in U_0, or equivalently,

$$\sqrt{\det g} \leq s_c(r)^{n-1}/r^{n-1}. \tag{11.31}$$

Suppose U contains a point q where $r \geq \pi R$, and let $\gamma : [0, b] \to U$ be the unit-speed radial geodesic from p to q. Because $s_c(\pi R) = 0$, (11.31) shows that $\det g(\gamma(t)) \to 0$ as $t \nearrow \pi R$, and therefore by continuity $\det g = 0$ at $\gamma(b)$, which contradicts the fact that $\det g > 0$ in every coordinate neighborhood. The upshot is that no normal neighborhood can include points where $r \geq \pi R$.

Now suppose $\gamma : [0, b] \to M$ is a unit-speed geodesic with $b \geq \pi R$, and assume for the sake of contradiction that γ has no conjugate points. Let $p = \gamma(0)$ and $v = \gamma'(0)$, so $\gamma(t) = \exp_p(tv)$ for $t \in [0, b]$. As in the proof of Theorem 11.9, because γ has no conjugate points, we can choose a star-shaped open subset $W \subseteq T_p M$ containing the set $L = \{tv : 0 \leq t < b_0\} \subseteq T_p M$ on which $\exp_p$ is a local diffeomorphism, and let $\tilde{g}$ be the pulled-back metric $\exp_p^* g$ on W, which satisfies the same curvature estimates as g. Then $\tilde{\gamma}(t) = tv$ is a radial $\tilde{g}$-geodesic in W of length greater than or equal to πR, which contradicts the argument in the preceding paragraph. $\square$

Corollary 11.17 (Injectivity Radius Comparison). *Let (M, g) be a Riemannian n-manifold, and suppose there is a positive constant $c = 1/R^2$ such that the Ricci curvature of M satisfies $Rc(v, v) \geq (n-1)c$ for all unit vectors v. Then for every point $p \in M$, we have $\mathrm{inj}(p) \leq \pi R$.*

Proof. Every radial geodesic segment in a geodesic ball is minimizing, but the preceding theorem shows that no geodesic segment of length πR or greater is minimizing. Thus no geodesic ball has radius greater than πR. $\square$

Corollary 11.18 (Diameter Comparison). *Let (M, g) be a complete, connected Riemannian n-manifold, and suppose there is a positive constant $c = 1/R^2$ such that the Ricci curvature of M satisfies $Rc(v, v) \geq (n-1)c$ for all unit vectors v. Then the diameter of M is less than or equal to πR.*

Proof. This follows from the fact that any two points of M can be connected by a minimizing geodesic segment, and the conjugate point comparison theorem implies that no such segment can have length greater than πR. $\square$

Our final comparison theorem is a powerful volume estimate under the assumption of a lower bound on the Ricci curvature. We will use it in the proof of Theorem 12.28 in the next chapter, and it plays a central role in many of the more advanced results of Riemannian geometry.

A weaker version of this result was proved by Paul Günther in 1960 [Gün60] for balls within the injectivity radius under the assumption of a lower bound on sectional curvature; it was improved by Richard L. Bishop in 1963 (announced in [Bis63], with a proof in [BC64]) to require only a lower Ricci curvature bound; and then it was extended by Misha Gromov in 1981 [Gro07] to cover all metric balls in the complete case, not just those inside the injectivity radius.

Theorem 11.19 (Bishop–Gromov Volume Comparison). *Let (M, g) be a connected Riemannian n-manifold, and suppose there is a constant c such that the Ricci curvature of M satisfies $Rc(v, v) \geq (n-1)c$ for all unit vectors v. Let $p \in M$ be given, and for every positive number δ, let $V_g(\delta)$ denote the volume of the metric ball of radius δ about p in (M, g), and let $V_c(\delta)$ denote the volume of a metric ball of radius δ in the n-dimensional Euclidean space, hyperbolic space, or sphere with constant sectional curvature c. Then for every $0 < \delta \leq \mathrm{inj}(p)$, we have*

$$V_g(\delta) \leq V_c(\delta), \tag{11.32}$$

and the quotient $V_g(\delta)/V_c(\delta)$ is a nonincreasing function of δ that approaches 1 as $\delta \searrow 0$. If (M, g) is complete, the same is true for all positive δ, not just $\delta \leq \mathrm{inj}(p)$. In either case, if equality holds in (11.32) for some δ, then g has constant sectional curvature on the entire metric ball of radius δ about p.

Proof. First consider $\delta \leq \mathrm{inj}(p)$, in which case a metric ball of radius δ in M is actually a geodesic ball. With the exception of the first and last paragraphs, the proof of Theorem 11.14 goes through with all of the inequalities reversed, and with the first Laplacian comparison theorem replaced by its counterpart Theorem 11.15, to show that $V_g(\delta)/V_c(\delta)$ is a nonincreasing function of δ that approaches 1 as $\delta \searrow 0$, and (11.32) follows.

In case M is complete, $\exp_p$ is defined and smooth on all of $T_p M$. Theorem 10.34 shows that $\mathrm{Cut}(p)$ has measure zero in M and $\exp_p$ maps the open subset $\mathrm{ID}(p) \subseteq T_p M$ diffeomorphically onto the complement of $\mathrm{Cut}(p)$ in M. Therefore, for every $\delta > 0$, the metric ball of radius δ is equal to $\exp_p\big(B_\delta(0) \cap \mathrm{ID}(p)\big)$ up to a set of measure zero, where $B_\delta(0)$ denotes the δ-ball about 0 in $T_p M$. Using an orthonormal basis to identify $(T_p M, g_p)$ with $(\mathbb{R}^n, \bar{g})$, we can compute the volume of a metric δ-ball as

$$V_g(\delta) = \int_{\mathrm{ID}(p) \cap B_\delta(0)} \exp_p^* dV_g = \int_{\mathrm{ID}(p) \cap B_\delta(0)} \sqrt{\det g}\, dV_{\bar{g}},$$

where $\det g$ denotes the determinant of the matrix of g in the normal coordinates determined by the choice of basis.

Let $\Phi: \mathbb{R}^+ \times \mathbb{S}^{n-1} \to T_p M \smallsetminus \{0\} \cong \mathbb{R}^n \smallsetminus \{0\}$ be the map $\Phi(\rho, \omega) = \rho\omega$ as in Corollary 10.17, and define $\tilde{s}_c: \mathbb{R}^+ \to [0, \infty)$ by $\tilde{s}_c(\rho) = s_c(\rho)$ if $c \leq 0$, while in the case $c = 1/R^2 > 0$,

$$\tilde{s}_c(\rho) = \begin{cases} s_c(\rho), & \rho < \pi R, \\ 0, & \rho \geq \pi R. \end{cases}$$

Corollary 11.18 shows that the cut time of every unit vector in $T_p M$ is less than or equal to πR. Thus $\tilde{s}_c(\rho) > 0$ whenever $\Phi(\rho, \omega) \in \mathrm{ID}(p)$, and we can define $\tilde{\lambda} \colon \mathbb{R}^+ \times \mathbb{S}^{n-1} \to \mathbb{R}$ by

$$\tilde{\lambda}(\rho, \omega) = \begin{cases} \dfrac{\left(\sqrt{\det g}\right) \circ \Phi(\rho, \omega) \rho^{n-1}}{\tilde{s}_c(\rho)^{n-1}}, & \Phi(\rho, \omega) \in \mathrm{ID}(p), \\ 0, & \Phi(\rho, \omega) \notin \mathrm{ID}(p), \end{cases}$$

and just as in the proof of Theorem 11.14, we can write

$$V_g(\delta) = \int_{\mathbb{S}^{n-1}} \int_0^\delta \tilde{\lambda}(\rho, \omega) \tilde{s}_c(\rho)^{n-1} \, d\rho \, dV_{\mathring{g}},$$

$$V_c(\delta) = \int_{\mathbb{S}^{n-1}} \int_0^\delta \tilde{s}_c(\rho)^{n-1} \, d\rho \, dV_{\mathring{g}}.$$

The arguments of Theorem 11.14 (with inequalities reversed) show that for each $\omega \in \mathbb{S}^{n-1}$, the function $\tilde{\lambda}(\rho, \omega)$ is nonincreasing in ρ for all positive ρ, and it follows just as in that proof that $V_g(\delta)/V_c(\delta)$ (now interpreted as a ratio of volumes of *metric* balls) is nonincreasing for all $\delta > 0$ and approaches 1 as $\delta \searrow 0$, and (11.32) follows.

Finally, suppose $V_g(\delta) = V_c(\delta)$ for some $\delta > 0$, and assume first that $\delta \leq \mathrm{inj}(p)$. An argument exactly analogous to the one at the end of the proof of Theorem 11.14 shows that $\lambda(\rho, \omega) \equiv 1$ everywhere on the set where $0 < \rho < \delta$. Combined with Lemma 11.13, this implies that

$$\Delta r = (n-1) \frac{s_c'(r)}{s_c(r)} \tag{11.33}$$

everywhere on $B_\delta(p) \smallsetminus \{p\}$. This means that along each unit-speed radial geodesic γ, the function $u(t) = (\Delta r)|_{\gamma(t)}/(n-1) = s_c'(t)/s_c(t)$ satisfies $u' + u^2 + c \equiv 0$ by direct computation. Comparing this to (11.30), we conclude that

$$\frac{\mathrm{tr}\left(\mathring{\mathscr{H}}_r^2\right) + Rc(\gamma', \gamma')}{n-1} \equiv c$$

on $B_\delta(p) \smallsetminus \{p\}$. Since $\mathrm{tr}\left(\mathring{\mathscr{H}}_r^2\right) \geq 0$ and $Rc(\gamma', \gamma') \geq (n-1)c$ everywhere, this is possible only if $\mathrm{tr}\left(\mathring{\mathscr{H}}_r^2\right)$ vanishes identically there. Because $\mathring{\mathscr{H}}_r^2$ is positive semidefinite and its trace is zero, it must vanish identically, which by definition of $\mathring{\mathscr{H}}_r$ means

$$\mathcal{H}_r = \frac{\Delta r}{n-1}\pi_r = \frac{s_c'(r)}{s_c(r)}\pi_r.$$

Proposition 11.3 then shows that g has constant sectional curvature c on $B_\delta(p)$.

Now suppose (M,g) is complete. The argument of Theorem 11.14 then shows that $\tilde{\lambda}(\rho,\omega) \equiv 1$ everywhere on the set where $0 < \rho < \delta$ and $\tilde{s}_c(\rho) > 0$. In view of the definition of $\tilde{\lambda}$, this implies in particular that $\mathrm{ID}(p) \cap B_\delta(0)$ contains all of the points in $B_\delta(0)$ where $\tilde{s}_c(r) > 0$. In case $c \leq 0$, $\tilde{s}_c(r) = s_c(r) > 0$ everywhere, so $\mathrm{ID}(p) \cap B_\delta(0) = B_\delta(0)$ and therefore the metric ball of radius δ around p is actually a geodesic ball, and the argument above applies.

In case $c = 1/R^2 > 0$, if $\delta \leq \pi R$, then $\tilde{s}_c(r) = s_c(r) > 0$ on $B_\delta(0)$, and once again we conclude that the metric δ-ball is a geodesic ball. On the other hand, if $\delta > \pi R$, then the diameter comparison theorem (Cor. 11.18) shows that the metric ball of radius δ is actually the entire manifold. The fact that the volume ratio is nonincreasing implies that $V_g(\pi R) = V_c(\pi R)$, and the argument above shows that g has constant sectional curvature c on the metric ball of radius πR. Since the closure of that ball is all of M, the result follows by continuity. $\square$

The next corollary is immediate.

Corollary 11.20. *Suppose (M,g) is a compact Riemannian manifold and there is a positive constant $c = 1/R^2$ such that the Ricci curvature of M satisfies $\mathrm{Rc}(v,v) \geq (n-1)c$ for all unit vectors v. Then the volume of M is no greater than the volume of the n-sphere of radius R with its round metric, and if equality holds, then (M,g) has constant sectional curvature c.* $\square$

(For explicit formulas for the volumes of n-spheres, see Problem 10-4.)

Problems

11-1. Let (M,g) be a Riemannian manifold, and let U be a normal neighborhood of $p \in M$. Use Corollary 6.42 to show that in every choice of polar normal coordinates $(\theta^1,\dots,\theta^{n-1},r)$ on a subset of $U \smallsetminus \{p\}$ (see Example 6.45), the covariant Hessian of r is given by

$$\nabla^2 r = \frac{1}{2}\sum_{\alpha,\beta=1}^{n-1}(\partial_r g_{\alpha\beta})d\theta^\alpha\,d\theta^\beta.$$

11-2. Prove the following extension to Proposition 11.2: Suppose P is an embedded submanifold of a Riemannian manifold (M,g), U is a normal neighborhood of P in M, and r is the radial distance function for P in U (see Prop. 6.37). If $\gamma: [0,b] \to U$ is a geodesic segment with $\gamma(0) \in P$ and $\gamma'(0)$ normal to P, and J is a Jacobi field along γ that is transverse to P in the sense of Problem 10-14, then $D_t J(t) = \mathcal{H}_r(J(t))$ for all $t \in [0,b]$.

11-3. Let (M,g) be a Riemannian manifold, and let f be any smooth local distance function defined on an open subset $U \subseteq M$. Let $F = \operatorname{grad} f$ (so the integral curves of F are unit-speed geodesics), and let $\mathcal{H}_f = \nabla F$ (the Hessian operator of f). Show that $\mathcal{H}_f$ satisfies the following Riccati equation along each integral curve γ of F:

$$D_t \mathcal{H}_f + \mathcal{H}_f^2 + R_{\gamma'} = 0, \tag{11.34}$$

where $R_{\gamma'}(w) = R(w, \gamma')\gamma'$. [Hint: Let W be any smooth vector field on U, and evaluate $\nabla_F \nabla_W F$ in two different ways.]

11-4. Let (M,g) be a compact Riemannian manifold. Prove that if R, L are positive numbers such that all sectional curvatures of M are less than or equal to $1/R^2$ and all closed geodesics have lengths greater than or equal to L, then

$$\operatorname{inj}(M) \geq \min\left(\pi R, \tfrac{1}{2} L\right).$$

[Hint: Assume not, and use the result of Problem 10-23(b).]

11-5. TRANSVERSE JACOBI FIELD COMPARISON THEOREM: Let P be an embedded hypersurface in a Riemannian manifold (M,g). Suppose $\gamma: [0,b] \to M$ is a unit-speed geodesic segment with $\gamma(0) \in P$ and $\gamma'(0)$ normal to P, and J is a normal Jacobi field along γ that is transverse to P. Let $\lambda = h(J(0), J(0))$, where h is the scalar second fundamental form of P with respect to the normal $-\gamma'(0)$. Let c be a real number, and let $u : \mathbb{R} \to \mathbb{R}$ be the unique solution to the initial value problem

$$\begin{aligned} u''(t) + cu(t) &= 0, \\ u(0) &= 1, \\ u'(0) &= \lambda. \end{aligned} \tag{11.35}$$

In the following statements, the principal curvatures of P are computed with respect to the normal $-\gamma'(0)$.

(a) If all sectional curvatures of M are bounded above by c, all principal curvatures of P at $\gamma(0)$ are bounded below by λ, and $u(t) \neq 0$ for $t \in (0,b)$, then $|J(t)| \geq u(t)|J(0)|$ for all $t \in [0,b]$.

(b) If all sectional curvatures of M are bounded below by c, all principal curvatures of P at $\gamma(0)$ are bounded above by λ, and $J(t) \neq 0$ for $t \in (0,b)$, then $|J(t)| \leq u(t)|J(0)|$ for all $t \in [0,b]$.

[Hint: Mimic the proof of Theorem 11.9, using the results of Problems 11-3 and 11-2.]

11-6. Suppose P is an embedded hypersurface in a Riemannian manifold (M,g) and N is a unit normal vector field along P. Suppose the principal curvatures of P with respect to $-N$ are bounded below by a constant λ, and the sectional curvatures of M are bounded above by $-\lambda^2$. Prove that P has no focal points along any geodesic segment with initial velocity N_p for $p \in P$.

11-7. Suppose P is an embedded hypersurface in a Riemannian manifold (M, g) and N is a unit normal vector field along P. Suppose the sectional curvatures of M are bounded below by a constant c, and the principal curvatures of P with respect to $-N$ are bounded above by a constant λ. Let u be the solution to the initial value problem (11.35). Prove that if b is a positive real number such that $u(b) = 0$, then P has a focal point along every geodesic segment with initial velocity N_p for some $p \in P$ and with length greater than or equal to b.

Chapter 12
Curvature and Topology

In this final chapter, we bring together most of the tools we have developed so far to prove some significant local-to-global theorems relating curvature and topology of Riemannian manifolds.

We focus first on constant-curvature manifolds. The main result here is the Killing–Hopf theorem, which shows that every complete, simply connected manifold with constant sectional curvature is isometric to one of our frame-homogeneous model spaces. A corollary of the theorem then shows that the ones that are not simply connected are just quotients of the models by discrete groups of isometries. The technique used to prove this result also leads to a global characterization of symmetric spaces in terms of parallel curvature tensors.

Next we turn to nonpositively curved manifolds. The primary result is the Cartan–Hadamard theorem, which topologically characterizes complete, simply connected manifolds with nonpositive sectional curvature: they are all diffeomorphic to $\mathbb{R}^n$. After proving the main result, we prove two other theorems, due to Cartan and Preissman, that place severe restrictions on the fundamental groups of complete manifolds with nonpositive or negative curvature.

Finally, we address the case of positive curvature. The main theorem is Myers's theorem, which says that a complete manifold with Ricci curvature bounded below by a positive constant must be compact and have a finite fundamental group and diameter no larger than that of the sphere with the same Ricci curvature. The borderline case is addressed by Cheng's maximal diameter theorem, which says that the diameter bound is achieved only by a round sphere. These results are supplemented by theorems of Milnor and Synge, which further restrict the topology of manifolds with nonnegative Ricci and positive sectional curvature, respectively.

Manifolds of Constant Curvature

Our first major local-to-global theorem is a global characterization of complete manifolds of constant curvature. If (M, g) has constant sectional curvature, Corollary

© Springer International Publishing AG 2018
J. M. Lee, *Introduction to Riemannian Manifolds*, Graduate Texts
in Mathematics 176, https://doi.org/10.1007/978-3-319-91755-9_12

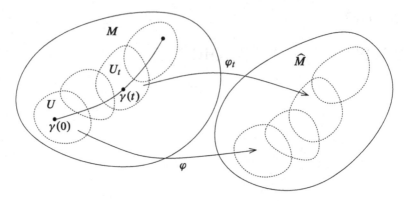

Fig. 12.1: Analytic continuation of a local isometry

10.15 shows that each point of M has a neighborhood that is isometric to an open subset of one of the constant-curvature model spaces of Chapter 3. In order to turn that into a global result, we use a technique modeled on the theory of analytic continuation in complex analysis.

Suppose (M, g) and $(\widehat{M}, \widehat{g})$ are Riemannian manifolds of the same dimension and $\varphi : U \to \widehat{M}$ is a local isometry defined on some connected open subset $U \subseteq M$. If $\gamma : [0, 1] \to M$ is a continuous path such that $\gamma(0) \in U$, then an *analytic continuation of φ along γ* is a family of pairs $\{(U_t, \varphi_t) : t \in [0, 1]\}$, where U_t is a connected neighborhood of $\gamma(t)$ and $\varphi_t : U_t \to \widehat{M}$ is a local isometry, such that $\varphi_0 = \varphi$ on $U_0 \cap U$, and for each $t \in [0, 1]$ there exists $\delta > 0$ such that $|t - t_1| < \delta$ implies that $\gamma(t_1) \in U_t$ and that φ_t agrees with φ_{t_1} on $U_t \cap U_{t_1}$ (see Fig. 12.1). Note that we require φ_t and φ_{t_1} to agree where they overlap only if t and t_1 are sufficiently close; in particular, if γ is not injective, the values of the analytic continuation may differ at different times at which γ returns to the same point. However, as the next lemma shows, all analytic continuations along the same path will end up with the same value.

Lemma 12.1 (Uniqueness of Analytic Continuation). *With (M, g), $(\widehat{M}, \widehat{g})$, and φ as above, if $\{(U_t, \varphi_t)\}$ and $\{(U'_t, \varphi'_t)\}$ are two analytic continuations of φ along the same path $\gamma : [0, 1] \to M$, then $\varphi_1 = \varphi'_1$ on a neighborhood of $\gamma(1)$.*

Proof. Let $\mathcal{T}$ be the set of all $t \in [0, 1]$ such that $\varphi_t = \varphi'_t$ on a neighborhood of $\gamma(t)$. Then $0 \in \mathcal{T}$, because both φ_0 and φ'_0 agree with φ on a neighborhood of $\gamma(0)$. We will show that $\mathcal{T}$ is open and closed in $[0, 1]$, from which it follows that $1 \in \mathcal{T}$, which proves the lemma.

To see that it is open, suppose $t \in \mathcal{T}$. By the definition of analytic continuation, there is some $\delta > 0$ such that if $t_1 \in [0, 1]$ and $|t_1 - t| < \delta$, then φ_{t_1} and φ'_{t_1} both agree with φ_t on a neighborhood of $\gamma(t)$, so $[0, 1] \cap (t - \delta, t + \delta) \subseteq \mathcal{T}$. Thus $\mathcal{T}$ is open.

To see that it is closed, suppose t is a limit point of $\mathcal{T}$. There is a sequence $t_i \to t$ such that $t_i \in \mathcal{T}$, which means that $\varphi_{t_i} = \varphi'_{t_i}$ on a neighborhood of t_i for each i. By definition of analytic continuation, for i large enough, we have $\gamma(t_i) \in U_t \cap U'_t$

and $\varphi_t = \varphi_{t_i} = \varphi'_{t_i} = \varphi'_t$ on a neighborhood of $\gamma(t_i)$. In particular, this means that $\varphi_t(\gamma(t_i)) = \varphi'_t(\gamma(t_i))$ and $d(\varphi_t)_{\gamma(t_i)} = d(\varphi'_t)_{\gamma(t_i)}$ for each such i. By continuity, therefore, $\varphi_t(\gamma(t)) = \varphi'_t(\gamma(t))$ and $d(\varphi_t)_{\gamma(t)} = d(\varphi'_t)_{\gamma(t)}$. Proposition 5.22 shows that $\varphi_t = \varphi'_t$ on a neighborhood of $\gamma(t)$, so $t \in \mathcal{T}$. $\square$

The previous lemma shows that all analytic continuations along the same path end up with the same value. But when we consider analytic continuations along *different* paths, this may not be the case. The next theorem gives a sufficient condition for analytic continuations along two different paths to result in the same value.

Theorem 12.2 (Monodromy Theorem for Local Isometries). *Let (M, g) and $(\widehat{M}, \widehat{g})$ be connected Riemannian manifolds. Suppose U is a neighborhood of a point $p \in M$ and $\varphi : U \to \widehat{M}$ is a local isometry that can be analytically continued along every path starting at p. If $\gamma_0, \gamma_1 : [0, 1] \to M$ are path-homotopic paths from p to a point $q \in M$, then the analytic continuations along γ_0 and γ_1 agree in a neighborhood of q.*

Proof. Suppose γ_0 and γ_1 are path-homotopic, and let $H : [0,1] \times [0,1] \to M$ be a path homotopy from γ_0 to γ_1. Write $H_s(t) = H(t, s)$, so that $H_0 = \gamma_0$, $H_1 = \gamma_1$, and $H_s(0) = p$ and $H_s(1) = q$ for all $s \in [0, 1]$. The hypothesis implies that for each $s \in [0, 1]$, there is an analytic continuation of φ along H_s, which we denote by $\{(U_t^{(s)}, \varphi_t^{(s)}) : t \in [0, 1]\}$.

Consider the function $P : [0, 1] \to \widehat{M}$ given by $P(s) = \varphi_1^{(s)}(q)$. Given $s \in [0, 1]$, if s' is sufficiently close to s, it follows from compactness of $H_s([0, 1])$ that the same family $\{(U_t^{(s)}, \varphi_t^{(s)}) : t \in [0, 1]\}$ serves as an analytic continuation along $H_{s'}$. This means that $P(s)$ is locally constant as a function of s, so $\varphi_1^{(1)}(q) = \varphi_1^{(0)}(q)$. Since a path from p to q can easily be modified near q to yield a path from p to any point q' sufficiently nearby, and the same analytic continuation works for the modified path, it follows from the same argument that $\varphi_1^{(1)}(q') = \varphi_1^{(0)}(q')$ for all q' in a neighborhood of q. $\square$

Corollary 12.3. *Let (M, g) and $(\widehat{M}, \widehat{g})$ be simply connected, complete Riemannian manifolds. Suppose U is a connected neighborhood of a point $p \in M$ and $\varphi : U \to \widehat{M}$ is a local isometry that can be analytically continued along every path starting at p. Then there is a global isometry $\Phi : M \to \widehat{M}$ whose restriction to U agrees with φ.*

Proof. Let $q \in M$ be arbitrary. We wish to define $\Phi(q)$ to be the value of an analytic continuation of φ along a path from p to q. The fact that M is simply connected means that all paths from p to q are path-homotopic, so the monodromy theorem implies that all such analytic continuations agree in a neighborhood of q; thus Φ is well defined. Because it is a globally defined local isometry and M is complete, it follows from Theorem 6.23 that Φ is a Riemannian covering map. Since $\widehat{M}$ is simply connected, every covering of $\widehat{M}$ is bijective by Corollary A.59, so Φ is a bijective local isometry and thus a global isometry. $\square$

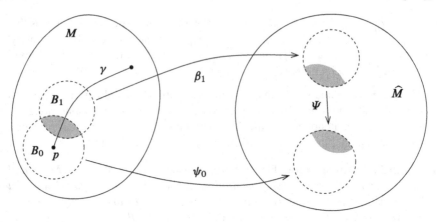

Fig. 12.2: Proof of the Killing–Hopf theorem

The next theorem, due to Wilhelm Killing and Heinz Hopf, is the main result of this section.

Theorem 12.4 (Killing–Hopf). *Let (M,g) be a complete, simply connected Riemannian n-manifold with constant sectional curvature, $n \geq 2$. Then M is isometric to one of the model spaces $\mathbb{R}^n$, $\mathbb{S}^n(R)$, and $\mathbb{H}^n(R)$.*

Proof. Given (M,g) satisfying the hypotheses, let $(\widehat{M},\widehat{g})$ be the Euclidean space, sphere, or hyperbolic space with the same constant sectional curvature as M. Note that $\widehat{M}$ is simply connected in each case. Let $p \in M$ be arbitrary. Corollary 10.15 shows that there exist an open subset $V \subseteq M$ containing p, an open subset $\widehat{V} \subseteq \widehat{M}$, and an isometry $\varphi : V \to \widehat{V}$. If we can show that φ can be analytically continued along every path starting at p, then we can conclude from Corollary 12.3 that there exists a global isometry $\Phi : M \to \widehat{M}$, thus proving the theorem.

To that end, let $\gamma : [0,1] \to M$ be a path starting at p. Corollary 10.15 shows that for each $t \in [0,1]$, there is a neighborhood of $\gamma(t)$ that is isometric to an open subset of $\widehat{M}$. After shrinking the neighborhoods if necessary, we may assume by Theorem 6.17 that each such neighborhood is a convex geodesic ball. By compactness, we can choose a partition $(t_0,\ldots,t_{k+1})$ of $[0,1]$ and a finite sequence of such balls $B_0,\ldots,B_k$ such that $\gamma([t_i,t_{i+1}]) \subseteq B_i$ for each $i = 0,\ldots,k$. After shrinking B_0 and adding more points to the partition if necessary, we can also assume $B_0 \subseteq V$.

We wish to construct local isometries $\psi_i : B_i \to \widehat{M}$ such that successive ones agree on overlaps. Begin with $\psi_0 = \varphi|_{B_0}$. By our choice of B_1, there is some isometry β_1 from B_1 to an open subset of $\widehat{M}$. Note that $B_0 \cap B_1$ is an intersection of geodesically convex sets and thus is connected, and it is nonempty because it contains $\gamma(t_1)$. The composition $\psi_0 \circ \beta_1^{-1}$ is an isometry from $\beta_1(B_0 \cap B_1)$ to $\psi_0(B_0 \cap B_1)$, both of which are connected open subsets of $\widehat{M}$, so by Problem 5-11(c), it is the restriction of a global isometry $\Psi : \widehat{M} \to \widehat{M}$ (Fig. 12.2). Let $\psi_1 = \Psi \circ \beta_1 : B_1 \to \widehat{M}$. Our choice of Ψ guarantees that $\psi_1|_{B_0 \cap B_1} = \psi_0|_{B_0 \cap B_1}$.

Proceeding similarly by induction, we can define $\psi_i : B_i \to \widehat{M}$ for $i = 2, \ldots, k$, such that ψ_i agrees with ψ_{i-1} on $B_i \cap B_{i-1}$.

Now we define the analytic continuation of φ as follows: start with $(U_t, \varphi_t) = (B_0, \psi_0)$ for $0 \le t \le t_1$, and thereafter let $(U_t, \varphi_t) = (B_i, \psi_i)$ for $t_i < t \le t_{i+1}$. It is easy to verify that this family of maps does the job. $\qquad\square$

Corollary 12.5 (Characterization of Constant-Curvature Manifolds). *The complete, connected, n-dimensional Riemannian manifolds of constant sectional curvature are, up to isometry, exactly the Riemannian quotients of the form $\widetilde{M}/\Gamma$, where $\widetilde{M}$ is one of the constant-curvature model spaces $\mathbb{R}^n$, $\mathbb{S}^n(R)$, or $\mathbb{H}^n(R)$, and Γ is a discrete subgroup of $\mathrm{Iso}\big(\widetilde{M}\big)$ that acts freely on $\widetilde{M}$.*

Proof. First suppose (M, g) is a complete, connected Riemannian n-manifold with constant sectional curvature, and let $\pi : \widetilde{M} \to M$ be its universal covering manifold with the pullback metric $\widetilde{g} = \pi^* g$. The preceding theorem shows that $\big(\widetilde{M}, \widetilde{g}\big)$ is isometric to one of the model spaces, so we might as well take it to be one of the models. Proposition C.20 shows that the group Γ of covering automorphisms acts freely and properly on $\widetilde{M}$, and Corollary 2.33 shows that M is isometric to $\widetilde{M}/\Gamma$. Moreover, if φ is any covering automorphism, then $\pi \circ \varphi = \pi$, and so $\varphi^* \widetilde{g} = \varphi^* \pi^* g = \pi^* g = \widetilde{g}$; thus Γ acts isometrically, so it is a subgroup of $\mathrm{Iso}\big(\widetilde{M}\big)$.

To show that Γ is discrete, suppose $\{\varphi_i\} \subseteq \Gamma$ is an infinite set with an accumulation point in $\mathrm{Iso}(\widetilde{M})$. (Note that Problem 5-11 shows that $\mathrm{Iso}\big(\widetilde{M}, \widetilde{g}\big)$ is a Lie group acting smoothly on $\widetilde{M}$ in each case, so it makes sense to talk about the topology of the isometry group.) Since the action of Γ is free, for every point $\widetilde{p} \in \widetilde{M}$ the set $\{\varphi_i(\widetilde{p})\}$ is infinite, and by continuity of the action it has an accumulation point in $\widetilde{M}$. But this is impossible, since the points $\{\varphi_i(\widetilde{p})\}$ all project to the same point in M under the covering map π, and so form a closed discrete set. Thus Γ is discrete in $\mathrm{Iso}(\widetilde{M})$.

Conversely, suppose $\widetilde{M}$ is one of the model spaces and Γ is a discrete subgroup of $\mathrm{Iso}\big(\widetilde{M}\big)$ that acts freely on $\widetilde{M}$. Then Γ is a closed Lie subgroup of $\mathrm{Iso}\big(\widetilde{M}\big)$ by Proposition C.9, so the result of Problem 6-28 shows that it acts properly on $\widetilde{M}$. It then follows from Proposition 2.32 that $\widetilde{M}/\Gamma$ is a smooth manifold and has a unique Riemannian metric such that the quotient map $\pi : \widetilde{M} \to \widetilde{M}/\Gamma$ is a Riemannian covering, and $\widetilde{M}/\Gamma$ is complete by Corollary 6.24. Since a Riemannian covering is in particular a local isometry, $\widetilde{M}/\Gamma$ has constant sectional curvature. $\qquad\square$

A complete, connected Riemannian manifold with constant sectional curvature is called a ***space form***. A space form is called ***spherical***, ***Euclidean***, or ***hyperbolic***, depending on whether its sectional curvature is positive, zero, or negative, respectively. The preceding corollary, combined with the characterization of the isometry groups of the simply connected model spaces given in Problem 5-11, essentially reduces the classification of space forms to the group-theoretic problem of classifying the discrete subgroups of $\mathrm{E}(n)$, $\mathrm{O}(n + 1)$, and $\mathrm{O}^+(n, 1)$ that act without fixed points. Nevertheless, the group-theoretic problem is still far from easy.

The spherical space forms were classified in 1972 by Joseph Wolf [Wol11]; the proof is intimately connected with the representation theory of finite groups. Although in even dimensions there are only spheres and projective spaces (see Problem

12-2), the situation in odd dimensions is far more complicated. Already in dimension 3 there are many interesting examples: some notable ones are the **lens spaces** obtained as quotients of $\mathbb{S}^3 \subseteq \mathbb{C}^2$ by cyclic groups rotating the two complex coordinates through different angles, and the quotients of SO(3) (which is diffeomorphic to $\mathbb{RP}^3$ and is therefore already a quotient of $\mathbb{S}^3$) by the **dihedral groups**, the symmetry groups of regular 3-dimensional polyhedra.

The complete classification of Euclidean space forms is known only in low dimensions. Problem 12-3 shows that every compact 2-dimensional Euclidean space form is diffeomorphic to the torus or the Klein bottle. It turns out that there are 10 classes of nondiffeomorphic compact Euclidean space forms of dimension 3, and 75 classes in dimension 4. The fundamental groups of compact Euclidean space forms are examples of **crystallographic groups**—discrete groups of Euclidean isometries with compact quotients, which have been studied extensively by physicists as well as geometers. (A quotient of $\mathbb{R}^n$ by a crystallographic group is a space form, provided it is a manifold, which is true whenever the group is torsion-free.) In higher dimensions, the classification is still elusive. The main things that are known are two results proved by Ludwig Bieberbach in the early twentieth century: (1) every Euclidean space form is a quotient of a flat torus by a finite group of isometries, and (2) in each dimension, there are only finitely many diffeomorphism classes of Euclidean space forms. See [Wol11] for proofs of the Bieberbach theorems and a complete survey of the state of the art.

In addition to the question of classifying Euclidean space forms up to diffeomorphism, there is also the question of classifying the different flat Riemannian metrics on a given manifold up to isometry. Problem 12-5 shows, for example, that there is a three-parameter family of nonisometric flat Riemannian metrics on the 2-torus.

Finally, the study of hyperbolic space forms is a vast and rich subject in its own right. A good introduction is the book [Rat06].

Symmetric Spaces Revisited

The technique of analytic continuation of local isometries can also be used to derive a fundamental global result about symmetric spaces. Theorem 10.19 showed that locally symmetric spaces are characterized by having parallel curvature. The next theorem is a global version of that result.

Theorem 12.6. *Suppose (M, g) is a complete, simply connected Riemannian manifold with parallel Riemann curvature tensor. Then M is a symmetric space.*

Proof. Let p be an arbitrary point of M. Theorem 10.19 shows that M is locally symmetric, so there is a point reflection $\varphi : U \to U$ defined on some connected neighborhood U of p. If we can show that φ can be analytically continued along every path starting at p, then Corollary 12.3 shows that φ extends to a global isometry $\Phi : M \to M$; because it agrees with φ on U, it satisfies $d\Phi_p = d\varphi_p = -\mathrm{Id}$, so it is a point reflection.

Let $\gamma : [0,1] \to M$ be a path starting at p. By compactness, we can find a partition $(t_0, \ldots, t_{k+1})$ of $[0,1]$ and convex geodesic balls $B_0, \ldots, B_k$ such that $\gamma([t_i, t_{i+1}]) \subseteq B_i$ for each $i = 0, \ldots, k$, with B_0 chosen small enough that $B_0 \subseteq U$. As in the proof of Theorem 12.4, we will construct local isometries $\psi_i : B_i \to M$ such that successive ones agree on overlaps.

Begin with $\psi_0 = \varphi|_{B_0}$, and let $p_1 = \gamma(t_1)$ and $q_1 = \psi_0(p_1)$. Because ψ_0 is a local isometry, the linear map $d\psi_0|_{p_1} : T_{p_1}M \to T_{q_1}M$ satisfies $(d\psi_0|_{p_1})^* g_{q_1} = g_{p_1}$ and $(d\psi_0|_{p_1})^* Rm_{q_1} = Rm_{p_1}$. Lemma 10.18 then shows that there is a local isometry $\psi_1 : B_1 \to M$ that satisfies $\psi_1(p_1) = \psi_0(p_1)$ and $d\psi_1|_{p_1} = d\psi_0|_{p_1}$, and it follows from Proposition 5.22 that ψ_1 and ψ_0 agree on the connected open set $B_0 \cap B_1$. Proceeding by induction, we obtain local isometries $\psi_i : B_i \to M$ satisfying $\psi_i|_{B_{i-1} \cap B_i} = \psi_{i-1}|_{B_{i-1} \cap B_i}$. Then the analytic continuation of φ is defined just as in the last paragraph of the proof of Theorem 12.4, thus completing the proof. $\square$

Corollary 12.7. *Suppose (M, g) is a complete, connected Riemannian manifold. The following are equivalent:*

(a) *M has parallel curvature tensor.*
(b) *M is a locally symmetric space.*
(c) *M is isometric to a Riemannian quotient $\widetilde{M}/\Gamma$, where $\widetilde{M}$ is a (globally) symmetric space and Γ is a discrete Lie group that acts freely, properly, and isometrically on $\widetilde{M}$.*

Proof. Theorem 10.19 shows that (a) $\Leftrightarrow$ (b). To show that (a) $\Rightarrow$ (c), assume that M has parallel curvature tensor, and let $(\widetilde{M}, \widetilde{g})$ be its universal covering manifold with the pullback metric. Since the covering map $\pi : \widetilde{M} \to M$ is a local isometry, $\widetilde{M}$ also has parallel curvature tensor, and Corollary 6.24 shows that it is complete. Then Theorem 12.6 shows that $\widetilde{M}$ is a symmetric space. If we let Γ denote the covering automorphism group with the discrete topology, then Proposition C.20 shows that Γ is a discrete Lie group acting smoothly, freely, and properly on $\widetilde{M}$, and it acts isometrically because the pullback metric is invariant under all covering automorphisms. Finally, Corollary 2.33 shows that M is isometric to $\widetilde{M}/\Gamma$.

Finally, we show that (c) $\Rightarrow$ (a). If M is isometric to a Riemannian quotient of the form $\widetilde{M}/\Gamma$ as in (c), then the quotient map $\widetilde{M} \to \widetilde{M}/\Gamma$ is a Riemannian covering by Proposition 2.32. Since $\widetilde{M}$ has parallel curvature tensor by the result of Problem 7-3, so does M. $\square$

An extensive treatment of the structure and classification of symmetric spaces can be found in [Hel01].

The analytic continuation technique used in this section was introduced in 1926 by Élie Cartan, in the same paper [Car26] in which he first defined symmetric spaces and proved many of their properties. The technique was generalized in 1956 by Warren Ambrose [Amb56] to allow for isometries between more general Riemannian manifolds with varying curvature; and then in 1959 it was generalized further by Noel Hicks [Hic59] to manifolds with connections in their tangent bundles, not necessarily Riemannian ones. The resulting theorem is known as the *Cartan–Ambrose–Hicks theorem*. We do not pursue it further, but a statement and proof can be found in [CE08].

Manifolds of Nonpositive Curvature

Our next major local-to-global theorem provides a complete topological characterization of complete, simply connected manifolds of nonpositive sectional curvature. This was proved in 1928 by Élie Cartan, generalizing earlier proofs for surfaces by Hans Carl Friedrich von Mangoldt and Jacques Hadamard. As with many other such theorems, tradition has given it a title that is not entirely historically accurate, but is generally recognized by mathematicians.

Theorem 12.8 (Cartan–Hadamard). *If (M, g) is a complete, connected Riemannian manifold with nonpositive sectional curvature, then for every point $p \in M$, the map $\exp_p : T_p M \to M$ is a smooth covering map. Thus the universal covering space of M is diffeomorphic to $\mathbb{R}^n$, and if M is simply connected, then M itself is diffeomorphic to $\mathbb{R}^n$.*

Proof. By Theorem 11.12, the assumption of nonpositive curvature guarantees that p has no conjugate points along any geodesic. Therefore, by Proposition 10.20, $\exp_p$ is a local diffeomorphism on all of $T_p M$.

Let $\tilde{g}$ be the (variable-coefficient) 2-tensor field $\exp_p^* g$ defined on $T_p M$. Because $\exp_p$ is a local diffeomorphism, $\tilde{g}$ is a Riemannian metric, and $\exp_p$ is a local isometry from $(T_p M, \tilde{g})$ to (M, g). Note that each line $t \mapsto tv$ in $T_p M$ is a $\tilde{g}$-geodesic, so $(T_p M, \tilde{g})$ is complete by Corollary 6.20. It then follows from Theorem 6.23 that $\exp_p$ is a smooth covering map. The remaining statements of the theorem follow immediately from uniqueness of the universal covering space. $\quad\square$

Because of this theorem, a complete, simply connected Riemannian manifold with nonpositive sectional curvature is called a ***Cartan–Hadamard manifold***. The basic examples are Euclidean and hyperbolic spaces. The next two propositions show that Cartan–Hadamard manifolds share many basic geometric properties with these model spaces. A ***line*** in a Riemannian manifold is the image of a nonconstant geodesic that is defined on all of $\mathbb{R}$ and restricts to a minimizing segment between any two of its points.

Proposition 12.9 (Basic Properties of Cartan–Hadamard Manifolds). *Suppose (M, g) is a Cartan–Hadamard manifold.*

- *(a) The injectivity radius of M is infinite.*
- *(b) The image of every nonconstant maximal geodesic in M is a line.*
- *(c) Any two distinct points in M are contained in a unique line.*
- *(d) Every open or closed metric ball in M is a geodesic ball.*
- *(e) For every point $q \in M$, the function $r(x) = d_g(q, x)$ is smooth on $M \smallsetminus \{q\}$ and $r(x)^2$ is smooth on all of M.*

Proof. These properties follow from Lemma 6.8, Proposition 6.11, and Corollaries 6.12 and 6.13, together with the fact that M is a normal neighborhood of each of its points. $\quad\square$

Suppose A, B, C are three points in a Cartan–Hadamard manifold that are **non-collinear** (meaning that they are not all contained in a single line). The **geodesic triangle** $\triangle ABC$ determined by the points is the union of the images of the (unique) geodesic segments connecting the three points. (In the present context, we do not require that a geodesic triangle bound a two-dimensional region, as we did in Chapter 9.) If $\triangle ABC$ is a geodesic triangle, we denote the angle in $T_A M$ formed by the initial velocities of the geodesic segments from A to B and A to C by $\angle A$ (or $\angle CAB$ if necessary to avoid ambiguity), and similarly for the other angles.

Proposition 12.10. *Suppose $\triangle ABC$ is a geodesic triangle in a Cartan–Hadamard manifold (M, g), and let a, b, c denote the lengths of the segments opposite the vertices A, B, and C, respectively. The following inequalities hold:*

(a) $c^2 \geq a^2 + b^2 - 2ab \cos \angle C$.
(b) $\angle A + \angle B + \angle C \leq \pi$.

If the sectional curvatures of g are all strictly negative, then strict inequality holds in both cases.

Proof. Problem 12-7. $\qquad\square$

Another consequence of the Cartan–Hadamard theorem is that there are stringent topological restrictions on which manifolds can carry metrics of nonpositive sectional curvature. The next corollary is immediate.

Corollary 12.11. *No simply connected compact manifold admits a metric of nonpositive sectional curvature.* $\qquad\square$

With a little more work, we can obtain a much more powerful result.

Corollary 12.12. *Suppose M and N are positive-dimensional compact, connected smooth manifolds, at least one of which is simply connected. Then $M \times N$ does not admit any Riemannian metric of nonpositive sectional curvature.*

Proof. Suppose for the sake of contradiction that M is simply connected and g is a metric on $M \times N$ with nonpositive sectional curvature. If $\widetilde{N}$ is the universal covering manifold of N, then there is a universal covering map $\pi : M \times \widetilde{N} \to M \times N$, and $\pi^* g$ is a complete metric of nonpositive sectional curvature on $M \times \widetilde{N}$. The Cartan–Hadamard theorem shows that $M \times \widetilde{N}$ is diffeomorphic to a Euclidean space.

Since M is simply connected, it is orientable (Cor. B.19). Let μ be a smooth orientation m-form for M, where $m = \dim M$. Then μ is closed because there are no nontrivial $(m + 1)$-forms on M, but Stokes's theorem shows that μ is not exact, because $\int_M \mu > 0$. On the other hand, if we let $p : M \times \widetilde{N} \to M$ denote the projection on the first factor, then the form $p^* \mu$ is exact on $M \times \widetilde{N}$, because every closed m-form on a Euclidean space is exact.

Choose an arbitrary point $y_0 \in \widetilde{N}$ and define $\sigma : M \to M \times \widetilde{N}$ by $\sigma(x) = (x, y_0)$; it is a diffeomorphism onto its image $\sigma(M) = M \times \{y_0\}$, which is a compact embedded submanifold of $M \times \widetilde{N}$. By diffeomorphism invariance of the integral,

$$\int_{\sigma(M)} p^*\mu = \int_M \sigma^*(p^*\mu) = \int_M (p \circ \sigma)^*\mu = \int_M \mu > 0,$$

because $p \circ \sigma = \text{Id}$. But $p^*\mu$ is exact on $M \times \tilde{N}$, which implies $\int_{\sigma(M)} p^*\mu = 0$, a contradiction. $\qquad\square$

Example 12.13. By the preceding corollary, every metric on $\mathbb{S}^m \times \mathbb{S}^n$ must have positive sectional curvature somewhere, provided $m \geq 2$ and $n \geq 1$, because both spheres are compact and $\mathbb{S}^m$ is simply connected. $\qquad /\!/$

With a little more algebraic topology, one can obtain more information. A topological space whose higher homotopy groups $\pi_k(M)$ all vanish for $k > 1$ is called **aspherical**.

Corollary 12.14. *If M is a smooth manifold that admits a metric of nonpositive sectional curvature, then M is aspherical.*

Proof. Suppose M admits a nonpositively curved metric, so its universal covering space is diffeomorphic to $\mathbb{R}^n$ by Cartan–Hadamard. Since $\mathbb{R}^n$ is contractible, it is aspherical. Since covering maps induce isomorphisms on π_k for $k > 1$ (see [Hat02], Prop. 4.1]), it follows that M is aspherical as well. $\qquad\square$

Cartan's Torsion Theorem

In addition to the topological restrictions on nonpositively curved manifolds imposed directly by the Cartan–Hadamard theorem, it turns out that there is a stringent restriction on the fundamental groups of such manifolds (Corollary 12.18 below). In preparation for the proof, we need the following two lemmas, both of which are interesting in their own right.

Lemma 12.15. *Suppose (M, g) is a Cartan–Hadamard manifold. Given $q \in M$, let $f : M \to [0, \infty)$ be the function $f(x) = \frac{1}{2}d_g(x, q)^2$. Then f is strictly geodesically convex, in the sense that for every geodesic segment $\gamma : [0, 1] \to M$, the following inequality holds for all $t \in (0, 1)$:*

$$f(\gamma(t)) < (1-t)f(\gamma(0)) + tf(\gamma(1)). \tag{12.1}$$

Proof. We can write $f(x) = \frac{1}{2}r(x)^2$, where r is the radial distance function from q with respect to any normal coordinates, and Proposition 12.9 shows that f is smooth on all of M. The Hessian comparison theorem implies that $\mathcal{H}_r \geq (1/r)\pi_r$ on $M \smallsetminus \{q\}$, and therefore

$$\mathcal{H}_f = \nabla(\text{grad } f) = \nabla(r \text{ grad} r) = \text{grad} r \otimes dr + r\mathcal{H}_r \geq \text{grad} r \otimes dr + \pi_r.$$

The expression on the right-hand side above is actually equal to the identity operator, as can be checked by applying it separately to $\partial_r = \text{grad} r$ and an arbitrary vector perpendicular to ∂_r. Therefore $\langle \mathcal{H}_f(v), v \rangle \geq |v|^2$ for all tangent vectors to $M \smallsetminus \{q\}$, and by continuity, the same holds at q as well.

Now suppose $\gamma : [0,1] \to M$ is a geodesic segment. By the chain rule,

$$\frac{d}{dt} f(\gamma(t)) = df_{\gamma(t)}(\gamma'(t)) = \langle \text{grad } f|_{\gamma(t)}, \gamma'(t) \rangle,$$

and thus (using the fact that γ' is parallel along γ)

$$\frac{d^2}{dt^2} f(\gamma(t)) = \langle \nabla_{\gamma'(t)} \text{grad } f|_{\gamma(t)}, \gamma'(t) \rangle = \langle \mathcal{H}_f(\gamma'(t)), \gamma'(t) \rangle > 0.$$

Therefore, the function $f \circ \gamma$ is strictly convex on $[0,1]$, and (12.1) follows. $\square$

Lemma 12.16. *Suppose (M,g) is a Cartan–Hadamard manifold and S is a compact subset of M containing more than one point. Then there is a unique closed ball of minimum radius containing S.*

Proof. Because S is compact, it is bounded. Let $\delta = \text{diam}(S)$, and let q_0 be any point of S, so that $S \subseteq \bar{B}_\delta(q_0)$. Let c_0 be the infimum of the radii of closed metric balls containing S, and let $\big(\bar{B}_{c_i}(p_i)\big)$ be a sequence of closed balls containing S such that $c_i \searrow c_0$ as $i \to \infty$. For i large enough that $c_i < 2\delta$, the fact that $q_0 \in S \subseteq \bar{B}_{c_i}(p_i)$ implies $d_g(p_i, q_0) \leq c_i < 2\delta$, so each of the points p_i lies in the compact ball $\bar{B}_{2\delta}(q_0)$. After passing to a subsequence, we may assume that p_i converges to some point $p_0 \in M$.

To show that $S \subseteq \bar{B}_{c_0}(p_0)$, suppose $q \in S$, and let $\varepsilon > 0$ be arbitrary. If i is large enough that $d_g(p_i, p_0) \leq \varepsilon/2$ and $c_i \leq c_0 + \varepsilon/2$, then

$$d_g(q, p_0) \leq d_g(q, p_i) + d_g(p_i, p_0) \leq c_i + \varepsilon/2 \leq c_0 + \varepsilon.$$

Since ε was arbitrary, this shows that $d_g(q, p_0) \leq c_0$, and thus $S \subseteq \bar{B}_{c_0}(p_0)$.

Now we need to show that $\bar{B}_{c_0}(p_0)$ is the unique closed ball of radius c_0 containing S. Suppose to the contrary that $\bar{B}_{c_0}(p'_0)$ is another such ball. Let $\gamma : [0,1] \to M$ be the geodesic segment from p_0 to p'_0, and let $m = \gamma\big(\tfrac{1}{2}\big)$ be its midpoint. Let $q \in S$ be arbitrary, and let $f : M \to [0,\infty)$ be the function $f(x) = \tfrac{1}{2} d_g(q,x)^2$. Lemma 12.15 implies that

$$\tfrac{1}{2} d_g(q,m)^2 = f(m) < \tfrac{1}{2} f(p_0) + \tfrac{1}{2} f(p'_0) \leq \tfrac{1}{2}\big(\tfrac{1}{2}c_0^2\big) + \tfrac{1}{2}\big(\tfrac{1}{2}c_0^2\big) = \tfrac{1}{2}c_0^2.$$

Thus for every $q \in S$, we have $d_g(q,m) < c_0$. Since S is compact, the continuous function $d_g(\cdot, m)$ takes on a maximum value $b_0 < c_0$ on S, showing that $S \subseteq \bar{B}_{b_0}(m)$. This contradicts the fact that c_0 is the minimum radius of a ball containing S. $\square$

Let us call the center of the smallest enclosing ball the **1-*center*** of the set S. (The term is borrowed from optimization theory, where the "1-center problem" is the problem of finding a location for a single production facility that minimizes the maximum distance to any client. Some Riemannian geometry texts refer to the 1-center of a set $S \subseteq M$ as its "circumcenter" or "center of gravity," but these terms seem inappropriate, because the 1-center does not coincide with the classical meaning of either term for finite subsets of the Euclidean plane.)

Theorem 12.17 (Cartan's Fixed-Point Theorem). *Suppose (M, g) is a Cartan–Hadamard manifold and G is a compact Lie group acting smoothly and isometrically on M. Then G has a fixed point in M, that is, a point $p_0 \in M$ such that $\varphi \cdot p_0 = p_0$ for all $\varphi \in G$.*

Proof. Let $q_0 \in M$ be arbitrary, and let $S = G \cdot q_0$ be the orbit of q_0. If S contains only the point q_0, then q_0 is a fixed point, so we may assume that the orbit contains at least two points. Because S is the image of the continuous map from G to M given by $\varphi \mapsto \varphi \cdot q_0$, it is compact, so by Lemma 12.16, there is a unique smallest closed geodesic ball containing S. Let p_0 be the center of this ball (the 1-center of S), and let c_0 be its radius.

Now let $\varphi_0 \in G$ be arbitrary, and note that $\varphi_0 \cdot S$ is the set of all points of the form $\varphi_0 \cdot \varphi \cdot q_0$ for $\varphi \in G$. As φ ranges over all of G, so does $\varphi_0 \varphi$, so this image set is exactly the orbit of q_0. In other words, $\varphi_0 \cdot S = S$.

Because G acts by isometries, $\varphi_0 \cdot \bar{B}_{c_0}(p_0) = \bar{B}_{c_0}(\varphi_0 \cdot p_0)$. Thus $\bar{B}_{c_0}(\varphi_0 \cdot p_0)$ is a closed ball of the same radius c_0 containing $\varphi_0 \cdot S = S$, so by uniqueness of the 1-center we must have $\varphi_0 \cdot p_0 = p_0$. Since φ_0 was arbitrary, this shows that p_0 is a fixed point of G. $\qquad\qquad\square$

Cartan's fixed-point theorem has important applications to Lie theory (where it is used to prove that all maximal compact subgroups of a Lie group are conjugate to each other). But our interest in the theorem is that it leads to the next corollary, also due to Cartan. If G is a group, an element $\varphi \in G$ is called a ***torsion element*** if $\varphi^k = 1$ for some integer $k > 1$. A group is said to be ***torsion-free*** if the only torsion element is the identity.

Corollary 12.18 (Cartan's Torsion Theorem). *Suppose (M, g) is a complete, connected Riemannian manifold with nonpositive sectional curvature. Then $\pi_1(M)$ is torsion-free.*

Proof. Let $\left(\widetilde{M}, \widetilde{g}\right)$ be the universal covering manifold of M with the pullback metric. Then $\left(\widetilde{M}, \widetilde{g}\right)$ is a Cartan–Hadamard manifold, and M is isometric to a Riemannian quotient $\widetilde{M}/\Gamma$, where Γ is subgroup of $\mathrm{Iso}\left(\widetilde{M}, \widetilde{g}\right)$, isomorphic to $\pi_1(M)$ and acting freely on $\widetilde{M}$.

Suppose φ is a torsion element of Γ. Then the cyclic group generated by φ is finite, and thus is a compact 0-dimensional Lie group with the discrete topology, acting smoothly on $\widetilde{M}$ because isometries are smooth. Cartan's fixed-point theorem shows that it has a fixed point. Because Γ acts freely, this implies that φ is the identity. Thus Γ is torsion-free, and the same is true of $\pi_1(M)$. $\qquad\square$

Preissman's Theorem

Taken together, the Cartan–Hadamard theorem and Cartan's torsion theorem put severe restrictions on the possible topologies of manifolds that can carry nonpositively curved metrics. If we make the stronger assumption of strictly *negative* sectional curvature, we can restrict the topology even further. The following theorem was first proved in 1943 by Alexandre Preissman [Pre43].

Theorem 12.19 (Preissman). *If (M, g) is a compact, connected Riemannian manifold with strictly negative sectional curvature, then every nontrivial abelian subgroup of $\pi_1(M)$ is isomorphic to $\mathbb{Z}$.*

Before proving Preissman's theorem, we need two more lemmas. Suppose (M, g) is a complete Riemannian manifold and $\varphi : M \to M$ is an isometry. A geodesic $\gamma : \mathbb{R} \to M$ is called an **axis for φ** if φ restricts to a nontrivial translation along γ, that is, if there is a nonzero constant c such that $\varphi(\gamma(t)) = \gamma(t + c)$ for all $t \in \mathbb{R}$. An isometry with no fixed points that has an axis is said to be **axial**.

Example 12.20 (Axial Isometries). In $\mathbb{R}^2$ with its Euclidean metric, every nontrivial translation $(x, y) \mapsto (x + a, y + b)$ is axial, and every line parallel to the vector $a\partial_x + b\partial_y$ is an axis. In the upper half-plane model of the hyperbolic plane, every dilation $(x, y) \mapsto (cx, cy)$ for $c \neq 1$ is axial, and the geodesic $\gamma(t) = (0, e^t)$ is an axis. On the other hand, the horizontal translations $(x, y) \mapsto (x + a, y)$ are not axial for the hyperbolic plane, because no geodesic is invariant under such an isometry. //

As the next lemma shows, there is a close relation between axes and closed geodesics. This lemma will also prove useful in the proof of Synge's theorem later in the chapter.

Lemma 12.21. *Suppose (M, g) is a compact, connected Riemannian manifold, and $\pi : \widetilde{M} \to M$ is its universal covering manifold endowed with the metric $\widetilde{g} = \pi^* g$. Then every covering automorphism of π has an axis, which restricts to a lift of a closed geodesic in M that is the shortest admissible path in its free homotopy class.*

Proof. Let φ be any nontrivial covering automorphism of π, so it is also an isometry of $(\widetilde{M}, \widetilde{g})$. Every continuous path $\widetilde{\sigma} : [0, 1] \to \widetilde{M}$ from a point $\widetilde{x} \in \widetilde{M}$ to its image $\varphi(\widetilde{x})$ projects to a loop $\sigma = \pi \circ \widetilde{\sigma}$ in M. We begin by showing that the set of all loops obtained from φ in this way is exactly one free homotopy class. Suppose $\sigma_0, \sigma_1 : [0, 1] \to M$ are two such loops, so that $\sigma_i = \pi \circ \widetilde{\sigma}_i$ for each i, where $\widetilde{\sigma}_i$ is a path in $\widetilde{M}$ from $\widetilde{x}_i$ to $\varphi(\widetilde{x}_i)$ (Fig. 12.3). For $i = 0, 1$, let $x_i = \pi(\widetilde{x}_i)$, so σ_i is a loop based at x_i. Choose a path $\widetilde{\alpha}$ from $\widetilde{x}_0$ to $\widetilde{x}_1$, and let $\widetilde{\beta} = \varphi \circ \widetilde{\alpha}$, which is a path from $\varphi(\widetilde{x}_0)$ to $\varphi(\widetilde{x}_1)$, and $\alpha = \pi \circ \widetilde{\alpha} = \pi \circ \widetilde{\beta}$, a path from x_0 to x_1 in M. Because $\widetilde{M}$ is simply connected, the two paths $\widetilde{\alpha}^{-1} \cdot \widetilde{\sigma}_0 \cdot \widetilde{\beta}$ and $\widetilde{\sigma}_1$ from $\widetilde{x}_1$ to $\varphi(\widetilde{x}_1)$ are path-homotopic, and therefore so are their images $\alpha^{-1} \cdot \sigma_0 \cdot \alpha$ and σ_1. An easy argument shows that the loops $\alpha^{-1} \cdot \sigma_0 \cdot \alpha$ and σ_0 are freely homotopic, by a homotopy that gradually shrinks the "tail" α to a point, so it follows that σ_0 and σ_1 are freely homotopic.

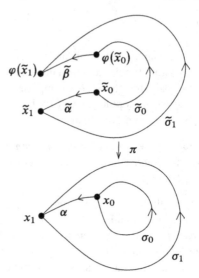

Fig. 12.3: Defining a free homotopy class

Conversely, let σ_0 and σ_1 be freely homotopic loops based at x_0 and x_1 in M, respectively, and suppose one of them, say σ_0, is of the form $\sigma_0 = \pi \circ \tilde{\sigma}_0$ for some path $\tilde{\sigma}_0$ from a point $\tilde{x}_0$ to $\varphi(\tilde{x}_0)$ in $\widetilde{M}$. We need to show that σ_1 is also of this form. The fact that the loops are freely homotopic means that there is a homotopy $H : [0,1] \times [0,1] \to M$ satisfying $H(s,0) = \sigma_0(s)$, $H(s,1) = \sigma_1(s)$, and $H(0,t) = H(1,t)$ for all t. Let $\alpha : [0,1] \to M$ be the path $\alpha(t) = H(0,t) = H(1,t)$ from x_0 to x_1. The existence of such a homotopy implies that the two loops $\alpha^{-1} \cdot \sigma_0 \cdot \alpha$ and σ_1 based at x_1 are path-homotopic (see [LeeTM, Lemma 7.17]). By the monodromy theorem (Prop. A.54(c)), their lifts starting at $\tilde{x}_1$ both end at the same point. Let $\tilde{\sigma}_1$ be the lift of σ_1 starting at the point $\tilde{x}_1 = \tilde{\alpha}(1)$, let $\tilde{\alpha}$ be the lift of α starting at $\tilde{x}_0$, and let $\tilde{\beta} = \varphi \circ \tilde{\alpha}$. Then $\tilde{\alpha}^{-1} \cdot \tilde{\sigma}_0 \cdot \tilde{\beta}$ is a lift of $\alpha^{-1} \cdot \sigma_0 \cdot \alpha$ starting at $\tilde{x}_1$, and it ends at $\tilde{\beta}(1) = \varphi(\tilde{x}_1)$. The monodromy theorem therefore implies that $\tilde{\sigma}_1$ is a path from $\tilde{x}_1$ to $\varphi(\tilde{x}_1)$, as claimed.

We have shown that the covering automorphism φ determines a unique free homotopy class in M. It is not the trivial class, because that class contains the constant loop, which cannot be the image of a path from a point $\tilde{x}$ to $\varphi(\tilde{x})$ because φ has no fixed points. Since M is compact, Proposition 6.28 shows that there is a closed geodesic $\gamma : [0,1] \to M$ that is the shortest admissible path in this free homotopy class. This means, in particular, that there is a lift $\tilde{\gamma}$ of γ that satisfies $\tilde{\gamma}(1) = \varphi(\tilde{\gamma}(0))$. Because $\widetilde{M}$ is complete, $\tilde{\gamma}$ extends to a geodesic in $\widetilde{M}$ defined on all of $\mathbb{R}$. Let $\tilde{\gamma}_0, \tilde{\gamma}_1 : \mathbb{R} \to \widetilde{M}$ be the geodesics defined by $\tilde{\gamma}_0(t) = \varphi \circ \tilde{\gamma}(t)$ and $\tilde{\gamma}_1(t) = \tilde{\gamma}(t+1)$. Then $\tilde{\gamma}_0(0) = \tilde{\gamma}(1) = \tilde{\gamma}_1(0)$, and

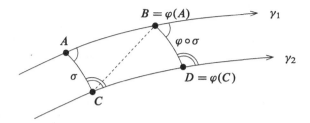

Fig. 12.4: Uniqueness of the axis in the negative curvature case

$$d\pi_{\tilde{\gamma}(1)}\big(\tilde{\gamma}_0'(0)\big) = d\pi_{\tilde{\gamma}(1)} \circ d\varphi_{\tilde{\gamma}(0)}\big(\tilde{\gamma}'(0)\big) \qquad \text{(definition of } \tilde{\gamma}_0\text{)}$$
$$= d\pi_{\tilde{\gamma}(0)}\big(\tilde{\gamma}'(0)\big) \qquad\qquad\quad \text{(because } \pi \circ \varphi = \pi\text{)}$$
$$= \gamma'(0) \qquad\qquad\qquad\qquad \text{(because } \pi \circ \tilde{\gamma} = \gamma\text{)}$$
$$= \gamma'(1) \qquad\qquad\qquad\qquad (\gamma \text{ is a closed geodesic)}$$
$$= d\pi_{\tilde{\gamma}(1)}\big(\tilde{\gamma}'(1)\big) \qquad\qquad\quad (\pi \circ \tilde{\gamma} = \gamma \text{ again)}$$
$$= d\pi_{\tilde{\gamma}(1)}\big(\tilde{\gamma}_1'(0)\big) \qquad\qquad\quad \text{(definition of } \tilde{\gamma}_1\text{)}.$$

The fact that $d\pi_{\tilde{\gamma}(1)}$ is injective then implies $\tilde{\gamma}_0'(0) = \tilde{\gamma}_1'(0)$, so by uniqueness of geodesics we have $\tilde{\gamma}_0 \equiv \tilde{\gamma}_1$, or in other words $\varphi\big(\tilde{\gamma}(t)\big) = \tilde{\gamma}(t+1)$ for all $t \in \mathbb{R}$. Thus $\tilde{\gamma}$ is an axis for φ. $\qquad\square$

In general, an axis of an isometry need not be unique, as we saw in the case of translations of $\mathbb{R}^2$ in Example 12.20. However, as the next lemma shows, the situation is different in the case of negatively curved Cartan–Hadamard manifolds.

Lemma 12.22. *Suppose (M, g) is a Cartan–Hadamard manifold with strictly negative sectional curvature. If $\varphi : M \to M$ is an axial isometry, then its axis is unique up to reparametrization.*

Proof. Let $\varphi : M \to M$ be an axial isometry, and suppose $\gamma_1, \gamma_2 : \mathbb{R} \to M$ are both axes for φ. After reparametrizing the geodesics if necessary, we can assume that φ translates both geodesics by time 1.

Suppose first that the two axes do not intersect. Then the points $A = \gamma_1(0)$, $B = \gamma_1(1) = \varphi(A)$, $C = \gamma_2(0)$, and $D = \gamma_2(1) = \varphi(C)$ are all distinct (Fig. 12.4). Let σ be the geodesic segment from A to C, so that $\varphi \circ \sigma$ is the geodesic segment from B to D, forming a "geodesic quadrilateral" $ABDC$. Because φ preserves angles, the angles $\angle CAB$ and $\angle ABD$ are supplementary, as are $\angle ACD$ and $\angle CDB$. Thus the angle sum of $ABDC$ is exactly 2π.

On the other hand, Proposition 12.10 shows that the geodesic triangles $\triangle ABC$ and $\triangle BCD$ have angle sums strictly less than π. The angle $\angle ACD$ is no larger than the sum $\angle ACB + \angle BCD$: one way to see this is to note that Problem 6-2 shows that the angle between two unit vectors $v, w \in T_C M$, namely $\arccos\langle v, w\rangle$, is exactly equal to the Riemannian distance between v and w regarded as points on the unit sphere in $T_C M$ with its round metric, so we can apply the triangle inequality for distances on the sphere. Similarly, $\angle ABD \le \angle ABC + \angle CBD$. Thus the sum of the four angles of $ABDC$ is strictly less than 2π, which is a contradiction.

It follows that γ_1 and γ_2 must intersect at some point, say $x = \gamma_1(t_1) = \gamma_2(t_2)$. But then $\varphi(x) = \gamma_1(t_1 + 1) = \gamma_2(t_2 + 1)$ is another intersection point, so Proposition 12.9(c) shows that γ_2 must be a reparametrization of γ_1. $\square$

Proof of Preissman's Theorem. Suppose (M, g) satisfies the hypotheses of the the- orem, and let $\pi : \widetilde{M} \to M$ be its universal covering, with the metric $\widetilde{g} = \pi^* g$. Then $(\widetilde{M}, \widetilde{g})$ is a Cartan–Hadamard manifold with strictly negative curvature. Because the fundamental group of M based at any point is isomorphic to the group of cov- ering automorphisms of π (see Prop. C.22), it suffices to show that every nontrivial abelian group of covering automorphisms is isomorphic to $\mathbb{Z}$. Let H be such a group.

Let φ be any non-identity element of H, and let $\gamma : \mathbb{R} \to \widetilde{M}$ be the axis for φ, which we may give a unit-speed parametrization. Thus there is a nonzero constant c such that $\varphi(\gamma(t)) = \gamma(t + c)$ for all $t \in \mathbb{R}$. If ψ is any other element of H, then for all $t \in \mathbb{R}$ we have

$$\varphi(\psi(\gamma(t))) = \psi(\varphi(\gamma(t))) = \psi(\gamma(t + c)),$$

which shows that $\psi \circ \gamma$ is also an axis for φ. Since axes are unique by Lemma 12.22, this means that $\psi \circ \gamma$ is a (unit-speed) reparametrization of γ. It cannot be a backward reparametrization (one satisfying $\varphi \circ \gamma(t) = \gamma(-t + a)$), because then ψ would be a nontrivial covering automorphism having $\gamma(a/2)$ as a fixed point, while the only covering automorphism with a fixed point is the identity. Thus there must be some $a \in \mathbb{R}$ such that $\psi(\gamma(t)) = \gamma(t + a)$ for all $t \in \mathbb{R}$.

Define a map $F : H \to \mathbb{R}$ as follows: for each $\psi \in H$, let $F(\psi)$ be the unique constant a such that $\psi(\gamma(t)) = \gamma(t + a)$ for all $t \in \mathbb{R}$. A simple computation shows that $F(\psi_1 \circ \psi_2) = F(\psi_1) + F(\psi_2)$, so F is a group homomorphism. It is injective because $F(\psi) = 0$ implies that ψ fixes every point in the image of γ, so it must be the identity. Thus $F(H)$ is an additive subgroup of $\mathbb{R}$ isomorphic to H.

Now let b be the infimum of the set of positive elements of $F(H)$. If $b = 0$, then there exists $\psi \in H$ such that $0 < F(\psi) < \operatorname{inj}(M)$, and then $\pi \circ \gamma$ is a unit-speed geodesic in M satisfying $\pi \circ \gamma(a) = \pi \circ \gamma(0)$, where $a = F(\psi)$, which contradicts the fact that $0 < a < \operatorname{inj}(M)$. Thus $b > 0$. If $b \notin F(H)$, there is a sequence of elements $\psi_i \in H$ such that $F(\psi_i) \searrow b$, and then for sufficiently large $i < j$ we have $0 < F(\psi_i \circ \psi_j^{-1}) < \operatorname{inj}(M)$, which leads to a contradiction as before. Thus b is the smallest positive element of $F(H)$, and then it is easy to verify that $F(H)$ consists exactly of all integral multiples of b, so it is isomorphic to $\mathbb{Z}$. $\square$

The most important consequence of Preissman's theorem is the following corol- lary.

Corollary 12.23. *No product of positive-dimensional connected compact manifolds admits a metric of strictly negative sectional curvature.*

Proof. Problem 12-8. $\square$

Manifolds of Positive Curvature

Finally, we consider manifolds with positive curvature. The most important fact about such manifolds is the following theorem, which was first proved in 1941 by Sumner B. Myers [Mye41], building on earlier work of Ossian Bonnet, Heinz Hopf, and John L. Synge.

Theorem 12.24 (Myers). *Let (M,g) be a complete, connected Riemannian n-manifold, and suppose there is a positive constant R such that the Ricci curvature of M satisfies $Rc(v,v) \geq (n-1)/R^2$ for all unit vectors v. Then M is compact, with diameter less than or equal to πR, and its fundamental group is finite.*

Proof. The diameter comparison theorem (Cor. 11.18) shows that the diameter of M is no greater than πR. To show that M is compact, we just choose a base point p and note that every point in M can be connected to p by a geodesic segment of length at most πR. Therefore, $\exp_p : \bar{B}_{\pi R}(0) \to M$ is surjective, so M is the continuous image of a compact set.

To prove the statement about the fundamental group, let $\pi : \widetilde{M} \to M$ denote the universal covering space of M, with the metric $\tilde{g} = \pi^* g$. Then $(\widetilde{M}, \tilde{g})$ is complete by Corollary 6.24, and $\tilde{g}$ also has Ricci curvature satisfying the same lower bound, so $\widetilde{M}$ is compact by the argument above. By Proposition A.61, for every $p \in M$ there is a one-to-one correspondence between $\pi_1(M, p)$ and the fiber $\pi^{-1}(p)$. If $\pi_1(M, p)$ were infinite, therefore, $\pi^{-1}(p)$ would be an infinite discrete closed subset of $\widetilde{M}$, contradicting the compactness of $\widetilde{M}$. Thus $\pi_1(M, p)$ is finite. $\square$

In case the manifold is already known to be compact, the hypothesis can be relaxed.

Corollary 12.25. *Suppose (M,g) is a compact, connected Riemannian n-manifold whose Ricci tensor is positive definite everywhere. Then M has finite fundamental group.*

Proof. Because the unit tangent bundle of M is compact by Proposition 2.9, the hypothesis implies that there is a positive constant c such that $Rc(v,v) \geq c$ for all unit tangent vectors v. Thus the hypotheses of Myers's theorem are satisfied with $(n-1)/R^2 = c$. $\square$

On the other hand, it is possible for complete, noncompact manifolds to have strictly positive Ricci or even sectional curvature, as long as the curvature gets arbitrarily close to zero, as the following example shows.

▶ **Exercise 12.26.** Let $n \geq 2$, and let $M \subseteq \mathbb{R}^{n+1}$ be the paraboloid $\{(x^1, \ldots, x^n, y) : y = |x|^2\}$ with the induced metric (see Problem 8-1). Show that M has strictly positive sectional and Ricci curvatures everywhere, but is not compact.

One of the most useful applications of Myers's theorem is to Einstein metrics.

Corollary 12.27. *If (M,g) is a complete Einstein manifold with positive scalar curvature, then M is compact.*

Proof. If g is Einstein with positive scalar curvature, then $Rc = \frac{1}{n}Sg$ satisfies the hypotheses of Myers's theorem with $(n-1)/R^2 = \frac{1}{n}S$. $\square$

In 1975, Shiu-Yuen Cheng proved the following theorem [Che75], showing that the diameter inequality in Myers's theorem is an equality only for a round sphere.

Theorem 12.28 (Cheng's Maximal Diameter Theorem). *Suppose (M,g) satisfies the hypotheses of Myers's theorem and* $\mathrm{diam}(M) = \pi R$. *Then (M,g) is isometric to* $\left(\mathbb{S}^n(R), \overset{\circ}{g}_R\right)$.

Proof. By Myers's theorem, M is compact, and thus by continuity of the distance function there are points $p_1, p_2 \in M$ such that $d_g(p_1, p_2) = \mathrm{diam}(M) = \pi R$. For $i = 1,2$ and $0 < \delta \leq \pi R$, let $B_\delta(p_i)$ denote the open *metric* ball in M of radius δ about p_i, and let $\overset{\circ}{B}_\delta$ denote a metric ball in $\left(\mathbb{S}^n(R), \overset{\circ}{g}_R\right)$ of the same radius. The Bishop–Gromov theorem shows that the ratio $\mathrm{Vol}(B_\delta(p_i))/\mathrm{Vol}(\overset{\circ}{B}_\delta)$ is a non-increasing function of δ for each i, which implies the following inequality for $0 < \delta < \pi R$:

$$\frac{\mathrm{Vol}(B_\delta(p_i))}{\mathrm{Vol}(M)} = \frac{\mathrm{Vol}(B_\delta(p_i))}{\mathrm{Vol}(B_{\pi R}(p_i))} \geq \frac{\mathrm{Vol}(\overset{\circ}{B}_\delta)}{\mathrm{Vol}(\overset{\circ}{B}_{\pi R})} = \frac{\mathrm{Vol}(\overset{\circ}{B}_\delta)}{\mathrm{Vol}(\mathbb{S}^n(R))}. \tag{12.2}$$

Now suppose δ_1, δ_2 are positive numbers such that $\delta_1 + \delta_2 = \pi R$. The fact that $d_g(p_1, p_2) = \pi R$ together with the triangle inequality implies that $B_{\delta_1}(p_1)$ and $B_{\delta_2}(p_2)$ are disjoint, so

$$\mathrm{Vol}(B_{\delta_1}(p_1)) + \mathrm{Vol}(B_{\delta_2}(p_2)) \leq \mathrm{Vol}(M). \tag{12.3}$$

On the other hand, $\mathrm{Vol}(\overset{\circ}{B}_{\delta_1}) + \mathrm{Vol}(\overset{\circ}{B}_{\delta_2}) = \mathrm{Vol}(\mathbb{S}^n(R))$ (think of balls centered at the north and south poles), so (12.2) and (12.3) imply

$$1 \geq \frac{\mathrm{Vol}(B_{\delta_1}(p_1)) + \mathrm{Vol}(B_{\delta_2}(p_2))}{\mathrm{Vol}(M)} \geq \frac{\mathrm{Vol}(\overset{\circ}{B}_{\delta_1}) + \mathrm{Vol}(\overset{\circ}{B}_{\delta_2})}{\mathrm{Vol}(\mathbb{S}^n(R))} = 1.$$

Thus the inequalities above are equalities, so in particular equality holds in (12.3). Moreover, the boundary of a metric ball has measure zero because it is contained in the image of a sphere under the exponential map, so we also have

$$\mathrm{Vol}(\overline{B}_{\delta_1}(p_1)) + \mathrm{Vol}(\overline{B}_{\delta_2}(p_2)) = \mathrm{Vol}(M). \tag{12.4}$$

Next we will show that $d_g(p_1, q) + d_g(p_2, q) = d_g(p_1, p_2) = \pi R$ for all $q \in M$. Suppose not: by the triangle inequality, the only way this can fail is if there is a point q such that $d_g(p_1, q) + d_g(p_2, q) > \pi R$. Choosing $\delta_1 < d_g(p_1, q)$ and $\delta_2 < d_g(p_2, q)$ such that $\delta_1 + \delta_2 = \pi R$, we see that q is not in the closed set $\overline{B}_{\delta_1}(p_1) \cup \overline{B}_{\delta_2}(p_2)$, so there is a neighborhood of q disjoint from that set (Fig. 12.5). But any such neighborhood would have positive volume, contradicting (12.4).

Suppose $\gamma : [0, \infty) \to M$ is a unit-speed geodesic starting at p_1, and choose $\varepsilon > 0$ small enough that $\gamma|_{[0,\varepsilon]}$ is minimizing. Then $d_g(p_1, \gamma(\varepsilon)) = \varepsilon$, and there is also a

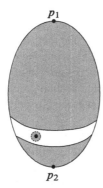

Fig. 12.5: A step in the proof of the maximal diameter theorem

minimizing unit-speed geodesic γ_2 from $\gamma(\varepsilon)$ to p_2, whose length is $d_g(p_2, \gamma(\varepsilon)) = \pi R - \varepsilon$. The piecewise regular curve consisting of $\gamma|_{[0,\varepsilon]}$ followed by a unit-speed reparametrization of γ_2 is an admissible curve from p_1 to p_2 of length πR, and is therefore minimizing; thus it is a (smooth) geodesic, and in fact coincides with γ, which is therefore minimizing on $[0, \pi R]$. This shows that the cut time of every unit-speed geodesic starting at p_1 is at least πR, so the injectivity radius $\mathrm{inj}(p_1)$ is at least πR; and then the injectivity radius comparison theorem (Cor. 11.17) shows that it is exactly πR. Thus $M \smallsetminus \{p_2\} = B_{\pi R}(p_1)$ is a geodesic ball around p_1. An analogous argument shows that $M \smallsetminus \{p_1\}$ is a geodesic ball around p_2.

The upshot of these observations is that the distance functions $r_1(q) = d_g(p_1, q)$ and $r_2(q) = d_g(p_2, q)$ are both smooth on $M \smallsetminus \{p_1, p_2\}$, and because their sum is constant, they satisfy $\Delta r_1 = -\Delta r_2$. Elementary trigonometric identities show that the function s_c defined by (10.8) with $c = 1/R$ satisfies $s_c(\pi R - t) = s_c(t)$ and $s_c'(\pi R - t) = -s_c'(t)$, so the Laplacian comparison theorem (Thm. 11.15) applied to r_1 and $r_2 = \pi R - r_1$ yields

$$\Delta r_1 \leq (n-1)\frac{s_c'(r_1)}{s_c(r_1)} = -(n-1)\frac{s_c'(r_2)}{s_c(r_2)} \leq -\Delta r_2 = \Delta r_1.$$

Thus both inequalities are equalities. Arguing exactly as in the last part of the proof of the Bishop–Gromov theorem, we conclude that (M, g) has constant sectional curvature $1/R^2$, and thus by Corollary 12.5 it admits a (finite-sheeted) Riemannian covering by $\big(\mathbb{S}^n(R), \mathring{g}_R\big)$. Since the injectivity domain of p_1 is the entire ball of radius πR, on which the metric has the form (10.17) in normal coordinates, it follows from Theorem 10.34 that the volume of M is equal to that of $\big(\mathbb{S}^n(R), \mathring{g}_R\big)$. Problem 2-14 then shows that the covering is one-sheeted, so it is a global isometry. □

If we weaken the hypothesis to require merely that the Ricci curvature be nonnegative instead of positive definite, there are still nontrivial topological consequences. Let Γ be a finitely generated group. We say that Γ has **polynomial growth** if there exist a finite generating set $S \subseteq \Gamma$ and positive real numbers C, k such that for every positive integer m, the number of distinct elements of Γ that can be expressed as products of at most m elements of S and their inverses is no larger than Cm^k. The **order of polynomial growth** of such a group is the infimum of such values of k. It is easy to check that $\mathbb{Z}^n$ has polynomial growth of order n; in fact, using the fundamental theorem on finitely generated abelian groups [DF04, p. 158], one can show that every finitely generated abelian group has polynomial growth of order equal to its free rank.

The following theorem, proved in 1968 by John Milnor [Mil68], shows in effect that fundamental groups of complete manifolds with nonnegative Ricci curvature are not too far from being abelian, at least when they are finitely generated. (Milnor conjectured in the same paper that the fundamental group of such a manifold is always finitely generated. This is true in the compact case, because compact manifolds are homotopy equivalent to finite CW complexes [Hat02, Cor. A.12], and these have finitely generated fundamental groups, as shown, for example, in [LeeTM, Thm. 10.15]. But the conjecture has not been proved in the noncompact case.) By contrast, Milnor proved in the same paper that the fundamental group of a negatively curved compact manifold never has polynomial growth, and that was extended in 1970 by André Avez [Ave70] to the case of a nonpositively curved compact manifold as long as it is not flat.

Theorem 12.29 (Milnor). *Suppose (M, g) is a complete, connected Riemannian n-manifold with nonnegative Ricci curvature. Every finitely generated subgroup of $\pi_1(M)$ has polynomial growth of order at most n.*

Proof. Problem 12-11. $\square$

Corollary 12.30. *If M is a connected smooth manifold whose fundamental group is free on more than one generator, then M admits no complete Riemannian metric with nonnegative Ricci curvature.*

▶ **Exercise 12.31.** Prove this corollary.

The corollary implies, for example, that a plane with two or more punctures does not admit a complete metric with nonnegative Gaussian curvature.

On the other hand, if we strengthen the hypothesis and assume positive *sectional curvature*, then we can reach a stronger conclusion. The following theorem was proved in 1936 by John L. Synge [Syn36].

Theorem 12.32 (Synge). *Suppose (M, g) is a compact, connected Riemannian n-manifold with strictly positive sectional curvature.*

(a) *If n is even and M is orientable, then M is simply connected.*
(b) *If n is odd, then M is orientable.*

Proof. Suppose for the sake of contradiction that the appropriate hypothesis holds but the conclusion is false. Let $\pi : \widetilde{M} \to M$ be the universal covering manifold of M with the pullback metric $\widetilde{g} = \pi^* g$. We can give $\widetilde{M}$ the pullback orientation in the even-dimensional case; in the odd-dimensional case, just choose an orientation on $\widetilde{M}$. In either case, there is a nontrivial covering automorphism $\varphi : \widetilde{M} \to \widetilde{M}$; in the even-dimensional case, it is orientation-preserving, and in the odd-dimensional case we can choose it to be orientation-reversing. By Lemma 12.21, there is a geodesic $\widetilde{\gamma} : \mathbb{R} \to \widetilde{M}$ that is an axis for φ, and $\gamma = \pi \circ \widetilde{\gamma}$ restricts to a closed geodesic in M that minimizes length in its free homotopy class. After reparametrizing if necessary, we may assume that $\varphi(\widetilde{\gamma}(t)) = \widetilde{\gamma}(t+1)$ for all $t \in \mathbb{R}$. Let $\widetilde{x}_0 = \widetilde{\gamma}(0)$, $\widetilde{x}_1 = \widetilde{\gamma}(1) = \varphi(\widetilde{x}_0)$, and $x = \pi(\widetilde{x}_0) = \pi(\widetilde{x}_1)$.

Let $\widetilde{P} : T_{\widetilde{x}_0}\widetilde{M} \to T_{\widetilde{x}_1}\widetilde{M}$ be parallel transport along $\widetilde{\gamma}|_{[0,1]}$, and let $P : T_x M \to T_x M$ be parallel transport along $\gamma|_{[0,1]}$. Because local isometries preserve parallelism, the following diagram commutes, and all four maps are linear isometries:

$$
\begin{array}{ccc}
T_{\widetilde{x}_0}\widetilde{M} & \xrightarrow{\ \widetilde{P}\ } & T_{\widetilde{x}_1}\widetilde{M} \\
\Big\downarrow{\scriptstyle d\pi_{\widetilde{x}_0}} & & \Big\downarrow{\scriptstyle d\pi_{\widetilde{x}_1}} \\
T_x M & \xrightarrow{\ P\ } & T_x M.
\end{array}
$$

Because $dV_{\widetilde{g}}$ applied to a parallel frame is constant, $\widetilde{P}$ is orientation-preserving. In the even-dimensional case, the vertical maps are both orientation-preserving, so P is too. In the odd-dimensional case, the fact that $d\pi_{\widetilde{x}_0} = d\pi_{\widetilde{x}_1} \circ d\varphi_{\widetilde{x}_0}$ implies that the two vertical maps induce opposite orientations on $T_x M$, so P is orientation-reversing.

The upshot is that in each case, P is a linear isometry of $T_x M$ whose determinant is $(-1)^n$. Note that the determinant is the product of the eigenvalues of P, counted with multiplicities. Since the real eigenvalues of a linear isometry are ± 1, and the others come in conjugate pairs whose product is 1, the equation $\det P = (-1)^n$ implies that the multiplicity of -1 has the same parity as n, and therefore the multiplicity of $+1$ must be even. We know that $\gamma'(0)$ is a $+1$-eigenvector, so there must be another independent $+1$-eigenvector $v \in T_x M$, which we can take to be orthogonal to $\gamma'(0)$. Extending v by parallel transport yields a nontrivial parallel normal vector field $V \in \mathfrak{X}(\gamma)$ satisfying $V(0) = V(1)$.

Consider the variation of γ defined by $\Gamma(s,t) = \exp_{\gamma(t)}(sV(t))$. Because $V(0) = V(1)$, this is a variation through admissible loops. The first variation formula (6.1) shows that the first derivative of $L_g(\Gamma_s)$ at $s = 0$ is zero, because γ is a geodesic and the boundary terms cancel. We can then apply the general version of the second variation formula given in Problem 10-9, with the image of the geodesic $s \mapsto \exp_x(sv)$ playing the role of M_1 and M_2. Because this is a totally geodesic submanifold, its second fundamental form vanishes, and we obtain

$$
\frac{d^2}{ds^2}\bigg|_{s=0} L_g(\Gamma_s) = \int_a^b \Big(|D_t V|^2 - Rm(V, \gamma', \gamma', V) \Big)\, dt.
$$

Because V is parallel along γ and $\sec(V,\gamma') > 0$ for all t, this second derivative is strictly negative, which implies that $L_g(\Gamma_s) < L_g(\gamma)$ for sufficiently small s. But this contradicts the fact that γ minimizes length in its free homotopy class. □

The next corollary shows that there are only two possibilities for fundamental groups of positively curved compact manifolds in the even-dimensional case. (Compare this to Problem 12-2, which shows that the only even-dimensional manifolds that admit complete metrics of *constant* positive curvature are spheres and real projective spaces.)

Corollary 12.33. *Suppose (M,g) is a compact, connected, even-dimensional Riemannian manifold with strictly positive sectional curvature. If M is orientable, then $\pi_1(M)$ is trivial, and if not, then $\pi_1(M) \cong \mathbb{Z}/2$.*

Proof. The orientable case is part of Synge's theorem. If M is nonorientable, it has a two-sheeted orientable covering manifold $\widetilde{M}$; Synge's theorem shows that $\widetilde{M}$ is simply connected, so it is the universal covering space of M. Because each fiber of the universal covering has the same cardinality as $\pi_1(M)$, it follows that $\pi_1(M)$ is isomorphic to the two-element group $\mathbb{Z}/2$. □

The preceding corollary implies, for example, that $\mathbb{RP}^n \times \mathbb{RP}^n$ has no metric of positive sectional curvature for any $n \geq 2$, because its fundamental group is isomorphic to $\mathbb{Z}/2 \times \mathbb{Z}/2$.

For odd-dimensional compact manifolds, not many topological obstructions to the existence of positively curved metrics are known, apart from orientability and finiteness of π_1. And in the simply connected case, almost nothing is known. There are many examples of simply connected compact manifolds that admit metrics of nonnegative sectional curvature (for example, products of nonnegatively curved, simply connected compact manifolds, by the result of Problem 8-10); but there is not a single known example of a simply connected compact manifold that admits a metric of nonnegative sectional curvature but not one of positive sectional curvature. It was conjectured by Heinz Hopf in the early 1930s that $\mathbb{S}^2 \times \mathbb{S}^2$ admits no positively curved metric, and it is reasonable to extend the conjecture to every product of positive-dimensional simply connected compact manifolds that both admit nonnegatively curved metrics; but nobody has come up with an example of such a product for which the conjecture can be proved or disproved.

Further Results

We end the book with a brief look at some other local-to-global theorems about manifolds with positive or nonnegative curvature, whose proofs are beyond our scope.

Some of the most powerful applications of comparison theory have been to prove "pinching theorems." Given a positive real number δ, a Riemannian manifold is said to be δ-*pinched* if there exists a positive constant c such that all sectional curvatures satisfy

$$\delta c \le \sec(\Pi) \le c.$$

It is said to be *strictly δ-pinched* if at least one of the inequalities is strict. The following celebrated theorem was originally proved by Marcel Berger and Wilhelm Klingenberg in the early 1960s [Ber60, Kli61].

Theorem 12.34 (The Sphere Theorem). *A complete, simply connected, Riemannian n-manifold that is strictly $\frac{1}{4}$-pinched is homeomorphic to $\mathbb{S}^n$.*

This result is sharp, at least in even dimensions, because the Fubini–Study metrics on complex projective spaces are $\frac{1}{4}$-pinched (Problem 8-13).

For noncompact manifolds, there is the following remarkable theorem of Jeff Cheeger and Detlef Gromoll [CG72].

Theorem 12.35 (The Soul Theorem). *If (M,g) is a complete, connected, noncompact Riemannian manifold with nonnegative sectional curvature, then there exists a compact totally geodesic submanifold $S \subseteq M$ (called a **soul** of M) such that M is diffeomorphic to the normal bundle of S in M. If M has strictly positive sectional curvature, then S is a point and M is diffeomorphic to a Euclidean space.*

Even if we assume only nonnegative Ricci curvature, something nontrivial can be said in the noncompact case. The next theorem is also due to Cheeger and Gromoll [CG71]. Recall that a *line* in a Riemannian manifold is the image of a nonconstant geodesic defined on $\mathbb{R}$ whose restriction to each compact subinterval is minimizing.

Theorem 12.36 (The Splitting Theorem). *If (M,g) is a complete, connected, noncompact Riemannian manifold with nonnegative Ricci curvature, and M contains a line, then there is a Riemannian manifold (N,h) with nonnegative Ricci curvature such that M is isometric to the Riemannian product $N \times \mathbb{R}$.*

The proofs of all three of the above theorems, which can be found, for example, in [Pet16], are elaborate applications of comparison theory.

All of our comparison theorems, and indeed most of the things we have proved about Riemannian manifolds, are based on the analysis of ordinary differential equations—the geodesic equation, the parallel transport equation, the Jacobi equation, and the Riccati equation. Using techniques of *partial* differential equations can lead to much stronger conclusions in some cases. For instance, in 1982, Richard Hamilton [Ham82] proved the following striking result about 3-manifolds.

Theorem 12.37 (Hamilton's 3-Manifold Theorem). *Suppose M is a simply connected compact Riemannian 3-manifold with positive definite Ricci curvature. Then M is diffeomorphic to $\mathbb{S}^3$.*

He followed it up four years later [Ham86] with an analogous result about 4-manifolds, with Ricci curvature replaced by the *curvature operator* described in Problem 8-33.

Theorem 12.38 (Hamilton's 4-Manifold Theorem). *Suppose M is a simply connected compact Riemannian 4-manifold with positive curvature operator. Then M is diffeomorphic to $\mathbb{S}^4$.*

Much more recently, Christoph Böhm and Burkhard Wilking [BW08] extended this result to all dimensions.

Theorem 12.39 (Böhm–Wilking). *Every simply connected compact Riemannian manifold with positive curvature operator is diffeomorphic to a sphere.*

The method of proof of these three sphere theorems was to start with an initial metric g_0, and then define a one-parameter family of metrics $\{g_t : t \geq 0\}$ evolving according to the following partial differential equation, called the ***Ricci flow***:

$$\frac{\partial}{\partial t} g_t = -2Rc_t,$$

where Rc_t is the Ricci curvature of g_t. Under the appropriate curvature assumption on g_0, the metric g_t can be rescaled to converge to a metric with constant positive sectional curvature as $t \to \infty$, and then it follows from the Killing–Hopf theorem that the original manifold must be diffeomorphic to a sphere. Since Hamilton first introduced it, this technique has been vastly generalized by Hamilton and others, culminating in 2003 in a proof by Grigory Perelman of the Thurston geometrization conjecture described in Chapter 3.

The Ricci flow has also been used to improve the original $\frac{1}{4}$-pinched sphere theorem significantly. The techniques used to prove the original theorem were not strong enough to prove that M is *diffeomorphic* to $\mathbb{S}^n$, leaving open the possibility that there might exist strictly $\frac{1}{4}$-pinched metrics on exotic spheres (topological spheres with nonstandard smooth structures). In 2009, Simon Brendle and Richard Schoen [BS09] were able to use Ricci-flow techniques to close this gap. They also were able to prove the theorem under a weaker hypothesis: a Riemannian metric g is said to be ***pointwise δ-pinched*** if for every $p \in M$, there exists a positive number $c(p)$ such that

$$\delta c(p) \leq \sec(\Pi) \leq c(p)$$

for every 2-plane $\Pi \subseteq T_p M$, and ***strictly pointwise δ-pinched*** if at least one of the inequalities is strict at each point.

Theorem 12.40 (Differentiable Sphere Theorem). *Let (M, g) be a compact, simply connected Riemannian manifold of dimension $n \geq 4$. If M is strictly pointwise $\frac{1}{4}$-pinched, then it is diffeomorphic to $\mathbb{S}^n$ with its standard smooth structure.*

For a nice exposition of the proof, see the recent book [Bre10].

Problems

12-1. Suppose (M, g) is a simply connected (but not necessarily complete) Riemannian n-manifold with constant sectional curvature. Prove that there exists an isometric immersion from M into one of the model spaces $\mathbb{R}^n$, $\mathbb{S}^n(R)$, $\mathbb{H}^n(R)$.

12-2. Prove that if n is even, then every orientation-preserving orthogonal linear map from $\mathbb{R}^{n+1}$ to itself fixes at least one point of the unit sphere, and use this to prove that every even-dimensional spherical space form is isometric to either a round sphere or a real projective space with a constant multiple of the metric defined in Example 2.34. (*Used on p. 349.*)

12-3. Show that every compact 2-dimensional Euclidean space form is diffeomorphic to the torus or the Klein bottle, and every noncompact one is diffeomorphic to $\mathbb{R}^2$, $\mathbb{S}^1 \times \mathbb{R}^2$, or the open Möbius band (see Example 2.35).

12-4. A *lattice* in $\mathbb{R}^n$ is an additive subgroup $\Lambda \subseteq \mathbb{R}^n$ of the form

$$\Lambda = \{m_1 v_1 + \cdots + m_n v_n : m_1, \ldots, m_n \in \mathbb{Z}\},$$

for some basis $(v_1, \ldots, v_n)$ of $\mathbb{R}^n$. Let $\mathbb{T}^n$ denote the n-torus.

(a) Show that if g is a flat Riemannian metric on $\mathbb{T}^n$, then $(\mathbb{T}^n, g)$ is isometric to a Riemannian quotient of the form $\mathbb{R}^n / \Lambda$ for some lattice Λ, acting on $\mathbb{R}^n$ by vector addition.

(b) Show that two such quotients $\mathbb{R}^n / \Lambda_1$ and $\mathbb{R}^n / \Lambda_2$ are isometric if and only if $\Lambda_2 = A(\Lambda_1)$ for some $A \in O(n)$.

12-5. CLASSIFICATION OF FLAT TORI: Let $\mathbb{T}^2 = \mathbb{S}^1 \times \mathbb{S}^1$ denote the 2-torus. Show that if g is a flat Riemannian metric on $\mathbb{T}^2$, then $(\mathbb{T}^2, g)$ is isometric to one and only one Riemannian quotient $\mathbb{R}^2 / \Lambda$, where Λ is a lattice generated by a basis of the form $((a, 0), (b, c))$, with $a > 0$, $0 \le b \le a/2$, $c > 0$, and $b^2 + c^2 \ge a^2$. [Hint: Given a lattice $\Lambda \subseteq \mathbb{R}^2$, let v_1 be an element of $\Lambda \smallsetminus \{(0, 0)\}$ of minimal norm; let v_2 be an element of $\Lambda \smallsetminus \langle v_1 \rangle$ of minimal norm (where $\langle v_1 \rangle$ is the cyclic subgroup generated by v_1), chosen so that the angle between v_1 and v_2 is less than or equal to $\pi/2$; and then apply a suitable orthogonal transformation.]

12-6. Suppose (M, g) is a complete, connected Riemannian manifold, and $p, q \in M$. Proposition 6.25 showed that every path-homotopy class of paths from p to q contains a geodesic segment γ. Show that if M has nonpositive sectional curvature, then γ is the unique geodesic segment in the given path homotopy class.

12-7. Prove Proposition 12.10 (inequalities for triangles in Cartan–Hadamard manifolds). [Hint: Compare with appropriate triangles in Euclidean space.]

12-8. Prove Corollary 12.23 (nonexistence of negatively curved metrics on compact product manifolds).

12-9. Prove the following generalization of the Cartan–Hadamard theorem, due to Robert Hermann [Her63]: Suppose (M, g) is a complete, connected Riemannian manifold with nonpositive sectional curvature, and $P \subseteq M$ is a connected, properly embedded, totally geodesic submanifold. If for some $x \in P$, the homomorphism $\pi_1(P, x) \to \pi_1(M, x)$ induced by the inclusion $P \hookrightarrow M$ is surjective, then the normal exponential map $E : NP \to M$ is a diffeomorphism. [Hint: Mimic the proof of the Cartan–Hadamard theorem, using the result of Problem 10-20.]

12-10. Give a counterexample to show that Myers's theorem is false in dimensions greater than 2 if we replace the lower bound on Ricci curvature by a positive lower bound on scalar curvature.

12-11. Prove Theorem 12.29 (Milnor's theorem on the fundamental group of a manifold with nonnegative Ricci curvature) as follows: Let $\pi : \widetilde{M} \to M$ be the universal covering space of M, and give $\widetilde{M}$ the pullback metric $\widetilde{g} = \pi^* g$. Because the covering automorphism group is isomorphic to $\pi_1(M)$ (Prop. C.22), it suffices to prove the result for every finitely generated subgroup of the automorphism group. Let Γ be such a subgroup, and let $S = \{\varphi_1, \dots, \varphi_N\}$ be a finite generating set. Choose a base point $p \in \widetilde{M}$. For each $i = 1, \dots, N$, let d_i be the Riemannian distance from p to $\varphi_i(p)$, and let $D = \max(d_1, \dots, d_n)$. For each nonidentity element $\psi \in \Gamma$, define the **length of** ψ to be the smallest integer m such ψ can be expressed as a product of m elements of S and their inverses.

(a) Show that there exists $\varepsilon > 0$ such that for any two distinct elements $\psi_1, \psi_2 \in \Gamma$, the metric balls $B_\varepsilon(\psi_1(p))$ and $B_\varepsilon(\psi_2(p))$ are disjoint.

(b) Show that if $\psi \in \Gamma$ has length m, then $B_\varepsilon(\psi(p)) \subseteq B_{mD+\varepsilon}(p)$.

(c) Use the Bishop–Gromov theorem to prove that the number of distinct elements of Γ of length at most m is bounded by a constant times m^n.

12-12. Prove that there is no compact manifold that admits both a metric of positive definite Ricci curvature and a metric of nonpositive sectional curvature.

12-13. Suppose (M, g) is a complete Riemannian manifold. Recall that a *line* in M is the image of a nonconstant geodesic defined on all of $\mathbb{R}$ that is minimizing between every pair of its points.

(a) Show that if (M, g) has strictly positive sectional curvature, then M contains no lines. [Hint: Given a geodesic $\gamma : \mathbb{R} \to M$, let $\alpha : \mathbb{R} \to \mathbb{R}$ be a smooth positive function such that $\alpha(t)$ is smaller than the minimum sectional curvature at $\gamma(t)$ for each t, and let $u : \mathbb{R} \to \mathbb{R}$ be the solution to the initial value problem $u'' + \alpha u = 0$ with $u(0) = 1$ and $u'(0) = 0$. Prove that there are numbers $a < 0 < b$ such that $u(a) = u(b) = 0$, and evaluate $I(V, V)$ for a vector field of the form $V(t) = u(t) E(t)$ along $\gamma|_{[a,b]}$, where E is a parallel unit normal vector field.]

(b) Give an example of a complete nonflat Riemannian manifold with nonnegative sectional curvature that contains a line.

Appendix A

Review of Smooth Manifolds

This book is designed for readers who already have a solid understanding of basic topology and smooth manifold theory, at roughly the level of [LeeTM] and [LeeSM]. In this appendix, we summarize the main ideas of these subjects that are used throughout the book. It is included here as a review, and to establish our notation and conventions.

Topological Preliminaries

An *n-dimensional topological manifold* (or simply an *n-manifold*) is a second-countable Hausdorff topological space that is *locally Euclidean of dimension n*, meaning that every point has a neighborhood homeomorphic to an open subset of $\mathbb{R}^n$. Given an n-manifold M, a *coordinate chart for M* is a pair (U, φ), where $U \subseteq M$ is an open set and $\varphi \colon U \to \hat{U}$ is a homeomorphism from U to an open subset $\hat{U} \subseteq \mathbb{R}^n$. If $p \in M$ and (U, φ) is a chart such that $p \in U$, we say that (U, φ) is a *chart containing p*.

We also occasionally need to consider manifolds with boundary. An n-dimensional *topological manifold with boundary* is a second-countable Hausdorff space in which every point has a neighborhood homeomorphic either to an open subset of $\mathbb{R}^n$ or to a (relatively) open subset of the half-space $\mathbb{R}^n_+ = \{(x^1, \dots, x^n) \in \mathbb{R}^n : x^n \geq 0\}$. A pair (U, φ), where U is an open subset of M and φ is a homeomorphism from U to an open subset of $\mathbb{R}^n$ or $\mathbb{R}^n_+$, is called an *interior chart* if $\varphi(U)$ is an open subset of $\mathbb{R}^n$ or an open subset of $\mathbb{R}^n_+$ that does not meet $\partial \mathbb{R}^n_+ = \{x \in \mathbb{R}^n_+ : x^n = 0\}$; and it is called a *boundary chart* if $\varphi(U)$ is an open subset of $\mathbb{R}^n_+$ with $\varphi(U) \cap \partial \mathbb{R}^n_+ \neq \varnothing$. A point $p \in M$ is called an *interior point of M* if it is in the domain of an interior chart, and it is a *boundary point of M* if it is in the domain of a boundary chart taking p to a point of $\partial \mathbb{R}^n_+$.

Notice that our convention is that a *manifold* without further qualification is always assumed to be a manifold without boundary, and the term "manifold with boundary" encompasses ordinary manifolds as a special case. But for clarity, we

© Springer International Publishing AG 2018

J. M. Lee, *Introduction to Riemannian Manifolds*, Graduate Texts in Mathematics 176, https://doi.org/10.1007/978-3-319-91755-9

sometimes use redundant phrases such as "manifold without boundary" for an ordinary manifold, and "manifold with or without boundary" when we wish to emphasize that the discussion applies equally well in either case.

Proposition A.1. [LeeTM, Props. 4.23 & 4.64] *Every topological manifold with or without boundary is locally path-connected and locally compact.*

Proposition A.2. [LeeTM, Thm. 4.77] *Every topological manifold with or without boundary is paracompact.*

If M and N are topological spaces, a map $F: M \to N$ is said to be a *closed map* if for each closed subset $K \subseteq M$, the image set $F(K)$ is closed in N, and an *open map* if for each open subset $U \subseteq M$, the image set $F(U)$ is open in N. It is a *quotient map* if it is surjective and $V \subseteq N$ is open if and only if $F^{-1}(V)$ is open.

Proposition A.3. [LeeTM, Prop. 3.69] *Suppose M and N are topological spaces, and $F: M \to N$ is a continuous map that is either open or closed.*

(a) *If F is surjective, it is a quotient map.*
(b) *If F is injective, it is a topological embedding.*
(c) *If F is bijective, it is a homeomorphism.*

The next lemma often gives a convenient shortcut for proving that a map is closed.

Lemma A.4 (Closed Map Lemma). [LeeTM, Lemma 4.50] *If F is a continuous map from a compact space to a Hausdorff space, then F is a closed map.*

Here is another condition that is frequently useful for showing that a map is closed. A map $F: M \to N$ between topological spaces is called a *proper map* if for each compact subset $K \subseteq N$, the preimage $F^{-1}(K)$ is compact.

Proposition A.5 (Proper Continuous Maps Are Closed). [LeeTM, Thm. 4.95] *Suppose M is a topological space, N is a locally compact Hausdorff space, and $F: M \to N$ is continuous and proper. Then F is a closed map.*

Fundamental Groups

Many of the deepest theorems in Riemannian geometry express relations between local geometric properties and global topological properties. Because the global properties are frequently expressed in terms of fundamental groups, we summarize some basic definitions and properties here.

Suppose M and N are topological spaces. If $F_0, F_1: M \to N$ are continuous maps, a *homotopy from F_0 to F_1* is a continuous map $H: M \times [0,1] \to N$ that satisfies $H(x,0) = F_0(x)$ and $H(x,1) = F_1(x)$ for all $x \in M$. If there exists such a homotopy, we say that F_0 and F_1 are *homotopic*.

We are most interested in homotopies in the following situation. If M is a topological space, a *path in M* is a continuous map $\alpha: [0,1] \to M$. If $p = \alpha(0)$ and $q = \alpha(1)$, we say that α is a *path from p to q*. If α_0 and α_1 are both paths from p to

q, a *path homotopy from α_0 to α_1* is a continuous map $H : [0,1] \times [0,1] \to M$ that satisfies

$$H(s,0) = \alpha_0(s) \text{ and } H(s,1) = \alpha_1(s) \text{ for all } s \in [0,1],$$
$$H(0,t) = p \text{ and } H(1,t) = q \text{ for all } t \in [0,1].$$

If there exists such a path homotopy, then α_0 and α_1 are said to be *path-homotopic*; this is an equivalence relation on the set of all paths in M from p to q. The equivalence class of a path α, called its *path class*, is denoted by $[\alpha]$.

Paths α and β that satisfy $\alpha(1) = \beta(0)$ are said to be *composable*; in this case, their *product* is the path $\alpha \cdot \beta : [0,1] \to M$ defined by

$$\alpha \cdot \beta(s) = \begin{cases} \alpha(2s), & 0 \le s \le \frac{1}{2}, \\ \beta(2s-1), & \frac{1}{2} \le s \le 1. \end{cases}$$

Path products are well defined on path classes: if $[\alpha_0] = [\alpha_1]$ and $[\beta_0] = [\beta_1]$, and α_0 and β_0 are composable, then so are α_1 and β_1, and $[\alpha_0 \cdot \beta_0] = [\alpha_1 \cdot \beta_1]$. Thus we obtain a well-defined product of path classes by $[\alpha] \cdot [\beta] = [\alpha \cdot \beta]$.

A path from a point p to itself is called a *loop based at p*. If M is a topological space and $p \in M$, then any two loops based at p are composable, and the set of path classes of loops based at p is a group under path class products, called the *fundamental group of M based at p* and denoted by $\pi_1(M, p)$. The class of the constant path $c_p(s) \equiv p$ is the identity element, and the class of the reverse path $\alpha^{-1}(s) = \alpha(1 - s)$ is the inverse of $[\alpha]$. (Although multiplication of path classes is associative, path products themselves are not, so a multiple product such as $\alpha \cdot \beta \cdot \gamma$ is to be interpreted as $(\alpha \cdot \beta) \cdot \gamma$.)

Proposition A.6 (Induced Homomorphisms). [LeeTM, Prop. 7.24 & Cor. 7.26] *Suppose M and N are topological spaces and $p \in M$. If $F : M \to N$ is a continuous map, then the map $F_* : \pi_1(M, p) \to \pi_1(N, F(p))$ defined by $F_*([\alpha]) = [F \circ \alpha]$ is a group homomorphism called the homomorphism induced by F. If F is a homeomorphism, then F_* is an isomorphism.*

A topological space M is said to be *simply connected* if it is path-connected and for some (hence every) point $p \in M$, the fundamental group $\pi_1(M, p)$ is trivial.

▶ **Exercise A.7.** Show that if M is a simply connected topological space and p, q are any two points in M, then all paths in M from p to q are path-homotopic.

A continuous map $F : M \to N$ is said to be a *homotopy equivalence* if there is a continuous map $G : N \to M$ such that $G \circ F$ and $F \circ G$ are both homotopic to identity maps.

Proposition A.8 (Homotopy Invariance of the Fundamental Group). [LeeTM, Thm. 7.40] *Suppose M and N are topological spaces and $F : M \to N$ is a homotopy equivalence. Then for every point $p \in M$, the induced homomorphism $F_* : \pi_1(M, p) \to \pi_1(N, F(p))$ is an isomorphism.*

Smooth Manifolds and Smooth Maps

The setting for the study of Riemannian geometry is provided by *smooth manifolds*, which are topological manifolds endowed with an extra structure that allows us to differentiate functions and maps. First, we note that when U is an open subset of $\mathbb{R}^n$, a map $F : U \to \mathbb{R}^k$ is said to be *smooth* (or *of class C^∞*) if all of its component functions have continuous partial derivatives of all orders. More generally, if the domain U is an arbitrary subset of $\mathbb{R}^n$, not necessarily open (such as a relatively open subset of $\mathbb{R}^n_+$), then F is said to be smooth if for each $x \in U$, F has a smooth extension to a neighborhood of x in $\mathbb{R}^n$. A *diffeomorphism* is a bijective smooth map whose inverse is also smooth.

If M is a topological n-manifold with or without boundary, then two coordinate charts (U, φ), (V, ψ) for M are said to be *smoothly compatible* if both of the *transition maps* $\psi \circ \varphi^{-1}$ and $\varphi \circ \psi^{-1}$ are smooth where they are defined (on $\varphi(U \cap V)$ and $\psi(U \cap V)$, respectively). Since these maps are inverses of each other, it follows that both transition maps are in fact diffeomorphisms. An *atlas for M* is a collection of coordinate charts whose domains cover M. It is called a *smooth atlas* if any two charts in the atlas are smoothly compatible. A *smooth structure on M* is a smooth atlas that is *maximal*, meaning that it is not properly contained in any larger smooth atlas. A *smooth manifold* is a topological manifold endowed with a specific smooth structure; and similarly, a *smooth manifold with boundary* is a topological manifold with boundary endowed with a smooth structure. (We usually just say "M is a smooth manifold," or "M is a smooth manifold with boundary," with the smooth structure understood from the context.) If M is a set, a *smooth manifold structure* on M is a second countable, Hausdorff, locally Euclidean topology together with a smooth structure making it into a smooth manifold.

▶ **Exercise A.9.** Let M be a topological manifold with or without boundary. Show that every smooth atlas for M is contained in a unique maximal smooth atlas, and two smooth atlases determine the same smooth structure if and only if their union is a smooth atlas.

The manifolds in this book are always assumed to be smooth. As in most parts of differential geometry, the theory still works under weaker differentiability assumptions (such as k times continuously differentiable), but such considerations are usually relevant only when one is treating questions of hard analysis that are beyond our scope.

If M is a smooth manifold with or without boundary, then every coordinate chart in the given maximal atlas is called a *smooth coordinate chart for M* or just a *smooth chart*. The set U is called a *smooth coordinate domain*; if its image is an open ball in $\mathbb{R}^n$, it is called a *smooth coordinate ball*. If in addition φ extends to a smooth coordinate map $\varphi' : U' \to \mathbb{R}^n$ on an open set $U' \supseteq \bar{U}$ such that $\varphi'(\bar{U})$ is the closure of the open ball $\varphi(U)$ in $\mathbb{R}^n$, then U is called a *regular coordinate ball*. In this case, $\bar{U}$ is diffeomorphic to a closed ball and is thus compact.

Proposition A.10. [LeeSM, Prop. 1.19] *Every smooth manifold has a countable basis of regular coordinate balls.*

Given a smooth chart (U, φ) for M, the component functions of φ are called *local coordinates for M*, and are typically written as $(x^1, \ldots, x^n)$, (x^i), or x, depending on context. Although, formally speaking, a coordinate chart is a map from an open subset $U \subseteq M$ to $\mathbb{R}^n$, it is common when one is working in the domain of a specific chart to use the coordinate map to *identify* U with its image in $\mathbb{R}^n$, and to identify a point in U with its coordinate representation $(x^1, \ldots, x^n) \in \mathbb{R}^n$.

In this book, we always write coordinates with upper indices, as in (x^i), and expressions with indices are interpreted using the *Einstein summation convention*: if in some monomial term the same index name appears exactly twice, once as an upper index and once as a lower index, then that term is understood to be summed over all possible values of that index (usually from 1 to the dimension of the space). Thus the expression $a^i v_i$ is to be read as $\sum_i a^i v_i$. As we will see below, index positions for other sorts of objects such as vectors and covectors are chosen whenever possible in such a way that summations that make mathematical sense obey the rule that each repeated index appears once up and once down in each term to be summed.

Because of the result of Exercise A.9, to define a smooth structure on a manifold, it suffices to exhibit a single smooth atlas for it. For example, $\mathbb{R}^n$ is a topological n-manifold, and it has a smooth atlas consisting of the single chart $(\mathbb{R}^n, \mathrm{Id}_{\mathbb{R}^n})$. More generally, if V is an n-dimensional vector space, then every basis $(b_1, \ldots, b_n)$ for V yields a linear *basis isomorphism* $B \colon \mathbb{R}^n \to V$ by $B(x^1, \ldots, x^n) = x^i b_i$, whose inverse is a global coordinate chart, and it is easy to check that all such charts are smoothly compatible. Thus every finite-dimensional vector space has a natural smooth structure, which we call its *standard smooth structure*. We always consider $\mathbb{R}^n$ or any other finite-dimensional vector space to be a smooth manifold with this structure unless otherwise specified.

If M is a smooth n-manifold with or without boundary and $W \subseteq M$ is an open subset, then W has a natural smooth structure consisting of all smooth charts (U, φ) for M such that $U \subseteq W$, so every open subset of a smooth n-manifold is a smooth n-manifold in a natural way, and similarly for manifolds with boundary. If $M_1, \ldots, M_k$ are smooth manifolds of dimensions $n_1, \ldots, n_k$, respectively, then their Cartesian product $M_1 \times \cdots \times M_k$ has a natural smooth atlas consisting of charts of the form $(U_1 \times \cdots \times U_k, \varphi_1 \times \cdots \times \varphi_k)$, where (U_i, φ_i) is a smooth chart for M_i; thus the product space is a smooth manifold of dimension $n_1 + \cdots + n_k$.

Suppose M and N are smooth manifolds with or without boundary. A map $F \colon M \to N$ is said to be *smooth* if for every $p \in M$, there exist smooth charts (U, φ) for M containing p and (V, ψ) for N containing $F(p)$ such that $F(U) \subseteq V$ and the composite map $\psi \circ F \circ \varphi^{-1}$ is smooth from $\varphi(U)$ to $\psi(V)$. In particular, if N is an open subset of $\mathbb{R}^k$ or $\mathbb{R}^k_+$ with its standard smooth structure, we can take ψ to be the identity map of N, and then smoothness of F just means that each point of M is contained in the domain of a chart (U, φ) such that $F \circ \varphi^{-1}$ is smooth. It is an easy consequence of the definition that identity maps, constant maps, and compositions of smooth maps are all smooth. A map $F \colon M \to N$ is said to be a *diffeomorphism* if it is smooth and bijective and $F^{-1} \colon N \to M$ is also smooth.

If $F \colon M \to N$ is smooth, and (U, φ) and (V, ψ) are any smooth charts for M and N respectively, the map $\widehat{F} = \psi \circ F \circ \varphi^{-1}$ is a smooth map between appropriate subsets of $\mathbb{R}^n$ or $\mathbb{R}^n_+$, called a ***coordinate representation of*** F. Often, in keeping with the practice of using local coordinates to identify an open subset of a manifold with an open subset of $\mathbb{R}^n$, we identify the coordinate representation of F with (the restriction of) F itself, and write things like "F is given in local coordinates by $F(x, y) = (xy, x - y)$," when we really mean that F is given by this formula for suitable choices of charts in the domain and codomain.

We let $C^\infty(M, N)$ denote the set of all smooth maps from M to N, and $C^\infty(M)$ the vector space of all smooth functions from M to $\mathbb{R}$. For every function $f : M \to \mathbb{R}$ or $\mathbb{R}^k$, we define the ***support of*** f, denoted by supp f, to be the closure of the set $\{x \in M : f(x) \neq 0\}$. If $A \subseteq M$ is a closed subset and $U \subseteq M$ is an open subset containing A, then a ***smooth bump function for A supported in U*** is a smooth function $f : M \to \mathbb{R}$ satisfying $0 \leq f(x) \leq 1$ for all $x \in M$, $f|_A \equiv 1$, and supp $f \subseteq U$.

If $\mathcal{U} = \{U_\alpha\}_{\alpha \in A}$ is an indexed open cover of M, then a ***smooth partition of unity subordinate to*** $\mathcal{U}$ is an indexed family $\{\psi_\alpha\}_{\alpha \in A}$ of functions $\psi_\alpha \in C^\infty(M)$ satisfying

- $0 \leq \psi_\alpha(x) \leq 1$ for all $\alpha \in A$ and all $x \in M$.
- supp $\psi_\alpha \subseteq U_\alpha$ for each $\alpha \in A$.
- The family of supports $\{\text{supp } \psi_\alpha\}_{\alpha \in A}$ is ***locally finite***: every point of M has a neighborhood that intersects supp ψ_α for only finitely many values of α.
- $\sum_{\alpha \in A} \psi_\alpha(x) = 1$ for all $x \in M$.

Proposition A.11 (Existence of Partitions of Unity). [LeeSM, Thm. 2.23] *If M is a smooth manifold with or without boundary and $\mathcal{U} = \{U_\alpha\}_{\alpha \in A}$ is an indexed open cover of M, then there exists a smooth partition of unity subordinate to $\mathcal{U}$.*

Proposition A.12 (Existence of Smooth Bump Functions). [LeeSM, Prop. 2.25] *If M is a smooth manifold with or without boundary, $A \subseteq M$ is a closed subset, and $U \subseteq M$ is an open subset containing A, then there exists a smooth bump function for A supported in U.*

Tangent Vectors

Let M be a smooth manifold with or without boundary. There are various equivalent ways of defining tangent vectors on M. The following definition is the most convenient to work with in practice. For every point $p \in M$, a ***tangent vector at p*** is a linear map $v : C^\infty(M) \to \mathbb{R}$ that is a ***derivation at p***, meaning that for all $f, g \in C^\infty(M)$ it satisfies the product rule

$$v(fg) = f(p)vg + g(p)vf. \tag{A.1}$$

The set of all tangent vectors at p is denoted by $T_p M$ and called the ***tangent space at p***.

Suppose M is n-dimensional and $\varphi : U \to \hat{U} \subseteq \mathbb{R}^n$ is a smooth coordinate chart on some open subset $U \subseteq M$. Writing the coordinate functions of φ as $(x^1, \ldots, x^n)$, we define the **coordinate vectors** $\partial/\partial x^1|_p, \ldots, \partial/\partial x^n|_p$ by

$$\frac{\partial}{\partial x^i}\bigg|_p f = \frac{\partial}{\partial x^i}\bigg|_{\varphi(p)} \left(f \circ \varphi^{-1}\right). \tag{A.2}$$

These vectors form a basis for T_pM, which therefore has dimension n. When there can be no confusion about which coordinates are meant, we usually abbreviate $\partial/\partial x^i|_p$ by the notation $\partial_i|_p$. Thus once a smooth coordinate chart has been chosen, every tangent vector $v \in T_pM$ can be written uniquely in the form

$$v = v^i \partial_i|_p = v^1 \partial_1|_p + \cdots + v^n \partial_n|_p, \tag{A.3}$$

where the components $v^1, \ldots, v^n$ are obtained by applying v to the coordinate functions: $v^i = v(x^i)$.

On a finite-dimensional vector space V with its standard smooth manifold structure, there is a natural (basis-independent) identification of each tangent space T_pV with V itself, obtained by identifying a vector $v \in V$ with the derivation $D_v|_p$, defined by

$$D_v\big|_p(f) = \frac{d}{dt}\bigg|_{t=0} f(p+tv).$$

In terms of the coordinates (x^i) induced on V by any basis, this is the correspondence $(v^1, \ldots, v^n) \leftrightarrow \sum_i v^i \partial_i|_p$.

If $F: M \to N$ is a smooth map and p is any point in M, we define a linear map $dF_p: T_pM \to T_{F(p)}N$, called the **differential of F at p**, by

$$dF_p(v)f = v(f \circ F), \quad v \in T_pM.$$

▶ **Exercise A.13.** Given a smooth map $F: M \to N$ and a point $p \in M$, show that dF_p is a well-defined linear map from T_pM to $T_{F(p)}N$. Show that the differential of the identity is the identity, and the differential of a composition is given by $d(G \circ F)_p = dG_{F(p)} \circ dF_p$.

Once we have chosen local coordinates (x^i) for M and (y^j) for N, we find by unwinding the definitions that the coordinate representation of the differential is given by the **Jacobian matrix** of the coordinate representation of F, which is its matrix of first-order partial derivatives:

$$dF_p\left(v^i \frac{\partial}{\partial x^i}\bigg|_p\right) = \frac{\partial \widehat{F}^j}{\partial x^i}(p) v^i \frac{\partial}{\partial y^j}\bigg|_{F(p)}.$$

For every $p \in M$, the dual vector space to T_pM is called the **cotangent space at p**. This is the space $T_p^*M = (T_pM)^*$ of real-valued linear functionals on T_pM, called **(tangent) covectors at p**. For every $f \in C^\infty(M)$ and $p \in M$, there is a covector $df_p \in T_p^*M$ called the **differential of f at p**, defined by

$$df_p(v) = vf \quad \text{for all } v \in T_pM. \tag{A.4}$$

Submanifolds

The theory of submanifolds is founded on the inverse function theorem and its corollaries.

Theorem A.14 (Inverse Function Theorem for Manifolds). [LeeSM, Thm. 4.5]
Suppose M and N are smooth manifolds and $F : M \to N$ is a smooth map. If the linear map dF_p is invertible at some point $p \in M$, then there exist connected neighborhoods U_0 of p and V_0 of $F(p)$ such that $F|_{U_0} : U_0 \to V_0$ is a diffeomorphism.

The most useful consequence of the inverse function theorem is the following. A smooth map $F : M \to N$ is said to have **constant rank** if the linear map dF_p has the same rank at every point $p \in M$.

Theorem A.15 (Rank Theorem). [LeeSM, Thm. 4.12] *Suppose M and N are smooth manifolds of dimensions m and n, respectively, and $F : M \to N$ is a smooth map with constant rank r. For each $p \in M$ there exist smooth charts (U, φ) for M centered at p and (V, ψ) for N centered at $F(p)$ such that $F(U) \subseteq V$, in which F has a coordinate representation of the form*

$$\hat{F}\left(x^1, \ldots, x^r, x^{r+1}, \ldots, x^m\right) = \left(x^1, \ldots, x^r, 0, \ldots, 0\right). \tag{A.5}$$

Here are the most important types of constant-rank maps. In all of these definitions, M and N are smooth manifolds, and $F : M \to N$ is a smooth map.

- F is a **submersion** if its differential is surjective at each point, or equivalently if it has constant rank equal to $\dim N$.
- F is an **immersion** if its differential is injective at each point, or equivalently if it has constant rank equal to $\dim M$.
- F is a **local diffeomorphism** if every point $p \in M$ has a neighborhood U such that $F|_U$ is a diffeomorphism onto an open subset of N, or equivalently if F is both a submersion and an immersion.
- F is a **smooth embedding** if it is an injective immersion that is also a topological embedding (a homeomorphism onto its image, endowed with the subspace topology).

Theorem A.16 (Local Embedding Theorem). [LeeSM, Thm. 4.25] *Every smooth immersion is locally an embedding: if $F : M \to N$ is a smooth immersion, then for every $p \in M$, there is a neighborhood U of p in M such that $F|_U$ is an embedding.*

Theorem A.17 (Local Section Theorem). [LeeSM, Thm. 4.26] *Suppose M and N are smooth manifolds and $\pi : M \to N$ is a smooth map. Then π is a smooth submersion if and only if every point of M is in the image of a smooth **local section** of π (a map $\sigma : U \to M$ defined on some open subset $U \subseteq N$, with $\pi \circ \sigma = \mathrm{Id}_U$).*

Proposition A.18 (Properties of Submersions). [LeeSM, Prop. 4.28] *Let M and N be smooth manifolds and let $\pi : M \to N$ be a smooth submersion.*

(a) π is an open map.

(b) If π is surjective, it is a quotient map.

Proposition A.19 (Uniqueness of Smooth Quotients). [LeeSM, Thm. 4.31] *Suppose M, N_1, and N_2 are smooth manifolds, and $\pi_1 : M \to N_1$ and $\pi_2 : M \to N_2$ are surjective smooth submersions that are constant on each other's fibers. Then there exists a unique diffeomorphism $F : N_1 \to N_2$ such that $F \circ \pi_1 = \pi_2$.*

Theorem A.20 (Global Rank Theorem). [LeeSM, Thm. 4.14] *Suppose M and N are smooth manifolds, and $F : M \to N$ is a smooth map of constant rank.*

(a) *If F is surjective, then it is a smooth submersion.*

(b) *If F is injective, then it is a smooth immersion.*

(c) *If F is bijective, then it is a diffeomorphism.*

Suppose $\widetilde{M}$ is a smooth manifold with or without boundary. An **immersed n-dimensional submanifold of M** is a subset $M \subseteq \widetilde{M}$ endowed with a topology that makes it into an n-dimensional topological manifold and a smooth structure such that the inclusion map $M \hookrightarrow \widetilde{M}$ is a smooth immersion. It is called an **embedded submanifold** if the inclusion is a smooth embedding, or equivalently if the given topology on M is the subspace topology. Immersed and embedded **submanifolds with boundary** are defined similarly. In each case, the **codimension of M** is the difference $\dim \widetilde{M} - \dim M$. A submanifold of codimension 1 is known as a **hypersurface**. Without further qualification, the word *submanifold* means an immersed submanifold, which includes an embedded one as a special case.

An embedded submanifold with or without boundary is said to be **properly embedded** if the inclusion map is proper.

Proposition A.21. [LeeSM, Prop. 5.5] *If $\widetilde{M}$ is a smooth manifold with or without boundary, then an embedded submanifold $M \subseteq \widetilde{M}$ is properly embedded if and only if it is a closed subset of $\widetilde{M}$.*

A properly embedded codimension-0 submanifold with boundary in $\widetilde{M}$ is called a **regular domain**.

Proposition A.22 (Slice Coordinates). [LeeSM, Thms. 5.8 & 5.51] *Let $\widetilde{M}$ be a smooth m-manifold without boundary and let $M \subseteq \widetilde{M}$ be an embedded n-dimensional submanifold with or without boundary. Then for each $p \in M$ there exist a neighborhood $\widetilde{U}$ of p in $\widetilde{M}$ and smooth coordinates $\left(x^1, \ldots, x^m\right)$ for $\widetilde{M}$ on $\widetilde{U}$ such that $M \cap \widetilde{U}$ is a set of one of the following forms:*

$$M \cap \widetilde{U} = \begin{cases} \{x \in \widetilde{U} : x^{n+1} = \cdots = x^m = 0\} & \text{if } p \notin \partial M, \\ \{x \in \widetilde{U} : x^{n+1} = \cdots = x^m = 0 \text{ and } x^n \geq 0\} & \text{if } p \in \partial M. \end{cases}$$

In such a chart, $\left(x^1, \ldots, x^n\right)$ form smooth local coordinates for M.

Coordinates satisfying either of these conditions are called *slice coordinates*; when it is necessary to distinguish them, coordinates satisfying the second condition are referred to as *boundary slice coordinates*.

If M is an immersed submanifold, then Theorem A.16 shows that every point of M has a neighborhood U in M that is embedded in $\widetilde{M}$, so we can construct slice coordinates for U (but they need not be slice coordinates for M).

▶ **Exercise A.23.** Suppose $\widetilde{M}$ is a smooth manifold and $M \subseteq \widetilde{M}$ is an embedded sub-manifold. Show that every smooth function $f : M \to \mathbb{R}$ can be extended to a smooth function on a neighborhood of M in $\widetilde{M}$ whose restriction to M is f; and if M is also closed in $\widetilde{M}$, then the neighborhood can be taken to be all of $\widetilde{M}$. [Hint: Extend f locally in slice coordinates by letting it be independent of $(x^{n+1}, \ldots, x^m)$, and patch together using a partition of unity.]

Here is the way that most submanifolds are presented. Suppose $\Phi: \widetilde{M} \to N$ is any map. Every subset of the form $\Phi^{-1}(\{y\}) \subseteq \widetilde{M}$ for some $y \in N$ is called a *level set of Φ*, or the *fiber of Φ over y*. The simpler notation $\Phi^{-1}(y)$ is also used for a level set when there is no likelihood of ambiguity.

Theorem A.24 (Constant-Rank Level Set Theorem). [LeeSM, Thm. 5.12] *Suppose $\widetilde{M}$ and N are smooth manifolds, and $\Phi: \widetilde{M} \to N$ is a smooth map with constant rank r. Every level set of Φ is a properly embedded submanifold of codimension r in $\widetilde{M}$.*

Corollary A.25 (Submersion Level Set Theorem). [LeeSM, Cor. 5.13] *Suppose $\widetilde{M}$ and N are smooth manifolds, and $\Phi: \widetilde{M} \to N$ is a smooth submersion. Every level set of Φ is a properly embedded submanifold of $\widetilde{M}$, whose codimension is equal to $\dim N$.*

In fact, a map does not have to be a submersion, or even to have constant rank, for its level sets to be embedded submanifolds. If $\Phi: \widetilde{M} \to N$ is a smooth map, a point $p \in \widetilde{M}$ is called a *regular point of Φ* if the linear map $d\Phi_p: T_p\widetilde{M} \to T_{\Phi(p)}N$ is surjective, and p is called a *critical point of Φ* if it is not. A point $c \in N$ is called a *regular value of Φ* if every point of $\Phi^{-1}(c)$ is a regular point of Φ, and a *critical value* otherwise. A level set $\Phi^{-1}(c)$ is called a *regular level set of Φ* if c is a regular value of Φ.

Corollary A.26 (Regular Level Set Theorem). [LeeSM, Cor. 5.14] *Let $\widetilde{M}$ and N be smooth manifolds, and let $\Phi: \widetilde{M} \to N$ be a smooth map. Every regular level set of Φ is a properly embedded submanifold of $\widetilde{M}$ whose codimension is equal to $\dim N$.*

Suppose $\widetilde{M}$ is a smooth manifold and $M \subseteq \widetilde{M}$ is an embedded codimension-k submanifold. A smooth map $\Phi: \widetilde{M} \to \mathbb{R}^k$ is called a *defining function for M* if M is a regular level set of Φ. More generally, if $\Phi: U \to \mathbb{R}^k$ is a smooth map defined on an open subset $U \subseteq \widetilde{M}$ such that $M \cap U$ is a regular level set of Φ, then Φ is called a *local defining function for M*. The following is a partial converse to the submersion level set theorem.

Proposition A.27 (Existence of Local Defining Functions). [LeeSM, Prop. 5.16] *Suppose $\widetilde{M}$ is a smooth manifold and $M \subseteq \widetilde{M}$ is an embedded submanifold of codimension k. For each point $p \in M$, there exist a neighborhood U of p in $\widetilde{M}$ and a smooth submersion $\Phi \colon U \to \mathbb{R}^k$ such that $M \cap U$ is a level set of Φ.*

If $M \subseteq \widetilde{M}$ is a submanifold (immersed or embedded), the inclusion map $\iota \colon M \hookrightarrow \widetilde{M}$ induces an injective linear map $d\iota_p \colon T_p M \to T_p \widetilde{M}$ for each $p \in M$. It is standard practice to identify $T_p M$ with its image $d\iota_p(T_p M) \subseteq T_p \widetilde{M}$, thus regarding $T_p M$ as a subspace of $T_p \widetilde{M}$. The next proposition gives two handy ways to identify this subspace in the embedded case.

Proposition A.28 (Tangent Space to a Submanifold). [LeeSM, Props. 5.37 & 5.38] *Suppose $\widetilde{M}$ is a smooth manifold, $M \subseteq \widetilde{M}$ is an embedded submanifold, and $p \in M$. As a subspace of $T_p \widetilde{M}$, the tangent space $T_p M$ is characterized by*

$$T_p M = \{ v \in T_p \widetilde{M} : vf = 0 \text{ whenever } f \in C^\infty(\widetilde{M}) \text{ and } f|_M = 0 \}.$$

If Φ is a local defining function for M on a neighborhood of p, then

$$T_p M = \operatorname{Ker} d\Phi_p.$$

Proposition A.29 (Lagrange Multiplier Rule). *Let $\widetilde{M}$ be a smooth m-manifold and $M \subseteq \widetilde{M}$ be an embedded hypersurface. Suppose $f \in C^\infty(\widetilde{M})$, and $p \in M$ is a point at which f attains a local maximum or minimum value among points in M. If $\Phi \colon U \to \mathbb{R}$ is a local defining function for M on a neighborhood U of p, then there is a real number λ (called a **Lagrange multiplier**) such that $df_p = \lambda d\Phi_p$.*

▶ **Exercise A.30.** Prove the preceding proposition. [Hint: One way to proceed is to begin by showing that there are smooth functions $x^1, \dots, x^{m-1}$ on a neighborhood of p such that $(x^1, \dots, x^{m-1}, \Phi)$ are local slice coordinates for M.]

In treating manifolds with boundary, many constructions can be simplified by viewing a manifold with boundary as being embedded in a larger manifold without boundary. Here is one way to do that.

If M is a topological manifold with nonempty boundary, the **double of M** is the quotient space $D(M)$ formed from the disjoint union of two copies of M (say M_1 and M_2), by identifying each point in ∂M_1 with the corresponding point in ∂M_2.

Proposition A.31. [LeeSM, Example 9.32] *Suppose M is a smooth manifold with nonempty boundary. Then $D(M)$ is a topological manifold without boundary, and it has a smooth structure such that the natural maps $M \to M_1 \to D(M)$ and $M \to M_2 \to D(M)$ are smooth embeddings. Moreover, $D(M)$ is compact if and only if M is compact, and connected if and only if M is connected.*

Vector Bundles

Suppose M is a smooth manifold with or without boundary. The ***tangent bundle of M***, denoted by TM, is the disjoint union of the tangent spaces at all points of M: $TM = \coprod_{p \in M} T_p M$. This set can be thought of both as a union of vector spaces and as a manifold in its own right. This kind of structure, called a *vector bundle*, is extremely common in differential geometry, so we take some time to review the definitions and basic properties of vector bundles.

For any positive integer k, a ***smooth vector bundle of rank k*** is a pair of smooth manifolds E and M with or without boundary, together with a smooth surjective map $\pi \colon E \to M$ satisfying the following conditions:

(i) For each $p \in M$, the set $E_p = \pi^{-1}(p)$ is endowed with the structure of a k-dimensional real vector space.

(ii) For each $p \in M$, there exist a neighborhood U of p and a diffeomorphism $\Phi \colon \pi^{-1}(U) \to U \times \mathbb{R}^k$ such that $\pi_U \circ \Phi = \pi$, where $\pi_U \colon U \times \mathbb{R}^k \to U$ is the projection onto the first factor; and for each $q \in U$, Φ restricts to a linear isomorphism from E_q to $\{q\} \times \mathbb{R}^k \cong \mathbb{R}^k$.

The space M is called the ***base*** of the bundle, E is called its ***total space***, and π is its ***projection***. Each set $E_p = \pi^{-1}(p)$ is called the ***fiber of E over p***, and each diffeomorphism $\Phi \colon \pi^{-1}(U) \to U \times \mathbb{R}^k$ described above is called a ***smooth local trivialization***.

It frequently happens that we are presented with a collection of vector spaces (such as tangent spaces), one for each point in a manifold, that we would like to "glue together" to form a vector bundle. There is a shortcut for showing that such a collection forms a vector bundle without first constructing a smooth manifold structure on the total space. As the next lemma shows, all we need to do is to exhibit the maps that we wish to regard as local trivializations and check that they overlap correctly.

Lemma A.32 (Vector Bundle Chart Lemma). [LeeSM, Lemma 10.6] *Let M be a smooth manifold with or without boundary, and suppose that for each $p \in M$ we are given a real vector space E_p of some fixed dimension k. Let $E = \coprod_{p \in M} E_p$ (the disjoint union of all the spaces E_p), and let $\pi \colon E \to M$ be the map that takes each element of E_p to the point p. Suppose furthermore that we are given*

(i) *an indexed open cover $\{U_\alpha\}_{\alpha \in A}$ of M;*

(ii) *for each $\alpha \in A$, a bijective map $\Phi_\alpha \colon \pi^{-1}(U_\alpha) \to U_\alpha \times \mathbb{R}^k$ whose restriction to each E_p is a linear isomorphism from E_p to $\{p\} \times \mathbb{R}^k \cong \mathbb{R}^k$;*

(iii) *for each $\alpha, \beta \in A$ such that $U_\alpha \cap U_\beta \neq \varnothing$, a smooth map $\tau_{\alpha\beta} \colon U_\alpha \cap U_\beta \to \mathrm{GL}(k, \mathbb{R})$ such that the composite map $\Phi_\alpha \circ \Phi_\beta^{-1}$ from $(U_\alpha \cap U_\beta) \times \mathbb{R}^k$ to itself has the form*

$$\Phi_\alpha \circ \Phi_\beta^{-1}(p, v) = (p, \tau_{\alpha\beta}(p)v). \tag{A.6}$$

Then E has a unique smooth manifold structure making it a smooth vector bundle of rank k over M, with π as projection and the maps Φ_α as smooth local trivializations.

The smooth $GL(k, \mathbb{R})$-valued maps $\tau_{\alpha\beta}$ of this lemma are called **transition functions for E**.

If $\pi\colon E \to M$ is a smooth vector bundle over M, then a **section of E** is a continuous map $\sigma\colon M \to E$ such that $\pi \circ \sigma = \mathrm{Id}_M$, or equivalently, $\sigma(p) \in E_p$ for all p. If $U \subseteq M$ is an open set, then a **local section of E over U** is a continuous map $\sigma\colon U \to E$ satisfying the analogous equation $\pi \circ \sigma = \mathrm{Id}_U$. A **smooth (local or global) section of E** is just a section that is smooth as a map between smooth manifolds. If we need to speak of "sections" that are not necessarily continuous, we use the following terminology: a **rough (local) section of E** is a map $\sigma\colon U \to E$ defined on some open set $U \subseteq M$, not necessarily smooth or even continuous, such that $\pi \circ \sigma = \mathrm{Id}_U$.

A **local frame** for E is an ordered k-tuple $(\sigma_1, \ldots, \sigma_k)$ of local sections over an open set U whose values at each $p \in U$ constitute a basis for E_p. It is called a **global frame** if $U = M$. If τ is a (local or global) section of E and $(\sigma_1, \ldots, \sigma_k)$ is a local frame for E over $U \subseteq M$, then the value of τ at each $p \in U$ can be written

$$\tau(p) = \tau^i(p)\sigma_i(p)$$

for some functions $\tau^1, \ldots, \tau^k\colon U \to \mathbb{R}$, called the **component functions of σ** with respect to the given frame.

The next lemma gives a criterion for smoothness that is usually easy to verify in practice.

Lemma A.33 (Local Frame Criterion for Smoothness). [LeeSM, Prop. 10.22] *Let $\pi\colon E \to M$ be a smooth vector bundle, and let $\tau\colon M \to E$ be a rough section. If (σ_i) is a smooth local frame for E over an open subset $U \subseteq M$, then τ is smooth on U if and only if its component functions with respect to (σ_i) are smooth.*

If $E \to M$ is a smooth vector bundle, then the set of smooth sections of E, denoted by $\Gamma(E)$, is a vector space (usually infinite-dimensional) under pointwise addition and multiplication by constants. Its zero element is the **zero section** ζ defined by $\zeta(p) = 0 \in E_p$ for all $p \in M$. For every section σ of E, the **support of σ** is the closure of the set $\{p \in M : \sigma(p) \neq 0\}$.

Given a smooth vector bundle $\pi_E\colon E \to M$, a **smooth subbundle of E** is a smooth vector bundle $\pi_D\colon D \to M$, in which D is an embedded submanifold (with or without boundary) of E and $\pi_D = \pi_E|_D$, such that for each $p \in M$, the subset $D_p = D \cap E_p$ is a linear subspace of E_p, and the vector space structure on D_p is the one inherited from E_p. Given a collection of subspaces $D_p \subseteq E_p$, one for each $p \in M$, the following lemma gives a convenient condition for checking that their union is a smooth subbundle.

Lemma A.34 (Local Frame Criterion for Subbundles). [LeeSM, Lemma 10.32] *Let $\pi\colon E \to M$ be a smooth vector bundle, and suppose that for each $p \in M$ we are given a k-dimensional linear subspace $D_p \subseteq E_p$. Suppose further that each $p \in M$ has a neighborhood U on which there are smooth local sections $\sigma_1, \ldots, \sigma_k\colon U \to E$ such that $\sigma_1(q), \ldots, \sigma_k(q)$ form a basis for D_q at each $q \in U$. Then $D = \bigcup_{p \in M} D_p \subseteq E$ is a smooth subbundle of E.*

Suppose $\widetilde{M}$ is a smooth manifold with or without boundary, and $M \subseteq \widetilde{M}$ is an immersed or embedded submanifold with or without boundary. If $E \to \widetilde{M}$ is any smooth rank-k vector bundle over $\widetilde{M}$, we obtain a smooth vector bundle $\pi|_M : E|_M \to M$ of rank k over M, called the **restriction of E to M**, whose total space is $E|_M = \pi^{-1}(M)$, whose fiber at each $p \in M$ is exactly the fiber of E, and whose local trivializations are the restrictions of the local trivializations of E. (See [LeeSM, Example 10.8].)

Every smooth section of E restricts to a smooth section of $E|_M$. Conversely, the next exercise shows that in most cases, smooth sections of $E|_M$ extend to smooth sections of E, at least locally near M.

▶ **Exercise A.35.** Suppose $\widetilde{M}$ is a smooth manifold, $E \to \widetilde{M}$ is a smooth vector bundle, and $M \subseteq \widetilde{M}$ is an embedded submanifold. Show that every smooth section of the restricted bundle $E|_M$ can be extended to a smooth section of E on a neighborhood of M in $\widetilde{M}$; and if M is closed in $\widetilde{M}$, the neighborhood can be taken to be all of $\widetilde{M}$.

The Tangent Bundle and Vector Fields

We continue to assume that M is a smooth manifold with or without boundary. In this section, we see how the theory of vector bundles applies to the particular case of the tangent bundle.

Given any smooth local coordinate chart $(U, (x^i))$ for M, we can define a local trivialization of TM over U by letting $\Phi : \pi^{-1}(U) \to U \times \mathbb{R}^n$ be the map sending $v \in T_p M$ to $(p, (v^1, \dots, v^n))$, where $v^i \partial/\partial x^i|_p$ is the coordinate representation for v. Given another such chart $(\widetilde{U}, (\widetilde{x}^i))$ for M, we obtain a corresponding local trivialization $\widetilde{\Phi}(v) = (p, (\widetilde{v}^1, \dots, \widetilde{v}^n))$, where $v = \widetilde{v}^i \partial/\partial \widetilde{x}^i|_p$. The transformation law for coordinate vectors shows that

$$\widetilde{v}^j = \frac{\partial \widetilde{x}^j}{\partial x^i}(p) v^i,$$

so the transition function between two such charts is given by the smooth matrix-valued function $(\partial \widetilde{x}^j/\partial x^i)$. Thus the chart lemma shows that these local trivializations turn TM into a smooth vector bundle.

A smooth coordinate chart (x^i) for M also yields smooth local coordinates $(x^1, \dots, x^n, v^1, \dots, v^n)$ on $\pi^{-1}(U) \subseteq TM$, called **natural coordinates for TM**, by following the local trivialization Φ with the map $U \times \mathbb{R}^n \to \mathbb{R}^n \times \mathbb{R}^n$ that sends $(p, (v^1, \dots, v^n))$ to $(x^1(p), \dots, x^n(p), v^1, \dots, v^n)$.

A section of TM is called a **vector field on M**. To avoid confusion between the point $p \in M$ at which a vector field is evaluated and the action of the vector field on a function, we usually write the value of a vector field X at $p \in M$ as $X_p \in T_p M$, or, if it is clearer (for example if X itself has one or more subscripts), as $X|_p$.

For example, if $(U, (x^i))$ is a smooth coordinate chart for M, for each i we get a smooth **coordinate vector field** on U, denoted by $\partial/\partial x^i$ (and often abbreviated by

∂_i if no confusion results), whose value at $p \in U$ is the coordinate vector $\partial/\partial x^i|_p$ defined above. Each coordinate vector field is smooth because its coordinate representation in natural coordinates is $(x^1, \ldots, x^n) \mapsto (x^1, \ldots, x^n, 0, \ldots, 0, 1, 0, \ldots, 0)$. The vector fields $\partial_1, \ldots, \partial_n$ form a smooth local frame for TM, called a ***coordinate frame***.

It follows from Lemma A.33 that a vector field is smooth if and only if its components are smooth with respect to some smooth local frame (such as a coordinate frame) in a neighborhood of each point. It is standard to use the notation $\mathfrak{X}(M)$ as another name for $\Gamma(TM)$, the space of all smooth vector fields on M.

Now suppose M and N are smooth manifolds with or without boundary, and $F: M \to N$ is a smooth map. We obtain a smooth map $dF: TM \to TN$, called the ***global differential of F***, whose restriction to each tangent space T_pM is the linear map dF_p defined above. In general, the global differential does not take vector fields to vector fields. In the special case that $X \in \mathfrak{X}(M)$ and $Y \in \mathfrak{X}(N)$ are vector fields such that $dF(X_p) = Y_{F(p)}$ for all $p \in M$, we say that the vector fields X and Y are ***F-related***.

Lemma A.36. [LeeSM, Prop. 8.19 & Cor. 8.21] *Let $F: M \to N$ be a diffeomorphism between smooth manifolds with or without boundary. For every $X \in \mathfrak{X}(M)$, there is a unique vector field $F_*X \in \mathfrak{X}(N)$, called the **pushforward of X**, that is F-related to X. For every $f \in C^\infty(N)$, it satisfies*

$$((F_*X)f) \circ F = X(f \circ F). \tag{A.7}$$

Suppose $X \in \mathfrak{X}(M)$. Given a real-valued function $f \in C^\infty(M)$, applying X to f yields a new function $Xf \in C^\infty(M)$ by $Xf(p) = X_p f$. The defining equation (A.1) for tangent vectors translates into the following product rule for vector fields:

$$X(fg) = f\, Xg + g\, Xf. \tag{A.8}$$

A map $X: C^\infty(M) \to C^\infty(M)$ is called a ***derivation of $C^\infty(M)$*** (as opposed to a *derivation at a point*) if it is linear over $\mathbb{R}$ and satisfies (A.8) for all $f, g \in C^\infty(M)$.

Lemma A.37. [LeeSM, Prop. 8.15] *Let M be a smooth manifold with or without boundary. A map $D: C^\infty(M) \to C^\infty(M)$ is a derivation if and only if it is of the form $Df = Xf$ for some $X \in \mathfrak{X}(M)$.*

Given smooth vector fields $X, Y \in \mathfrak{X}(M)$, define a map $[X,Y]: C^\infty(M) \to C^\infty(M)$ by

$$[X,Y]f = X(Yf) - Y(Xf).$$

A straightforward computation shows that $[X,Y]$ is a derivation of $C^\infty(M)$, and thus by Lemma A.37 it defines a smooth vector field, called the ***Lie bracket of X and Y***.

Proposition A.38 (Properties of Lie Brackets). [LeeSM, Prop. 8.28] *Let M be a smooth manifold with or without boundary and $X, Y, Z \in \mathfrak{X}(M)$.*

(a) $[X, Y]$ is bilinear over $\mathbb{R}$ as a function of X and Y.

(b) $[X, Y] = -[Y, X]$ (antisymmetry).

(c) $[X, [Y, Z]] + [Y, [Z, X]] + [Z, [X, Y]] = 0$ (Jacobi identity).

(d) For $f, g \in C^\infty(M)$, $[fX, gY] = fg[X, Y] + (fXg)Y - (gYf)X$.

Proposition A.39 (Naturality of Lie Brackets). [LeeSM, Prop. 8.30 & Cor. 8.31] *Let* $F \colon M \to N$ *be a smooth map between manifolds with or without boundary, and let* $X_1, X_2 \in \mathfrak{X}(M)$ *and* $Y_1, Y_2 \in \mathfrak{X}(N)$ *be vector fields such that* X_i *is* F-*related to* Y_i *for* $i = 1, 2$. *Then* $[X_1, X_2]$ *is* F-*related to* $[Y_1, Y_2]$. *In particular, if* F *is a diffeomorphism, then* $F_*[X_1, X_2] = [F_*X_1, F_*X_2]$.

Now suppose $\widetilde{M}$ is a smooth manifold with or without boundary and $M \subseteq \widetilde{M}$ is an immersed or embedded submanifold with or without boundary. The bundle $T\widetilde{M}|_M$, obtained by restricting $T\widetilde{M}$ to M, is called the **ambient tangent bundle**. It is a smooth bundle over M whose rank is equal to the dimension of $\widetilde{M}$. The tangent bundle TM is naturally viewed as a smooth subbundle of $T\widetilde{M}|_M$, and smooth vector fields on M can also be viewed as smooth sections of $T\widetilde{M}|_M$. A vector field $X \in \mathfrak{X}(\widetilde{M})$ always restricts to a smooth section of $T\widetilde{M}|_M$, and it restricts to a smooth section of TM if and only if it is **tangent to** M, meaning that $X_p \in T_p M \subseteq T_p \widetilde{M}$ for each $p \in M$.

Corollary A.40 (Brackets of Vector Fields Tangent to Submanifolds). [LeeSM, Cor. 8.32] *Let* $\widetilde{M}$ *be a smooth manifold and let* M *be an immersed submanifold with or without boundary in* $\widetilde{M}$. *If* Y_1 *and* Y_2 *are smooth vector fields on* $\widetilde{M}$ *that are tangent to* M, *then* $[Y_1, Y_2]$ *is also tangent to* M.

▶ **Exercise A.41.** Let $\widetilde{M}$ be a smooth manifold with or without boundary and let $M \subseteq \widetilde{M}$ be an embedded submanifold with or without boundary. Show that a vector field $X \in \mathfrak{X}(\widetilde{M})$ is tangent to M if and only if $(Xf)|_M = 0$ whenever $f \in C^\infty(\widetilde{M})$ is a function that vanishes on M.

Integral Curves and Flows

A **curve** in a smooth manifold M (with or without boundary) is a continuous map $\gamma \colon I \to M$, where $I \subseteq \mathbb{R}$ is some interval. If γ is smooth, then for each $t_0 \in I$ we obtain a vector $\gamma'(t_0) = d\gamma_{t_0}(d/dt|_{t_0})$, called the **velocity of** γ **at time** t_0. It acts on functions by

$$\gamma'(t_0)f = (f \circ \gamma)'(t_0).$$

In any smooth local coordinates, the coordinate expression for $\gamma'(t_0)$ is exactly the same as it would be in $\mathbb{R}^n$: the components of $\gamma'(t_0)$ are the ordinary t-derivatives of the components of γ.

If $X \in \mathfrak{X}(M)$, then a smooth curve $\gamma \colon I \to M$ is called an **integral curve of** X if its velocity at each point is equal to the value of X there: $\gamma'(t) = X_{\gamma(t)}$ for each $t \in I$.

The fundamental fact about vector fields (at least in the case of manifolds without boundary) is that there exists a unique maximal integral curve starting at each point, varying smoothly as the point varies. These integral curves are all encoded into a global object called a *flow*, which we now define.

Given a smooth manifold M (without boundary), a **flow domain** for M is an open subset $\mathcal{D} \subseteq \mathbb{R} \times M$ with the property that for each $p \in M$, the set $\mathcal{D}^{(p)} = \{t \in \mathbb{R} : (t, p) \in \mathcal{D}\}$ is an open interval containing 0. Given a flow domain $\mathcal{D}$ and a map $\theta : \mathcal{D} \to M$, for each $t \in \mathbb{R}$ we let $M_t = \{p \in M : (t, p) \in \mathcal{D}\}$, and we define maps $\theta_t : M_t \to M$ and $\theta^{(p)} : \mathcal{D}^{(p)} \to M$ by $\theta_t(p) = \theta^{(p)}(t) = \theta(t, p)$.

A **flow** on M is a continuous map $\theta : \mathcal{D} \to M$, where $\mathcal{D} \subseteq \mathbb{R} \times M$ is a flow domain, that satisfies

$$\theta_0 = \mathrm{Id}_M,$$
$$\theta_t \circ \theta_s(p) = \theta_{t+s}(p) \quad \text{wherever both sides are defined.}$$

If θ is a smooth flow, we obtain a smooth vector field $X \in \mathfrak{X}(M)$ defined by $X_p = \left(\theta^{(p)}\right)'(0)$, called the **infinitesimal generator of θ**.

Theorem A.42 (Fundamental Theorem on Flows). [LeeSM, Thm. 9.12] *Let X be a smooth vector field on a smooth manifold M (without boundary). There is a unique smooth maximal flow $\theta : \mathcal{D} \to M$ whose infinitesimal generator is X. This flow has the following properties:*

(a) *For each $p \in M$, the curve $\theta^{(p)} : \mathcal{D}^{(p)} \to M$ is the unique maximal integral curve of X starting at p.*

(b) *If $s \in \mathcal{D}^{(p)}$, then $\mathcal{D}^{(\theta(s,p))}$ is the interval $\mathcal{D}^{(p)} - s = \{t - s : t \in \mathcal{D}^{(p)}\}$.*

(c) *For each $t \in \mathbb{R}$, the set M_t is open in M, and $\theta_t : M_t \to M_{-t}$ is a diffeomorphism with inverse θ_{-t}.*

Although the fundamental theorem guarantees only that each point lies on an integral curve that exists for a short time, the next lemma can often be used to prove that a particular integral curve exists for all time.

Lemma A.43 (Escape Lemma). *Suppose M is a smooth manifold and $X \in \mathfrak{X}(M)$. If $\gamma : I \to M$ is a maximal integral curve of X whose domain I has a finite least upper bound b, then for every $t_0 \in I$, $\gamma\big([t_0, b)\big)$ is not contained in any compact subset of M.*

▶ **Exercise A.44.** Prove the escape lemma.

Proposition A.45 (Canonical Form for a Vector Field). [LeeSM, Thm. 9.22] *Let X be a smooth vector field on a smooth manifold M, and suppose $p \in M$ is a point where $X_p \neq 0$. There exist smooth coordinates (x^i) on some neighborhood of p in which X has the coordinate representation $\partial/\partial x^1$.*

The fundamental theorem on flows leads to a way of differentiating one vector field along the flow of another. Suppose M is a smooth manifold, $X, Y \in \mathfrak{X}(M)$, and θ is the flow of X. The **Lie derivative of Y with respect to X** is the vector field $\mathcal{L}_X Y$ defined by

$$(\mathscr{L}_X Y)_p = \frac{d}{dt}\bigg|_{t=0} d(\theta_{-t})_{\theta_t(p)}\big(Y_{\theta_t(p)}\big) = \lim_{t \to 0} \frac{d(\theta_{-t})_{\theta_t(p)}\big(Y_{\theta_t(p)}\big) - Y_p}{t}.$$

This formula is useless for computation, however, because typically the flow of a vector field is difficult or impossible to compute explicitly. Fortunately, there is another expression for the Lie derivative that is much easier to compute.

Proposition A.46. [LeeSM, Thm. 9.38] *Suppose M is a smooth manifold and $X, Y \in \mathfrak{X}(M)$. The Lie derivative of Y with respect to X is equal to the Lie bracket $[X, Y]$.*

One of the most important applications of the Lie derivative is as an obstruction to invariance under a flow. If θ is a smooth flow, we say that a vector field Y is *invariant under θ* if $(\theta_t)_* Y = Y$ wherever the left-hand side is defined.

Proposition A.47. [LeeSM, Thm. 9.42] *Let M be a smooth manifold and $X \in \mathfrak{X}(M)$. A smooth vector field is invariant under the flow of X if and only if its Lie derivative with respect to X is identically zero.*

A k-tuple of vector fields $X_1, \ldots, X_k$ is said to *commute* if $[X_i, X_j] = 0$ for each i and j.

Proposition A.48 (Canonical Form for Commuting Vector Fields). [LeeSM, Thm. 9.46] *Let M be a smooth n-manifold, and let $(X_1, \ldots, X_k)$ be a linearly independent k-tuple of smooth commuting vector fields on an open subset $W \subseteq M$. For each $p \in W$, there exists a smooth coordinate chart $(U, (x^i))$ centered at p such that $X_i = \partial/\partial x^i$ for $i = 1, \ldots, k$.*

Smooth Covering Maps

A *covering map* is a surjective continuous map $\pi : \widetilde{M} \to M$ between connected and locally path-connected topological spaces, for which each point of M has connected neighborhood U that is *evenly covered*, meaning that each connected component of $\pi^{-1}(U)$ is mapped homeomorphically onto U by π. It is called a *smooth covering map* if $\widetilde{M}$ and M are smooth manifolds with or without boundary and each component of $\pi^{-1}(U)$ is mapped *diffeomorphically* onto U. For every evenly covered open set $U \subseteq M$, the components of $\pi^{-1}(U)$ are called the *sheets of the covering over U*.

Here are the main properties of covering maps that we need.

Proposition A.49 (Elementary Properties of Smooth Covering Maps). [LeeSM, Prop. 4.33]

(a) *Every smooth covering map is a local diffeomorphism, a smooth submersion, an open map, and a quotient map.*

(b) *An injective smooth covering map is a diffeomorphism.*
(c) *A topological covering map is a smooth covering map if and only if it is a local diffeomorphism.*

Proposition A.50 *A covering map is a proper map if and only if it is finite-sheeted.*

▶ **Exercise A.51.** Prove the preceding proposition.

Example A.52 (A Covering of the n-Torus). The ***n-torus*** is the manifold $\mathbb{T}^n = \mathbb{S}^1 \times \cdots \times \mathbb{S}^1$, regarded as the subset of $\mathbb{R}^{2n}$ defined by $(x^1)^2 + (x^2)^2 = \cdots = (x^{2n-1})^2 + (x^{2n})^2 = 1$. The smooth map $X \colon \mathbb{R}^n \to \mathbb{T}^n$ given by $X(u^1, \ldots, u^n) = (\cos u^1, \sin u^1, \ldots, \cos u^n, \sin u^n)$ is a smooth covering map.   //

If $\pi \colon \widetilde{M} \to M$ is a covering map and $F \colon B \to M$ is a continuous map from a topological space B into M, then a ***lift of*** F is a continuous map $\widetilde{F} \colon B \to \widetilde{M}$ such that $\pi \circ \widetilde{F} = F$.

Proposition A.53 (Lifts of Smooth Maps are Smooth). *If $\pi \colon \widetilde{M} \to M$ is a smooth covering map, B is a smooth manifold with or without boundary, and $F \colon B \to M$ is a smooth map, then every lift of F is smooth.*

Proof. Since π is a smooth submersion, every lift $\widetilde{F} \colon B \to \widetilde{M}$ can be written locally as $\widetilde{F} = \sigma \circ F$, where σ is a smooth local section of π (see Thm. A.17).   □

Proposition A.54 (Lifting Properties of Covering Maps). *Suppose $\pi \colon \widetilde{M} \to M$ is a covering map.*

(a) UNIQUE LIFTING PROPERTY [LeeTM, Thm. 11.12] *If B is a connected topological space and $F \colon B \to M$ is a continuous map, then any two lifts of F that agree at one point are identical.*
(b) PATH LIFTING PROPERTY [LeeTM, Cor. 11.14] *Suppose $f \colon [0,1] \to M$ is a continuous path. For every $\widetilde{p} \in \pi^{-1}(f(0))$, there exists a unique lift $\widetilde{f} \colon [0,1] \to \widetilde{M}$ of f such that $\widetilde{f}(0) = \widetilde{p}$.*
(c) MONODROMY THEOREM [LeeTM, Thm. 11.15] *Suppose $f, g \colon [0,1] \to M$ are path-homotopic paths and $\widetilde{f}, \widetilde{g} \colon [0,1] \to \widetilde{M}$ are their lifts starting at the same point. Then $\widetilde{f}$ and $\widetilde{g}$ are path-homotopic and $\widetilde{f}(1) = \widetilde{g}(1)$.*

Theorem A.55 (Injectivity Theorem). [LeeTM, Thm. 11.16] *If $\pi \colon \widetilde{M} \to M$ is a covering map, then for each point $\widetilde{x} \in \widetilde{M}$, the induced fundamental group homomorphism $\pi_* \colon \pi_1(\widetilde{M}, \widetilde{x}) \to \pi_1(M, \pi(\widetilde{x}))$ is injective.*

Theorem A.56 (Lifting Criterion). [LeeTM, Thm. 11.18] *Suppose $\pi \colon \widetilde{M} \to M$ is a covering map, B is a connected and locally path-connected topological space, and $F \colon B \to M$ is a continuous map. Given $b \in B$ and $\widetilde{x} \in \widetilde{M}$ such that $\pi(\widetilde{x}) = F(b)$, the map F has a lift to $\widetilde{M}$ if and only if $F_*\big(\pi_1(B, b)\big) \subseteq \pi_*\big(\pi_1(\widetilde{M}, \widetilde{x})\big)$.*

Corollary A.57 (Lifting Maps from Simply Connected Spaces). [LeeTM, Cor. 11.19] *Suppose* $\pi: \widetilde{M} \to M$ *and* $F: B \to M$ *satisfy the hypotheses of Theorem A.56, and in addition* B *is simply connected. Then every continuous map* $F: B \to M$ *has a lift to* $\widetilde{M}$*. Given any* $b \in B$*, the lift can be chosen to take* b *to any point in the fiber over* $F(b)$*.*

Corollary A.58 (Covering Map Homeomorphism Criterion). *A covering map* $\pi: \widetilde{M} \to M$ *is a homeomorphism if and only if the induced homomorphism* $\pi_*: \pi_1(\widetilde{M}, \widetilde{x}) \to \pi_1(M, \pi(\widetilde{x}))$ *is surjective for some (hence every)* $\widetilde{x} \in \widetilde{M}$*. A smooth covering map is a diffeomorphism if and only if the induced homomorphism is surjective.*

Proof. By Theorem A.56, the hypothesis implies that the identity map $\mathrm{Id}: M \to M$ has a lift $\widetilde{\mathrm{Id}}: M \to \widetilde{M}$, which in this case is a continuous inverse for π. If π is a smooth covering map, then the lift is also smooth.   □

Corollary A.59 (Coverings of Simply Connected Spaces). [LeeTM, Cor. 11.33] *If* M *is a simply connected manifold with or without boundary, then every covering of* M *is a homeomorphism, and if* M *is smooth, every smooth covering is a diffeomorphism.*

Proposition A.60 (Existence of a Universal Covering Manifold). [LeeSM, Cor. 4.43] *If* M *is a connected smooth manifold, then there exist a simply connected smooth manifold* $\widetilde{M}$*, called the **universal covering manifold of** M, and a smooth covering map* $\pi: \widetilde{M} \to M$*. It is unique in the sense that if* $\widetilde{M}'$ *is any other simply connected smooth manifold that admits a smooth covering map* $\pi': \widetilde{M}' \to M$*, then there exists a diffeomorphism* $\Phi: \widetilde{M} \to \widetilde{M}'$ *such that* $\pi' \circ \Phi = \pi$*.*

Proposition A.61. [LeeTM, Cor. 11.31] *With* $\pi: \widetilde{M} \to M$ *as in the previous proposition, each fiber of* π *has the same cardinality as the fundamental group of* M*.*

▶ **Exercise A.62.** Suppose $\pi: \widetilde{M} \to M$ is a covering map. Show that $\widetilde{M}$ is compact if and only if M is compact and π is a finite-sheeted covering.

We will revisit smooth covering maps at the end of Appendix C.

Appendix B
Review of Tensors

Of all the constructions in smooth manifold theory, the ones that play the most fundamental roles in Riemannian geometry are *tensors* and *tensor fields*. Most of the technical machinery of Riemannian geometry is built up using tensors; indeed, Riemannian metrics themselves are tensor fields. This appendix offers a brief review of their definitions and properties. For a more detailed exposition of the material summarized here, see [LeeSM].

Tensors on a Vector Space

We begin with tensors on a finite-dimensional vector space. There are many ways of constructing tensors on a vector space, but for simplicity we will stick with the most concrete description, as real-valued multilinear functions. The simplest tensors are just linear functionals, also called *covectors*.

Covectors

Let V be an n-dimensional vector space (all of our vector spaces are assumed to be real). When we work with bases for V, it is usually important to consider **ordered bases**, so will assume that each basis comes endowed with a specific ordering. We use parentheses to denote ordered k-tuples, and braces to denote unordered ones, so an ordered basis is designated by either $(b_1, \ldots, b_n)$ or (b_i), and the corresponding unordered basis by $\{b_1, \ldots, b_n\}$ or $\{b_i\}$.

The **dual space of V**, denoted by V^*, is the space of linear maps from V to $\mathbb{R}$. Elements of the dual space are called **covectors** or **linear functionals on V**. Under the operations of pointwise addition and multiplication by constants, V^* is a vector space.

Suppose $(b_1, \ldots, b_n)$ is an ordered basis for V. For each $i = 1, \ldots, n$, define a covector $\beta^i \in V^*$ by

© Springer International Publishing AG 2018
J. M. Lee, *Introduction to Riemannian Manifolds*, Graduate Texts in Mathematics 176, https://doi.org/10.1007/978-3-319-91755-9

$$\beta^i(b_j) = \delta^i_j,$$

where δ^i_j is the **Kronecker delta symbol**, defined by

$$\delta^i_j = \begin{cases} 1 & \text{if } i = j, \\ 0 & \text{if } i \neq j. \end{cases} \tag{B.1}$$

It is a standard exercise to prove that $(\beta^1, \ldots, \beta^n)$ is a basis for V^*, called the **dual basis to (b_i)**. Therefore, V^* is finite-dimensional, and its dimension is the same as that of V. Every covector $\omega \in V^*$ can thus be written in terms of the dual basis as

$$\omega = \omega_j \beta^j, \tag{B.2}$$

where the components ω_j are defined by $\omega_j = \omega(b_j)$. The action of ω on an arbitrary vector $v = v^i b_i \in V$ is then given by

$$\omega(v) = \omega_j v^j. \tag{B.3}$$

(Here and throughout the rest of this appendix we use the Einstein summation convention; see p. 375.)

Every vector $v \in V$ uniquely determines a linear functional on V^*, by $\omega \mapsto \omega(v)$. Because we are assuming V to be finite-dimensional, it is straightforward to show that this correspondence gives a canonical (basis-independent) isomorphism between V and V^{**} (the dual space of V^*). Given $\omega \in V^*$ and $v \in V$, we can denote the natural action of ω on v either by the traditional functional notation $\omega(v)$, or by either of the more symmetric notations $\langle \omega, v \rangle$ or $\langle v, \omega \rangle$. The latter notations are meant to emphasize that it does not matter whether we think of the resulting number as the effect of the linear functional ω acting on the vector v, or as the effect of $v \in V^{**}$ acting on the covector ω. Note that when applied to a vector and a covector, this pairing makes sense without any choice of an inner product on V.

Higher-Rank Tensors on a Vector Space

Now we generalize from linear functionals to multilinear ones. If $V_1, \ldots, V_k$ and W are vector spaces, a map $F : V_1 \times \cdots \times V_k \to W$ is said to be **multilinear** if it is linear as a function of each variable separately, when all the others are held fixed:

$$F(v_1, \ldots, av_i + a'v_i', \ldots, v_k) = aF(v_1, \ldots, v_i, \ldots, v_k) + a'F(v_1, \ldots, v_i', \ldots, v_k).$$

Given a finite-dimensional vector space V, a **covariant k-tensor on V** is a multilinear map

$$F : \underbrace{V \times \cdots \times V}_{k \text{ copies}} \to \mathbb{R}.$$

Similarly, a **contravariant k-tensor on V** is a multilinear map

$$F : \underbrace{V^* \times \cdots \times V^*}_{k \text{ copies}} \to \mathbb{R}.$$

We often need to consider tensors of mixed types as well. A **mixed tensor of type** **(k, l)**, also called a **k-contravariant, l-covariant tensor**, is a multilinear map

$$F : \underbrace{V^* \times \cdots \times V^*}_{k \text{ copies}} \times \underbrace{V \times \cdots \times V}_{l \text{ copies}} \to \mathbb{R}.$$

Actually, in many cases it is necessary to consider real-valued multilinear functions whose arguments consist of k covectors and l vectors, but not necessarily in the order implied by the definition above; such an object is still called a tensor of type (k, l). For any given tensor, we will make it clear which arguments are vectors and which are covectors.

The spaces of tensors on V of various types are denoted by

$$T^k(V^*) = \{\text{covariant } k\text{-tensors on } V\};$$

$$T^k(V) = \{\text{contravariant } k\text{-tensors on } V\};$$

$$T^{(k,l)}(V) = \{\text{mixed } (k,l)\text{-tensors on } V\}.$$

The **rank** of a tensor is the number of arguments (vectors and/or covectors) it takes. By convention, a 0-tensor is just a real number. (You should be aware that the notation conventions for describing the spaces of covariant, contravariant, and mixed tensors are not universally agreed upon. Be sure to look closely at each author's conventions.)

Tensor Products

There is a natural product, called the **tensor product**, linking the various tensor spaces over V: if $F \in T^{(k,l)}(V)$ and $G \in T^{(p,q)}(V)$, the tensor $F \otimes G \in T^{(k+p,l+q)}(V)$ is defined by

$$F \otimes G(\omega^1, \ldots, \omega^{k+p}, v_1, \ldots, v_{l+q})$$
$$= F(\omega^1, \ldots, \omega^k, v_1, \ldots, v_l) G(\omega^{k+1}, \ldots, \omega^{k+p}, v_{l+1}, \ldots, v_{l+q}).$$

The tensor product is associative, so we can unambiguously form tensor products of three or more tensors on V. If (b_i) is a basis for V and (β^j) is the associated dual basis, then a basis for $T^{(k,l)}(V)$ is given by the set of all tensors of the form

$$b_{i_1} \otimes \cdots \otimes b_{i_k} \otimes \beta^{j_1} \otimes \cdots \otimes \beta^{j_l}, \tag{B.4}$$

as the indices i_p, j_q range from 1 to n. These tensors act on basis elements by

$$b_{i_1} \otimes \cdots \otimes b_{i_k} \otimes \beta^{j_1} \otimes \cdots \otimes \beta^{j_l} (\beta^{s_1}, \ldots, \beta^{s_k}, b_{r_1}, \ldots, b_{r_l}) = \delta_{i_1}^{s_1} \cdots \delta_{i_k}^{s_k} \delta_{r_1}^{j_1} \cdots \delta_{r_l}^{j_l}.$$

It follows that $T^{(k,l)}(V)$ has dimension n^{k+l}, where $n = \dim V$. Every tensor $F \in T^{(k,l)}(V)$ can be written in terms of this basis (using the summation convention) as

$$F = F^{i_1 \ldots i_k}_{j_1 \ldots j_l} \, b_{i_1} \otimes \cdots \otimes b_{i_k} \otimes \beta^{j_1} \otimes \cdots \otimes \beta^{j_l}, \tag{B.5}$$

where

$$F^{i_1 \ldots i_k}_{j_1 \ldots j_l} = F\left(\beta^{i_1}, \ldots, \beta^{i_k}, b_{j_1}, \ldots, b_{j_l}\right).$$

If the arguments of a mixed tensor F occur in a nonstandard order, then the horizontal as well as vertical positions of the indices are significant and reflect which arguments are vectors and which are covectors. For example, if A is a $(1,2)$-tensor whose first argument is a vector, second is a covector, and third is a vector, its basis expression would be written

$$A = A_i{}^j{}_k \, \beta^i \otimes b_j \otimes \beta^k,$$

where

$$A_i{}^j{}_k = A\left(b_i, \beta^j, b_k\right). \tag{B.6}$$

There are obvious identifications among some of these tensor spaces:

$$\begin{aligned}
T^{(0,0)}(V) &= T^0(V) = T^0(V^*) = \mathbb{R}, \\
T^{(1,0)}(V) &= T^1(V) = V, \\
T^{(0,1)}(V) &= T^1(V^*) = V^*, \\
T^{(k,0)}(V) &= T^k(V), \\
T^{(0,k)}(V) &= T^k(V^*).
\end{aligned} \tag{B.7}$$

A less obvious, but extremely important, identification is the following:

$$T^{(1,1)}(V) \cong \operatorname{End}(V),$$

where $\operatorname{End}(V)$ denotes the space of linear maps from V to itself (also called the **endomorphisms of V**). This is a special case of the following proposition.

Proposition B.1. *Let V be a finite-dimensional vector space. There is a natural (basis-independent) isomorphism between $T^{(k+1,l)}(V)$ and the space of multilinear maps*

$$\underbrace{V^* \times \cdots \times V^*}_{k \text{ copies}} \times \underbrace{V \times \cdots \times V}_{l \text{ copies}} \to V.$$

▶ **Exercise B.2.** Prove Proposition B.1. [Hint: In the special case $k = 0, l = 1$, consider the map $\Phi \colon \operatorname{End}(V) \to T^{(1,1)}(V)$ defined by letting ΦA be the $(1,1)$-tensor defined by $\Phi A(\omega, v) = \omega(Av)$. The general case is similar.]

We can use the result of Proposition B.1 to define a natural operation called *trace* or **contraction**, which lowers the rank of a tensor by 2. In one special case, it is easy to describe: the operator $\mathrm{tr}\colon T^{(1,1)}(V) \to \mathbb{R}$ is just the trace of F when it is regarded as an endomorphism of V, or in other words the sum of the diagonal entries of any matrix representation of F. Since the trace of a linear endomorphism is basis-independent, this is well defined. More generally, we define $\mathrm{tr}\colon T^{(k+1,l+1)}(V) \to T^{(k,l)}(V)$ by letting $(\mathrm{tr}\, F)(\omega^1,\dots,\omega^k,v_1,\dots,v_l)$ be the trace of the $(1,1)$-tensor

$$F\left(\omega^1,\dots,\omega^k,\cdot,v_1,\dots,v_l,\cdot\right) \in T^{(1,1)}(V).$$

In terms of a basis, the components of $\mathrm{tr}\, F$ are

$$(\mathrm{tr}\, F)^{i_1\dots i_k}_{j_1\dots j_l} = F^{i_1\dots i_k m}_{j_1\dots j_l m}.$$

In other words, just set the last upper and lower indices equal and sum. Even more generally, we can contract a given tensor on any pair of indices as long as one is contravariant and one is covariant. There is no general notation for this operation, so we just describe it in words each time it arises. For example, we can contract the tensor A with components given by (B.6) on its first and second indices to obtain a covariant 1-tensor B whose components are $B_k = A_i{}^i{}_k$.

▶ **Exercise B.3.** Show that the trace on any pair of indices (one upper and one lower) is a well-defined linear map from $T^{(k+1,l+1)}(V)$ to $T^{(k,l)}(V)$.

Symmetric Tensors

There are two classes of tensors that play particularly important roles in differential geometry: the symmetric and alternating tensors. Here we discuss the symmetric ones; we will take up alternating tensors later in this appendix when we discuss differential forms.

If V is a finite-dimensional vector space, a covariant tensor $F \in T^k(V^*)$ is said to be **symmetric** if its value is unchanged by interchanging any pair of arguments:

$$F(v_1,\dots,v_i,\dots,v_j,\dots,v_k) = F(v_1,\dots,v_j,\dots,v_i,\dots,v_k)$$

whenever $1 \le i < j \le k$. It follows immediately that the value is unchanged under *every* rearrangement of the arguments. If we denote the components of F with respect to a basis by $F_{i_1\dots i_k}$, then F is symmetric if and only if its components are unchanged by every permutation of the indices $i_1,\dots,i_k$. Every 0-tensor and every 1-tensor are vacuously symmetric. The set of symmetric k-tensors on V is a linear subspace of $T^k(V^*)$, which we denote by $\Sigma^k(V^*)$.

A tensor product of symmetric tensors is generally not symmetric. However, there is a natural product of symmetric tensors that does yield symmetric tensors. Given a covariant k-tensor F, the **symmetrization of F** is the k-tensor $\mathrm{Sym}\, F$ defined by

$$(\operatorname{Sym}F)(v_1,\ldots,v_k) = \frac{1}{k!}\sum_{\sigma\in S_k} F\left(v_{\sigma(1)},\ldots,v_{\sigma(k)}\right),$$

where S_k is the group of all permutations of $\{1,\ldots,k\}$, called the **symmetric group on k elements**. It is easy to check that $\operatorname{Sym}F$ is symmetric, and $\operatorname{Sym}F = F$ if and only if F is symmetric. Then if F and G are symmetric tensors on V, of ranks k and l, respectively, their **symmetric product** is defined to be the $(k+l)$-tensor FG (denoted by juxtaposition without an explicit product symbol) given by

$$FG = \operatorname{Sym}(F \otimes G).$$

The most important special case is the symmetric product of two 1-tensors, which can be characterized as follows: if ω and η are covectors, then

$$\omega\eta = \tfrac{1}{2}(\omega\otimes\eta + \eta\otimes\omega).$$

(To prove this, just evaluate both sides on an arbitrary pair of vectors (v,w), and use the definition of $\omega\eta$.) If ω is any 1-tensor, the notation ω^2 means the symmetric product $\omega\omega$, which in turn is equal to $\omega\otimes\omega$.

Tensor Bundles and Tensor Fields

On a smooth manifold M with or without boundary, we can perform the same linear-algebraic constructions on each tangent space T_pM that we perform on any vector space, yielding tensors at p. The disjoint union of tensor spaces of a particular type at all points of the manifold yields a vector bundle, called a **tensor bundle**.

The most fundamental tensor bundle is the **cotangent bundle**, defined as

$$T^*M = \coprod_{p\in M} T^*_p M.$$

More generally, the **bundle of (k,l)-tensors** on M is defined as

$$T^{(k,l)}TM = \coprod_{p\in M} T^{(k,l)}(T_pM).$$

As special cases, the **bundle of covariant k-tensors** is denoted by $T^kT^*M = T^{(0,k)}TM$, and the **bundle of contravariant k-tensors** is denoted by $T^kTM = T^{(k,0)}TM$. Similarly, the **bundle of symmetric k-tensors** is

$$\Sigma^kT^*M = \coprod_{p\in M} \Sigma^k(T^*_pM).$$

There are the usual identifications among these bundles that follow from (B.7): for example, $T^1TM = T^{(1,0)}TM = TM$ and $T^1T^*M = T^{(0,1)}TM = \Sigma^1T^*M = T^*M$.

▶ **Exercise B.4.** Show that each tensor bundle is a smooth vector bundle over M, with a local trivialization over every open subset that admits a smooth local frame for TM.

A **tensor field on M** is a section of some tensor bundle over M (see p. 383 for the definition of a section of a bundle). A section of $T^1T^*M = T^{(0,1)}TM$ (a covariant 1-tensor field) is also called a **covector field**. As we do with vector fields, we write the value of a tensor field F at $p \in M$ as F_p or $F|_p$. Because covariant tensor fields are the most common and important tensor fields we work with, we use the following shorthand notation for the space of all smooth covariant k-tensor fields:

$$\mathcal{T}^k(M) = \Gamma(T^kT^*M).$$

The space of smooth 0-tensor fields is just $C^\infty(M)$.

Let $(E_i) = (E_1, \dots, E_n)$ be any smooth local frame for TM over an open subset $U \subseteq M$. Associated with such a frame is the **dual coframe**, which we typically denote by $(\varepsilon^1, \dots, \varepsilon^n)$; these are smooth covector fields satisfying $\varepsilon^i(E_j) = \delta^i_j$. For example, given a coordinate frame $(\partial/\partial x^1, \dots, \partial/\partial x^n)$ over some open subset $U \subseteq M$, the dual coframe is $(dx^1, \dots, dx^n)$, where dx^i is the differential of the coordinate function x^i.

In terms of any smooth local frame (E_i) and its dual coframe (ε^i), the tensor fields $E_{i_1} \otimes \cdots \otimes E_{i_k} \otimes \varepsilon^{j_1} \otimes \cdots \otimes \varepsilon^{j_l}$ form a smooth local frame for $T^{(k,l)}(T^*M)$. In particular, in local coordinates (x^i), a (k,l)-tensor field F has a coordinate expression of the form

$$F = F^{i_1 \dots i_k}_{j_1 \dots j_l} \, \partial_{i_1} \otimes \cdots \otimes \partial_{i_k} \otimes dx^{j_1} \otimes \cdots \otimes dx^{j_l}, \tag{B.8}$$

where each coefficient $F^{i_1 \dots i_k}_{j_1 \dots j_l}$ is a smooth real-valued function on U.

▶ **Exercise B.5.** Suppose $F: M \to T^{(k,l)}TM$ is a rough (k,l)-tensor field. Show that F is smooth on an open set $U \subseteq M$ if and only if whenever $\omega^1, \dots, \omega^k$ are smooth covector fields and $X_1, \dots, X_l$ are smooth vector fields defined on U, the real-valued function $F(\omega^1, \dots, \omega^k, X_1, \dots, X_l)$, defined on U by

$$F(\omega^1, \dots, \omega^k, X_1, \dots, X_l)(p) = F_p(\omega^1|_p, \dots, \omega^k|_p, X_1|_p, \dots, X_l|_p),$$

is smooth.

An important property of tensor fields is that they are multilinear over the space of smooth functions. Suppose $F \in \Gamma(T^{(k,l)}TM)$ is a smooth tensor field. Given smooth covector fields $\omega^1, \dots, \omega^k \in \mathcal{T}^1(M)$ and smooth vector fields $X_1, \dots, X_l \in \mathcal{X}(M)$, Exercise B.5 shows that the function $F(\omega^1, \dots, \omega^k, X_1, \dots, X_l)$ is smooth, and thus F induces a map

$$\tilde{F}: \underbrace{\mathcal{T}^1(M) \times \cdots \times \mathcal{T}^1(M)}_{k \text{ factors}} \times \underbrace{\mathfrak{X}(M) \times \cdots \times \mathfrak{X}(M)}_{l \text{ factors}} \to C^\infty(M).$$

It is easy to check that this map is **multilinear over $C^\infty(M)$**, that is, for all functions $u, v \in C^\infty(M)$ and smooth vector or covector fields α, β,

$$\tilde{F}(\ldots, u\alpha + v\beta, \ldots) = u\tilde{F}(\ldots, \alpha, \ldots) + v\tilde{F}(\ldots, \beta, \ldots).$$

Even more important is the converse: as the next lemma shows, every such map that is multilinear over $C^\infty(M)$ defines a tensor field. (This lemma is stated and proved in [LeeSM] for covariant tensor fields, but the same argument works in the case of mixed tensors.)

Lemma B.6 (Tensor Characterization Lemma). [LeeSM, Lemma 12.24] *A map*

$$\mathcal{F}: \underbrace{\mathcal{T}^1(M) \times \cdots \times \mathcal{T}^1(M)}_{k \text{ factors}} \times \underbrace{\mathfrak{X}(M) \times \cdots \times \mathfrak{X}(M)}_{l \text{ factors}} \to C^\infty(M)$$

is induced by a smooth (k, l)-tensor field as above if and only if it is multilinear over $C^\infty(M)$. Similarly, a map

$$\mathcal{F}: \underbrace{\mathcal{T}^1(M) \times \cdots \times \mathcal{T}^1(M)}_{k \text{ factors}} \times \underbrace{\mathfrak{X}(M) \times \cdots \times \mathfrak{X}(M)}_{l \text{ factors}} \to \mathfrak{X}(M)$$

is induced by a smooth $(k + 1, l)$-tensor field as in Proposition B.1 if and only if it is multilinear over $C^\infty(M)$.

Because of this result, it is common to use the same symbol for both a tensor field and the multilinear map on sections that it defines, and to refer to either of these objects as a tensor field.

Pullbacks of Tensor Fields

Suppose $F: M \to N$ is a smooth map and A is a covariant k-tensor field on N. For every $p \in M$, we define a tensor $dF_p^*(A) \in T^k(T_p^*M)$, called the **pointwise pullback of A by F at p**, by

$$dF_p^*(A)(v_1, \ldots, v_k) = A\big(dF_p(v_1), \ldots, dF_p(v_k)\big)$$

for $v_1, \ldots, v_k \in T_pM$; and we define the **pullback of A by F** to be the tensor field F^*A on M defined by

$$(F^*A)_p = dF_p^*\big(A_{F(p)}\big).$$

Proposition B.7 (Properties of Tensor Pullbacks). [LeeSM, Prop. 12.25] *Suppose $F: M \to N$ and $G: P \to M$ are smooth maps, A and B are covariant tensor fields on N, and f is a real-valued function on N.*

(a) $F^*(fB) = (f \circ F)F^*B$.
(b) $F^*(A \otimes B) = F^*A \otimes F^*B$.
(c) $F^*(A + B) = F^*A + F^*B$.
(d) F^*B is a (continuous) tensor field, and it is smooth if B is smooth.
(e) $(F \circ G)^*B = G^*(F^*B)$.
(f) $(\mathrm{Id}_N)^*B = B$.

If f is a continuous real-valued function (i.e., a 0-tensor field), the pullback F^*f is just the composition $f \circ F$.

The following proposition shows how pullbacks are computed in local coordinates.

Proposition B.8. [LeeSM, Cor. 12.28] *Let $F : M \to N$ be smooth, and let B be a covariant k-tensor field on N. If $p \in M$ and (y^i) are smooth coordinates for N on a neighborhood of $F(p)$, then F^*B has the following expression in a neighborhood of p:*

$$F^* \left(B_{i_1 \ldots i_k} dy^{i_1} \otimes \cdots \otimes dy^{i_k} \right) = \left(B_{i_1 \ldots i_k} \circ F \right) d\left(y^{i_1} \circ F\right) \otimes \cdots \otimes d\left(y^{i_k} \circ F\right).$$

Lie Derivatives of Tensor Fields

We can extend the notion of Lie derivatives to tensor fields. This can be done for mixed tensor fields of any rank, but for simplicity we restrict attention to covariant tensor fields. Suppose X is a smooth vector field on M and θ is its flow. If A is a smooth covariant tensor field on M, the **Lie derivative of A with respect to X** is the smooth covariant tensor field $\mathcal{L}_X A$ defined by

$$(\mathcal{L}_X A)_p = \frac{d}{dt}\bigg|_{t=0} (\theta_t^* A)_p = \lim_{t \to 0} \frac{d(\theta_t)_p^* \left(A_{\theta_t(p)} \right) - A_p}{t}.$$

Proposition B.9. [LeeSM, Thm. 12.32] *Suppose M is a smooth manifold, $X \in \mathfrak{X}(M)$, and A is a smooth covariant k-tensor field on M. For all smooth vector fields $Z_1, \ldots, Z_k$,*

$$(\mathcal{L}_X A)(Z_1, \ldots, Z_k) = X\left(A(Z_1, \ldots, Z_k)\right)$$
$$- A(\mathcal{L}_X Z_1, Z_2, \ldots, Z_k) - A(Z_1, \mathcal{L}_X Z_2, \ldots, Z_k) - \cdots - A(Z_1, \ldots, \mathcal{L}_X Z_k).$$

As we did for vector fields, we say that a covariant tensor field A is **invariant under a flow** θ if $(\theta_t)^*A = A$ wherever it is defined. The next proposition is a tensor analogue of Proposition A.47.

Proposition B.10. [LeeSM, Thm. 12.37] *Let M be a smooth manifold and $X \in \mathfrak{X}(M)$. A smooth covariant tensor field is invariant under the flow of X if and only if its Lie derivative with respect to X is identically zero.*

Differential Forms and Integration

In addition to symmetric tensors, the other class of tensors that play a special role in differential geometry is that of *alternating tensors*, which we now define. Let V be a finite-dimensional vector space. If F is a covariant k-tensor on V, we say that F is **alternating** if its value changes sign whenever two different arguments are interchanged:

$$F(v_1, \ldots, v_i, \ldots, v_j, \ldots, v_k) = -F(v_1, \ldots, v_j, \ldots, v_i, \ldots, v_k).$$

The set of alternating covariant k-tensors on V is a linear subspace of $T^k(V^*)$, denoted by $\Lambda^k(V^*)$. If $F \in \Lambda^k(V^*)$, then the effect of an arbitrary permutation of its arguments is given by

$$F(v_{\sigma(1)}, \ldots, v_{\sigma(k)}) = (\operatorname{sgn} \sigma) F(v_1, \ldots, v_k),$$

where sgn σ represents the **sign** of the permutation $\sigma \in S_k$, which is $+1$ if σ is even (i.e., can be written as a composition of an even number of transpositions), and -1 if σ is odd. The components of F with respect to any basis change sign similarly under a permutation of the indices. All 0-tensors and 1-tensors are alternating.

Analogously to the symmetrization operator defined above, if F is any covariant k-tensor on V, we define the **alternation of F** by

$$(\operatorname{Alt} F)(v_1, \ldots, v_k) = \frac{1}{k!} \sum_{\sigma \in S_k} (\operatorname{sgn} \sigma) F\left(v_{\sigma(1)}, \ldots, v_{\sigma(k)}\right). \tag{B.9}$$

Again, it is easy to check that Alt F is alternating, and Alt $F = F$ if and only if F is alternating. Given $\omega \in \Lambda^k(V^*)$ and $\eta \in \Lambda^l(V^*)$, we define their **wedge product** by

$$\omega \wedge \eta = \frac{(k+l)!}{k! \, l!} \operatorname{Alt}(\omega \otimes \eta).$$

It is immediate that $\omega \wedge \eta$ is an alternating $(k+l)$-tensor. The wedge product is easily seen to be bilinear and **anticommutative**, which means that

$$\omega \wedge \eta = (-1)^{kl} \eta \wedge \omega, \qquad \text{for } \omega \in \Lambda^k(V^*) \text{ and } \eta \in \Lambda^l(V^*).$$

It is also the case, although not so easy to see, that it is associative; see [LeeSM, Prop. 14.11] for a proof. If (b_i) is a basis for V and (β^i) is the dual basis, the wedge products $\beta^{i_1} \wedge \cdots \wedge \beta^{i_k}$, as $(i_1, \ldots, i_k)$ range over strictly increasing multi-indices, form a basis for $\Lambda^k(V^*)$, which therefore has dimension $\binom{n}{k} = n!/(k!(n-k)!)$, where $n = \dim V$. In terms of these basis elements, the wedge product satisfies

$$\beta^{i_1} \wedge \cdots \wedge \beta^{i_k}(b_{j_1}, \ldots, b_{j_k}) = \det\left(\delta^{i_p}_{j_q}\right).$$

More generally, if $\omega^1, \ldots, \omega^k$ are any covectors and $v_1, \ldots, v_k$ are any vectors, then

$$\omega^1 \wedge \cdots \wedge \omega^k (v_1, \ldots, v_k) = \det\left(\omega^i(v_j)\right). \tag{B.10}$$

The wedge product can be defined analogously for *contravariant* alternating tensors $\alpha \in \Lambda^p(V)$, $\beta \in \Lambda^q(V)$, simply by regarding α, β, and $\alpha \wedge \beta$ as alternating multilinear functionals on V^*.

The convention we use for the wedge product is referred to in [LeeSM] as the **determinant convention**. There is another convention that is also in common use, the **Alt convention**, which amounts to multiplying the right-hand side of (B.10) by a factor of $1/k!$. The choice of which definition to use is a matter of taste, though there are various reasons to justify each choice depending on the context. The determinant convention is most common in introductory differential geometry texts, and is used, for example, in [Boo86, Cha06, dC92, LeeJeff09, LeeSM, Pet16, Spi79, Tu11]. The Alt convention is used in [KN96] and is more common in complex differential geometry.

Given an alternating k-tensor $\omega \in \Lambda^k(V^*)$ and a vector $v \in V$, we define an alternating $(k-1)$-tensor $v \lrcorner \omega$ by

$$(v \lrcorner \omega)(w_1, \ldots, w_{k-1}) = \omega(v, w_1, \ldots, w_{k-1}).$$

The operation $\omega \mapsto v \lrcorner \omega$ is known as **interior multiplication by v**, and is also denoted by $\omega \mapsto i_v \omega$. By convention, $v \lrcorner \omega = 0$ when ω is a 0-tensor. Direct computation shows that $i_v \circ i_v = 0$, and i_v satisfies the following product rule for an alternating k-tensor ω and an alternating l-tensor η:

$$i_v(\omega \wedge \eta) = (i_v \omega) \wedge \eta + (-1)^k \omega \wedge (i_v \eta). \tag{B.11}$$

If M is a smooth manifold with or without boundary, the subbundle of $T^k T^* M$ consisting of alternating tensors is denoted by $\Lambda^k T^* M$, and an alternating tensor field on M is called a **differential k-form**, or just a **k-form**. The space of all smooth k-forms is denoted by $\Omega^k(M) = \Gamma(\Lambda^k T^* M)$. Because every 1-tensor field is alternating, $\Omega^1(M)$ is the same as the space $\mathcal{T}^1(M)$ of smooth covector fields.

Exterior Derivatives

The most important operation on differential forms is the *exterior derivative*, defined as follows. Suppose M is a smooth n-manifold with or without boundary, and (x^i) are any smooth local coordinates on M. A smooth k-form ω can be expressed in these coordinates as

$$\omega = \sum_{j_1 < \cdots < j_k} \omega_{j_1 \ldots j_k} \, dx^{j_1} \wedge \cdots \wedge dx^{j_k},$$

and then we define the **exterior derivative of ω**, denoted by $d\omega$, to be the $(k+1)$-form defined in coordinates by

$$d\omega = \sum_{j_1 < \cdots < j_k} \sum_{i=1}^{n} \frac{\partial \omega_{j_1 \ldots j_k}}{\partial x^i} dx^i \wedge dx^{j_1} \wedge \cdots \wedge dx^{j_k}.$$

For a 0-form f (a smooth real-valued function), this reduces to

$$df = \sum_{i=1}^{n} \frac{\partial f}{\partial x^i} dx^i,$$

which is exactly the *differential of f* defined by (A.4).

Proposition B.11 (Properties of Exterior Differentiation). [LeeSM, Thm. 14.24]
Suppose M is a smooth n-manifold with or without boundary.

(a) *For each $k = 0, \ldots, n$, the operator $d : \Omega^k(M) \to \Omega^{k+1}(M)$ is well defined, independently of coordinates.*
(b) *d is linear over $\mathbb{R}$.*
(c) *$d \circ d \equiv 0$.*
(d) *If $\omega \in \Omega^k(M)$ and $\eta \in \Omega^l(M)$, then*

$$d(\omega \wedge \eta) = d\omega \wedge \eta + (-1)^k \omega \wedge d\eta.$$

Proposition B.12 (Exterior Derivative of a 1-form). [LeeSM, Prop. 14.29] *If ω is a smooth 1-form and X, Y are vector fields, then*

$$d\omega(X, Y) = X(\omega(Y)) - Y(\omega(X)) - \omega([X, Y]).$$

Proposition B.13 (Naturality of the Exterior Derivative). [LeeSM, Prop. 14.26] *If $F : M \to N$ is a smooth map, then for each k, the pullback map $F^* : \Omega^k(N) \to \Omega^k(M)$ commutes with d: for all $\omega \in \Omega^k(N)$,*

$$F^*(d\omega) = d(F^*\omega). \tag{B.12}$$

A smooth differential form $\omega \in \Omega^k(M)$ is *closed* if $d\omega = 0$, and *exact* if there exists a smooth $(k-1)$-form η on M such that $\omega = d\eta$. The fact that $d \circ d = 0$ implies that every exact form is closed, but the converse is not true in general. However, the next lemma gives an important special case in which it is true. If V is a vector space, a subset $S \subseteq V$ is said to be *star-shaped* with respect to a point $x \in S$ if for every $y \in S$, the line segment from x to y is contained in S.

Lemma B.14 (The Poincaré Lemma). [LeeSM, Thm. 17.14] *If U is a star-shaped open subset of $\mathbb{R}^n$, then for $k \geq 1$, every closed k-form on U is exact.*

Orientations

If V is a finite-dimensional vector space, an *orientation of V* is an equivalence class of ordered bases for V, where two ordered bases are considered equivalent if the transition matrix that expresses one basis in terms of the other has positive determinant. Every vector space has exactly two orientations. Once an orientation is chosen, a basis is said to be *positively oriented* if it belongs to the chosen orientation, and *negatively oriented* if not. The *standard orientation of* $\mathbb{R}^n$ is the one determined by the *standard basis* $(e_1,\dots,e_n)$, where e_i is the vector $(0,\dots,1,\dots,0)$ with 1 in the ith place and zeros elsewhere.

If M is a smooth manifold, an *orientation for* M is a choice of orientation for each tangent space that is continuous in the sense that in a neighborhood of every point there is a (continuous) local frame that determines the given orientation at each point of the neighborhood. If there exists an orientation for M, we say that M is *orientable*. An *oriented manifold* is a smooth orientable manifold together with a choice of orientation. If M is an oriented n-manifold, then a smooth coordinate chart $(U,(x^i))$ is said to be an *oriented chart* if the coordinate frame $(\partial/\partial x^1,\dots,\partial/\partial x^n)$ is positively oriented at each point. Exactly the same definitions apply to manifolds with boundary.

Proposition B.15 (Orientation Determined by an n-Form). [LeeSM, Prop. 15.5] *Let M be an n-manifold with or without boundary. Every nonvanishing n-form $\mu \in \Omega^n(M)$ determines a unique orientation of M by declaring an ordered basis $(b_1,\dots,b_n)$ for T_pM to be positively oriented if and only if $\mu_p(b_1,\dots,b_n) > 0$. Conversely, if M is oriented, there is a smooth nonvanishing n-form that determines the orientation in this way.*

Because of this proposition, a nonvanishing n-form on a smooth n-manifold is called an *orientation form*. If in addition M is oriented and μ determines the given orientation, we say that μ is *positively oriented*.

Suppose M and N are both smooth n-manifolds with or without boundary and $F: M \to N$ is a local diffeomorphism. If N is oriented, we define the ***pullback orientation on M induced by F*** to be the orientation determined by $F^*\mu$, where μ is any positively oriented orientation form for N. If both M and N are oriented, we say that F is *orientation-preserving* if the pullback orientation is equal to the given orientation on M, and *orientation-reversing* if the pullback orientation is the opposite orientation.

Proposition B.16 (Orientation of a Hypersurface). [LeeSM, Prop. 15.21] *Suppose M is an oriented smooth n-manifold with or without boundary, $S \subseteq M$ is a smooth immersed hypersurface, and N is a continuous vector field along S (i.e., a continuous map $N: S \to TM$ such that $N_p \in T_pM$ for each $p \in S$). If N is nowhere tangent to S, then S has a unique orientation determined by the $(n-1)$-form $\iota^*(N \lrcorner \mu)$, where $\iota: S \hookrightarrow M$ is inclusion and μ is any positively oriented orientation form for M.*

A special case of the preceding proposition occurs when M is a manifold with boundary and the hypersurface in question is the boundary of M. In that case we declare a vector field N along ∂M to be *outward-pointing* if for each $p \in \partial M$, there is a smooth curve $\gamma: (-\varepsilon,0] \to M$ such that $\gamma(0) = p$ and $\gamma'(0) = N_p$, and $N_p \notin T_p(\partial M)$. We can always construct a global smooth outward-pointing vector field by taking $-\partial/\partial x^n$ in boundary coordinates in a neighborhood of each boundary point, and gluing together with a partition of unity. Since an outward-pointing vector field is nowhere tangent to ∂M, we have the following proposition.

Proposition B.17 (Induced Orientation on a Boundary). [LeeSM, Prop. 15.24] *If M is an oriented smooth manifold with boundary, then ∂M is orientable, and it has a canonical orientation (called the **induced orientation** or **Stokes orientation**) determined by $\iota^*(N \lrcorner \mu)$, where $\iota: \partial M \hookrightarrow M$ is inclusion, N is any outward-pointing vector field along ∂M, and μ is any positively oriented orientation form for M.*

Proposition B.18 (Orientation Covering Theorem). [LeeSM, Thm. 15.41] *If M is a connected, nonorientable smooth manifold, then there exist an oriented smooth manifold $\widehat{M}$ and a two-sheeted smooth covering map $\widehat{\pi}: \widehat{M} \to M$, called the **orientation covering** of M.*

Corollary B.19. *Every simply connected smooth manifold is orientable.*

Proof. Corollary A.59 shows that a simply connected manifold does not admit two-sheeted covering maps. □

Integration of Differential Forms

Suppose first that ω is a (continuous) n-form on an open subset $U \subseteq \mathbb{R}^n$ or $\mathbb{R}^n_+$. It can be written $\omega = f \, dx^1 \wedge \cdots \wedge dx^n$ for some continuous real-valued function f, and we define the ***integral of ω over U*** to be the ordinary multiple integral

$$\int_U \omega = \int_U f\left(x^1, \ldots, x^n\right) dx^1 \cdots dx^n,$$

provided the integral is well defined. This will always be the case, for example, if f is continuous and has compact support in U.

In general, if ω is a compactly supported n-form on a smooth n-manifold M with or without boundary, we define the ***integral of ω over M*** by choosing finitely many oriented smooth coordinate charts $\{U_i\}_{i=1}^k$ whose domains cover the support of ω, together with a smooth partition of unity $\{\psi_1\}_{i=1}^k$ subordinate to this cover, and defining

$$\int_M \omega = \sum_{i=1}^k \int_{\widehat{U}_i} \left(\varphi_i^{-1}\right)^*(\psi_i \omega),$$

where $\widehat{U}_i = \varphi_i(U_i)$, and the integrals on the right-hand side are defined as above. Proposition 16.5 in [LeeSM] shows that this definition does not depend on the choice of oriented charts or partition of unity.

Proposition B.20 (Properties of Integrals of Forms). [LeeSM, Prop. 16.6] *Suppose M and N are nonempty oriented smooth n-manifolds with or without boundary, and ω, η are compactly supported n-forms on M.*

(a) LINEARITY: *If $a,b \in \mathbb{R}$, then*

$$\int_M a\omega + b\eta = a \int_M \omega + b \int_M \eta.$$

(b) ORIENTATION REVERSAL: *If $-M$ denotes M with the opposite orientation, then*

$$\int_{-M} \omega = -\int_M \omega.$$

(c) POSITIVITY: *If ω is a positively oriented orientation form, then $\int_M \omega > 0$.*

(d) DIFFEOMORPHISM INVARIANCE: *If $F : N \to M$ is an orientation-preserving or orientation-reversing diffeomorphism, then*

$$\int_M \omega = \begin{cases} \displaystyle\int_N F^*\omega & \text{if } F \text{ is orientation-preserving,} \\ -\displaystyle\int_N F^*\omega & \text{if } F \text{ is orientation-reversing.} \end{cases}$$

Theorem B.21 (Stokes's Theorem). [LeeSM, Thm. 16.11] *If M is an oriented smooth n-manifold with boundary and ω is a compactly supported smooth $(n-1)$-form on M, then*

$$\int_M d\omega = \int_{\partial M} \omega. \tag{B.13}$$

In the statement of this theorem, ∂M is understood to have the Stokes orientation, and the right-hand side is interpreted as the integral of the pullback of ω to ∂M.

The following special case is frequently useful.

Corollary B.22. [LeeSM, Cor. 16.13] *Suppose M is a compact oriented smooth n-manifold (without boundary). Then the integral of every exact n-form on M is zero.*

Densities

On an oriented n-manifold with or without boundary, n-forms are the natural objects to integrate. But in order to integrate on a nonorientable manifold, we need closely related objects called *densities*.

If V is an n-dimensional real vector space, a ***density on*** V is a function

$$\mu : \underbrace{V \times \cdots \times V}_{n \text{ copies}} \to \mathbb{R}$$

satisfying the following formula for every linear map $T : V \to V$:

$$\mu(Tv_1, \ldots, Tv_n) = |\det T| \, \mu(v_1, \ldots, v_n). \tag{B.14}$$

A density μ is said to be **positive** if $\mu(v_1,\dots,v_n) > 0$ whenever $(v_1,\dots,v_n)$ is a basis of V; it is clear from (B.14) that if this is true for some basis, then it is true for every one. Every nonzero alternating n-tensor μ determines a positive density $|\mu|$ by the formula

$$|\mu|(v_1,\dots,v_n) = |\mu(v_1,\dots,v_n)|.$$

The set $\mathcal{D}(V)$ of all densities on V is a 1-dimensional vector space, spanned by $|\mu|$ for any nonzero alternating n-tensor μ.

When M is a smooth manifold with or without boundary, the set

$$\mathcal{D}M = \coprod_{p \in M} \mathcal{D}(T_p M)$$

is called the **density bundle of M**. It is a smooth rank-1 vector bundle, with $\left| dx^1 \wedge \cdots \wedge dx^n \right|$ as a smooth local frame over any smooth coordinate chart. A **density on M** is a (smooth) section μ of $\mathcal{D}M$; in any local coordinates, it can be written

$$\mu = u \left| dx^1 \wedge \cdots \wedge dx^n \right|$$

for some locally defined smooth function u.

Under smooth maps, densities pull back in the same way as differential forms: Suppose $F \colon M \to N$ is a smooth map between n-manifolds with or without boundary, and μ is a density on N. The **pullback of μ** is the density $F^* \mu$ on M defined by

$$(F^* \mu)_p(v_1,\dots,v_n) = \mu_{F(p)}(dF_p(v_1),\dots,dF_p(v_n)).$$

In coordinates, this satisfies

$$F^* \left(u \left| dy^1 \wedge \cdots \wedge dy^n \right| \right) = (u \circ F) \left| \det DF \right| \left| dx^1 \wedge \cdots \wedge dx^n \right|, \qquad \text{(B.15)}$$

where DF represents the matrix of partial derivatives of F in these coordinates.

Because the pullback formula (B.15) is exactly analogous to the change of variables formula for multiple integrals, we can define the integral of a compactly supported density on a smooth manifold with or without boundary in exactly the same way as the integral of an n-form is defined on an oriented manifold, except that now there is no need to have an orientation. Thus if $\mu = u \left| dx^1 \cdots dx^n \right|$ is a smooth density that is compactly supported in an open set $V \subseteq \mathbb{R}^n$, we define the integral of μ by

$$\int_V \mu = \int_V u \, dx^1 \cdots dx^n.$$

It follows from (B.15) that the value of this integral is diffeomorphism-invariant, so we can define the integral of a compactly supported density on a smooth manifold with or without boundary by breaking it up with a partition of unity and integrating each term in coordinates as above.

Appendix C

Review of Lie Groups

Lie groups play many important roles in Riemannian geometry, both as examples of Riemannian manifolds and as isometry groups of other manifolds. In this appendix, we summarize the main facts about Lie groups that are used in this book. For details, consult [LeeSM], especially Chapters 7, 8, 20, and 21.

Definitions and Properties

A *Lie group* is a smooth manifold G that is also a group in the algebraic sense, with the property that the multiplication map $m: G \times G \to G$ given by $m(\varphi_1, \varphi_2) = \varphi_1 \varphi_2$ and the inversion map $i: G \to G$ given by $i(\varphi) = \varphi^{-1}$ are smooth. For generic Lie groups, we usually denote the identity element by e, unless there is a more specific common notation for a particular group.

Given a Lie group G, each $\varphi \in G$ defines a map $L_\varphi: G \to G$, called *left translation*, by $L_\varphi(\varphi') = \varphi \varphi'$. It is a diffeomorphism with inverse $L_{\varphi^{-1}}$. Similarly, *right translation* $R_\varphi: G \to G$ is the diffeomorphism $R_\varphi(\varphi') = \varphi' \varphi$, with inverse $R_{\varphi^{-1}}$.

A subgroup $H \subseteq G$ that is endowed with a topology and smooth structure making it into a Lie group and an immersed or embedded submanifold of G is called a *Lie subgroup of G*.

If G and H are Lie groups, a group homomorphism $F: G \to H$ that is also a smooth map is called a *Lie group homomorphism*. It is a *Lie group isomorphism* if it has an inverse that is also a Lie group homomorphism. A Lie group isomorphism from G to itself is called an *automorphism of G*.

Proposition C.1. [LeeSM, Thms. 7.5 & 21.27] *Suppose $F: G \to H$ is a Lie group homomorphism.*

(a) *F has constant rank.*
(b) *The kernel of F is an embedded Lie subgroup of G.*
(c) *The image of F is an immersed Lie subgroup of H.*

© Springer International Publishing AG 2018
J. M. Lee, *Introduction to Riemannian Manifolds*, Graduate Texts in Mathematics 176, https://doi.org/10.1007/978-3-319-91755-9

Example C.2 (Lie Groups).

(a) Every countable group with the discrete topology is a zero-dimensional Lie group, called a ***discrete Lie group***.

(b) $\mathbb{R}^n$ and $\mathbb{C}^n$ are Lie groups under addition.

(c) The sets $\mathbb{R}^*$, $\mathbb{R}^+$, and $\mathbb{C}^*$ of nonzero real numbers, positive real numbers, and nonzero complex numbers, respectively, are Lie groups under multiplication.

(d) The unit circle $\mathbb{S}^1 \subseteq \mathbb{C}$ is a 1-dimensional Lie group under complex multiplication, called the ***circle group***.

(e) Given Lie groups $G_1, \dots, G_k$, their ***direct product group*** is the Lie group whose underlying manifold is $G_1 \times \cdots \times G_k$, and whose multiplication is given by $(g_1, \dots, g_k)(g_1', \dots, g_k') = (g_1 g_1', \dots, g_k g_k')$. For example, the ***n-torus*** is the n-fold product group $\mathbb{T}^n = \mathbb{S}^1 \times \cdots \times \mathbb{S}^1$. The 2-torus is often simply called ***the torus***.

(f) The ***general linear groups*** $\mathrm{GL}(n, \mathbb{R})$ and $\mathrm{GL}(n, \mathbb{C})$, consisting of all invertible $n \times n$ real or complex matrices, respectively, are Lie groups under matrix multiplication. The identity element in both cases is the $n \times n$ identity matrix, denoted by I_n. More generally, if V is an n-dimensional real or complex vector space, the group $\mathrm{GL}(V)$ of invertible linear maps from V to itself is a Lie group isomorphic to $\mathrm{GL}(n, \mathbb{R})$ or $\mathrm{GL}(n, \mathbb{C})$.

(g) The real and complex ***special linear groups*** are the subgroups $\mathrm{SL}(n, \mathbb{R}) \subseteq \mathrm{GL}(n, \mathbb{R})$ and $\mathrm{SL}(n, \mathbb{C}) \subseteq \mathrm{GL}(n, \mathbb{C})$ consisting of matrices of determinant 1.

(h) The ***orthogonal group*** $\mathrm{O}(n) \subseteq \mathrm{GL}(n, \mathbb{R})$ is the subgroup consisting of orthogonal matrices, those that satisfy $A^T A = I_n$, where A^T is the transpose of A. The ***special orthogonal group*** is the subgroup $\mathrm{SO}(n) = \mathrm{O}(n) \cap \mathrm{SL}(n, \mathbb{R})$.

(i) Similarly, the ***unitary group*** $\mathrm{U}(n) \subseteq \mathrm{GL}(n, \mathbb{C})$ is the subgroup consisting of complex unitary matrices, those that satisfy $A^* A = I_n$, where $A^* = \bar{A}^T$ is the conjugate transpose of A, called its ***adjoint***. The ***special unitary group*** is $\mathrm{SU}(n) = \mathrm{U}(n) \cap \mathrm{SL}(n, \mathbb{C})$.   //

If G is a Lie group and V is a finite-dimensional vector space, a Lie group homomorphism $\rho \colon G \to \mathrm{GL}(V)$ is called a (finite-dimensional) ***representation of G***. It is said to be ***faithful*** if it is injective.

The Lie Algebra of a Lie Group

Suppose G is a Lie group. A vector field X on G is said to be ***left-invariant*** if it is invariant under all left translations, meaning that $(L_g)_* X = X$ for every $g \in G$. Similarly, X is ***right-invariant*** if it is invariant under all right translations.

We denote the set of all smooth left-invariant vector fields on G by $\mathrm{Lie}(G)$. It is a vector subspace of $\mathfrak{X}(G)$ that is closed under Lie brackets (see [LeeSM, Prop. 8.33]). Thus it is an example of a ***Lie algebra***: a vector space $\mathfrak{g}$ endowed with a bilinear operation $[\cdot, \cdot] \colon \mathfrak{g} \times \mathfrak{g} \to \mathfrak{g}$, called the ***bracket***, that is antisymmetric (meaning $[X, Y] = -[Y, X]$) and satisfies the ***Jacobi identity***:

$$[X,[Y,Z]] + [Y,[Z,X]] + [Z,[X,Y]] = 0.$$

With the Lie bracket structure that it inherits from $\mathfrak{X}(G)$, the Lie algebra $\mathrm{Lie}(G)$ is called the *Lie algebra of G*.

Proposition C.3. [LeeSM, Thm. 8.37] *Let G be a Lie group with identity e. The evaluation map $X \mapsto X_e$ is a vector space isomorphism from $\mathrm{Lie}(G)$ to T_eG, so $\mathrm{Lie}(G)$ has the same dimension as G itself.*

If $\mathfrak{g}$ is a Lie algebra, a *Lie subalgebra* of $\mathfrak{g}$ is a vector subspace $\mathfrak{h} \subseteq \mathfrak{g}$ that is closed under Lie brackets, and is thus a Lie algebra with the restriction of the same bracket operation. Given Lie algebras $\mathfrak{g}_1$ and $\mathfrak{g}_2$, a *Lie algebra homomorphism* is a linear map $F: \mathfrak{g}_1 \to \mathfrak{g}_2$ that preserves brackets: $F([X,Y]) = [F(X), F(Y)]$. A *Lie algebra isomorphism* is an invertible Lie algebra homomorphism, and a *Lie algebra automorphism* is a Lie algebra isomorphism from a Lie algebra to itself.

Example C.4 (Lie Algebras).

(a) If M is a smooth positive-dimensional manifold, the space $\mathfrak{X}(M)$ of all smooth vector fields on M is an infinite-dimensional Lie algebra under the Lie bracket.

(b) The vector space $\mathfrak{gl}(n,\mathbb{R})$ of all $n \times n$ real matrices is an n^2-dimensional Lie algebra under the *commutator bracket* $[A, B] = AB - BA$. The canonical identification of both $\mathfrak{gl}(n,\mathbb{R})$ and $\mathrm{Lie}(\mathrm{GL}(n,\mathbb{R}))$ with the tangent space to $\mathrm{GL}(n,\mathbb{R})$ at the identity yields a Lie algebra isomorphism $\mathfrak{gl}(n,\mathbb{R}) \cong \mathrm{Lie}(\mathrm{GL}(n,\mathbb{R}))$ [LeeSM, Prop. 8.41].

(c) Similarly, the space $\mathfrak{gl}(n,\mathbb{C})$ of $n \times n$ complex matrices is a $2n^2$-dimensional (real) Lie algebra under the commutator bracket, isomorphic to the Lie algebra of $\mathrm{GL}(n,\mathbb{C})$ [LeeSM, Prop. 8.48].

(d) The space $\mathfrak{o}(n)$ of skew-symmetric $n \times n$ real matrices is a Lie subalgebra of $\mathfrak{gl}(n,\mathbb{R})$, isomorphic to the Lie algebra of $\mathrm{O}(n)$ [LeeSM, Example 8.47].

(e) If V is a vector space, the space $\mathfrak{gl}(V)$ of all linear maps from V to itself is a Lie algebra under the commutator bracket $[A, B] = A \circ B - B \circ A$. In case V is finite-dimensional, $\mathfrak{gl}(V)$ is isomorphic to the Lie algebra of the Lie group $\mathrm{GL}(V)$.

(f) A Lie algebra in which all brackets are zero is called an *abelian Lie algebra*. Every vector space can be made into an abelian Lie algebra by defining all brackets to be zero. The Lie algebra of a connected Lie group is abelian if and only if the group is abelian (see [LeeSM, Problems 8-25 and 20-7]). //

The following proposition shows that every Lie group homomorphism induces a Lie algebra homomorphism between the respective Lie algebras.

Proposition C.5 (The Induced Lie Algebra Homomorphism). [LeeSM, Thm. 8.44] *Let G and H be Lie groups, and let $\mathfrak{g}$ and $\mathfrak{h}$ be their Lie algebras. Given a Lie group homomorphism $F: G \to H$, there is a unique Lie algebra homomorphism $F_*: \mathfrak{g} \to \mathfrak{h}$, called the **induced Lie algebra homomorphism of F**, with the property that for each $X \in \mathfrak{g}$, the vector field $F_*X \in \mathfrak{h}$ is F-related to X.*

Proposition C.6 (Properties of Induced Homomorphisms). [LeeSM, Prop. 8.45]
Let G, H, K be Lie groups.

(a) $(\mathrm{Id}_G)_* = \mathrm{Id}_{\mathrm{Lie}(G)} \colon \mathrm{Lie}(G) \to \mathrm{Lie}(G)$.
(b) *If $F_1 \colon G \to H$ and $F_2 \colon H \to K$ are Lie group homomorphisms, then*

$$(F_2 \circ F_1)_* = (F_2)_* \circ (F_1)_* \colon \mathrm{Lie}(G) \to \mathrm{Lie}(K).$$

(c) *Isomorphic Lie groups have isomorphic Lie algebras.*

The Exponential Map of a Lie Group

If G is a Lie group, a Lie group homomorphism from $\mathbb{R}$ (with its additive group structure) to G is called a ***one-parameter subgroup of G***. (Note that the term "one-parameter subgroup" refers, confusingly, to a homomorphism, not to a Lie subgroup of G. But the *image* of a one-parameter subgroup is always a Lie subgroup of dimension at most 1.) Theorem 20.1 of [LeeSM] shows that for every $X \in \mathrm{Lie}(G)$, the integral curve of X starting at the identity is a one-parameter subgroup, called the ***one-parameter subgroup generated by X***, and every one-parameter subgroup is of this form.

The ***exponential map of G*** is the map $\exp^G \colon \mathrm{Lie}(G) \to G$ defined by setting $\exp^G(X) = \gamma(1)$, where γ is the integral curve of X starting at e. (The exponential map is more commonly denoted simply by $\exp$, but we use the notation $\exp^G$ to distinguish the Lie group exponential map from the Riemannian exponential map, introduced in Chapter 5.)

Proposition C.7 (Properties of the Lie Group Exponential Map). [LeeSM, Props. 20.5 & 20.8] *Let G be a Lie group and let $\mathfrak{g}$ be its Lie algebra.*

(a) $\exp^G \colon \mathfrak{g} \to G$ *is smooth.*
(b) *For every $X \in \mathfrak{g}$, the map $\gamma \colon \mathbb{R} \to G$ defined by $\gamma(t) = \exp^G(tX)$ is the one-parameter subgroup generated by X.*
(c) *The differential $\left(d \exp^G\right)_0 \colon T_0\mathfrak{g} \to T_eG$ is the identity map, under the canonical identifications of $T_0\mathfrak{g}$ and T_eG with $\mathfrak{g}$.*

The exponential map is the key ingredient in the proof of the following fundamental result.

Theorem C.8 (Closed Subgroup Theorem). [LeeSM, Thm. 20.12 & Cor. 20.13] *Suppose G is a Lie group and $H \subseteq G$ is a subgroup in the algebraic sense. Then H is an embedded Lie subgroup of G if and only if it is closed in the topological sense.*

An important special case of the closed subgroup theorem is that of a ***discrete subgroup***, that is, a subgroup that is discrete in the subspace topology.

Proposition C.9. [LeeSM, Prop. 21.28] *Every discrete subgroup of a Lie group is a closed Lie subgroup of dimension zero.*

Adjoint Representations

Let G be a Lie group and $\mathfrak{g}$ its Lie algebra. For every $\varphi \in G$, conjugation by φ gives a Lie group automorphism $C_\varphi \colon G \to G$, called an **inner automorphism**, by $C_\varphi(\psi) = \varphi\psi\varphi^{-1}$. Let $\mathrm{Ad}(\varphi) = (C_\varphi)_* \colon \mathfrak{g} \to \mathfrak{g}$ be the induced Lie algebra automorphism. It follows from the definition that $C_{\varphi_1} \circ C_{\varphi_2} = C_{\varphi_1\varphi_2}$, and therefore $\mathrm{Ad}(\varphi_1) \circ \mathrm{Ad}(\varphi_2) = \mathrm{Ad}(\varphi_1\varphi_2)$; in other words, $\mathrm{Ad} \colon G \to \mathrm{GL}(\mathfrak{g})$ is a representation, called the **adjoint representation of G**. Proposition 20.24 in [LeeSM] shows that Ad is a smooth map.

Example C.10 (Adjoint Representations of Lie Groups).

(a) If G is an abelian Lie group, then C_φ is the identity map for every $\varphi \in G$, so the adjoint representation is trivial: $\mathrm{Ad}(\varphi) = \mathrm{Id}_\mathfrak{g}$ for all $\varphi \in G$.

(b) Suppose G is a Lie subgroup of $\mathrm{GL}(n, \mathbb{R})$. Then for each $A \in G$, the conjugation map $C_A(B) = ABA^{-1}$ is the restriction to G of a linear map from the space $\mathrm{M}(n, \mathbb{R})$ of $n \times n$ matrices to itself. Its differential, therefore, is given by the same formula: $\mathrm{Ad}(A)X = AXA^{-1}$ for every $A \in G$ and $X \in \mathrm{Lie}(G) \subseteq \mathfrak{gl}(n, \mathbb{R})$. //

There is also an adjoint representation for Lie algebras. If $\mathfrak{g}$ is any Lie algebra, a **representation of $\mathfrak{g}$** is a Lie algebra homomorphism from $\mathfrak{g}$ to $\mathfrak{gl}(V)$, the Lie algebra of all linear endomorphisms of some vector space V. For each $X \in \mathfrak{g}$, define a map $\mathrm{ad}(X) \colon \mathfrak{g} \to \mathfrak{g}$ by $\mathrm{ad}(X)Y = [X, Y]$. This defines ad as a map from $\mathfrak{g}$ to $\mathfrak{gl}(\mathfrak{g})$, and a straightforward computation shows that it is a representation of $\mathfrak{g}$, called the **adjoint representation of $\mathfrak{g}$**.

The next proposition shows how the two adjoint representations are related.

Proposition C.11. [LeeSM, Thm. 20.27] *Let G be a Lie group and $\mathfrak{g}$ its Lie algebra, and let $\mathrm{Ad} \colon G \to \mathrm{GL}(\mathfrak{g})$ and $\mathrm{ad} \colon \mathfrak{g} \to \mathfrak{gl}(\mathfrak{g})$ be their respective adjoint representations. The induced Lie algebra representation $\mathrm{Ad}_* \colon \mathfrak{g} \to \mathfrak{gl}(\mathfrak{g})$ is given by $\mathrm{Ad}_* = \mathrm{ad}$.*

Group Actions on Manifolds

The most important applications of Lie groups in differential geometry involve their actions on other manifolds.

First we consider actions by abstract groups, not necessarily endowed with a smooth manifold structure or even a topology. Let G be a group and M a set. A **left action of G on M** is a map $G \times M \to M$, usually written $(\varphi, p) \mapsto \varphi \cdot p$, satisfying $\varphi_1 \cdot (\varphi_2 \cdot p) = (\varphi_1\varphi_2) \cdot p$ and $e \cdot p = p$ for all $\varphi_1, \varphi_2 \in G$ and $p \in M$, where e is the identity element of G. Similarly, a **right action of G on M** is a map $M \times G \to M$ satisfying $(p \cdot \varphi_1) \cdot \varphi_2 = p \cdot (\varphi_1\varphi_2)$ and $p \cdot e = p$. In some cases, it will be important to give a name to an action such as $\theta \colon G \times M \to M$, in which case we write $\theta_\varphi(p)$ in place of $\varphi \cdot p$, with a similar convention for right actions. Since right actions can be converted to left actions and vice versa by setting $\varphi \cdot p = p \cdot \varphi^{-1}$, for most purposes we lose no real generality by restricting attention to left actions. (But there are also situations in which right actions arise naturally.)

Given an action of G on M, for each $p \in M$ the *isotropy subgroup at* p is the subgroup $G_p \subseteq G$ consisting of all elements that fix p: that is, $G_p = \{\varphi \in G : \varphi \cdot p = p\}$. The group action is said to be *free* if $\varphi \cdot p = p$ for some $\varphi \in G$ and $p \in M$ implies $\varphi = e$, or in other words, if $G_p = \{e\}$ for every p. It is said to be *effective* if $\varphi_1 \cdot p = \varphi_2 \cdot p$ for all p if and only if $\varphi_1 = \varphi_2$, or equivalently, if the only element of G that fixes every element of M is the identity. For effective actions, each element of G is uniquely determined by the map $p \mapsto \varphi \cdot p$. In such cases, we will sometimes use the same notation φ to denote either the element $\varphi \in G$ or the map $p \mapsto \varphi \cdot p$. The action is said to be *transitive* if for every pair of points $p, q \in M$, there exists $\varphi \in G$ such that $\varphi \cdot p = q$.

If M is a smooth manifold, an action by a group G on M is said to be an *action by diffeomorphisms* if for each $\varphi \in G$, the map $p \mapsto \varphi \cdot p$ is a diffeomorphism of M. If G is a Lie group, an action of G on a smooth manifold M is said to be a *smooth action* if the defining map $G \times M \to M$ is smooth. In this case, it is also an action by diffeomorphisms, because each map $p \mapsto \varphi \cdot p$ is smooth and has $p \mapsto \varphi^{-1} \cdot p$ as an inverse. If G is any countable group, an action by G on M is an action by diffeomorphisms if and only if it is a smooth action when G is regarded as a 0-dimensional Lie group with the discrete topology.

Example C.12 (Semidirect Products). Group actions provide an important way to construct Lie groups out of other Lie groups. Suppose H and N are Lie groups and $\theta \colon H \times N \to N$ is a smooth left action of H on N by automorphisms of N, meaning that for each $h \in H$, the map $\theta_h \colon N \to N$ given by $\theta_h(n) = h \cdot n$ is a Lie group automorphism. The *semidirect product of* H *and* N determined by θ, denoted by $N \rtimes_\theta H$, is the Lie group whose underlying manifold is the Cartesian product $N \times H$, and whose group multiplication is $(n, h)(n', h') = (n\theta_h(n'), hh')$. //

▶ **Exercise C.13.** Verify that $N \rtimes_\theta H$ is indeed a Lie group with the multiplication defined above, and with (e, e) as identity and $(n, h)^{-1} = (\theta_{h^{-1}}(n^{-1}), h^{-1})$.

Now suppose G is a Lie group, and M and N are smooth manifolds endowed with G-actions (on the left, say). A map $F \colon M \to N$ is said to be *equivariant* with respect to the given G actions if $F(\varphi \cdot x) = \varphi \cdot F(x)$ for all $\varphi \in G$ and $x \in M$.

Theorem C.14 (Equivariant Rank Theorem). [LeeSM, Thm. 7.25] *Suppose* M, N *are smooth manifolds, and* G *is a Lie group acting smoothly and transitively on* M *and smoothly on* N. *If* $F \colon M \to N$ *is a smooth* G-*equivariant map, then* F *has constant rank. Thus, if* F *is surjective, it is a smooth submersion; if it is injective, it is a smooth immersion; and if it is bijective, it is a diffeomorphism.*

A smooth action of G on M is said to be a *proper action* if the map $G \times M \to M \times M$ defined by $(\varphi, x) \mapsto (\varphi \cdot x, x)$ is a proper map, meaning that the preimage of every compact set is compact. The following characterization is usually the easiest way to prove that a given action is proper.

Proposition C.15. [LeeSM, Prop. 21.5] *Suppose G is a Lie group acting smoothly on a smooth manifold M. The action is proper if and only if the following condition is satisfied: if (p_i) is a sequence in M and (φ_i) is a sequence in G such that both (p_i) and $(\varphi_i \cdot p_i)$ converge, then a subsequence of (φ_i) converges.*

Corollary C.16. *Every smooth action by a compact Lie group on a smooth manifold is proper.*

Proof. If G is a compact Lie group, then *every* sequence in G has a convergent subsequence, so every smooth G-action is proper by the preceding proposition. □

The next theorem is the most important application of proper actions. If a group G acts on a manifold M, then each $p \in M$ determines a subset $G \cdot p = \{\varphi \cdot p : \varphi \in G\}$, called the **orbit of p**. Because two orbits are either identical or disjoint, the orbits form a partition of G. The set of orbits is denoted by M/G, and with the quotient topology it is called the **orbit space** of the action.

Theorem C.17 (Quotient Manifold Theorem). [LeeSM, Thm. 21.10] *Suppose G is a Lie group acting smoothly, freely, and properly on a smooth manifold M. Then the orbit space M/G is a topological manifold whose dimension is equal to the difference $\dim M - \dim G$, and it has a unique smooth structure with the property that the quotient map $\pi : M \to M/G$ is a smooth submersion.*

Example C.18 (Real Projective Spaces). For each nonnegative integer n, the n-dimensional **real projective space**, denoted by $\mathbb{RP}^n$, is defined as the set of one-dimensional linear subspaces of $\mathbb{R}^{n+1}$. It can be identified with the orbit space of $\mathbb{R}^{n+1} \smallsetminus \{0\}$ under the action of the group $\mathbb{R}^*$ of nonzero real numbers given by scalar multiplication: $\lambda \cdot (x^1, \dots, x^{n+1}) = (\lambda x^1, \dots, \lambda x^{n+1})$. It is easy to check that this action is smooth and free. To see that it is proper, we use Proposition C.15. Suppose (x_i) is a sequence in $\mathbb{R}^{n+1} \smallsetminus \{0\}$ and (λ_i) is a sequence in $\mathbb{R}^*$ such that $x_i \to x \in \mathbb{R}^{n+1} \smallsetminus \{0\}$ and $\lambda_i x_i \to y \in \mathbb{R}^{n+1} \smallsetminus \{0\}$. Then $|\lambda_i| = |\lambda_i x_i|/|x_i|$ converges to the nonzero real number $|y|/|x|$. Thus the numbers λ_i all lie in a compact set of the form $\{\lambda : 1/C \le |\lambda| \le C\}$ for some positive number C, so a subsequence converges to a nonzero real number. Therefore, by the quotient manifold theorem, $\mathbb{RP}^n$ has a unique structure as a smooth n-dimensional manifold such that the quotient map $\mathbb{R}^{n+1} \smallsetminus \{0\} \to \mathbb{RP}^n$ is a smooth submersion. //

Example C.19 (Complex Projective Spaces). Similarly, the n-dimensional **complex projective space**, denoted by $\mathbb{CP}^n$, is the set of 1-dimensional complex subspaces of $\mathbb{C}^{n+1}$, identified with the orbit space of $\mathbb{C}^{n+1} \smallsetminus \{0\}$ under the $\mathbb{C}^*$-action given by $\lambda \cdot z = \lambda z$. The same argument as in the preceding example shows that this action is smooth, free, and proper, so $\mathbb{CP}^n$ is a smooth $2n$-dimensional manifold and the quotient map $\pi : \mathbb{C}^{n+1} \smallsetminus \{0\} \to \mathbb{CP}^n$ is a smooth submersion. //

Group Actions and Covering Spaces

There is a close connection between smooth covering maps and smooth group actions. To begin, suppose $\widetilde{M}$ and M are smooth manifolds and $\pi : \widetilde{M} \to M$

is a smooth covering map. A *covering automorphism of* π is a diffeomorphism $\varphi \colon \widetilde{M} \to \widetilde{M}$ such that $\pi \circ \varphi = \pi$; the set of all covering automorphisms is a group under composition, called the *covering automorphism group* and denoted by $\mathrm{Aut}_\pi(\widetilde{M})$.

Proposition C.20. [LeeSM, Prop. 21.12] *Let* $\pi \colon \widetilde{M} \to M$ *be a smooth covering map. With the discrete topology,* $\mathrm{Aut}_\pi(\widetilde{M})$ *is a discrete Lie group acting smoothly, freely, and properly on* $\widetilde{M}$.

Because of the requirement that a covering automorphism φ satisfy $\pi \circ \varphi = \pi$, it follows that φ restricts to an action on each fiber of π. The next proposition describes the conditions in which this action is transitive. A covering map $\pi \colon \widetilde{M} \to M$ is called a *normal covering* if the image of the homomorphism $\pi_* \colon \pi_1(\widetilde{M}, \widetilde{x}) \to \pi_1(M, \pi(\widetilde{x}))$ is a normal subgroup of $\pi_1(M, \pi(\widetilde{x}))$. (Note that every universal covering is a normal covering, because the trivial subgroup is always normal.)

Proposition C.21. [LeeTM, Cor. 12.5] *If* $\pi \colon \widetilde{M} \to M$ *is a smooth covering map, then the automorphism group of* π *acts transitively on each fiber of* π *if and only if* π *is a normal covering.*

The universal covering is the most important special case.

Proposition C.22 (Automorphisms of the Universal Covering). [LeeTM, Cor. 12.9] *Suppose* M *is a connected smooth manifold and* $\pi \colon \widetilde{M} \to M$ *is its universal covering. Then* $\mathrm{Aut}_\pi(\widetilde{M})$ *is isomorphic to the fundamental group of* M, *and it acts transitively on each fiber of* π.

The next theorem, which is an application of the quotient manifold theorem, is a partial converse to Proposition C.20.

Proposition C.23. [LeeSM, Thm. 21.13] *Suppose* $\widetilde{M}$ *is a connected smooth manifold and* Γ *is a discrete Lie group acting smoothly, freely, and properly on* $\widetilde{M}$. *Then the orbit space* $\widetilde{M}/\Gamma$ *has a unique smooth manifold structure such that the quotient map* $\widetilde{M} \to \widetilde{M}/\Gamma$ *is a smooth normal covering map.*

Example C.24 (The Universal Covering of $\mathbb{RP}^n$). The two-element group $\Gamma = \{\pm 1\}$ acts smoothly, freely, and properly on $\mathbb{S}^n$ by multiplication, and thus $\mathbb{S}^n/\Gamma$ is a smooth manifold and the quotient map $\pi \colon \mathbb{S}^n \to \mathbb{S}^n/\Gamma$ is a two-sheeted smooth normal covering map. Note also that the quotient map $\mathbb{R}^{n+1} \smallsetminus \{0\} \to \mathbb{RP}^n$ defined in Example C.18 restricts to a surjective smooth map $q \colon \mathbb{S}^n \to \mathbb{RP}^n$, which is a submersion because each point of $\mathbb{S}^n$ is in the image of a smooth local section. Since q and π are constant on each other's fibers, Proposition A.19 shows that there is a diffeomorphism $\mathbb{S}^n/\Gamma \to \mathbb{RP}^n$, and q is equal to the composition $\mathbb{S}^n \to \mathbb{S}^n/\Gamma \to \mathbb{RP}^n$. Since this is a smooth normal covering map followed by a diffeomorphism, q is also a smooth normal covering map. Thus $\mathbb{S}^n$ is the universal covering space of $\mathbb{RP}^n$, and the fundamental group of $\mathbb{RP}^n$ is isomorphic to $\{\pm 1\}$.   //

References

[AP94]   Marco Abate and Giorgio Patrizio, *Finsler Metrics, a Global Approach: with Appli-cations to Geometric Function Theory*, Springer, New York, 1994.

[Amb56]   Warren Ambrose, *Parallel translation of Riemannian curvature*, Ann. of Math. (2) **64** (1956), 337–363.

[Ave70]   André Avez, *Variétés riemanniennes sans points focaux*, C. R. Acad. Sci. Paris Sér. A-B **270** (1970), A188–A191 (French).

[BCS00]   David Bao, Shiing-Shen Chern, and Zhongmin Shen, *An Introduction to Riemann–Finsler Geometry*, Graduate Texts in Mathematics, vol. 200, Springer, New York, 2000.

[Ber60]   M. Berger, *Les variétés Riemanniennes (1/4)-pincées*, Ann. Scuola Norm. Sup. Pisa (3) **14** (1960), 161–170 (French).

[Ber03]   Marcel Berger, *A Panoramic View of Riemannian Geometry*, Springer-Verlag, Berlin, 2003.

[Bes87]   Arthur L. Besse, *Einstein Manifolds*, Springer, Berlin, 1987.

[BBBMP]   Laurent Bessières, Gérard Besson, Michel Boileau, Sylvain Maillot, and Joan Porti, *Geometrisation of 3-Manifolds*, EMS Tracts in Mathematics, vol. 13, European Mathematical Society (EMS), Züurich, 2010.

[Bis63]   Richard L. Bishop, *A relation between volume, mean curvature, and diameter*, Notices Amer. Math. Soc. **10** (1963), 364.

[BC64]   Richard L. Bishop and Richard J. Crittenden, *Geometry of Manifolds*, Pure and Applied Mathematics, Vol. XV, Academic Press, New York-London, 1964.

[BW08]   Christoph Böhm and Burkhard Wilking, *Manifolds with positive curvature operators are space forms*, Ann. of Math. (2) **167** (2008), no. 3, 1079–1097.

[Boo86]   William M. Boothby, *An Introduction to Differentiable Manifolds and Riemannian Geometry*, 2nd ed., Academic Press, Orlando, FL, 1986.

[Bre10]   Simon Brendle, *Ricci Flow and the Sphere Theorem*, Graduate Studies in Mathematics, vol. 111, American Mathematical Society, Providence, RI, 2010.

[BS09]   Simon Brendle and Richard Schoen, *Manifolds with 1/4-pinched curvature are space forms*, J. Amer. Math. Soc. **22** (2009), 287–307.

[Car26]   Élie Cartan, *Sur une classe remarquable d'espaces de Riemann*, Bull. Soc. Math. France **54** (1926), 214–264 (French).

[Cha06]   Isaac Chavel, *Riemannian Geometry*, 2nd ed., Cambridge Studies in Advanced Mathematics, vol. 98, Cambridge University Press, Cambridge, 2006. A modern introduction.

[CE08]   Jeff Cheeger and David G. Ebin, *Comparison Theorems in Riemannian Geometry*, AMS Chelsea Publishing, Providence, 2008. Revised reprint of the 1975 original.

[CG71]   Jeff Cheeger and Detlef Gromoll, *The splitting theorem for manifolds of nonnegative Ricci curvature*, J. Differential Geometry **6** (1971), 119–128.

© Springer International Publishing AG 2018

J. M. Lee, *Introduction to Riemannian Manifolds*, Graduate Texts
in Mathematics 176, https://doi.org/10.1007/978-3-319-91755-9

[CG72] Jeff Cheeger and Detlef Gromoll. *On the structure of complete manifolds of nonnegative curvature*, Ann. Math. (2) **96** (1972), 413–443.

[Che75] Shiu-Yuen Cheng, *Eigenvalue comparison theorems and its geometric applications*, Math. Z. **143** (1975), no. 3, 289–297.

[Che55] Shiing-Shen Chern, *An elementary proof of the existence of isothermal parameters on a surface*, Proc. Amer. Math. Soc. **6** (1955), 771–782.

[CB09] Yvonne Choquet-Bruhat, *General Relativity and the Einstein Equations*, Oxford Mathematical Monographs, Oxford University Press, Oxford, 2009.

[Cla70] Chris J. S. Clarke, *On the Global Isometric Embedding of Pseudo-Riemannian Manifolds*, Proc. R. Soc. Lond. Ser. A Math. Phys. Eng. Sci. **314** (1970), no. 1518, 417–428.

[CM11] Tobias H. Colding and William P. Minicozzi II, *A Course in Minimal Surfaces*, Graduate Studies in Mathematics, vol. 121, American Mathematical Society, Providence, RI, 2011.

[dC92] Manfredo Perdigão do Carmo, *Riemannian Geometry*, Mathematics: Theory & Applications, Birkhäuser Boston, Inc., Boston, MA, 1992. Translated from the second Portuguese edition by Francis Flaherty.

[DF04] David S. Dummit and Richard M. Foote, *Abstract Algebra*, 3rd ed., John Wiley & Sons, Inc., Hoboken, NJ, 2004.

[Fra61] Theodore Frankel, *Manifolds with positive curvature*, Pacific J. Math. **11** (1961), 165–174.

[GHL04] Sylvestre Gallot, Dominique Hulin, and Jacques Lafontaine, *Riemannian Geometry*, 3rd ed., Universitext, Springer-Verlag, Berlin, 2004.

[Gan73] David Gans, *An Introduction to Non-Euclidean Geometry*, Academic Press, New York, 1973.

[Gau65] Carl F. Gauss, *General Investigations of Curved Surfaces*, Raven Press, New York, 1965.

[Gra82] Alfred Gray, *Comparison theorems for the volumes of tubes as generalizations of the Weyl tube formula*, Topology **21** (1982), no. 2, 201–228.

[Gra04] Alfred Gray, *Tubes*, 2nd ed., Progress in Mathematics, vol. 221, Birkhäuser Verlag, Basel, 2004. With a preface by Vicente Miquel.

[Gre93] Marvin J. Greenberg, *Euclidean and Non-Euclidean Geometries: Development and History*, W. H. Freeman, New York, 1993.

[Gre70] Robert E. Greene, *Isometric Embeddings of Riemannian and Pseudo-Riemannian Manifolds*, Memoirs of the Amer. Math. Soc., No. 97, Amer. Math. Soc., Providence, 1970.

[Gro07] Misha Gromov, *Metric Structures for Riemannian and Non-Riemannian Spaces*, Reprint of the 2001 English edition, Modern Birkhäuser Classics, Birkhäuser Boston, Inc., Boston, MA, 2007. Based on the 1981 French original; With appendices by M. Katz, P. Pansu and S. Semmes; Translated from the French by Sean Michael Bates.

[Gün60] Paul Günther, *Einige Sätze über das Volumenelement eines Riemannschen Raumes*, Publ. Math. Debrecen **7** (1960), 78–93 (German).

[Ham82] Richard S. Hamilton, *Three-manifolds with positive Ricci curvature*, J. Differential Geom. **17** (1982), 255–306.

[Ham86] Richard S. Hamilton, *Four-manifolds with positive curvature operator*, J. Differential Geom. **24** (1986), no. 2, 153–179.

[Hat02] Allen Hatcher, *Algebraic Topology*, Cambridge University Press, Cambridge, 2002.

[HE73] Stephen W. Hawking and George F. R. Ellis, *The Large-Scale Structure of Space-Time*, Cambridge University Press, Cambridge, 1973.

[Hel01] Sigurdur Helgason, *Differential Geometry, Lie Groups, and Symmetric Spaces*, Graduate Studies in Mathematics, vol. 34, American Mathematical Society, Providence, RI, 2001. Corrected reprint of the 1978 original.

[Her63] Robert Hermann, *Homogeneous Riemannian manifolds of non-positive sectional curvature*, Nederl. Akad. Wetensch. Proc. Ser. A 66 = Indag. Math. **25** (1963), 47–56.

[Hic59] Noel J. Hicks, *A theorem on affine connexions*, Illinois J. Math. **3** (1959), 242–254.

[Hil71]   David Hilbert, *The Foundations of Geometry*, Open Court Publishing Co., Chicago, 1971. Translated from the tenth German edition by Leo Unger.

[Hop89]   Heinz Hopf, *Differential Geometry in the Large*, 2nd ed., Lecture Notes in Mathematics, vol. 1000, Springer, 1989. Notes taken by Peter Lax and John W. Gray; with a preface by S. S. Chern; with a preface by K. Voss.

[Ive92]   Birger Iversen, *Hyperbolic Geometry*, London Mathematical Society Student Texts, vol. 25, Cambridge University Press, Cambridge, 1992.

[Jos17]   Jürgen Jost, *Riemannian Geometry and Geometric Analysis*, 7th ed., Universitext, Springer, Cham, 2017.

[JP13]   Marek Jarnicki and Peter Pflug, *Invariant Distances and Metrics in Complex Analysis*, Second extended edition, De Gruyter Expositions in Mathematics, vol. 9, Walter de Gruyter GmbH & Co. KG, Berlin, 2013.

[Kaz85]   Jerry Kazdan, *Prescribing the Curvature of a Riemannian Manifold*, CBMS Regional Conf. Ser. in Math., Amer. Math. Soc., Providence, 1985.

[KL08]   Bruce Kleiner and John Lott, *Notes on Perelman's papers*, Geom. Topol. **12** (2008), no. 5, 2587–2855.

[Kli61]   Wilhelm P. A. Klingenberg, *Über Riemannsche Mannigfaltigkeiten mit positiver Krümmung*, Comment. Math. Helv. **35** (1961), 47–54 (German).

[Kli95]   Wilhelm P. A. Klingenberg, *Riemannian Geometry*, 2nd ed., De Gruyter Studies in Mathematics, vol. 1, Walter de Gruyter & Co., Berlin, 1995.

[Kob72]   Shoshichi Kobayashi, *Transformation Groups in Differential Geometry*, Springer, Berlin, 1972.

[KN96]   Shoshichi Kobayashi and Katsumi Nomizu, *Foundations of Differential Geometry*, Wiley Classics Library, vol. I–II, John Wiley & Sons, Inc., New York, 1996. Reprint of the 1963 and 1969 originals; A Wiley-Interscience Publication.

[LeeJeff09]   Jeffrey M. Lee, *Manifolds and Differential Geometry*, Graduate Studies in Mathematics, vol. 107, American Mathematical Society, Providence, RI, 2009.

[LeeAG]   John M. Lee, *Axiomatic Geometry*, Pure and Applied Undergraduate Texts, vol. 21, American Mathematical Society, Providence, RI, 2013.

[LeeSM]   John M. Lee, *Introduction to Smooth Manifolds*, 2nd ed., Graduate Texts in Mathematics, vol. 218, Springer, New York, 2013.

[LeeTM]   John M. Lee, *Introduction to Topological Manifolds*, 2nd ed., Graduate Texts in Mathematics, vol. 202, Springer, New York, 2011.

[LP87]   John M. Lee and Thomas H. Parker, *The Yamabe problem*, Bull. Amer. Math. Soc. (N.S.) **17** (1987), 37–91.

[MP11]   William H. Meeks III and Joaqun Pérez, *The classical theory of minimal surfaces*, Bull. Amer. Math. Soc. (N.S.) **48** (2011), 325–407.

[Mil63]   John W. Milnor, *Morse Theory*, based on lecture notes by M. Spivak and R. Wells. Annals of Mathematics Studies, No. 51, Princeton University Press, Princeton, N.J., 1963.

[Mil68]   John W. Milnor, *A note on curvature and fundamental group*, J. Differential Geometry **2** (1968), 1–7.

[Mil76]   John W. Milnor, *Curvatures of left invariant metrics on Lie groups*, Advances in Math. **21** (1976), no. 3, 293–329.

[Mor98]   Frank Morgan, *Riemannian Geometry: A Beginner's Guide*, 2nd ed., A K Peters Ltd., Wellesley, MA, 1998.

[MF10]   John W. Morgan and Frederick Tsz-Ho Fong, *Ricci Flow and Geometrization of 3-Manifolds*, University Lecture Series, vol. 53, American Mathematical Society, Providence, RI, 2010.

[MT14]   John W. Morgan and Gang Tian, *The Geometrization Conjecture*, Clay Mathematics Monographs, vol. 5, American Mathematical Society, Providence, RI; Clay Mathematics Institute, Cambridge, MA, 2014.

[Mun56]   James Raymond Munkres, *Some Applications of Triangulation Theorems*, ProQuest LLC, Ann Arbor, MI, 1956. Thesis (Ph.D.)-University of Michigan.

[Mye41] Sumner B. Myers, *Riemannian manifolds with positive mean curvature*, Duke Math.
         J. **8** (1941), 401–404.

[MS39]   Sumner B. Myers and Norman E. Steenrod, *The group of isometries of a Riemannian
         manifold*, Ann. of Math. (2) **40** (1939), no. 2, 400–416.

[Nas56] John Nash, *The imbedding problem for Riemannian manifolds*, Ann. Math. **63** (1956),
         20–63.

[O'N83] Barrett O'Neill, *Semi-Riemannian Geometry with Applications to General Relativity*,
         Academic Press, New York, 1983.

[Pet16] Peter Petersen, *Riemannian Geometry*, 3rd ed., Graduate Texts in Mathematics, vol.
         171, Springer, Cham, 2016.

[Poo81] Walter A. Poor, *Differential Geometric Structures*, McGraw-Hill, 1981.

[Pre43] Alexandre Preissman, *Quelques propriétés globales des espaces de Riemann*, Com-
         ment. Math. Helv. **15** (1943), 175–216 (French).

[Rad25] Tibor Radó, *Über den Begriff der Riemannschen Fläche*, Acta Litt. Sci. Szeged. **2**
         (1925), 101–121 (German).

[Rat06] John G. Ratcliffe, *Foundations of Hyperbolic Manifolds*, 2nd ed., Graduate Texts in
         Mathematics, vol. 149, Springer, New York, 2006.

[Sco83] Peter Scott, *The Geometries of 3-Manifolds*, Bull. London Math. Soc. **15** (1983), 401–
         487.

[Spi79] Michael Spivak, *A Comprehensive Introduction to Differential Geometry*, Vol. I–V,
         Publish or Perish, Berkeley, 1979.

[Str86] Robert S. Strichartz, *Sub-Riemannian geometry*, J. Differential Geom. **24** (1986), no.
         2, 221–263.

[Syn36] John L. Synge, *On the connectivity of spaces of positive curvature*, Q. J. Math. **os-7**
         (1936), 316–320.

[Thu97] William P. Thurston, *Three-Dimensional Geometry and Topology. Vol. 1*, Princeton
         Mathematical Series, vol. 35, Princeton University Press, Princeton, NJ, 1997. Edited
         by Silvio Levy.

[Tu11]   Loring W. Tu, *An Introduction to Manifolds*, 2nd ed., Universitext, Springer, New
         York, 2011.

[Wol11] Joseph A. Wolf, *Spaces of Constant Curvature*, 6th ed., AMS Chelsea Publishing,
         Providence, RI, 2011.

Notation Index

Subject Index

Symbols
(1, 1)-Hessian, 320
1-center, 355
2-point homogeneous, 189, 261
3-point homogeneous, 261

A
abelian Lie algebra, 409
absolute derivative, 102
acceleration
 in $\mathbb{R}^n$, 86
 of a curve in a manifold, 103
 of a plane curve, 2
 tangential, 86
action, *see* group action
adapted frame, 16
adjoint matrix, 408
adjoint representation
 of a Lie algebra, 411
 of a Lie group, 68, 411
admissible curve, 34
admissible family, 152
admissible loop, 150
admissible partition
 for a curve, 34
 for a family of curves, 152
 for a vector field along a curve, 108
affine connection, 91
aims at a point, 167
algebraic Bianchi identity, 203
algebraic curvature tensor, 212
Alt convention for wedge product, 401
alternating tensor, 400
alternation, 400
ambient manifold, 226
ambient tangent bundle, 386
Ambrose, Warren, 351

analytic continuation of a local isometry, 346
angle
 between vectors, 10, 12
 of a geodesic triangle, 353
 tangent, 264, 265
angle excess, 281
angle-sum theorem, 2, 271, 275
anti-de Sitter space, 80
 universal, 83
anti-self-dual, 50
anticommutative, 400
antipodal points, 144
arc length, parametrization by, 35
arc-length function, 35
area-minimizing hypersurface, 239
area of a hypersurface, 239
aspherical, 354
asymptotically parallel, 282
atlas, 374
automorphism of a Lie group, 407
 inner, 411
Avez, André, 364
axial isometry, 357
axis for an isometry, 357

B
Böhm, Christoph, 368
Böhm–Wilking theorem, 368
backward reparametrization, 34
ball
 geodesic, 158, 163
 metric, 163
 Poincaré, 62
 regular coordinate, 374
 smooth coordinate, 374
 volume of, 314

© Springer International Publishing AG 2018
J. M. Lee, *Introduction to Riemannian Manifolds*, Graduate Texts
in Mathematics 176, https://doi.org/10.1007/978-3-319-91755-9

Printed in the United States
by Baker & Taylor Publisher Services